Other Titles in This Series

(Continued in the back of this publication)

Translations of
MATHEMATICAL MONOGRAPHS

Volume 145

Modular Forms and Hecke Operators

A. N. Andrianov
V. G. Zhuravlev

American Mathematical Society
Providence, Rhode Island

А. Н. Андрианов, В. Г. Журавлев

МОДУЛЯРНЫЕ ФОРМЫ И ОПЕРАТОРЫ ГЕККЕ

Translated from the Russian by Neal Koblitz

1991 *Mathematics Subject Classification.* Primary 11Fxx; Secondary 11E45.

ABSTRACT. The book contains the exposition of the theory of Hecke operators in the spaces of modular forms of an arbitrary degree. The main consideration is given to the study of multiplicative properties of the Fourier coefficients of modular forms.

The book can be used by researchers and graduate students working in algebra and number theory, to learn about theta-series, theory of modular forms in one and several variables, theory of Hecke operators, and about recent developments in arithmetic of modular forms.

Library of Congress Cataloging-in-Publication Data

Andrianov, A. N. (Anatoliĭ Nikolaevich)
   [Moduli͡arnye formy i operatory Gekke. English]
   Modular forms and Hecke operators / A. N. Andrianov, V. G. Zhuravlev.
     p. cm. — (Translations of mathematical monographs, ISSN 0065-9282; v. 145)
   Includes bibliographical references.
   ISBN 0-8218-0277-1
   1. Forms, Modular. 2. Hecke operators. I. Zhuravlev, V. G. (Vladimir Georgievich) II. Title.
III. Series.
QA243.A52813   1995
512.9′44—dc20
                                                      95-30915
                                                        CIP

Contents

Introduction

Throughout the history of number theory, a problem that has attracted and continues to attract the interest of researchers is that of studying the number $r(q, a)$ of integer solutions to equations of the form

$$q(x_1, \ldots, x_m) = a,$$

where q is a quadratic form. The classical theory gave many exact formulas for the functions $r(q, a)$ which revealed remarkable multiplicative properties of these numbers. For example, Jacobi's formula

$$r(x_1^2 + \cdots + x_4^2, a) = 8\sigma_1(a)$$

for the number of representations of an odd integer a as a sum of four squares and Ramanujan's formula

$$r(x_1^2 + \cdots + x_{24}^2, a) = \frac{16}{691}\sigma_{11}(a) + \frac{33152}{691}\tau(a)$$

for the number of representations of an odd integer a as a sum of 24 squares, where $\sigma_k(a)$ denotes the sum of the kth powers of the positive divisors of a and $\tau(a)$ is defined as the coefficients in the power series

$$x\{(1-x)(1-x^2)\ldots\}^{24} = \sum_{a=1}^{\infty} \tau(a)x^a,$$

involve the multiplicative functions $\sigma_k(a)$ with multiplication table

$$\sigma_k(a) \cdot \sigma_k(b) = \sum_{d|a,b} d^k \sigma_k(ab/d^2)$$

and the Ramanujan function $\tau(a)$, which follows the same multiplication rule as $\sigma_{11}(a)$. In 1937, Hecke explained why this phenomenon occurs. In particular, from Hecke's theory it follows that, given a positive definite integral quadratic form q in an even number of variables, the function $r(q, a)$ is a linear combination of multiplicative functions whose values can be interpreted as eigenvalues of certain invariantly defined linear operators (called "Hecke operators") on the spaces of modular forms. In subsequent years, the work of Eichler, Sato, Deligne and others uncovered fundamental relations between Hecke operators and algebraic geometry. In particular, their eigenvalues were interpreted in terms of the roots of the zeta-functions of suitable algebraic varieties over finite fields. Another line of development, initiated by Selberg and then greatly expanded by Langlands, considers Hecke operators from the point of view of the representation theory of locally compact groups and hopes to find a prominent place for them in a future noncommutative class field theory.

1

A natural generalization of the problem of representing numbers by quadratic forms is the problem of representing quadratic forms by quadratic forms. If q and a are two quadratic forms in m and n variables, respectively, then we might want to study the number $r(q, a)$ of changes of variables (by means of $m \times n$ integer matrices) which take the form q to the form a. For example, if a is an integer, then $r(q, ax^2) = r(q, a)$. In 1935–1937, Siegel laid the groundwork for the arithmetic–analytic study of $r(q, a)$ and began constructing a theory of modular forms in several variables. Neither he nor later investigators were able to find much in the way of "exact formulas" (in the sense of the classical theory) for these functions $r(q, a)$. Moreover, since quadratic forms generally do not have any composition rule, it is not at all clear in what sense one can speak of multiplicative functions of the type $r(q, a)$. Thus, there were no arithmetic motives for trying to carry over the theory of Hecke operators to Siegel modular forms. But the concept of Hecke operators was so simple and natural that, soon after Hecke's work, attempts were made to develop a Hecke theory for such modular forms.

As this theory developed, the Hecke operators on spaces of modular forms in several variables were found to have arithmetic meaning. In particular, the theory provided a framework for discovering certain multiplicative properties of the number of integer representations of quadratic forms by quadratic forms.

The theory has now reached a sufficient level of maturity, and the time has come for a detailed and systematic exposition of its fundamental methods and results. The purpose of this book is, starting with the basics and ending with the latest results, to explain the current status of the theory of Hecke operators on spaces of holomorphic modular forms of integer and half-integer weight for congruence-subgroups of integral symplectic groups. In the spirit of Hecke's original approach, we consider Hecke operators principally as an instrument for studying the multiplicative properties of the Fourier coefficients of modular forms. We do not discuss other directions of the theory, such as the connection of Hecke operators with algebraic geometry, representation theory, and Galois theory, since in the case of several variables the study of these connections is far from complete.

The book can also be used as an introduction to the theory of modular forms in one or several variables and the theory of theta-series.

The book is intended for those who plan to work in the arithmetic theory of quadratic forms and modular functions, those who already are working in this area, and those who merely want some familiarity with the field. The reader can get an idea of the book's contents from the chapter and section headings and the introductory remarks at the beginning of each chapter. Here we shall only make some general comments. The first three chapters are independent of one another, except for a few general lemmas. Chapter 4 relies upon Chapters 2 and 3. Most of the exercises are not standard drill problems, but rather indicate interesting branches of the theory which for one reason or another did not fit into the main text. All of the prerequisites from algebra and number theory that go beyond the standard university courses are given at the end of the book in three Appendices. The most important references to the literature are concentrated in the Remarks.

We hope that this book will help attract the attention of young researchers to a beautiful and mysterious realm of number theory, and will make the path easier for all who wish to enter.

CHAPTER 1

Theta-Series

In this chapter we look at a fundamental instrument in the analytic study of quadratic Diophantine equations and systems—the theta-series of quadratic forms.

§1. Definition of theta-series

1. Representations of quadratic forms by quadratic forms. Suppose $q(x_1, \ldots, x_m)$ and $a(y_1, \ldots, y_n)$ are two quadratic forms with coefficients in some commutative ring K'. By a *representation of the form a by the form q over a subring K of K'* we mean a matrix $C = (c_{ij})$ in the set $M_{m,n}(K)$ of all $m \times n$-matrices with entries in K such that the change of variables

$$x_i = \sum_{j=1}^{n} c_{ij} y_j \quad (i = 1, \ldots, m)$$

takes the form q to the form a. We let $r_K(q, a)$ denote the number of representations of a by q over the ring K. In the case $n = 1$, i.e., $a(y_1) = a \cdot y_1^2$, the definition says that a column C with components $c_1, \ldots, c_m \in K$ is a representation of a by q if and only if $(c_1, \ldots, c_m)$ is a solution of the equation

$$(1.1) \qquad\qquad q(x_1, \ldots, x_m) = a,$$

and hence $r_K(q, a \cdot y_1^2)$ is equal to the number $r_K(q, a)$ of solutions in K to the equation (1.1). Similarly, in the general case $r_K(q, a)$ can be interpreted as the number of solutions of a certain system of equations of degree two.

If the number $2 = 2 \cdot 1_{K'}$ is not a zero divisor in K', then it is convenient to use matrix language. To every quadratic form

$$(1.2) \qquad\qquad q(x_1, \ldots, x_m) = \sum_{1 \leqslant \alpha, \beta \leqslant m} q_{\alpha\beta} x_\alpha x_\beta$$

we associate the *even* symmetric matrix

$$(1.3) \qquad\qquad Q = Q(q) = (q_{\alpha\beta}) + {}^t(q_{\alpha\beta}),$$

where t denotes the transpose. We call Q the *matrix of the form q* (it is more traditional but less convenient to call Q the matrix of the form $2q$). Then q can be written in terms of its matrix as follows:

$$(1.4) \qquad\qquad q(x_1, \ldots, x_m) = \frac{1}{2} {}^t x Q x,$$

where x is the column with components $x_1, \ldots, x_m$. Using these definitions and notation, we immediately see that $C \in M_{m,n}(K)$ is a representation of a form a in n variables by the form q if and only if $X = C$ is a solution of the matrix equation

$$(1.5) \qquad\qquad {}^t X Q X = A,$$

where A is the matrix of the form a and X is an $m \times n$-matrix. In particular, $r_K(q, a)$ is equal to the number $r_K(Q, A)$ of solutions over K of the equation (1.5).

The methods of studying $r_K(q, a)$ and even the formulation of the questions naturally depend upon the nature of the ring K and the properties of the quadratic forms under consideration. For now we shall limit ourselves to a simple but useful observation.

Two quadratic forms q and q' in the same number m of variables are said to be *equivalent* over K (or K-*equivalent*) if there exists a representation of one form by the other that lies in the group $GL_m(K)$ of invertible $m \times m$-matrices over K. In this case we write $q \sim_K q'$. The set $\{q\}_K$ of all forms that are equivalent over K to a given form q is called the *class* of q over K. If Q and Q' are the matrices of the forms q and q', then $q \sim_K q'$ means that

$$(1.6) \qquad\qquad {}^t V Q V = Q' \quad \text{with } V \in GL_m(K).$$

In this case we say that Q and Q' are *equivalent* over K (or K-*equivalent*), and we write $Q \sim_K Q'$. We let $\{Q\}_K$ denote the K-*equivalence class* of the matrix Q.

For any fixed matrices $V \in GL_m(K)$ and $V' \in GL_n(K)$, the map $C \to VCV'$ is obviously a one-to-one correspondence from $M_{m,n}(K)$ to itself. From this obvious fact and the definitions we have

PROPOSITION 1.1. *The function $r_K(Q, A)$ depends only on the K-equivalence classes of Q and A (or the K-equivalence classes of the corresponding quadratic forms).*

The history of quadratic forms over rings is almost as old and colorful as that of mathematics itself. The questions asked and the approaches to answering them vary greatly from one ring to another and from one type of quadratic form to another. In this book we are interested in the development of analytic methods in the simplest nontrivial situation, and so we take the ring of rational integers $\mathbf{Z}$ as our ground ring. Other rings will play only an auxiliary role. In addition, as a rule we will be considering only representations by positive definite forms. If q is such a form, the number $r_{\mathbf{Z}}(q, a)$ of integral representations by q is always finite, and the theory studies various properties of the function $a \to r_{\mathbf{Z}}(q, a)$.

It is the theory of modular forms which provides a natural language and an apparatus for studying this function. The theta-series of quadratic forms are the link between modular forms and quadratic forms.

PROBLEM 1.2. Let Q and A be symmetric $m \times m$- and $n \times n$-matrices, respectively, with coefficients in the field of real numbers $\mathbf{R}$, and let $Q > 0$ (i.e., Q is positive definite). Prove that the equation ${}^t X Q X = A$ is solvable in real $m \times n$-matrices X if and only if $A \geqslant 0$ (i.e., A is positive semidefinite) and rank $A \leqslant m$, and in this case the entries in any solution $X = C = (c_{ij})$ satisfy the inequality $\max_{i,j} |c_{ij}| \leqslant \lambda^{-1/2} \mu^{1/2}$, where λ is the smallest eigenvalue of the matrix Q and μ is the largest eigenvalue of the matrix A; in particular, $r_{\mathbf{Z}}(Q, A) < \infty$.

2. Definition of theta-series. We start with some notation. Let

(1.7) $$\mathbf{A}_n = \{A \in \mathbf{E}_n; A \geqslant 0\}$$

be the set of $n \times n$ integral even symmetric (i.e., with even numbers on the main diagonal) semidefinite matrices, and let

(1.8) $$\mathbf{A}_n^+ = \{A \in \mathbf{A}_n; A > 0\}.$$

We fix $Q \in \mathbf{A}_m^+$ and $n = 1, 2, \ldots$. For every $A \in \mathbf{A}_n$ we can define the finite number $r(Q, A) = r_Z(Q, A)$ of integral representations of the form with matrix A by the form with matrix Q. Thus, the matrix Q corresponds to a function $r(Q, \cdot): \mathbf{A}_n \to \mathbf{Z}$. In order to study this function analytically, it is natural to consider the following generating series, written as a power series,

$$\sum_{A=((1+e_{\alpha\beta})a_{\alpha\beta})\in\mathbf{A}_n} r(Q, A) \prod_{1\leqslant\alpha\leqslant\beta\leqslant n} t_{\alpha\beta}^{a_{\alpha\beta}}$$

in $\langle n \rangle = n(n+1)/2$ variables $t_{\alpha\beta}$, where $e_{\alpha\beta}$ are the coefficients of the identity matrix $E_n = (e_{\alpha\beta})$. Setting $t_{\alpha\beta} = \exp(2\pi i z_{\alpha\beta})$, we obtain the Fourier series

(1.9)
$$\theta^n(Z, Q) = \sum_{A=((1+e_{\alpha\beta})a_{\alpha\beta})\in\mathbf{A}_n} r(Q, A) \exp\left(\pi i 2 \sum_{1\leqslant\alpha\leqslant\beta\leqslant n} a_{\alpha\beta} z_{\alpha\beta}\right)$$
$$= \sum_{A\in\mathbf{A}_n} \mathbf{r}(Q, A) \exp(\pi i \sigma(AZ)),$$

where Z is an $n \times n$ symmetric matrix with coefficients $z_{\alpha\beta}$ on and above the main diagonal, and where $\sigma(M)$ denotes the trace of the matrix M. The last form of writing the generating series is the more convenient one in most situations, and in particular for finding the domain of convergence.

We write the matrix Z in the form $Z = X + iY$, where X and Y are real matrices and $i = \sqrt{-1}$. If the matrix Y does not satisfy the condition $Y \geqslant 0$, then there exists a row of integers $c = (c_1, \ldots, c_n)$ such that $cY{}^tc < 0$. (A real solution of this inequality exists by definition; a rational solution exists by continuity; and an integral solution can be obtained from the rational solution using homogeneity.) Let C denote the $m \times n$ integer matrix with c in the first row and zeros everywhere else. Then the matrices $A_d = d^2 \cdot {}^tCQC = {}^t(dC)Q(dC)$ with rational integers d belong to $\mathbf{A}_n$, they satisfy the condition $r(Q, A_d) \geqslant 1$, and we obviously have

$$|r(Q, A_d) \exp(\pi i \sigma(A_dZ))| = r(Q, A_d) \exp(-\pi d^2 \sigma(QCY{}^tC)) \to \infty \quad \text{as } d \to \infty.$$

Thus, in this case the general term in (1.9) does not approach zero, and the series diverges. Consequently, if the series (1.9) converges on some open subset of the $\langle n \rangle$-dimensional complex space of the variables $z_{\alpha\beta}$, then this subset must be contained in the region

(1.10) $$\mathbf{H}_n = \{Z = X + iY \in S_n(\mathbf{C}); Y > 0\},$$

called the *Siegel upper half-plane of degree n*. The region $\mathbf{H}_n$ is obviously an open subset of $\langle n \rangle$-dimensional complex space.

PROPOSITION 1.3. *The series $\theta^n(Z, Q)$, where $Q \in A_m^+$ and $n \in \mathbf{N}$, converges absolutely and uniformly on any subset of $\mathbf{H}_n$ of the form*

$$(1.11) \qquad \mathbf{H}_n(\varepsilon) = \{Z = X + iY \in \mathbf{H}_n; Y \geqslant \varepsilon E_n\},$$

where $\varepsilon > 0$ and E_n is the $n \times n$ identity matrix.

PROOF. Let ε_1 denote the smallest eigenvalue of the matrix Q. Then for any $N \in M_{m,n}(\mathbf{R})$ we have the inequality ${}^tNQN \geqslant \varepsilon_1 \, {}^tNN$. Consequently, on the set $\mathbf{H}_n(\varepsilon)$ the series

$$(1.12) \qquad \sum_{N \in M_{m,n}} \exp(\pi i \sigma({}^tNQNZ))$$

is majorized by the convergent numerical series

$$\sum_{N \in M_{m,n}} \exp(-\pi \varepsilon \varepsilon_1 \sigma({}^tNN)) = \left(\sum_{t \in \mathbf{Z}} \exp(-\pi \varepsilon \varepsilon_1 t^2) \right)^{mn},$$

and hence it converges absolutely and uniformly on this set. If we gather together all of the terms in (1.12) for which ${}^tNQN$ is equal to a fixed matrix $A \in A_n$, we see that the number of such terms is $r(Q, A)$, and thus the series (1.12) is equal to $\theta^n(Z, Q)$ in any region of absolute convergence. $\square$

The series

$$(1.13) \qquad \theta^n(Z, Q) = \sum_{A \in A_n} r(Q, A) \exp(\pi i \sigma(AZ)) = \sum_{N \in M_{m,n}} \exp(\pi i \sigma({}^tNQNZ))$$

is called the *theta-series of degree n for the matrix Q* (or the *corresponding quadratic form*). Proposition 1.3 immediately implies

THEOREM 1.4. *The theta-series $\theta^n(Z, Q)$ of degree n for the matrix $Q \in A_m^+$ determines a holomorphic function on $\mathbf{H}_n$; the function $\theta^n(Z, Q)$ is bounded on every subset $\mathbf{H}_n(\varepsilon) \subset \mathbf{H}_n$ with $\varepsilon > 0$.*

§2. Symplectic transformations

1. The symplectic group. On the Siegel upper half-plane $\mathbf{H}_n$, which arose in §1 as the domain of definition of theta-series, we have an action of a large group of biholomorphic one-to-one transformations. An obvious example of such transformations is the set of maps

$$U(V): Z \to {}^tV^{-1}ZV^{-1}, \qquad T(S): Z \to Z + S,$$

where $V \in GL_n(\mathbf{R})$ and $S \in S_n(\mathbf{R})$. Another such transformation is the map $J: Z \to -Z^{-1}$. To see that this has the desired properties, we must check that $-Z^{-1}$ exists and belongs to $\mathbf{H}_n$ for any $Z = X + iY \in \mathbf{H}_n$. Since Y is a symmetric positive definite matrix, there exists $V \in GL_n(\mathbf{R})$ such that $VY \, {}^tV$ is the identity matrix. We set $T = VX \, {}^tV$. Then $VZ \, {}^tV = T + iE_n$. Since the matrix $T^2 + E_n$ is positive definite, it is invertible, and so $T + iE$ is also invertible, with $(T + iE)^{-1} = (T - iE)(T^2 + E)^{-1}$ (here $E = E_n$). Thus, Z is an invertible matrix, and

$$-Z^{-1} = -{}^tV(T + iE)^{-1}V = {}^tV(-T + iE)(T^2 + E)^{-1}V$$

is contained in $\mathbf{H}_n$.

We now examine the group of analytic automorphisms of $\mathbf{H}_n$ that is generated by $U(V)$, $T(S)$, and J. We first note that each of the generating transformations is a fractional-linear transformation of the form

$$(2.1) \qquad Z \to M\langle Z \rangle = (AZ + B)(CZ + D)^{-1} \quad (Z \in \mathbf{H}_n),$$

where A, B, C, D are $n \times n$-matrices, and in the three cases the $2n \times 2n$-matrix $M = \begin{pmatrix} A & B \\ C & D \end{pmatrix}$ is, respectively:

$$(2.2) \qquad U(V) = \begin{pmatrix} V^* & 0 \\ 0 & V \end{pmatrix}, \quad \text{where } V \in GL_n(\mathbf{R}), \quad V^* = {}^tV^{-1},$$

$$(2.3) \qquad T(S) = \begin{pmatrix} E & S \\ 0 & E \end{pmatrix}, \quad \text{where } E = E_n, \quad S \in S_n(\mathbf{R}),$$

in which $S_n(\mathbf{R})$ denotes the set of symmetric matrices in $M_n(\mathbf{R})$, and

$$(2.4) \qquad J = J_n = \begin{pmatrix} 0 & E \\ -E & 0 \end{pmatrix}.$$

Furthermore, it is easy to see that the composition of any two automorphisms of the form (2.1) is also an automorphism of the form (2.1) with matrix equal to the product of the original matrices. Thus, we obtain

PROPOSITION 2.1. *Let* $\mathbf{S}$ *be the subgroup of* $GL_{2n}(\mathbf{R})$ *generated by the matrices* (2.2)–(2.4). *Then for every* $M = \begin{pmatrix} A & B \\ C & D \end{pmatrix} \in \mathbf{S}$ *the matrix* $CZ + D$ *is invertible for all* $Z \in \mathbf{H}_n$, *and the map*

$$(2.5) \qquad f(M) \colon Z \to M\langle Z \rangle$$

is a holomorphic automorphism of $\mathbf{H}_n$. *The map* $M \to f(M)$ *gives a homomorphism from* $\mathbf{S}$ *to the group of holomorphic automorphisms of* $\mathbf{H}_n$.

In order to characterize $\mathbf{S}$ as an algebraic group, we first note that each generator (2.2)–(2.4) leaves invariant the skew-symmetric bilinear form with matrix (2.4), i.e., it satisfies the relation ${}^tMJ_nM = J_n$. Hence, $\mathbf{S}$ is contained in the group

$$(2.6) \qquad \mathrm{Sp}_n(\mathbf{R}) = \{M \in M_{2n}(\mathbf{R}); \, {}^tMJ_nM = J_n\},$$

which is called the *real symplectic group of degree n*. It follows from the definition that a $2n \times 2n$ real matrix $M = \begin{pmatrix} A & B \\ C & D \end{pmatrix}$ with $n \times n$-blocks A, B, C, D is *symplectic* (i.e., belongs to the symplectic group of degree n) if and only if

$$(2.7) \qquad {}^tAC = {}^tCA, \quad {}^tBD = {}^tDB, \quad {}^tAD - {}^tCB = E_n.$$

It is easy to see that a matrix M is symplectic if and only if the matrix ${}^tM = JM^{-1}J^{-1}$ is symplectic. This implies that the conditions (2.7) can be rewritten in the form

$$(2.8) \qquad A\,{}^tB = B\,{}^tA, \quad C\,{}^tD = D\,{}^tC, \quad A\,{}^tD - B\,{}^tC = E_n.$$

Finally, we note that the inverse of a symplectic matrix $M = \begin{pmatrix} A & B \\ C & D \end{pmatrix}$ is

$$(2.9) \qquad M^{-1} = J^{-1}\,{}^tMJ = \begin{pmatrix} {}^tD & -{}^tB \\ -{}^tC & {}^tA \end{pmatrix}.$$

THEOREM 2.2. *The symplectic group of degree n is generated by the matrices (2.2)–(2.4). In other words,* $\mathbf{S} = \mathrm{Sp}_n(\mathbf{R})$.

PROOF. Let $M = \begin{pmatrix} A & B \\ C & D \end{pmatrix}$ be an arbitrary symplectic matrix. The upper-left block of the symplectic matrix $U(V)MU(V_1)$ is equal to $V^*AV_1^*$, and so by a suitable choice of V, $V_1 \in GL_n(\mathbf{R})$ this block can be brought to the form $\begin{pmatrix} E_r & 0 \\ 0 & 0 \end{pmatrix}$, where r is the rank of A and E_r is the $r \times r$ identity matrix. Thus, we may assume from the beginning that $A = \begin{pmatrix} E_r & 0 \\ 0 & 0 \end{pmatrix}$. Then, if $C = \begin{pmatrix} C_1 & C_2 \\ C_3 & C_4 \end{pmatrix}$ is the corresponding partition of C into blocks, the first relation in (2.7) shows that $C_2 = 0$ and $C_1 = {}^tC_1$. In addition, $\det C_4 \neq 0$, since otherwise the first n columns of M would be linearly dependent. If we now pass from M to the matrix $T(\lambda E_n)M$, where λ is a real number, we see that the new matrix has upper-left block equal to

$$A' = A + \lambda C = \begin{pmatrix} E_r + \lambda C_1 & 0 \\ \lambda C_3 & \lambda C_4 \end{pmatrix}$$

and so it has rank n for λ sufficiently small. We see that from the beginning we may assume, without loss of generality, that $A = E_n$. Now, by the first relation in (2.7), C is a symmetric matrix, and

$$J^{-1}T(C)JM = \begin{pmatrix} E_n & 0 \\ -C & E_n \end{pmatrix}\begin{pmatrix} E_n & B \\ C & D \end{pmatrix} = \begin{pmatrix} E_n & B \\ 0 & D_1 \end{pmatrix}.$$

The third and second relations in (2.7) show that $D_1 = E_n$ and ${}^tB = B$ in the last matrix, and hence it is equal to the matrix $T(B)$. □

PROBLEM 2.3. Suppose that the matrix $M = \begin{pmatrix} A & B \\ C & D \end{pmatrix} \in M_{2n}(\mathbf{R})$ satisfies the condition ${}^tMJ_nM = rJ_n$, $r \neq 0$, where J_n is the matrix (2.4). Show that the map (2.5) is defined and holomorphic on $\mathbf{H}_n$, maps $\mathbf{H}_n$ onto itself if $r > 0$, and maps $\mathbf{H}_n$ onto $\{\overline{Z} = X - Yi;\ Z = X + iY \in \mathbf{H}_n\}$ if $r < 0$.

2. The Siegel upper half-plane. We saw that the group $\mathrm{Sp}_n(\mathbf{R})$ acts as a group of holomorphic automorphisms of $\mathbf{H}_n$ according to the rule

$$(2.10) \qquad M = \begin{pmatrix} A & B \\ C & D \end{pmatrix} : Z \to M\langle Z\rangle = (AZ + B)(CZ + D)^{-1}.$$

PROPOSITION 2.4. *The action of the symplectic group on the upper half-plane is transitive.*

PROOF. Let $Z = X + iY \in \mathbf{H}_n$. Since $Y > 0$, there exists a matrix $A \in GL_n(\mathbf{R})$ with ${}^tAA = Y$. Then $M = T(X)U(A^{-1})$ is a symplectic matrix, and $M\langle iE_n\rangle = X + i\,{}^tAA = Z$. □

This transitivity implies that $\mathbf{H}_n$ can be identified with the quotient $\mathrm{Sp}_n(\mathbf{R})/\mathrm{St}(Z)$ of the symplectic group by the stabilizer of an arbitrary point $Z \in \mathbf{H}_n$. All of the stabilizers are obviously conjugate to the stabilizer of the point iE_n. The structure of the latter group is given by the next proposition.

PROPOSITION 2.5. *One has*

$$\{M \in \mathrm{Sp}_n(\mathbf{R}); M\langle iE_n\rangle = iE_n\} = \left\{M = \begin{pmatrix} A & B \\ -B & A \end{pmatrix}; u(M) = A + iB \in U(n)\right\},$$

where $U(n)$ is the unitary group of order n. The map $M \to u(M)$ is an isomorphism of $\mathrm{St}(iE_n)$ with the unitary group $U(n)$.

PROOF. The proposition follows easily from the definitions. $\square$

If $Z = X + iY$ and $Z_1 = X_1 + iY_1$ are any two matrices in $\mathbf{H}_n$ and $t \in \mathbf{R}$, $0 \leqslant t \leqslant 1$, then the matrix $tY + (1 - t)Y_1$ is obviously positive definite. Hence, $tZ + (1 - t)Z_1 \in \mathbf{H}_n$. This remark implies

PROPOSITION 2.6. $\mathbf{H}_n$ *is a convex and simply connected domain.*

The upper half-plane $\mathbf{H}_n$ is obviously an open subset of $n(n + 1)$-dimensional real space. We shall show that $\mathbf{H}_n$ has an $n(n + 1)$-dimensional volume element that is invariant under all symplectic transformations, and we shall find such an element. With this purpose in mind, we first examine what happens under symplectic transformations to the Euclidean volume element on $\mathbf{H}_n$.

LEMMA 2.7. *Let*

$$dZ = \prod_{1 \leqslant \alpha \leqslant \beta \leqslant n} dx_{\alpha\beta} dy_{\alpha\beta} \quad (Z = (x_{\alpha\beta} + iy_{\alpha\beta}) \in \mathbf{H}_n)$$

be the Euclidean volume element on $\mathbf{H}_n$. Then for any symplectic matrix $M = \begin{pmatrix} A & B \\ C & D \end{pmatrix}$ we have

$$dM\langle Z\rangle = |\det(CZ + D)|^{-2n-2} dZ.$$

PROOF. For $Z = (z_{\alpha\beta}) = (x_{\alpha\beta} + iy_{\alpha\beta})$ we set $Z' = (z'_{\gamma\delta}) = (x'_{\gamma\delta} + iy'_{\gamma\delta}) = M\langle Z\rangle$. To prove the lemma, we must find the Jacobian determinant of the variables $x'_{\gamma\delta}, y'_{\gamma\delta}$ with respect to the variables $x_{\alpha\beta}, y_{\alpha\beta}$, i.e., the determinant of the transition matrix from the $n(n + 1)$-vector whose components (in any order) are the differentials $dx'_{\gamma\delta}, dy'_{\gamma\delta}$, to the analogous vector with components $dx_{\alpha\beta}, dy_{\alpha\beta}$. It is actually simpler to work with the corresponding question for $\langle n\rangle$-vectors made up of the complex differentials $dz'_{\gamma\delta} = dx'_{\gamma\delta} + i\,dy'_{\gamma\delta}$ and $dz_{\alpha\beta} = dx_{\alpha\beta} + i\,dy_{\alpha\beta}$. If $Z_1, Z_2 \in \mathbf{H}_n$, then, taking into account the symmetry of Z'_2, we have

(2.11)
$$\begin{aligned} Z'_2 - Z'_1 &= (Z_2\,{}^tC + {}^tD)^{-1}(Z_2\,{}^tA + {}^tB) - (AZ_1 + B)(CZ_1 + D)^{-1} \\ &= (Z_2\,{}^tC + {}^tD)^{-1}\{(Z_2\,{}^tA + {}^tB)(CZ_1 + D) \\ &\qquad\qquad - (Z_1\,{}^tC + D)(AZ_1 + B)\}(CZ_1 + D)^{-1} \\ &= (Z_2\,{}^tC + {}^tD)^{-1}(Z_2 - Z_1)(CZ_1 + D)^{-1}, \end{aligned}$$

where we used (2.7) in the last step. From (2.11) it follows that

(2.12)
$$DZ' = {}^t(CZ + D)^{-1}DZ(CZ + D)^{-1},$$

where $DZ = (dz_{\alpha\beta})$ and $DZ' = (dz'_{\gamma\delta})$ are symmetric matrices of complex differentials. We let ρ denote the $\langle n\rangle$-dimensional representation of $GL_n(\mathbf{C})$ which associates to every matrix U the linear transformation $(v_{\alpha\beta}) \to U(v_{\alpha\beta})\,{}^tU$ of the variables $v_{\alpha\beta} = v_{\beta\alpha}$,

$1 \leqslant \alpha, \beta \leqslant n$ (this is the *symmetric square* of the standard representation of GL_n). Then

$$(2.13) \qquad\qquad \det \rho(U) = (\det U)^{n+1}. \qquad \square$$

In fact, by replacing U by a matrix of the form $W^{-1}UW$ for a suitable W, without loss of generality we may assume that U is upper triangular (for example, a matrix in Jordan normal form). If $u_1, \ldots, u_n$ are the diagonal entries in U and the variables $v_{\alpha\beta}$, $1 \leqslant \alpha \leqslant \beta \leqslant n$, are ordered lexicographically, then it is not hard to see that $\rho(U)$ is also an upper-triangular matrix with diagonal entries $u_1 u_1, \ldots, u_1 u_n, \ldots, u_n u_n$; this implies (2.13). The relations (2.12)–(2.13) enable us to find the determinant of the transition matrix for the complex differentials under the map $Z \to Z'$. We return to the real differentials. Let $\mathbf{d}Z$ ($\mathbf{d}Z'$) be the $\langle n \rangle$-dimensional column with components $dz_{\alpha\beta}$ ($dz'_{\gamma\delta}$), arranged in any order. Then in the above notation we can write $\mathbf{d}Z' = \rho((CZ + D)^*)\mathbf{d}Z$. Setting $\rho((CZ+D)^*) = R + iS$ and separating the real and imaginary parts, we obtain the relations $\mathbf{d}X' = R\mathbf{d}X - S\mathbf{d}Y$, $\mathbf{d}Y' = S\mathbf{d}X + R\mathbf{d}Y$. Thus, the Jacobian matrix of the transformation $Z \to Z'$ is $\begin{pmatrix} R & -S \\ S & R \end{pmatrix}$, and

$$\det \begin{pmatrix} R & -S \\ S & R \end{pmatrix} = \det \begin{pmatrix} E_n & iE_n \\ 0 & E_n \end{pmatrix} \begin{pmatrix} R & -S \\ S & R \end{pmatrix} \begin{pmatrix} E_n & -iE_n \\ 0 & E_n \end{pmatrix}$$

$$= \det \begin{pmatrix} R + iS & 0 \\ 0 & R - iS \end{pmatrix}$$

$$= |\det(R + iS)|^2 = |\det(CZ + D)|^{-2n-2}.$$

LEMMA 2.8. *If $Z' = X' + iY' = M\langle Z \rangle$, where $Z = X + iY \in \mathbf{H}_n$ and $M = \begin{pmatrix} A & B \\ C & D \end{pmatrix} \in \mathrm{Sp}_n(\mathbf{R})$, then*

$$Y' = {}^t(C\overline{Z} + D)^{-1} Y(CZ + D)^{-1},$$

in particular,

$$(2.14) \qquad\qquad \det Y' = |\det(CZ + D)|^{-2} \det Y.$$

PROOF. If we compute $Y' = -(1/2i)(\overline{Z}' - Z')$ using equation (2.11), we obtain the lemma. $\qquad \square$

Combining Lemmas 2.7 and 2.8, we obtain

PROPOSITION 2.9. *The volume element on the Siegel upper half-plane $\mathbf{H}_n$ that is given by*

$$d^*Z = (\det Y)^{-n-1} dZ = \det(y_{\alpha\beta})^{-n-1} \prod_{1 \leqslant \alpha \leqslant \beta \leqslant n} dx_{\alpha\beta} \cdot dy_{\alpha\beta},$$

where $Z = X + iY = (x_{\alpha\beta}) + i(y_{\alpha\beta}) \in \mathbf{H}_n$, is invariant relative to all symplectic transformations, i.e., $d^*M\langle Z \rangle = d^*Z$ for $M \in \mathrm{Sp}_n(\mathbf{R})$.

PROBLEM 2.10. Prove that the Cayley map

$$Z \to W = (Z - iE_n)(Z + iE_n)^{-1} \quad (Z \in \mathbf{H}_n)$$

gives an analytic isomorphism of $\mathbf{H}_n$ with the bounded region $\{W \in S_n(\mathbf{C}); \overline{W} \cdot W < E_n\}$, where the inequality is understood in the sense of Hermitian matrices. Prove that the inverse map is given by the formula $W \to Z = i(E_n + W)(E_n - W)^{-1}$

PROBLEM 2.11. Prove that the volume element

$$d^*Y = (\det Y)^{-(n+1)/2} \prod_{1 \leqslant \alpha \leqslant \beta \leqslant n} dy_{\alpha\beta}$$

on the space $P_n = \{Y \in S_n(\mathbf{R}); Y > 0\}$ is invariant relative to all transformations of the form $Y \to {}^t g Y g$, where $g \in GL_n(\mathbf{R})$.

§3. Symplectic transformations of theta-series

1. Transformations of theta-series. The analytic and algebraic study of theta-series is based on the remarkable fact that the theta-series of integral quadratic forms transform according to certain simple rules under a rather large group of symplectic transformations. Usually, this group is the subgroup of the symplectic group that is generated by certain standard transformations whose action on theta-series can be found by a direct computation. However, in the multidimensional situation, where the generators and relations for these subgroups are often unknown, it is very difficult to find the transformation groups of arbitrary theta-series directly. Instead, we first express all theta-series in terms of the simplest ones—the "theta-functions". After determining the transformation groups of theta-functions, we can readily find the transformation groups of arbitrary theta-series.

We introduce some notation. If Q is a symmetric $k \times k$-matrix and N is a $k \times l$-matrix, we write

$$(3.1) \qquad\qquad Q[N] = {}^t NQN.$$

If $Z \in \mathbf{H}_k$, $W, W' \in M_{k,1}(\mathbf{C})$, then it is easy to see that the series

$$(3.2) \qquad \theta^k(Z; W, W') = \sum_{N \in M_{k,1}} \exp(\pi i (Z[N - W'] + 2 {}^t NW - {}^t W' \cdot W))$$

is absolutely convergent. Here, if $Z \in \mathbf{H}_n(\varepsilon)$, where $\varepsilon > 0$ (see (1.11)), and if W and W' belong to fixed compact subsets of $M_{k,1}(\mathbf{C})$, then the series (3.2) converges uniformly, as do the series that are obtained from it by taking partial derivatives. The holomorphic function on $\mathbf{H}_k \times M_{k,1}(\mathbf{C}) \times M_{k,1}(\mathbf{C})$ that is defined by the series (3.2) is called the *theta-function of degree n*.

Let $V \in \Lambda^k = GL_k(\mathbf{Z})$. If we replace Z by $Z[{}^tV]$ in (3.2) and take into account the absolute convergence of the series, we find by a simple computation that

$$\theta^k(Z[{}^tV]; W, W') = \theta^k(Z; V^{-1}W, {}^tVW'),$$

and hence, replacing W by VW and W' by V^*W', we obtain the identity

$$(3.3) \qquad \theta^k(Z[{}^tV]; VW, V^*W') = \theta^k(Z; W, W') \quad (V \in \Lambda^k).$$

Now let $S = (s_{\alpha\beta})$ be an integral symmetric $k \times k$-matrix. Substituting $Z + S$ in place of Z in (3.2), we obtain

$$\theta^k(Z + S; W, W') = \sum_{N \in M_{k,1}} \exp(\pi i(Z[N - W'] + S[N - W'] + 2\,{}^t NW - {}^t W'W)).$$

Since $S[N - W'] = S[N] - 2\,{}^t NSW' + S[W']$ and

$$S[N] = \sum_{\alpha,\beta} s_{\alpha\beta} n_\alpha n_\beta = \sum_\alpha s_{\alpha\alpha} n_\alpha^2 + 2 \sum_{\alpha<\beta} s_{\alpha\beta} n_\alpha n_\beta$$

$$\equiv \sum_\alpha s_{\alpha\alpha} n_\alpha = {}^t N \cdot \mathrm{dc}(S) \pmod 2,$$

where $\mathrm{dc}(S)$ is the column with components $s_{\alpha\alpha}$, it follows that

$$\theta^k(Z + S; W, W') = \exp\left(\frac{\pi i}{2}\,{}^t W' \cdot \mathrm{dc}(S)\right)\theta^k\left(Z; W - SW' + \frac{1}{2}\mathrm{dc}(S), W'\right),$$

from which, if we substitute $W \to W + SW' - \frac{1}{2}\mathrm{dc}(S)$ and divide both sides by $\exp(\frac{\pi i}{2}\,{}^t W' \cdot \mathrm{dc}(S))$, we obtain the identity

$$(3.4) \quad \exp\left(-\frac{\pi i}{2}\,{}^t W' \cdot \mathrm{dc}(S)\right)\theta^k\left(Z + S;\, W + SW' - \frac{1}{2}\mathrm{dc}(S), W'\right) = \theta^k(Z; W, W').$$

The last formula—and the only nontrivial formula among the basic transformation rules for theta-functions—is the *inversion formula*, which had its origin in the famous *Jacobi inversion formula*.

LEMMA 3.1 (Inversion formula for theta-functions). *One has the identity*

$$(3.5) \qquad \theta^k(-Z^{-1}; W', -W) = (\det(-iZ))^{1/2}\theta^k(Z; W, W'),$$

where the square root is positive for $Z = iY$ and is extended to arbitrary Z by analytic continuation (see Proposition 2.6).

PROOF. The function

$$\exp(-\pi i\,{}^t W' W)\theta^k(Z; W, W') = \sum_{N \in M_{k,1}} \exp(\pi i Z[N - W'] + 2\pi i\,{}^t(N - W')W)$$

obviously depends holomorphically on the components w'_t of the vector W' and is periodic of period 1 in each component. We introduce new variables u_t by setting $u_t = \exp(2\pi i w'_t)$, $w'_t = 1/(2\pi i)\log u_t$. Then our function is a single-valued analytic function in $u_1, \ldots, u_k$ in the region $0 \leqslant |u_1|, \ldots, |u_k| < \infty$. Consequently, it has an absolutely convergent Laurent expansion

$$(3.6) \quad \begin{aligned} \exp(-\pi i\,{}^t W' W)\theta^k(Z; W, W') &= \sum_{l_1,\ldots,l_k=-\infty}^{\infty} c(L)u_1^{l_1}\cdots u_k^{l_k} \\ &= \sum_{L \in M_{k,1}} c(L)\exp(2\pi i\,{}^t W'L), \end{aligned}$$

where the coefficients $c(L)$ depend only on Z, W, and L. This series converges uniformly if W' belongs to any set of the form $M_{k,1}(\mathbf{R}) + W_0$ with fixed W_0 (since the series is majorized by an absolutely convergent numerical series); and the series can be integrated term-by-term over subsets of such sets. We multiply both sides of (3.6) by $\exp(-2\pi i\, {}^tW'L)$, set $W' = H + W_0$ (where W_0 will be chosen later), and integrate term-by-term over the unit cube $C = \{H = (h_t) \in M_{k,1}(\mathbf{R}); \ 0 \leqslant h_t \leqslant 1\}$ with respect to the Euclidean measure $dH = dh_1 \cdots dh_k$. We obtain

$$c(L) = \int_C \theta^k(Z; W, H + W_0) \exp(-\pi i\, {}^t(H + W_0)W - 2\pi i\, {}^t(H + W_0)L)\, dH$$

$$= \int_C \sum_{N \in M_{k,1}} \exp(\pi i (Z[N - H - W_0])$$

$$+ 2\, {}^t(N - H - W_0)W + 2\, {}^t(N - H - W_0)L))\, dH$$

(note that the numbers ${}^tNL$ are integers). If we integrate the uniformly convergent series term-by-term, we obtain

$$c(L) = \sum_{N \in M_{k,1} - N + C} \int \exp(\pi i (Z[H + W_0] - 2\, {}^t(H + W_0)(W + L)))\, dH$$

$$= \int_{M_{k,1}(\mathbf{R})} \exp(\pi i (Z[H + W_0] - 2\, {}^t(H + W_0)(W + L)))\, dH.$$

Applying the obvious identity

$$Z[H + W_0] - 2\, {}^t(H + W_0)(W + L) = Z[H + W_0 - Z^{-1}(W + L)] - Z^{-1}[W + L]$$

and then setting $W_0 = Z^{-1}(W + L)$, we arrive at the formula

$$c(L) = \exp(-\pi i Z^{-1}[L + W]) \int_{M_{k,1}(\mathbf{R})} \exp(\pi i Z[H])\, dH.$$

If we substitute these expressions into (3.6), we find that

$$\theta^k(Z; W, W') = I(Z) \sum_{L \in M_{k,1}} \exp(\pi i (-Z^{-1}[L + W] + 2\, {}^tLW' + {}^tWW'))$$

$$= I(Z)\theta^k(-Z^{-1}; W', -W),$$

where

$$I(Z) = \int_{M_{k,1}(\mathbf{R})} \exp(\pi i Z[H])\, dH.$$

If $Z = iY$, then, making the change of variables $H = VH'$, where $V \in GL_k(\mathbf{R})$ and $Y[V] = E_k$, we obtain

$$I(Z) = \int_{M_{k,1}(\mathbf{R})} \exp(-\pi\, {}^tH' \cdot H')|\det V|\, dH'$$

$$= (\det Y)^{-1/2} \left(\int_{\mathbf{R}} \exp(-\pi h^2)\, dh \right)^k = \det(-iZ)^{-1/2}.$$

Since the left and right sides of this equality are holomorphic functions of each entry in the matrix $Z \in \mathbf{H}_k$, it follows by the principle of analytic continuation that the equality holds for all $Z \in \mathbf{H}_k$. $\qquad \square$

In order to determine the transformation group that is implicit in the functional equations (3.3)–(3.5), we introduce some notation. First of all, for $W, W' \in M_{k,1}(\mathbf{C})$ we set

$$\Omega = \begin{pmatrix} W \\ W' \end{pmatrix} \in M_{2k,1}(\mathbf{C}),$$

and we define

(3.7) $$\theta(Z, \Omega) = \theta^k(Z; W, W').$$

Next, we let

(3.8) $$\Gamma^k = \mathrm{Sp}_k(\mathbf{Z}) = \mathrm{Sp}_k(\mathbf{R}) \cap M_{2k}$$

denote the set of integral symplectic matrices. It follows from the definition that Γ^k is a semigroup. The relation (2.9) shows that Γ^k is a group. The group Γ^k is called the *Siegel modular group of degree k*. Given a matrix $M = \begin{pmatrix} A & B \\ C & D \end{pmatrix} \in \Gamma^k$, we let $\xi(M)$ and $\eta(M)$ denote the diagonal entries (arranged in a column) of the symmetric matrices $B\,{}^tA$ and $C\,{}^tD$, respectively, and we set

(3.9) $$\zeta(M) = \begin{pmatrix} \xi(M) \\ \eta(M) \end{pmatrix} = \begin{pmatrix} \mathrm{dc}(B\,{}^tA) \\ \mathrm{dc}(C\,{}^tD) \end{pmatrix} \in M_{2k,1}.$$

Finally, given $M = \begin{pmatrix} A & B \\ C & D \end{pmatrix} \in \Gamma^k$ and a function F on $\mathbf{H}_k \times M_{2k,1}(\mathbf{C})$, we define

(3.10)
$$
\begin{aligned}
(F|M)(Z, \Omega) = {}& \det(CZ + D)^{-1/2} \exp\left(-\frac{\pi i}{2}\,{}^t\zeta(M) J_k M \Omega \right) \\
& \times F\left(M\langle M \rangle, M\Omega - \frac{1}{2}\zeta(M) \right),
\end{aligned}
$$

where J_k is the matrix (2.4). The function $F|M$ is defined up to a sign, which depends on which root is chosen in the first factor.

REMARK. From now on, unless stated otherwise, the symbol $\varphi^{1/2}$, where $\varphi = \varphi(Z)$ is a certain nonvanishing holomorphic function on the upper half-plane $\mathbf{H}_k$, will denote one of the two single-valued (because $\mathbf{H}_k$ is simply-connected) holomorphic functions on $\mathbf{H}_k$ obtained by analytic continuation of any local element of the function $\pm\varphi^{1/2}$. The two possible choices differ by a sign.

Now the formulas (3.3)–(3.5) can be rewritten in the form

(3.11) $$(\theta|M)(Z, \Omega) = \chi(M)\theta(Z, \Omega),$$

where the matrix M in Γ^k is equal, respectively, to

(3.12) $$U(V), \quad T(S), \quad J_k,$$

where $V \in \Lambda^k$, $S \in S_k$, and the number $\chi(M)$ is chosen so that the product $\chi(M)\det(CZ + D)^{1/2}$ is equal to 1 in the first two cases and is equal to $\det(-iZ)^{1/2}$ in the third case.

PROPOSITION 3.2. *Let Γ' denote the subgroup of Γ^k that is generated by all elements of the form (3.12). Then (3.11) holds for any $M \in \Gamma'$. Here $\chi(M)$ is a certain eighth root of unity that depends upon the sign chosen for the root in (3.10).*

PROOF. Since (3.11) holds for the generators of the group Γ', the proposition follows if we verify that for any $M, M_1 \in \Gamma'$

$$(3.13) \qquad (\theta|M|M_1)(Z, \Omega) = \varepsilon \cdot (\theta|MM_1)(Z, \Omega),$$

where ε is an eighth root of unity that depends both on the matrices M and M_1 and on the choice of sign in the definitions of $\theta|M$, $\theta|M|M_1$, and $\theta|MM_1$.

First of all, it is not hard to check by a direct substitution that, when the vector $\zeta(M)$ in the definition of $F|M$ is replaced by any vector of the form $\zeta(M) + 2L$, where $L = \binom{L_1}{L_2} \in M_{2k,1}$, the expression for $F|M$ is multiplied by a fourth root of unity equal to

$$\exp\left(-\frac{\pi i}{2} \cdot {}^t L J_k \zeta(M) - \pi i \, {}^t L_1 L_2\right).$$

Furthermore, if $M = \begin{pmatrix} A & B \\ C & D \end{pmatrix}$ and $M_1 = \begin{pmatrix} A_1 & B_1 \\ C_1 & D_1 \end{pmatrix}$, then from the definitions we have

$$\begin{aligned}
(\theta|M|M_1)(Z, \Omega) = {} & \det(C_1 Z + D_1)^{-1/2} \exp\left(-\frac{\pi i}{2} \cdot {}^t\zeta(M_1) J M_1 \Omega\right) \\
& \times \det(C M_1\langle Z\rangle + D)^{-1/2} \\
& \times \exp\left(-\frac{\pi i}{2} \cdot {}^t\zeta(M) J M \left(M_1 \Omega - \frac{1}{2}\zeta(M_1)\right)\right) \\
& \times \theta\left(M\langle M_1\langle Z\rangle\rangle, M\left(M_1\Omega - \frac{1}{2}\zeta(M_1)\right) - \frac{1}{2}\zeta(M)\right).
\end{aligned}$$

Since

$$(3.14) \qquad \begin{aligned}
CM_1\langle Z\rangle + D &= (C(A_1 Z + B_1) + D(C_1 Z + D_1))(C_1 Z + D_1)^{-1} \\
&= (C_2 Z + D_2)(C_1 Z + D_1)^{-1},
\end{aligned}$$

if $MM_1 = \begin{pmatrix} A_2 & B_2 \\ C_2 & D_2 \end{pmatrix}$, and since the number $\exp(\frac{\pi i}{4} \cdot {}^t\zeta(M) J M \zeta(M_1))$ to the eighth power is 1, it follows that, up to an eighth root of unity, the last expression is equal to

$$\begin{aligned}
\det(C_2 Z + D_2)^{-1/2} \exp\left(-\frac{\pi i}{2} \cdot {}^t(M\zeta(M_1) + \zeta(M)) J MM_1 \Omega\right) \\
\times \theta\left(MM_1\langle Z\rangle, MM_1\Omega - \frac{1}{2}(M\zeta(M_1) + \zeta(M))\right),
\end{aligned}$$

where we used the equality $J[M] = J$ to transform the exponent in the first term. If we prove that $M\zeta(M_1) + \zeta(M) \equiv \zeta(MM_1) \pmod 2$, then it will follow that the last expression differs from $(\theta|MM_1)(Z, \Omega)$ only by a fourth root of unity. This will prove (3.13), and hence the proposition.

LEMMA 3.3. *For any* $M, M_1 \in \Gamma^k$ *one has* $\zeta(MM_1) \equiv M\zeta(M_1) + \zeta(M) \pmod 2$.

PROOF OF THE LEMMA. If $M = \begin{pmatrix} A & B \\ C & D \end{pmatrix}$ and $M_1 = \begin{pmatrix} A_1 & B_1 \\ C_1 & D_1 \end{pmatrix}$, then

$$MM_1 = \begin{pmatrix} AA_1 + BC_1 & AB_1 + BD_1 \\ CA_1 + DC_1 & CB_1 + DD_1 \end{pmatrix},$$

and, by definition,

$$\zeta(MM_1) = \begin{pmatrix} \mathrm{dc}(AB_1 + BD_1) \cdot {}^t(AA_1 + BC_1) \\ \mathrm{dc}(CA_1 + DC_1) \cdot {}^t(CB_1 + DD_1) \end{pmatrix}.$$

If M and S are square integer matrices of the same size, and if S is symmetric, then it is not hard to see that

$$\mathrm{dc}(MS\,{}^tM) \equiv M\,\mathrm{dc}(S)\,(\mathrm{mod}\,2).$$

From this congruence, the relations (2.8), and the fact that the diagonal does not change when taking the transpose, it follows that

$$\zeta(MM_1) = \begin{pmatrix} A \cdot \mathrm{dc}(B_1\,{}^tA_1) + B \cdot \mathrm{dc}(C_1\,{}^tD_1) + \mathrm{dc}(A(B_1\,{}^tC_1 + A_1\,{}^tD_1)\,{}^tB) \\ C \cdot \mathrm{dc}(B_1\,{}^tA_1) + D \cdot \mathrm{dc}(C_1\,{}^tD_1) + \mathrm{dc}(C(A_1\,{}^tD_1 + B_1\,{}^tC_1)\,{}^tD) \end{pmatrix}$$
$$\equiv M\zeta(M_1) + \zeta(M)\,(\mathrm{mod}\,2),$$

since, by (2.8), $A_1\,{}^tD_1 + B_1\,{}^tC_1 = E_k + 2B_1\,{}^tC_1 \equiv E_k\,(\mathrm{mod}\,2)$. This proves the lemma and Proposition 3.2. $\square$

PROBLEM 3.4. Show that the theta-function of degree k satisfies the relations

$$\theta^k(Z; W + L, W') = \exp(-\pi i\,{}^tLW')\theta^k(Z; W, W'),$$
$$\theta^k(Z; W + ZL, W') = \exp(\pi i(Z[L] - 2\,{}^tLW + {}^tLZW'))\theta^k(Z; W, W')$$

for any k-dimensional integer vector $L \in M_{k,1}$. Further show that any function $F(Z; W, W')$ that satisfies all of these relations and is holomorphic in W has the form $F(Z; W, W') = F_0(Z, W')\theta^k(Z; W, W')$, where F_0 depends only on Z and W'.

PROBLEM 3.5. Show that the theta-function of degree k satisfies the relations

$$\theta^k(Z; W, W' + L) = \exp(\pi i\,{}^tLW)\theta^k(Z; W, W'),$$
$$\theta^k(Z; W, W' - Z^{-1}L) = \exp(\pi i(-Z^{-1}[L] - 2\,{}^tLW' + {}^tLZW))\theta^k(Z; W, W')$$

for any k-dimensional integer vector $L \in M_{k,1}$. Further show that any function $F(Z; W, W')$ that satisfies all of these relations and is holomorphic in W' has the form $F(Z; W, W') = F_1(Z, W)\theta^k(-Z^{-1}; W', -W)$, where F_1 depends only on Z and W.

2. The Siegel modular group and the theta-group. In this subsection we show that the group Γ' that is generated by matrices of the form (3.12) is actually the entire Siegel group Γ^k. Thus, the functional equations (3.11) hold for all $M \in \Gamma^k$. These equations take a particularly simple form if M belongs to a certain subgroup of Γ^k called the theta-group.

THEOREM 3.6. *The Siegel modular group of degree k*

$$\Gamma^k = \mathrm{Sp}_k(Z)$$

is generated by matrices of the form (3.12).

For later use, we shall prove a more general fact.

PROPOSITION 3.7. *Let M be a $2k \times 2k$-matrix with entries in the rational number field $\mathbf{Q}$ which satisfies the condition*

$$(3.15) \qquad\qquad {}^{t}MJ_{k}M = rJ_{k},$$

where $r \neq 0$. Then there exists a matrix g in the group Γ' that is generated by matrices of the form (3.12) such that the product gM has a $k \times k$ block of zeros in the lower-left corner:

$$gM = \begin{pmatrix} A_{1} & B_{1} \\ 0 & D_{1} \end{pmatrix}.$$

We first note that Theorem 3.6 follows from Proposition 3.7. Namely, if $M \in \Gamma^{k}$, then, by Proposition 3.7, there exists $g \in \Gamma'$ such that the matrix $M_{1} = gM$ has the above form. Since M_{1} is a symplectic matrix, we have ${}^{t}A_{1}D_{1} = E_{k}$ and ${}^{t}B_{1}D_{1} = {}^{t}D_{1}B_{1}$ (see (2.7)). Since M_{1} is also an integer matrix, it follows that $A_{1}, D_{1} \in \Lambda^{k}$ and $S = B_{1}D_{1}^{-1} = {}^{t}(B_{1}D_{1}^{-1}) \in S_{k}$. Thus, $M_{1} = T(S)U(D_{1})$ and $M = g^{-1}M_{1} \in \Gamma'$.

Before proceeding to the proof of Proposition 3.7, we prove two useful lemmas.

LEMMA 3.8. *Let $k \geqslant 2$, and let $u = {}^{t}(u_{1}, \ldots, u_{k})$ be an arbitrary nonzero k-dimensional column of integers. Then there exists a matrix V in the group $SL_{k}(\mathbf{Z})$ of $k \times k$ integer matrices of determinant $+1$ such that*

$$(3.16) \qquad\qquad Vu = {}^{t}(d, 0, \ldots, 0),$$

where d is the greatest common divisor of $u_{1}, \ldots, u_{k}$.

PROOF. For $k = 2$ the lemma follows from the fact that the g.c.d. of two integers can be written as an integer linear combination of those integers. The general case follows by an obvious induction on k. □

LEMMA 3.9. *Let u be a nonzero $2k$-dimensional column of integers. Then there exists a matrix $g \in \Gamma'$ such that*

$$(3.17) \qquad\qquad gu = {}^{t}(d, 0, \ldots, 0),$$

where d is the greatest common divisor of the entries in u.

PROOF. We shall write u in the form $u = \binom{a}{c}$, where a and c are k-dimensional columns. We first prove that the set $\{gu;\ g \in \Gamma'\}$ contains a column $u' = \binom{a'}{c'}$ with $a' = 0$ or $c' = 0$. Assume the contrary. Then $a' \neq 0$ and $c' \neq 0$ for every $u' = gu$. Let α' and γ' denote the greatest common divisor of the entries of a' and c', respectively. We choose u' in such a way that the product $\alpha'\gamma'$ is minimal and $\alpha' \geqslant \gamma'$ (if $\gamma' > \alpha'$, we replace u' by $J_{k}u'$). By Lemma 3.8, by replacing u' by $U(V)u'$ with a suitable $V \in SL_{k}(\mathbf{Z})$, we may assume that $c' = {}^{t}(\gamma', 0, \ldots, 0)$. Replacing u' by $T(S)u'$ takes a' to $a' + Sc'$ and does not change c'. We can clearly choose $S \in S_{k}$ in such a way that all of the entries in the column $a' + Sc'$ belong to the set $\{0, 1, \ldots, \gamma' - 1\}$. Then the greatest common divisor δ of these entries satisfies the conditions: $\delta < \gamma' \leqslant \alpha'$ and $\delta\gamma' < \alpha'\gamma'$. This contradicts our choice of α' and γ'. Thus, our set contains a column u' with $a' = 0$ or $c' = 0$. If $a' = 0$, we replace u' by $J_{k}u'$. Hence, we may assume that $c' = 0$. It then follows by Lemma 3.8 that for suitable $V \in SL_{k}(\mathbf{Z})$ the column $U(V)u'$ has the form (3.17). □

PROOF OF PROPOSITION 3.7. Without loss of generality we may assume that M is an integer matrix. By Lemma 3.9, we may also assume that the first column of M has the form (3.17). In the case $k = 1$ this proves the proposition. Suppose that the proposition has already been proved for $2k' \times 2k'$-matrices for $k' < k$. The relation (3.15) for the matrix $M = \begin{pmatrix} A & B \\ C & D \end{pmatrix}$ is equivalent to the conditions

$$(3.18) \qquad {}^tAC = {}^tCA, \quad {}^tBD = {}^tDB, \quad {}^tAD - {}^tCB = r \cdot E_k.$$

By assumption, the matrices A and C have the form

$$(3.19) \qquad A = \begin{pmatrix} a_{11} & a_{12} & \cdots & a_{1k} \\ 0 & & & \\ \vdots & & A_0 & \\ 0 & & & \end{pmatrix}, \qquad C = \begin{pmatrix} 0 & c_{12} & \cdots & c_{1k} \\ 0 & & & \\ \vdots & & C_0 & \\ 0 & & & \end{pmatrix},$$

where $a_{11} \neq 0$. From the relation ${}^tAC = {}^tCA$ it follows that $c_{12} = \cdots = c_{1k} = 0$, ${}^tA_0C_0 = {}^tC_0A_0$. Since ${}^tAD = {}^tCB + rE_k$, we conclude that

$$D = \begin{pmatrix} d_{11} & 0 & \cdots & 0 \\ d_{21} & & & \\ \vdots & & D_0 & \\ d_{1k} & & & \end{pmatrix}, \qquad {}^tA_0D_0 - {}^tC_0B_0 = rE_{k-1},$$

where B_0 denotes the corresponding block of B. Finally, the relation ${}^tBD = {}^tDB$ implies the relation ${}^tB_0D_0 = {}^tD_0B_0$. From all of these relations it follows that the matrix $M_0 = \begin{pmatrix} A_0 & B_0 \\ C_0 & D_0 \end{pmatrix}$ satisfies the condition: ${}^tM_0J_{k-1}M_0 = rE_{k-1}$. By the induction assumption, there exists a matrix $g_0 \in \Gamma'_{k-1}$ such that $g_0M_0 = \begin{pmatrix} * & * \\ 0 & * \end{pmatrix}$. For an arbitrary $(2k-2) \times (2k-2)$-matrix $M' = \begin{pmatrix} A' & B' \\ C' & D' \end{pmatrix}$, $k \geqslant 2$, we define the $2k \times 2k$-matrix $\widehat{M'} = \begin{pmatrix} A_1 & B_1 \\ C_1 & D_1 \end{pmatrix}$ with blocks

$$A_1 = \begin{pmatrix} 1 & 0 \\ 0 & A' \end{pmatrix}, \quad \begin{pmatrix} 0 & 0 \\ 0 & B' \end{pmatrix}, \quad C_1 = \begin{pmatrix} 0 & 0 \\ 0 & C' \end{pmatrix}, \quad D_1 = \begin{pmatrix} 1 & 0 \\ 0 & D' \end{pmatrix}.$$

With this notation, the C-block of the matrix $\widehat{g_0}M$ consists of zeros. Thus, to prove the proposition it suffices to verify that the map $g_0 \to \widehat{g_0}$ takes Γ'_{k-1} to Γ'_k. This, in turn, follows if we show that the map takes all of the generators of Γ'_{k-1} to matrices in Γ'_k. This is obvious for all of the generators except J_{k-1}. For J_{k-1} we have

$$(3.20) \qquad \widehat{J}_{k-1} = \begin{pmatrix} E^1 & E_k - E^1 \\ E^1 - E_k & E^1 \end{pmatrix} = (-\widehat{E}_{k-1})(J_kT(E^1))^3 \in \Gamma'_k,$$

where $E^1 = E^1_k$ is the $k \times k$-matrix $\mathrm{diag}(1, 0, \ldots, 0)$. $\square$

From what we have proved it follows that the functional equation (3.11) holds for any matrix M in the modular group Γ^k. By the remark at the beginning of the proof of

Proposition 3.2, in the case when $\zeta(M) \equiv 0 \pmod 2$ we may suppose that $\zeta(M) = 0$. Then the functional equation (3.11) can be written in the form

$$(3.21) \qquad \det(CZ + D)^{-1/2}\theta(M\langle Z\rangle, M\Omega) = \chi(M)\theta(Z, \Omega),$$

where $\chi(M)$ is an eighth root of unity. From Lemma 3.3 it follows that the set

$$\Theta^k = \{M \in \Gamma^k; \zeta(M) \equiv 0 \pmod 2\}$$

is a subgroup of Γ^k. Returning to our original notation, we see that we have the following theorem.

THEOREM 3.10. *The set*

$$\Theta^k = \left\{M = \begin{pmatrix} A & B \\ C & D \end{pmatrix} \in \Gamma^k; \mathrm{dc}(B\,^tA) \equiv \mathrm{dc}(C\,^tD) \equiv 0 \pmod 2\right\}$$

is a subgroup of the modular group. For every $M = \begin{pmatrix} A & B \\ C & D \end{pmatrix} \in \Theta^k$ *the theta-function* $\theta^k(Z; W, W')$ *satisfies the functional equation*

$$\det(CZ + D)^{-1/2}\theta^k(M\langle Z\rangle; AW + BW', CW + DW') = \chi(M)\theta^k(Z; W, W'),$$

where $\chi(M)$ *is an eighth root of unity that depends on the choice of square root on the left.*

The group Θ^k is called the *theta-group of degree k.*

PROBLEM 3.11 (Witt). Prove that the theta-group of degree k is generated by the matrices $U(V)$ with $V \in \Lambda^k$, the matrices $T(S)$ with $S \in \mathbf{E}_k$ (where $\mathbf{E}_k$ is the set of even symmetric $k \times k$-matrices), the matrix J_k, and the matrices of the form

$$\begin{pmatrix} E^i & E_k - E^i \\ E^i - E_k & E^i \end{pmatrix} \quad \text{with } E^i = \mathrm{diag}(\underbrace{1, \ldots, 1}_{i}, \underbrace{0, \ldots, 0}_{k-i}) \text{ for } i = 1, \ldots, k-1.$$

3. Symplectic transformations of theta-series. We now examine the action of symplectic transformations on the theta-series of arbitrary positive definite integral quadratic forms. With a view toward the applications of theta-series (for example, to the problem of integral representations of quadratic forms by quadratic forms, where the representing matrix satisfies certain congruences), it is convenient to generalize the definition of theta-series by introducing some new parameters.

Let $Q \in S_m(\mathbf{R})$, $Q > 0$, $Z \in \mathbf{H}_n$, $W, W' \in M_{m,n}(\mathbf{C})$. By analogy with the proof of Proposition 1.3, it is not hard to see that the series

$$(3.22) \qquad \begin{aligned} \theta^n(Z, Q, (W, W')) &= \theta^n(Z, Q, \Omega) \\ &= \sum_{N \in M_{m,n}} e\{Q[N - W']Z + 2\,^tNQW - \,^tWQW'\}, \end{aligned}$$

where $\Omega = (W, W') \in M_{m,2n}(\mathbf{C})$, and for an arbitrary square matrix T we set

$$(3.23) \qquad e\{T\} = \exp(\pi i\sigma(T)),$$

where $\sigma(T)$ is, as usual, the trace of T, converges absolutely and uniformly if Ω belongs to a fixed compact subset of $M_{m,2n}(\mathbf{C})$ and $Z \in \mathbf{H}_n(\varepsilon)$ with $\varepsilon > 0$ (see (1.11)). Thus, this series determines a holomorphic function on the space $\mathbf{H}_n \times M_{m,2n}(\mathbf{C})$. The series (3.22) is called the *theta-function of degree n for the matrix Q* (or the *corresponding quadratic form*). If we set $W = 0$ and $W' = 0$ in (3.22), we obtain the theta-series $\theta^n(Z, Q)$.

If $m = 1$ and $Q = (1)$, then $\theta^n(Z, Q, (W, W'))$ obviously becomes the theta-function $\theta^n(Z; {}^tW, {}^tW')$ in (3.2). It turns out that, conversely, every theta-function (3.22) is a restriction of a suitable theta-function (3.2). This fact enables us to reduce the study of the action of symplectic transformations on general theta-functions to the case we have already examined.

We first recall the definition and the simplest properties of the tensor product of two matrices. If A and $B = (b_{\alpha\beta})$ are $m \times m$ and $n \times n$ matrices, respectively, over the field of complex numbers, we define their *tensor product* by setting

$$A \otimes B = (Ab_{\alpha\beta}) \in M_{mn}(\mathbf{C}).$$

It follows from the definition that the tensor product is linear in each argument, and

$$(A \otimes B)(A_1 \otimes B_1) = AA_1 \otimes BB_1.$$

From this relation it follows that the matrix $A \otimes B$ is invertible whenever A and B are invertible, and

$$(A \otimes B)^{-1} = A^{-1} \otimes B^{-1},$$

in addition,

$$\det(A \otimes B) = \det(A \otimes E_n) \cdot \det(E_m \otimes B) = (\det A)^n \cdot (\det B)^m.$$

Finally,

$${}^t(A \otimes B) = {}^tA \otimes {}^tB$$

and if A and B are real, symmetric, and positive definite, then $A \otimes B$ is also a positive definite matrix.

LEMMA 3.12. *Let $m, n \geqslant 1$, $Q \in S_m(\mathbf{R})$, $Q > 0$, $Z \in \mathbf{H}_n$, $W, W' \in M_{m,n}(\mathbf{C})$. Then*

$$(3.24) \qquad \theta^n(Z, Q, (W, W')) = \theta^{mn}(Q \otimes Z; c(QW), c(W')),$$

where the theta-function of (3.2) is on the right and the theta-function of (3.22) is on the left, and for every matrix $T = (t_1, \ldots, t_n) \in M_{m,n}(\mathbf{C})$ with columns t_α we set

$$c(T) = \begin{pmatrix} t_1 \\ \vdots \\ t_n \end{pmatrix} \in M_{mn,1}(\mathbf{C}).$$

In addition, for arbitrary $M = \begin{pmatrix} A & B \\ C & D \end{pmatrix} \in \mathrm{Sp}_n(\mathbf{R})$ the matrix

$$M_Q = \begin{pmatrix} A_Q & B_Q \\ C_Q & D_Q \end{pmatrix} = \begin{pmatrix} E_m \otimes A & Q \otimes B \\ Q^{-1} \otimes C & E_m \otimes D \end{pmatrix}$$

belongs to the symplectic group $\mathrm{Sp}_{mn}(\mathbf{R})$, and one has the following identities:

$$\theta^n(M\langle Z\rangle, Q, (W, W')\, {}^tM)$$

$$(3.25)$$

$$= \theta^{mn}(M_Q\langle Q \otimes Z\rangle; A_Q c(QW) + B_Q c(W'), C_Q c(QW) + D_Q c(W')),$$

$$(3.26) \qquad \det(CZ + D)^m = \det(C_Q(Q \otimes Z) + D_Q).$$

PROOF. First of all, in the above notation we have

$$(Q \otimes Z)[c(T)] = \sum_{\alpha,\beta=1}^{n} z_{\alpha\beta} \, {}^t t_\alpha Q t_\beta = \sigma(Q[T] \cdot Z).$$

Similarly, for $T, V \in M_{m,n}(\mathbf{C})$ we have

$$ {}^t c(T) c(V) = \sum_{\alpha=1}^{n} {}^t t_\alpha v_\alpha = \sigma({}^t TV).$$

We thus obtain:

$$\theta^n(Z, Q, (W, W')) = \sum_{c(N) \in M_{mn,1}} \exp(\pi i((Q \otimes Z)[c(N) - c(W')]$$
$$+ 2 \, {}^t c(N) c(QW) - {}^t c(W') c(QW)))$$
$$= \theta^{mn}(Q \otimes Z; c(QW), c(W')),$$

which proves the first part of the lemma.

To prove that $M_Q \in \mathrm{Sp}_{mn}(\mathbf{R})$, it suffices to verify that the blocks of M_Q satisfy (2.7) whenever the blocks of M satisfy these relations. Using the above properties of the tensor product, we obtain

$$ {}^t C_Q A_Q = (Q^{-1} \otimes {}^t C)(E_m \otimes A) = Q^{-1} \otimes {}^t CA$$
$$= Q^{-1} \otimes {}^t AC = {}^t(E_m \otimes A)(Q^{-1} \otimes C) = {}^t A_Q C_Q.$$

Similarly,

$$ {}^t B_Q D_Q = Q \otimes {}^t BD = Q \otimes {}^t DB = {}^t D_Q B_Q.$$

Finally,

$$ {}^t A_Q D_Q - {}^t C_Q B_Q = (E_m \otimes {}^t AD) - (E_m \otimes {}^t CB) = E_m \otimes E_n = E_{mn}.$$

To prove (3.25) it is now sufficient to verify that

(3.27) $$Q \otimes M\langle Z \rangle = M_Q \langle Q \otimes Z \rangle$$

and

$$c(Q(W \, {}^t A + W' \, {}^t B)) = A_Q c(QW) + B_Q c(W'),$$
$$c(W \, {}^t C + W' \, {}^t D) = C_Q c(QW) + D_Q c(W').$$

We have

$$A_Q(Q \otimes Z) + B_Q = (E_m \otimes A)(Q \otimes Z) + (Q \otimes B) = Q \otimes (AZ + B),$$
$$C_Q(Q \otimes Z) + D_Q = (Q^{-1} \otimes C)(Q \otimes Z) + (E_m \otimes D) = E_m \otimes (CZ + D),$$

from which (3.27) and (3.26) follow. Finally, if $A = (a_{\alpha\beta})$, $B = (b_{\alpha\beta})$, $W = (w_1, \ldots, w_n)$, and $W' = (w_1', \ldots, w_n')$, then obviously

$$
A_Q c(QW) + B_Q c(W') = (E_m a_{\alpha\beta}) \begin{pmatrix} Qw_1 \\ \vdots \\ Qw_n \end{pmatrix} + (Qb_{\alpha\beta}) \begin{pmatrix} w_1' \\ \vdots \\ w_n' \end{pmatrix}
$$

$$
= \begin{pmatrix} Q(w_1 a_{11} + \cdots + w_n a_{1n}) \\ \cdots \\ Q(w_1 a_{n1} + \cdots + w_n a_{nn}) \end{pmatrix} + \begin{pmatrix} Q(w_1' b_{11} + \cdots + w_n' b_{1n}) \\ \cdots \\ Q(w_1' b_{n1} + \cdots + w_n' b_{nn}) \end{pmatrix}
$$

$$
= c(Q(W \cdot {}^t A + W' \cdot {}^t B)).
$$

The second relation can be verified in the same way. $\qquad\square$

Now suppose that Q is the matrix of a nondegenerate integral quadratic form q in m variables, i.e., $Q \in \mathbf{E}_m$ and $\det Q \neq 0$. By the *level of the matrix Q* (or *the form q*) we mean the least positive integer $q = q(Q)$ such that $q \cdot Q^{-1} \in \mathbf{E}_m$.

THEOREM 3.13. *Suppose that $m, n \geqslant 1$, $Q \in \mathbf{A}_m^+$, and q is the level of Q. Then for every matrix $M = \begin{pmatrix} A & B \\ C & D \end{pmatrix}$ in the group*

$$
(3.28) \qquad \Gamma_0^n(q) = \left\{ \begin{pmatrix} A & B \\ C & D \end{pmatrix} \in \Gamma^n; C \equiv 0 (\operatorname{mod} q) \right\}
$$

the theta-function (3.22) of degree n for the matrix Q satisfies the functional equation

$$
(3.29) \qquad \det(CZ + D)^{-m/2} \theta^n(M\langle Z \rangle, Q, \Omega\,{}^t M) = \chi_Q^n(M) \theta^n(Z, Q, \Omega),
$$

where $\chi_Q^n(M)$ is a certain eighth root of unity that for odd m also depends on the choice of root of the determinant on the left. In particular, the theta-series (1.13) of degree n for the matrix Q satisfies the following functional equation for every $M = \begin{pmatrix} A & B \\ C & D \end{pmatrix} \in \Gamma_0^n(q)$:

$$
(3.30) \qquad \det(CZ + D)^{-m/2} \theta^n(M\langle Z \rangle, Q) = \chi_Q^n(M) \theta^n(Z, Q).
$$

PROOF. Let M_Q be the matrix that is constructed from M in Lemma 3.12. By Lemma 3.12, $M_Q \in \operatorname{Sp}_{mn}(\mathbf{R})$. From the definitions it follows that M_Q is an integer matrix, so that $M_Q \in \Gamma^{mn}$. Finally, all of the diagonal entries in the matrices $B_Q\,{}^t A_Q = Q \otimes B\,{}^t A$ and $C_Q\,{}^t D_Q = Q^{-1} \otimes C\,{}^t D = qQ^{-1} \otimes (q^{-1} C\,{}^t D)$ are even, because all of the diagonal entries in the first factors of the tensor products are even, and the second factors are integer matrices. Thus, M_Q is contained in the theta-group Θ^{mn}. Using Lemma 3.12 for the matrix M and Theorem 3.10 for the matrix M_Q, we obtain

$$
\det(CZ + D)^{-m/2} \theta^n(M\langle Z \rangle, Q, \Omega\,{}^t M)
$$
$$
= \det(C_Q(Q \otimes Z) + D_Q)^{-1/2} \cdot \theta^{mn}(M_Q\langle Q \otimes Z \rangle; A_Q \cdot c(QW)
$$
$$
+ B_Q \cdot c(W'), C_Q \cdot c(QW) + D_Q \cdot c(W'))
$$
$$
= \chi(M_Q) \theta^{mn}(Q \otimes Z; c(QW), c(W')) = \chi(M_Q) \theta^n(Z, Q, (W, W')),
$$

which proves (3.29) if we set $\chi_Q^n(M) = \chi(M_Q)$. (3.30) follows from (3.29) if we set $\Omega = 0$. $\qquad\square$

Theorem 3.13 answers (except for the computation of the factor $\chi_Q^n(M)$) our question about the action on $\theta^n(Z, Q)$ of symplectic transformations in the subgroup $\Gamma_0^n(q)$ of the Siegel modular group Γ^n. However, when studying certain properties of theta-series, such as their behavior near the boundary of $\mathbf{H}_n$, one needs to understand the action on theta-series of an arbitrary transformation in Γ^n. In general (when $q \neq 1$), such transformations do not take the theta-series $\theta^n(Z, Q)$ to itself (even modulo a multiplicative factor). But the theta-series does remain inside a certain finite-dimensional space that depends on n and q—the space of generalized theta-series of degree n for the matrix Q. Suppose that $Q \in \mathbf{A}_m^+$ and q is the level of Q. We consider the set of matrices

$$(3.31) \qquad T^n(Q) = \{T \in M_{m,n}; QT \equiv 0 (\mathrm{mod}\, q)\},$$

and for each $T \in T^n(Q)$ we define the *generalized theta-series of degree n for Q* by setting

$$(3.32) \qquad \theta^n(Z, Q | T) = \theta^n(Z, Q, (0, -q^{-1}T)) = \sum_{N \in M_{m,n}} e\{Q[N + q^{-1}T]Z\}.$$

It is clear that, as a function of T, the generalized theta-series depends only on T modulo q, and we have

$$\theta^n(Z, Q | T) = \theta^n(Z, Q), \quad \text{if } T \equiv 0 (\mathrm{mod}\, q).$$

Thus, the space spanned by all generalized theta-series of degree n for Q is finite dimensional and contains the theta-series $\theta^n(Z, Q)$.

PROPOSITION 3.14. *Under the action of the generators* (3.12) *of the modular group* Γ^n, *we have the following transformation formulas for generalized theta-series of degree n for a matrix* $Q \in \mathbf{A}_m^+$:

$$\theta^n(U(V^*)\langle Z \rangle; Q | T) = \theta^n(Z; Q | TV) \qquad (V \in \Lambda^n),$$
$$\theta^n(T(S)\langle Z \rangle, Q | T) = e\{q^{-2}Q[T]S\}\theta^n(Z, Q | T) \qquad (S \in S_n).$$

Finally,

$$\theta^n(J_n\langle Z \rangle, Q | T) = (\det Q)^{-n/2}(\det(-iZ))^{m/2}$$
$$\times \sum_{T' \in T^n(Q)/\mathrm{mod}\, q} e\{2q^{-2} \cdot {}^tTQT'\}\theta^n(Z, Q | T'),$$

where $(\det(-iZ))^{1/2}$ *is positive if* $Z = iY$, *and is uniquely determined by analytic continuation for arbitrary* $Z \in \mathbf{H}_n$.

We first prove the following generalization of Lemma 3.1, which is actually a corollary of that lemma.

LEMMA 3.15 (Inversion formula for theta-functions of degree n for Q). *The theta-function* (3.22) *satisfies the following identity*:

$$(3.33) \qquad \begin{aligned} &\theta^n(-Z^{-1}, Q^{-1}, (QW', -QW)) \\ &= (\det Q)^{n/2}(\det(-iZ))^{m/2}\theta^n(Z, Q, (W, W')), \end{aligned}$$

where $(\det(-iZ))^{1/2}$ *is the function defined in Proposition* 3.14.

PROOF. We derive (3.33) from (3.5), using the connection that is given in Lemma 3.12 between the theta-functions in the two formulas. If we use (3.24) and the properties of the tensor product of matrices (listed right before Lemma 3.12), we obtain

$$\theta^n(-Z^{-1}, Q^{-1}, (QW', -QW)) = \theta^{mn}(Q^{-1} \otimes (-Z^{-1}); c(W'), -c(QW)).$$

Since $Q^{-1} \otimes (-Z^{-1}) = -(Q \otimes Z)^{-1}$, by Lemma 3.1 the last expression is equal to

$$\det(-i(Q \otimes Z))^{1/2} \theta^{mn}(Q \otimes Z; c(QW), c(W'))$$

and hence, by (3.24), it is equal to the expression on the right in (3.33). $\qquad \square$

PROOF OF PROPOSITION 3.14. The first two formulas follow directly from the definitions:

$$\theta^n(Z['V], Q|T) = \sum_{N \in M_{m,n}} e\{'VQ[N + q^{-1}T]VZ\}$$

$$= \sum_{N \in M_{m,n}} e\{Q[NV + q^{-1}TV]Z\}$$

$$= \sum_{N' \in M_{m,n}} e\{Q[N' + q^{-1}TV]Z\},$$

since we have $M_{m,n}V = M_{m,n}$ for $V \in \Lambda^n$;

$$\theta^n(Z + S, Q|T) = \sum_{N \in M_{m,n}} e\{Q[N + q^{-1}T]Z\}e\{Q[q^{-1}T]S\},$$

since

$$\sigma(Q[N + q^{-1}T]S) = \sigma(Q[N]S) + 2\sigma('Nq^{-1}QTS) + \sigma(Q[q^{-1}T]S)$$
$$\equiv \sigma(Q[q^{-1}T]S) \pmod{2},$$

because $N \in M_{m,n}$, $S \in S_n$, and all of the diagonal entries in Q—and hence also in $Q[N]$—are even, while $q^{-1}QT$ is an integer matrix. To prove the third identity, we apply the inversion formula (3.33) (with Q replaced by Q^{-1}, $W' = 0$, and $W = 1/qQT$) to the theta-series $\theta^n(Z, Q|T) = \theta^n(Z, Q, (0, -q^{-1}T))$.

We obtain

$$\theta^n(-Z^{-1}, Q|T) = (\det Q)^{-n/2}(\det(-iZ))^{m/2}\theta^n(Z, Q^{-1}, (q^{-1}QT, 0))$$
$$= (\det Q)^{-n/2}(\det(-iZ))^{m/2} \sum_{N \in M_{m,n}} e\{Q^{-1}[N]Z + 2q^{-1}\,'NT\}.$$

For every $L \in M_{m,n}$ the matrix $qQ^{-1}L$ is obviously an integer matrix belonging to the set $T^n(Q)$ (see (3.31)). Conversely, any matrix $T' \in T^n(Q)$ is uniquely representable in the form $qQ^{-1}L$ $(L \in M_{m,n})$. Thus, the map $L \to qQ^{-1}L$ gives an isomorphism of the additive group $M_{m,n}$ with the group $T^n(Q)$. Under this isomorphism the subgroup $QM_{m,n}$ is obviously mapped onto $qM_{m,n} \subset T^n(Q)$, so that we obtain an isomorphism of quotient groups: $M_{m,n}/QM_{m,n} \xrightarrow{\sim} T^n(Q)/\mathrm{mod}\, q$. Continuing the above chain of equalities, we have

$$(\det Q)^{-n/2}(\det(-iZ))^{m/2}$$
$$\times \sum_{\substack{N \in M_{m,n} \\ L \in M_{m,n}/QM_{m,n}}} e\{Q^{-1}[QN + L]Z + 2q^{-1}\,'(QN + L)T\}$$

$$= (\det Q)^{-n/2}(\det(-iZ))^{m/2}$$

$$\times \sum_{N,L} e\left\{ Q[N + q^{-1}(qQ^{-1}L)]Z + \frac{2}{q}\,{}^t(N + q^{-1}(qQ^{-1}L))QT \right\}$$

$$= (\det Q)^{-n/2}(\det(-iZ))^{m/2}$$

$$\times \sum_{\substack{N \in M_{m,n} \\ T' \in T^n(Q)/\bmod q}} e\{Q[N + q^{-1}T']Z + 2q^{-2} \cdot {}^tT'QT\},$$

since $q^{-1} \cdot {}^tNQT = {}^tN(q^{-1}Q)T$ is an integer matrix. $\qquad\square$

PROBLEM 3.16. Prove that there are $(\det Q)^n$ elements in the set $T^n(Q)/\bmod q$.

PROBLEM 3.17. Suppose that $Q \in \mathbf{A}_m^+$, q is the level of Q, and $T \in T^n(Q)$. Prove that for every matrix $M = \begin{pmatrix} A & B \\ C & D \end{pmatrix}$ in the group $\Gamma_0^n(q)$ the theta-series $\theta^n(Z, Q|T)$ satisfies the functional equation

$$\det(CZ + D)^{-m/2}\theta^n(M\langle Z\rangle, Q|T) = \chi_Q^n(M)e\{q^{-2}A \cdot {}^tB \cdot Q[T]\}\theta^n(Z, Q|TA)$$

with the same scalar $\chi_Q^n(M)$ as in Theorem 3.13. Thus, if $M \equiv E_{2n}(\bmod q)$, then

$$\det(CZ + D)^{-m/2}\theta^n(M\langle Z\rangle, Q|T) = \chi_Q^n(M)\theta^n(Z, Q|T).$$

§4. Computation of the multiplier

The scalar factor $\chi_Q^n(M)$ that appears in the functional equations for theta-functions and theta-series is usually called the *multiplier of degree n for Q*. In this section we shall find an explicit expression for this multiplier.

1. Automorphy factors. Suppose that $Q \in \mathbf{A}_m^+$, q is the level of Q, and $n \in \mathbf{N}$. For $M \in \Gamma_0^n(q)$ and $Z \in \mathbf{H}_n$ we set

$$(4.1) \qquad j_Q^n(M, Z) = \theta^n(M\langle Z\rangle, Q)\theta^n(Z, Q)^{-1}.$$

By definition, the function $j_q^n \colon \Gamma_0^n(q) \times \mathbf{H}_n \to \mathbf{C}$ is meromorphic in the second argument. By Theorem 3.13, it can be written in the form

$$j_Q^n(M, Z) = \det(CZ + D)^{m/2}\chi_Q^n(M) \quad \left(M = \begin{pmatrix} A & B \\ C & D \end{pmatrix} \right),$$

from which it follows that, as a function of Z, it is holomorphic and nonzero on $\mathbf{H}_n$ for every $M \in \Gamma_0^n(q)$. Finally, from (4.1) we see that the following relation holds for any $M, M_1 \in \Gamma_0^n(q)$ and $Z \in \mathbf{H}_n$:

$$\begin{aligned} j_Q^n(MM_1, Z) &= \theta^n(MM_1\langle Z\rangle, Q)\theta^n(Z, Q)^{-1} \\ &= \theta^n(M\langle M_1\langle Z\rangle\rangle, Q)\theta^n(M_1\langle Z\rangle, Q)^{-1}\theta^n(M_1\langle Z\rangle, Q)\theta^n(Z, Q)^{-1} \\ &= j_Q^n(M, M_1\langle Z\rangle)j_Q^n(M_1, Z). \end{aligned}$$

Let S be a multiplicative semigroup acting on a set H, $S \ni g \colon h \to g(h)$, as a subsemigroup of the group of all one-to-one maps of H onto itself. A function φ on

$S \times H$ with values in a multiplicative group T will be called an *automorphy factor* of S on H with values in T if for all $g, g_1 \in S$ and $h \in H$ one has

$$\varphi(gg_1, h) = \varphi(g, g_1(h))\varphi(g_1, h).$$

LEMMA 4.1. *Let* $\varphi : S \times H \to T$ *be an automorphy factor of* S *on* H *with values in* T. *Then:*

(1) *If* $f : T \to T_1$ *is a homomorphism of groups, then the function* $(g, h) \to f(\varphi(g, h))$ *is an automorphy factor of* S *on* H *with values in* T_1.

(2) *If* $\chi : S \to T$ *is a map whose image is contained in the center of the group* T, *then the function* $(g, h) \to \chi(g)\varphi(g, h)$ *is an automorphy factor if and only if* χ *is a semigroup homomorphism.*

(3) *For every function* F *on* H *with values in a left* T-*module* V *and for every* $g \in S$ *define the function* $F|g : H \to V$ *by setting*

$$(F|g)(h) = (F|_\varphi g)(h) = \varphi(g, h)^{-1} F(g(h)).$$

Then for any $g, g_1 \in S$ *one has*

$$F|g|g_1 = F|gg_1.$$

PROOF. All three assertions follow directly from the definitions. $\square$

The discussion at the beginning of the section shows that the function j_Q^n is an automorphy factor of $\Gamma_0^n(q)$, where q is the level of the quadratic form Q, on the upper half-plane of degree n with values in the multiplicative group $\mathbf{C}^*$ of nonzero complex numbers. The next lemma gives other examples of automorphy factors.

LEMMA 4.2. *For any matrix* $M = \begin{pmatrix} A & B \\ C & D \end{pmatrix}$ *in the group*

$$(4.2) \qquad S_R^n = GSp_n^+(\mathbf{R}) = \{M \in M_{2n}(\mathbf{R}); J_n[M] = r(M)J_n, r(M) > 0\}$$

(*the general symplectic group of degree* n) *and any* $Z \in \mathbf{H}_n$ *the matrix* $CZ + D$ *is invertible. The correspondence that associates such an* M *to the map*

$$f_M : Z \to M\langle Z\rangle = (AZ + B)(CZ + D)^{-1}$$

gives a homomorphism of the group S_R^n *to the group of holomorphic automorphisms of the space* $\mathbf{H}_n$. *The functions*

$$(4.3) \qquad (M, Z) \to CZ + D, \quad j(M, Z)^k = \det(CZ + D)^k \quad (k \in \mathbf{Z})$$

are automorphy factors of S_R^n *on* $\mathbf{H}_n$ *with values in the groups* $GL_n(\mathbf{C})$ *and* $\mathbf{C}^*$, *respectively.*

PROOF. If $\lambda \in \mathbf{R}$, then obviously $r(\lambda M) = \lambda^2 r(M)$. Thus, $\lambda M \in Sp_n(\mathbf{R})$ if $\lambda = r(M)^{-1/2}$, and the first two statements in the lemma follow immediately from the corresponding statements for the group $Sp_n(\mathbf{R})$ (see Proposition 2.1 and Theorem 2.2). Furthermore, if $M = \begin{pmatrix} A & B \\ C & D \end{pmatrix}$, $M_1 = \begin{pmatrix} A_1 & B_1 \\ C_1 & D_1 \end{pmatrix}$, and $MM_1 = \begin{pmatrix} A_2 & B_2 \\ C_2 & D_2 \end{pmatrix}$, then

$$(C(A_1Z + B_1)(C_1Z + D_1)^{-1} + D)(C_1Z + D_1)$$
$$= (CA_1 + DC_1)Z + (CB_1 + DD_1) = C_2Z + D_2.$$

The analogous identity for $j(M, Z)^k$ follows from this, since the map $A \rightarrow (\det A)^k$ is a group homomorphism from $GL_n(\mathbf{C})$ to $\mathbf{C}^*$.   □

2. Quadratic forms of level 1.

PROPOSITION 4.3. *Let $Q \in \mathbf{A}_m^+$, where $m \in \mathbf{N}$. If the level of Q is 1, then $m \equiv 0 \pmod{8}$ and $\chi_Q^n(M) = 1$ for all $M \in \Gamma^n$ and $n \in \mathbf{N}$.*

PROOF. Because Q has level 1, it follows that Q^{-1} is an integer matrix whenever Q is. Since $\det Q \cdot \det Q^{-1} = 1$, we have $\det Q = \pm 1$. Since $Q > 0$, it follows that $\det Q = 1$.

By the inversion formula (3.33) with $W = W' = 0$, we have

$$\theta^n(-Z^{-1}, Q^{-1}) = \det(-iZ)^{m/2} \theta^n(Z, Q).$$

On the other hand, since Q^{-1} is a unimodular matrix, it follows that when we replace N by $Q^{-1}N$ in the definition (1.13) of the theta-series $\theta^n(Z', Q)$, we obtain $\theta^n(Z', Q^{-1}) = \theta^n(Z', Q)$. Thus,

$$(4.4) \qquad \theta^n(-Z^{-1}, Q) = \det(-iZ)^{m/2} \theta^n(Z, Q).$$

If $n = 1$, then the relations $\theta^1(z + 1, Q) = \theta^1(z, Q)$ and (4.4) imply that

$$\theta^1(-(z + 1)^{-1}, Q) = (-i(z + 1))^{m/2} \theta^1(z, Q).$$

We set

$$A = \begin{pmatrix} 0 & 1 \\ -1 & 0 \end{pmatrix} \begin{pmatrix} 1 & 1 \\ 0 & 1 \end{pmatrix} = \begin{pmatrix} 0 & 1 \\ -1 & -1 \end{pmatrix}.$$

Then for $z \in \mathbf{H}_1$ we have $A\langle z \rangle = -(z+1)^{-1}$, $A^2\langle z \rangle = -(z+1)z^{-1}$, $A^3\langle z \rangle = z$. Using these relations, we obtain

$$
\begin{aligned}
\theta^1(z, Q)^\nu &= \theta^1(A\langle A^2\langle z \rangle \rangle, Q)^\nu \\
&= (-i(-(z + 1)z^{-1} + 1))^{\nu m/2} \theta^1(A^2\langle z \rangle, Q)^\nu \\
&= (iz^{-1})^{\nu m/2}(-i(-(z + 1)^{-1} + 1))^{\nu m/2} \theta^1(A\langle z \rangle, Q)^\nu \\
&= (iz^{-1})^{\nu m/2}(-iz(z + 1)^{-1})^{\nu m/2}(-i(z + 1))^{\nu m/2} \theta^1(z, Q)^\nu,
\end{aligned}
$$

where we take $\nu = 1$ for m even and $\nu = 2$ for m odd. We choose a point $z_0 \in \mathbf{H}_1$ for which $\theta^1(z_0, Q) \neq 0$. We then have the equality

$$(iz_0^{-1})^{\nu m/2}(-iz_0(z_0 + 1)^{-1})^{\nu m/2}(-i(z_0 + 1))^{\nu m/2} = 1,$$

from which (since νm is even) it follows that

$$i^{\nu m/2} \cdot (-i)^{\nu m/2} \cdot (-i)^{\nu m/2} = (-i)^{\nu m/2} = 1,$$

and hence

$$(4.5) \qquad \nu m/2 \equiv 0 \pmod{4}.$$

Thus, $m \equiv 0 \pmod{4}$. But then $\nu = 1$, and the congruence (4.5) shows that $m \equiv 0 \pmod{8}$.

Since $m \equiv 0 \pmod{8}$, we can rewrite (4.4) in the form

$$\theta^n(J_n\langle Z \rangle, Q) = \det(-Z)^{m/2} \theta^n(Z, Q).$$

For the other generators of the modular group Γ^n we immediately find from the definitions that

$$\theta^n(M\langle Z\rangle, Q) = \theta^n(Z, Q) \quad \text{for } M = U(V) \text{ and } T(S).$$

This implies that the automorphy factor $j_Q^n(M, Z)$ for $M = J_n$, $U(V)$, and $T(S)$ is equal, respectively, to $\det(-Z)^{m/2}$, 1, and 1. The automorphy factor $j(M, Z)^{m/2}$ also takes these values on the generators. On the other hand, for any $M \in \Gamma^n$ we have, by Theorem 3.13,

$$j_Q^n(M, Z) = \chi_Q^n(M) j(M, Z)^{m/2},$$

and, by Lemma 4.1(2), the map $\chi_Q^n: \Gamma^n \to \mathbf{C}^*$ is a homomorphism of groups. The above discussion shows that this homomorphism is trivial on the generators of Γ^n, and hence on the entire group. $\square$

PROBLEM 4.4. Verify that the matrix Q_8 of the quadratic form

$$\frac{1}{2}\sum_{r=1}^{8} x_r^2 + \frac{1}{2}\left(\sum_{r=1}^{8} x_r\right)^2 - x_1 x_2 - x_2 x_8$$

is contained in $\mathbf{A}_8^+$ and satisfies the condition: $\det Q_8 = q(Q_8) = 1$. Conclude from this that for any natural number m divisible by 8 there exist matrices $Q_m \in \mathbf{A}_m^+$ with $\det Q_m = q(Q_m) = 1$.

3. The multiplier as a Gauss sum. We first fix the square root of a complex number $z \neq 0$ by setting

(4.6) $$z^{k/2} = (z^{1/2})^k = (|z|^{1/2} \exp(\pi i \varphi))^k,$$

where $|z|^{1/2} > 0$, $-\pi/2 < \varphi \leqslant \pi/2$, and k is any integer. Next, suppose that $\begin{pmatrix} A & B \\ C & D \end{pmatrix} \in S_{\mathbf{R}}^n$. By Lemma 4.2, the function $f(Z) = \det(CZ + D)$ is nonzero on $\mathbf{H}_n$. If $\det D \neq 0$, then of the two branches of $f(Z)^{1/2}$ that are holomorphic on $\mathbf{H}_n$ and differ from one another by a sign (see the remark in §3.1), we shall usually use the notation $f(Z)^{1/2}$ to denote the branch satisfying the condition

(4.7) $$\lim_{Z=i\lambda E_n, \lambda \to +0} \det(CZ + D)^{1/2} = (\det D)^{1/2},$$

where the right side is understood in the sense of (4.6). Finally, for any integer k we set

(4.8) $$\det(CZ + D)^{k/2} = (\det(CZ + D)^{1/2})^k.$$

PROPOSITION 4.5. *Suppose that, under the assumptions of Theorem 3.13, the level q is greater than 1. If the function $\det(CZ + D)^{-m/2}$ on the right in the functional equations (3.29) and (3.30) is understood in the sense of (4.7)–(4.8), then for any $M = \begin{pmatrix} A & B \\ C & D \end{pmatrix} \in \Gamma_0^n(q)$ the multiplier $\chi_Q^n(M)$ in these equations can be computed from the formula*

(4.9) $$\chi_Q^n(M) = (\det D)^{m/2} |\det D|^m G(-D^{-1}C, Q),$$

where the roots are understood in the sense of (4.6), *and* $G(S, Q)$ *denotes the following Gauss sum, for S an* $n \times n$ *symmetric matrix with rational entries and for Q an* $m \times m$ *symmetric integer matrix with even integers on the main diagonal:*

$$(4.10) \qquad G(S, Q) = d^{-mn} \sum_{L \in M_{m,n}(\mathbf{Z}/d\mathbf{Z})} e\{Q[L]S\},$$

where d is any positive integer such that $dS \in M_n$.

Note that $\det D$ is prime to q for $M = \begin{pmatrix} A & B \\ C & D \end{pmatrix} \in \Gamma_0^n(q)$ (for example, use the third relation in (2.7)), and hence it is nonzero if $q > 1$. Further note that the Gauss sum (4.10) does not depend on the choice of integer d satisfying the property $dS \in M_n$.

PROOF. We compute the limit

$$(4.11) \qquad \lim_{\lambda \to +0} \lambda^{mn/2} \det(i\lambda C + D)^{-m/2} \theta^n(M\langle i\lambda E \rangle, Q),$$

where $E = E_n$ is the $n \times n$ identity matrix, in two different ways. On the one hand, by Theorem 3.13 it is equal to

$$\lim_{\lambda \to +0} \lambda^{mn/2} \chi_Q^n(M) \theta^n(i\lambda E, Q),$$

which, by the inversion formula (3.33), can be written in the form

$$(4.12) \qquad \begin{aligned} &\lim_{\lambda \to +0} \lambda^{mn/2} \chi_Q^n(M) \det(\lambda E)^{-m/2} (\det Q)^{-n/2} \theta^n(i\lambda^{-1} E, Q^{-1}) \\ &= \chi_Q^n(M)(\det Q)^{-n/2} \lim_{\lambda \to +0} \theta^n(i\lambda^{-1} E, Q^{-1}) = \chi_Q^n(M)(\det Q)^{-n/2}. \end{aligned}$$

On the other hand, we set $M\langle i\lambda E \rangle = BD^{-1} + Z_0$. Then, applying (2.8), we find that

$$\begin{aligned} Z_0(i\lambda C + D) &= (i\lambda A + B) - BD^{-1}(i\lambda C + D) \\ &= i\lambda(A - BD^{-1}C) = i\lambda(A \cdot {}^t D - BD^{-1}C \cdot {}^t D)D^* \\ &= i\lambda(A \cdot {}^t D - B \cdot {}^t C)D^* = i\lambda D^*. \end{aligned}$$

Thus,

$$(4.13) \qquad Z_0 = i\lambda D^*(i\lambda C + D)^{-1}.$$

Substituting, we obtain

$$\theta^n(M\langle i\lambda E \rangle, Q) = \sum_{N \in M_{m,n}} e\{Q[N](BD^{-1} + Z_0)\}.$$

Let d be a positive integer for which dBD^{-1} is an integer matrix. We represent N in the form $N = L + dN_1$, where $L \in M_{m,n}/dM_{m,n}$, $N_1 \in M_{m,n}$. Since then

$$Q[N] = Q[L] + d^2 Q[N_1] + d \cdot {}^t LQN_1 + d \cdot {}^t N_1 QL$$

and since

$$(d^2 Q[N_1] + d \cdot {}^t LQN_1 + d \cdot {}^t N_1 QL)BD^{-1}$$

is an integer matrix with even trace, it follows that

$$\theta^n(M\langle i\lambda E\rangle, Q) = \sum_{L\in M_{m,n}/dM_{m,n}} e\{Q[L](BD^{-1} + Z_0)\}$$

$$\times \sum_{N_1\in M_{m,n}} e\{d^2 Q[N_1]Z_0 + 2d \cdot {}^t N_1 QLZ_0\}$$

$$= \sum_{L\in M_{m,n}/dM_{m,n}} e\{Q[L](BD^{-1} + Z_0)\}\theta^n(d^2Z_0, Q, (dLZ_0, 0)),$$

which, by the inversion formula (3.33), can be rewritten in the form

$$(\det Q)^{-n/2}(\det(-id^2Z_0))^{-m/2}$$

(4.14)

$$\times \sum_{L\in M_{m,n}/dM_{m,n}} e\{Q[L](BD^{-1} + Z_0)\}\theta^n(-(d^2Z_0)^{-1}, Q^{-1}, (0, -dQLZ_0)).$$

From (4.13) it follows that $Z_0 \to 0$ if $\lambda \to 0$; in addition, $-Z_0^{-1} = -C \cdot {}^t D + i\lambda^{-1}D \cdot {}^t D$. If we take into account that $\theta^n(Z, Q(W, W'))$ converges uniformly if W and W' are in fixed compact subsets and $Z \in \mathbf{H}_n(\varepsilon)$ with $\varepsilon > 0$, we see that in computing

$$\lim_{\lambda\to+0} \theta^n(-(d^2Z_0)^{-1}, Q^{-1}, (0, -dQLZ_0))$$

we can take the limit term by term. This implies that the limit exists and is equal to 1. Thus, from (4.14) it follows that the limit (4.11) is equal to

$$(\det Q)^{-n/2} \sum_{L\in M_{m,n}/dM_{m,n}} e\{Q[L]BD^{-1}\}$$

$$\times \lim_{\lambda\to+0} (\det(-i\lambda C + D))^{-m/2}(\det(-id^2Z_0))^{-m/2}$$

$$= (\det Q)^{-n/2}G(BD^{-1}, Q)(\det D)^{-m/2}$$

$$\times \lim_{\lambda\to+0} (\det(-i(iD^*(i\lambda C + D)^{-1})))^{-m/2}.$$

According to Lemma 3.15, the function inside the last limit is continuous in $Z = iD^*(i\lambda C + D)^{-1} \in \mathbf{H}_n$, and so the limit is $(\det(-i(iD^*D^{-1})))^{-m/2} = |\det D|^m$. Thus, the limit (4.11) is equal to

$$(\det Q)^{-n/2}(\det D)^{-m/2}|\det D|^m G(BD^{-1}, Q).$$

Comparing this expression with (4.12), we find that

$$(4.15) \qquad \chi_Q^n(M) = (\det D)^{-m/2}|\det D|^m G(BD^{-1}, Q),$$

and so to prove the proposition it remains to verify that

$$(4.16) \quad G(BD^{-1}, Q) = G(-D^{-1}C, Q), \quad \text{if } \begin{pmatrix} A & B \\ C & D \end{pmatrix} \in \Gamma^n \text{ and } \det D \neq 0.$$

In order to do this, we make a small modification in our definition (4.10) of Gauss sums. Suppose that S and Q are as before, and D satisfies the conditions

$$(4.17) \qquad\qquad D \in M_n, \quad \det D \neq 0, \quad DS \in M_n.$$

We then set

$$(4.18) \qquad G_D(S, Q) = |\det D|^{-m} \sum_{L \in M_{m,n}/M_{m,n}D} e\{Q[L]S\}.$$

It is easy to see that if D satisfies (4.17) and M is any nonsingular $n \times n$ integer matrix, then

$$G_{MD}(S, Q) = G_D(S, Q).$$

Thus, if D and D_1 are two matrices that satisfy (4.17), then because the matrix $D' = \det D \cdot \det D_1 \cdot E_n$ is divisible on the right by both D and D', it follows that

$$G_D(S, Q) = G_{D'}(S, Q) = G_{D_1}(S, Q),$$

so that $G_D(S, Q)$ does not depend on the choice of D. Then, taking D to be the matrix dE_n, where $d \in \mathbf{N}$ and $dS \in M_n$, we see that

$$(4.19) \qquad G_D(S, Q) = G(S, Q).$$

Returning to the proof of (4.16), we can write

$$G(-D^{-1}C, Q) = G_D(-D^{-1}C, Q), \quad G(BD^{-1}, Q) = G_{t_D}(BD^{-1}, Q)$$

(note that ${}^tDBD^{-1} = {}^tDD^* \cdot {}^tB = {}^tB$ is an integer matrix). Since ${}^tDB = {}^tBD$, it follows that the map $L \to LB$ gives a homomorphism of quotient groups

$$B: M_{m,n}/M_{m,n}\,{}^tD \to M_{m,n}/M_{m,n}D.$$

Similarly, the map $L \to L \cdot {}^tC$ gives a homomorphism

$${}^tC: M_{m,n}/M_{m,n}D \to M_{m,n}/M_{m,n}\,{}^tD.$$

Since $B\,{}^tC = A\,{}^tD - E_n$, it follows that the composition $B\,{}^tC$ of these two homomorphisms coincides with the automorphism of multiplying by $-E_n$. Since ${}^tCB = {}^tAD - E_n$, the homomorphism ${}^tCB$ also coincides with multiplication by $-E_n$. Hence, the maps ${}^tC$ and B are isomorphisms, and we can write

$$G_D(-D^{-1}C, Q) = \sum_{L \in M_{m,n}/M_{m,n}\,{}^tD} e\{Q[LB](-D^{-1}C)\}$$

$$= \sum_{L \in M_{m,n}/M_{m,n}\,{}^tD} e\{Q[L](-BD^{-1}C\,{}^tB)\} = G_{t_D}(BD^{-1}, Q),$$

since $C\,{}^tB = {}^t(A\,{}^tD - E_n) = D\,{}^tA - E_n$. □

PROBLEM 4.6. In the notation of the definition (4.10) of Gauss sums, let $S = d^{-1}S'$, where S' is a symmetric integer matrix. Show that the Gauss sum of degree n reduces to the usual Gauss sum modulo d of the quadratic form with matrix $Q \otimes S'$:

$$G(S, Q) = d^{-mn} \sum_{L \in M_{mn,1}(\mathbf{Z}/d\mathbf{Z})} \exp((\pi i/d)(Q \otimes S')[L]).$$

PROBLEM 4.7 (The Gauss sum as an "automorphic form"). For $n \in \mathbf{N}$ and $q > 1$ define the set

$$S = \left\{ S = BD^{-1}; \begin{pmatrix} A & B \\ C & D \end{pmatrix} \in \Gamma_0^n(q) \right\}.$$

Prove the following facts:

(1) If $M = \begin{pmatrix} A & B \\ C & D \end{pmatrix} \in \Gamma_0^n(q)$ and $S \in \mathbf{S}$, then $\det(CS + D) \neq 0$; and if to every such M we associate the map $S \to M\langle S \rangle = (AS + B) \cdot (CS + D)^{-1}$, we obtain a transitive action of the group $\Gamma_0^n(q)$ on the set $\mathbf{S}$.

(2) If m is even, $Q \in \mathbf{A}_m^+$, and q is the level of Q, then the Gauss sum (4.10), regarded as a function on $\mathbf{S}$, satisfies the functional equation

$$\det(CS + D)^{m/2} G(M\langle S \rangle, Q) = \chi_Q^n(M) G(S, Q)$$

for any $M = \begin{pmatrix} A & B \\ C & D \end{pmatrix} \in \Gamma_0^n(q)$ and $S \in \mathbf{S}$, where $\chi_Q^n(M)$ is the same multiplier as in the corresponding functional equation for the theta-series. [**Hint:** Use the fact that in this case χ_Q^n is a homomorphism of the group $\Gamma_0^n(q)$.]

4. Quadratic forms in an even number of variables. Suppose that $Q \in \mathbf{A}_m^+$, where $m = 2k$ is even. By Theorem 3.13, the automorphy factor $j_Q^n(M, Z)$ (see (4.1)) for $n \in \mathbf{N}$ and $M = \begin{pmatrix} A & B \\ C & D \end{pmatrix} \in \Gamma_0^n(q)$, where q is the level of Q, can be written in the form

$$j_Q^n(M, Z) = \chi_Q^n(M) \det(CZ + D)^k = \chi_Q^n(M) j(M, Z)^k,$$

where $j(M, Z)$ is the automorphy factor (4.3) for the group $S_{\mathbf{R}}^n$, and hence also for the group $\Gamma_0^n(q)$, and χ_Q^n is a function on $\Gamma_0^n(q)$ with values in the group of eighth roots of unity. According to Lemma 4.1(2), the map $\chi_Q = \chi_Q^n : \Gamma_0^n(q) \to \mathbf{C}^*$ is a group homomorphism:

$$(4.20) \qquad \chi_Q(MM_1) = \chi_Q(M)\chi_Q(M_1) \quad \text{for } M, M_1 \in \Gamma_0^n(q).$$

If $q = 1$, then χ_Q is trivial by Proposition 4.3. Hence we may assume that $q > 1$. In this case, by Proposition 4.5,

$$(4.21) \qquad \chi_Q(M) = (\det D)^k G(-D^{-1}C, Q).$$

We let K denote the subgroup of $\Gamma_0^n(q)$ generated by matrices of the form $U(V)$ (see (2.2)) for $V \in SL_n(\mathbf{Z})$, $T(S)$ (see (2.3)) for $S \in S_n$, and $^t T(S)$ for $S \in qS_n$. From (4.21) it immediately follows that the character χ_Q is trivial on all of the generators of K, and hence on all of K. Hence, χ_Q is constant on every double coset KMK with $M \in \Gamma_0^n(q)$.

LEMMA 4.8. *Every double coset KMK, where $M = \begin{pmatrix} A & B \\ C & D \end{pmatrix} \in \Gamma_0^n(q)$, has a representative of the form $M_0 = \begin{pmatrix} A_0 & B_0 \\ C_0 & D_0 \end{pmatrix}$, where*

$$A_0 = \begin{pmatrix} E_{n-1} & 0 \\ 0 & \alpha \end{pmatrix}, \quad B_0 = \begin{pmatrix} 0 & 0 \\ 0 & \beta \end{pmatrix},$$

$$C_0 = \begin{pmatrix} 0 & 0 \\ 0 & \gamma \end{pmatrix}, \quad D_0 = \begin{pmatrix} E_{n-1} & 0 \\ 0 & \delta \end{pmatrix}, \quad \text{and} \quad \begin{pmatrix} \alpha & \beta \\ \gamma & \delta \end{pmatrix} \in \Gamma_0^1(q).$$

Here

(4.22) $\delta \equiv \det D \,(\mathrm{mod}\, q)$.

PROOF. It suffices to prove that every double coset has a representative $M_1 = \begin{pmatrix} A_1 & B_1 \\ C_1 & D_1 \end{pmatrix}$ all of whose entries in the first rows and columns of A_1, B_1, C_1, D_1 are zero, except for the entries in the first row and first column of A_1 and D_1, which equal 1. Once that has been proved, the lemma will follow by induction on n.

When we pass from the matrix M to the matrix $M' = \begin{pmatrix} A' & B' \\ C' & D' \end{pmatrix} = MU(V)$, the block D goes to the block $D' = DV$. By Lemma 3.8, the matrix $V \in SL_n(\mathbf{Z})$ can be chosen in such a way that the first row of D' has the form $(d, 0, \ldots, 0)$, where d is the greatest common divisor of the entries in the first row of D. Let c be the greatest common divisor of the entries $c'_{11}, \ldots, c'_{1n}$ in the first row of C'. Then there exist integers $s_{21}, \ldots, s_{2n}$ such that $c'_{11} s_{21} + \cdots + c'_{n1} s_{2n} = c$. Let $S = (s_{\alpha\beta})$ denote any symmetric integer matrix whose second column is ${}^t(s_{21}, \ldots, s_{2n})$. Then in the matrix $M'' = \begin{pmatrix} A'' & B'' \\ C'' & D'' \end{pmatrix} = M'T(S)$ the first two entries in the first row of the block $D'' = C'S + D'$ are equal to $c'_{11} s_{11} + \cdots + c'_{1n} s_{1n} + d$ and c, respectively. Since c divides $c'_{11}, \ldots, c'_{n1}$, and c and d are relatively prime, it follows that these two entries are relatively prime. Thus, if we again multiply M'' on the right by a suitable matrix of the form $U(V')$, we may assume that the first row of the D-block of this matrix is $(1, 0, \ldots, 0)$. Since

$$\begin{pmatrix} 1 & 0 & \cdots & 0 \\ -d_{21} & 1 & \cdots & 0 \\ \cdots & \cdots & \cdots & \cdots \\ -d_{n1} & 0 & \cdots & 1 \end{pmatrix} \begin{pmatrix} 1 & 0 & \cdots & 0 \\ d_{21} & & & \\ \vdots & & * & \\ d_{n1} & & & \end{pmatrix} = \begin{pmatrix} 1 & 0 & \cdots & 0 \\ 0 & & & \\ \vdots & & * & \\ 0 & & & \end{pmatrix},$$

it follows that, after multiplying the above matrix on the left by a suitable matrix of the form $U(V'')$, we may assume that its D-block has the form

(4.23) $$\begin{pmatrix} 1 & 0 & \cdots & 0 \\ 0 & & & \\ \vdots & & * & \\ 0 & & & \end{pmatrix}.$$

If the block D in M already has the form (4.23), then we can pass from M to the matrix

$$M' = \begin{pmatrix} A' & B' \\ C' & D' \end{pmatrix} = T(S) M \, {}^t T(S_1) = \begin{pmatrix} A & B + SD \\ C + DS_1 & D \end{pmatrix},$$

where we can obviously choose symmetric integer matrices S and S_1, the second of which is divisible by q, so that the first column of the matrix $B + SD$ and the first row of the matrix $C + DS_1$ consist of zeros. Then the relations $C' \cdot {}^t D' = D' \cdot {}^t C'$ and ${}^t B' \cdot D' = {}^t D' \cdot B'$ (see (2.7), (2.8)) imply that the first column of C' and the first row of B' also consist of zeros. Finally, if we take into account the structure of the matrices B', C', D' and the relations ${}^t A' \cdot D' - {}^t C' \cdot B' = E_n$ and $A' \cdot {}^t D' - B' \cdot {}^t C' = E_n$ (see (2.7), (2.8)), we conclude that A' as well as D' has the form (4.23). The first part of the lemma is proved.

Since we obviously have $\det D' \equiv \det D \pmod q$ for any matrix

$$M' = \begin{pmatrix} A' & B' \\ C' & D' \end{pmatrix} \in K \begin{pmatrix} A & B \\ C & D \end{pmatrix} K,$$

the congruence (4.22) follows. $\square$

We proceed with the computation of the multiplier χ_Q. Since χ_Q is constant on K-double cosets, by (4.21) and Lemma 4.8 we find that for any $M \in KM_0K$

$$\chi_Q(M) = \chi_Q(M_0) = \delta^k G(-D_0^{-1}C_0, Q)$$

$$= \delta^k d^{-mn} \sum_{L = (l_1, \ldots, l_n) \in M_{m,n}(\mathbf{Z}/d\mathbf{Z})} e\left\{ \begin{pmatrix} 0 & 0 \\ 0 & -\delta^{-1}\gamma \end{pmatrix} \begin{pmatrix} * & * \\ * & Q[l_n] \end{pmatrix} \right\},$$

where $M_0 = \begin{pmatrix} A_0 & B_0 \\ C_0 & D_0 \end{pmatrix}$, d is any natural number divisible by δ, and $l_1, \ldots, l_n$ are the columns of the matrix L. By the definition of $e\{T\}$ (see (3.23)), the last expression is equal to

$$\delta^k d^{-m} \sum_{l \in M_{m,1}(\mathbf{Z}/d\mathbf{Z})} e\{-\delta^{-1}\gamma Q[l]\}.$$

By the formula (4.21) applied to $M_1 = \begin{pmatrix} \alpha & \beta \\ \gamma & \delta \end{pmatrix}$, the last expression can be written as $\chi_Q^1\left(\begin{pmatrix} \alpha & \beta \\ \gamma & \delta \end{pmatrix}\right)$; hence in the notation of Lemma 4.8 we obtain the relation

(4.24) $$\chi_Q^n(M) = \chi_Q^1\left(\begin{pmatrix} \alpha & \beta \\ \gamma & \delta \end{pmatrix}\right).$$

We have thereby reduced the calculation of the multiplier χ_Q^n for arbitrary n to the case $n = 1$.

PROPOSITION 4.9. *Let* $Q \in \mathbf{A}_m^+$, *where* $m = 2k$ *is even and the level* q *of* Q *is greater than* 1. *Then for any* $\begin{pmatrix} \alpha & \beta \\ \gamma & \delta \end{pmatrix} \in \Gamma_0^1(q)$ *one has*

(4.25) $$\chi_Q^1\left(\begin{pmatrix} \alpha & \beta \\ \gamma & \delta \end{pmatrix}\right) = \chi_Q(\delta),$$

where χ_Q *is the character of the quadratic form* Q, *i.e., it is the real Dirichlet character modulo* q *defined on integers* δ *prime to* q *by the formula*

(4.26) $$\chi_Q(\delta) = (\text{sign}\,\delta)^k |\delta|^{-k} \sum_{l \in M_{m,1}(\mathbf{Z}/\delta\mathbf{Z})} \exp((\pi i/\delta)Q[l]),$$

in particular

$$\chi_Q(-1) = (-1)^k.$$

If p *is an odd prime, then* $\chi_Q(p)$ *can be computed from the formula*

$$\chi_Q(p) = \left(\frac{(-1)^k \det Q}{p}\right) \qquad (\text{Legendre symbol}).$$

PROOF. The formula (4.21) shows that the number $\xi = \chi_Q^1\left(\begin{pmatrix} \alpha & \beta \\ \gamma & \delta \end{pmatrix}\right)$ belongs to the field $\mathbf{Q}_{|\delta|}$ of $|\delta|$th roots of unity. On the other hand, because χ_Q^1 is a character of the group $\Gamma_0^1(q)$ and $\chi_Q^1\left(\begin{pmatrix} 1 & b \\ 0 & 1 \end{pmatrix}\right) = 1$ for any $b \in \mathbf{Z}$, we obtain

$$\chi_Q^1\left(\begin{pmatrix} \alpha & \beta \\ \gamma & \delta \end{pmatrix}\right) = \chi_Q^1\left(\begin{pmatrix} \alpha & \beta \\ \gamma & \delta \end{pmatrix}\begin{pmatrix} 1 & b \\ 0 & 1 \end{pmatrix}\right) = \chi_Q^1\left(\begin{pmatrix} \alpha & \beta + b\alpha \\ \gamma & \delta + b\gamma \end{pmatrix}\right),$$

so that ξ also belongs to any of the fields $\mathbf{Q}_{|\delta + b\gamma|}$. But the arithmetic progression $\delta + b\gamma$ ($b \in \mathbf{Z}$) contains numbers that are relatively prime, and $\mathbf{Q}_{|a|} \cap \mathbf{Q}_{|b|} = \mathbf{Q}$ if a is prime to b (in this case the compositum of $\mathbf{Q}_{|a|}$ and $\mathbf{Q}_{|b|}$ is $\mathbf{Q}_{|ab|}$, and its degree over $\mathbf{Q}$ is the product of the degrees of $\mathbf{Q}_{|a|}$ and $\mathbf{Q}_{|b|}$). Hence, ξ is a rational number. Consequently, ξ does not change under any of the automorphisms $\exp(2\pi i/\delta) \to \exp(2\pi i t/\delta)$ of the field $\mathbf{Q}_{|\delta|}$ (here $(t, \delta) = 1$). Taking $t = -\beta$ and taking into account that $\gamma\beta \equiv 1 \pmod{\delta}$, we find that

$$\chi_Q^1\left(\begin{pmatrix} \alpha & \beta \\ \gamma & \delta \end{pmatrix}\right) = \delta^k d^{-m} \sum_{l \in M_{m,1}(\mathbf{Z}/d\mathbf{Z})} e\{\delta^{-1}Q[l]\}$$

depends only on δ and Q. If we set $d = |\delta|$ here, we obtain (4.26).

Given any integer δ prime to q, there exist integers a and b such that $a\delta - qb = 1$. Then $\begin{pmatrix} a & b \\ q & \delta \end{pmatrix} \in \Gamma_0^1(q)$, and

$$\chi_Q(\delta) = \chi_Q^1\left(\begin{pmatrix} a & b \\ q & \delta \end{pmatrix}\right) = \chi_Q^1\left(\begin{pmatrix} a & b \\ q & \delta \end{pmatrix}\begin{pmatrix} 1 & t \\ 0 & 1 \end{pmatrix}\right)$$

$$= \chi_Q^1\left(\begin{pmatrix} a & b + tq \\ q & \delta + tq \end{pmatrix}\right) = \chi_Q(\delta + tq)$$

for any $t \in \mathbf{Z}$. Thus, the function $\chi_Q(\delta)$ is defined for all δ prime to q, and it depends only on the residue class of δ modulo q. If δ_1 is also an integer prime to q and $\begin{pmatrix} a_1 & b_1 \\ q & \delta_1 \end{pmatrix} \in \Gamma_0^1(q)$, then

$$\chi_Q(\delta)\chi_Q(\delta_1) = \chi_Q^1\left(\begin{pmatrix} a & b \\ q & \delta \end{pmatrix}\begin{pmatrix} a_1 & \delta_1 \\ q & b_1 \end{pmatrix}\right)$$

$$= \chi_Q^1\left(\begin{pmatrix} * & * \\ * & qb_1 + \delta\delta_1 \end{pmatrix}\right) = \chi_Q(qb_1 + \delta\delta_1) = \chi_Q(\delta\delta_1).$$

Thus, χ_Q is a real Dirichlet character modulo q. Now let p be an odd prime not dividing q. If we set $\delta = p$ in (4.26), we can write

$$\chi_Q(p) = p^{-k} \sum_{l \in M_{m,1}(\mathbf{Z}/p\mathbf{Z})} \exp((2\pi i/p)((1/2)Q[l])).$$

If $M \in M_m$ and the determinant of M is prime to p, then the map $l \to Ml$ obviously gives a bijection of the set $M_{m,1}/pM_{m,1}$ with itself. Hence, for any such M we can write

$$\chi_Q(p) = p^{-k} \sum_{l \in M_{m,1}(\mathbf{Z}/p\mathbf{Z})} \exp((2\pi i/p)((1/2)Q[Ml])).$$

It is well known (see Appendix 1.1) that the matrix M can be chosen in such a way that the quadratic form $(1/2)Q[MX]$ is congruent modulo p to a diagonal quadratic form $a_1 x_1^2 + \cdots + a_m x_m^2$. Here we clearly have

$$a_1 \cdots a_m \equiv \det((1/2)Q[M]) = 2^{-m}(\det M)^2 \det Q \pmod{p}.$$

With this choice of M, the last formula for $\chi_Q(p)$ can be written in the form

$$\chi_Q(p) = p^{-k} G_p(a_1) \cdots G_p(a_m),$$

where $G_p(a)$ denotes the *Gauss sum*

$$(4.27) \qquad G_p(a) = \sum_{t \in \mathbf{F}_p = \mathbf{Z}/p\mathbf{Z}} \exp(2\pi i a t^2 / p).$$

If we use the definition and properties of the Legendre symbol modulo p (see Appendix 2.3), we find that

$$G_p(a) = \sum_{b \in \mathbf{F}_p} \left(1 + \left(\frac{b}{p}\right)\right) \exp(2\pi i a b / p)$$

$$(4.28) \qquad = \sum_{b \in \mathbf{F}_p} \left(\frac{b}{p}\right) \exp(2\pi i a b / p)$$

$$= \left(\frac{a}{p}\right) \sum_{b \in \mathbf{F}_p} \left(\frac{ab}{p}\right) \exp(2\pi i a b / p) = \left(\frac{a}{p}\right) G_p(1).$$

On the other hand, taking into account that p is an odd prime, we have

$$G_p(1)^2 = \left(\frac{-1}{p}\right) G_p(1) G_p(-1)$$

$$(4.29) \qquad = \left(\frac{-1}{p}\right) \sum_{t_1, t_2 \in \mathbf{F}_p} \exp(2\pi i (t_1 - t_2)(t_1 + t_2)/p)$$

$$= \left(\frac{-1}{p}\right) \sum_{a \in \mathbf{F}_p} \sum_{b \in \mathbf{F}_p} \exp(2\pi i a b / p) = \left(\frac{-1}{p}\right) p.$$

Returning to the calculation of $\chi_Q(p)$, from the above formulas and the properties of the Legendre symbol we obtain

$$\chi_Q(p) = p^{-k} \left(\frac{a_1 \cdots a_m}{p}\right) G_p(1)^m = \left(\frac{(-1)^k a_1 \cdots a_m}{p}\right)$$

$$= \left(\frac{(-1)^k 2^{-m}(\det M)^2 \det Q}{p}\right) = \left(\frac{(-1)^k \det Q}{p}\right). \qquad \square$$

The number $(-1)^k \det Q$ is called the *discriminant* of the quadratic form with matrix Q. The reader can easily verify that the discriminant of any integral quadratic form in an even number of variables is congruent to 0 or 1 modulo 4.

The next theorem summarizes our computation of the multiplier for theta-series of quadratic forms in an even number of variables.

THEOREM 4.10. *Suppose that $Q \in \mathbf{A}_m^+$, q is the level of Q, and $n \geqslant 1$. Further suppose that $m = 2k$ is even. Then for any matrix $M = \begin{pmatrix} A & B \\ C & D \end{pmatrix} \in \Gamma_0^n(q)$ the multiplier $\chi_Q^n(M)$ in the functional equations* (3.29) *and* (3.30) *of Theorem* 3.13 *is given by the following formulas:*

if $q = 1$, then

$$(4.30) \qquad\qquad \chi_Q^n(M) = 1,$$

if $q > 1$, then

$$(4.31) \qquad\qquad \chi_Q^n(M) = \chi_Q(\det D),$$

where χ_Q is the character of the quadratic form of Q, i.e., the real Dirichlet character modulo q that satisfies the conditions

$$(4.32) \qquad\qquad \chi_Q(-1) = (-1)^k,$$

$$(4.33) \qquad\qquad \chi_Q(p) = \left(\frac{(-1)^k \det Q}{p} \right) \qquad (\textit{Legendre symbol}),$$

if p is an odd prime not dividing q, and

$$(4.34) \qquad\qquad \chi_Q(2) = 2^{-k} \sum_{t \in M_{m,1}(\mathbf{Z}/2\mathbf{Z})} \exp(\pi i Q[t]/2),$$

if q is odd.

PROOF. Formula (4.30) was proved in Proposition 4.3, and (4.31) follows from (4.24), (4.25), and (4.22). Formulas (4.32)–(4.34) were proved in Proposition 4.9. □

5. Quadratic forms in an odd number of variables. We first prove the following useful proposition.

PROPOSITION 4.11. *Let Q be a nonsingular symmetric integer matrix with even entries on the main diagonal. Suppose that the order of Q is odd. Then its determinant $\det Q$ and its level $q = q(Q)$ satisfy the congruences*

$$(4.35) \qquad\qquad \det Q \equiv 0 \pmod 2, \quad q(Q) \equiv 0 \pmod 4.$$

PROOF. For brevity we shall use the term "even matrix" to refer to a symmetric integer matrix with even entries on the main diagonal. Recall that the level of a nonsingular even matrix Q is the least natural number q such that $q \cdot Q^{-1}$ is an even matrix. Let $Q = (a_{\alpha\beta})$ be a nonsingular $m \times m$ even matrix, where m is odd. We set $Q^{-1} = Q^* = (\det Q)^{-1} \cdot (A_{\alpha\beta})$. Then for every $\alpha = 1, \ldots, m$ we have the equality

$$\det Q = \sum_{\beta=1}^m a_{\alpha\beta} A_{\alpha\beta};$$

summing these equalities, we obtain

$$m \det Q = \sum_{\beta=1}^m a_{\alpha\alpha} A_{\alpha\alpha} + 2 \sum_{1 \leqslant \alpha < \beta \leqslant m} a_{\alpha\beta} A_{\alpha\beta}.$$

Since all of the coefficients $a_{\alpha\alpha}$ $(1 \leqslant \alpha \leqslant m)$ are even, it follows that $m \det Q$ is divisible by 2, and hence so is $\det Q$.

To prove the second congruence in (4.35), we first note that since $q(Q)Q^{-1}$ is an integer matrix, its determinant is an integer, i.e., $\det Q$ divides $q(Q)^m$. Thus, if m is odd, the level $q = q(Q)$ is divisible by 2. To show that q is divisible by 4, we use induction on the odd number m. The congruence is obvious if $m = 1$. Suppose that it has already been proved for all nonsingular even matrices of odd order less than m, where $m > 1$, and let Q be a nonsingular even matrix of order m. We consider two cases:

(1) All of the entries of Q are even, i.e., $Q = 2Q_1$, where Q_1 is an integer matrix. Then $Q_2 = qQ^{-1} = (q/2)Q_1^{-1}$ is a nonsingular even matrix of odd order, and hence has even determinant. Since this determinant divides $(q/2)^m$, it follows that $q/2$ is even, and hence q is divisible by 4.

(2) Not all of the entries in Q are even. In this case, if we make a suitable permutation of the rows of Q and the same permutation of its columns—i.e., for suitable $V \in GL_m(\mathbf{Z})$ we perform the transformation $Q \to Q[V]$, which does not change the level q and takes even matrices to even matrices—then we may suppose that the entry $a_{12} = a_{21}$ is odd. We divide Q into blocks $\begin{pmatrix} Q_{11} & Q_{12} \\ Q_{21} & Q_{22} \end{pmatrix}$, where $Q_{11} = \begin{pmatrix} a_{11} & a_{12} \\ a_{21} & a_{22} \end{pmatrix}$. Since

$$\det Q_{11} = a_{11}a_{22} - a_{12}^2 \equiv -a_{12}^2 \equiv 1 (\mathrm{mod}\, 2),$$

the matrix Q_{11} is invertible and the matrix Q_{11}^{-1} is a 2-integral matrix with even diagonal. From the obvious identity

$$Q[U] = \begin{pmatrix} Q_{11} & 0 \\ 0 & Q_{22} - Q_{11}^{-1}[Q_{12}] \end{pmatrix}, \quad \text{where } U = \begin{pmatrix} E_2 & -Q_{11}^{-1}Q_{12} \\ 0 & E_{m-2} \end{pmatrix},$$

we obtain

$$Q^{-1}[U] = \begin{pmatrix} Q_{11}^{-1} & 0 \\ 0 & (Q_{22} - Q_{11}^{-1}[Q_{12}])^{-1} \end{pmatrix}.$$

The matrix $Q_{22} - Q_{11}^{-1}[Q_{12}]$ is 2-integral, and we easily see that it has even diagonal. This implies that it can be written in the form $d^{-1}Q'$, where d is an odd natural number and Q' is an even matrix. By the last identity, $q \cdot d \cdot (Q')^{-1}$ is a 2-integral matrix with even diagonal. Consequently, by the induction assumption, the number qd is divisible by 4, and hence so is q. □

Proposition 4.11 shows that the level of any quadratic form in an odd number of variables—or, equivalently, the level of the corresponding matrix $Q \in \mathbf{A}_m^+$—is divisible by 4; hence, we have the inclusion $\Gamma_0^n(q) \subset \Gamma_0^n(4)$. According to Theorem 3.13, for any $M \in \Gamma_0^n(q)$ the theta-series $\theta^n(Z, Q)$ satisfies the functional equation (3.30), in which the multiplier χ_Q^n is not a character of the group $\Gamma_0^n(q)$ as in the case of theta-series of quadratic forms in an even number of variables (see (4.31)), but rather is more complicated.

On the other hand, the example of the simplest quadratic form with 1×1-matrix $(2) \in \mathbf{A}_1^+$ shows that there exist matrices of level 4. Using the notation in (4.1) and (4.10) and the formula (4.15), we can write the functional equation for the theta-series

$\theta^n(Z, (2))$ in the form

(4.36) $$\theta^n(M\langle Z\rangle, (2)) = j^n_{(2)}(M, Z)\theta^n(Z, (2)),$$

where $M = \begin{pmatrix} A & B \\ C & D \end{pmatrix} \in \Gamma^n_0(4)$ and

(4.37) $$j^n_{(2)}(M, Z) = \chi^n_{(2)}(M)\det(CZ + D)^{1/2},$$

in which the square root is determined from the condition (4.7);

(4.38) $$\chi^n_{(2)}(M) = \varepsilon^{-1}_{\text{sign}(\det D)}|\det D|^{1/2}G(BD^{-1}, (2)),$$

where for any odd integer d we determine ε_d from the formula

(4.39) $$\varepsilon_d = \begin{cases} 1, & \text{if } d \equiv 1 (\text{mod } 4), \\ i = +\sqrt{-1}, & \text{if } d \equiv -1 (\text{mod } 4), \end{cases}$$

(4.40) $$G(BD^{-1}, (2)) = |d|^{-n} \sum_{r \in M_{n,1}(\mathbf{Z}/d\mathbf{Z})} e\{2BD^{-1}[r]\},$$

where d is any nonzero integer such that $d \cdot D^{-1} \in M_n$, and $e\{\dots\}$ is the function (3.23).

Since $\Gamma^n_0(q) \subset \Gamma^n_0(4)$, and since the product of the theta-series for the matrices Q and (2) is the theta-series for a matrix of even order and the same level q, it follows that Theorem 4.10 enables us to obtain the functional equation for the theta-series $\theta^n(Z, Q)$ in terms of the automorphy factor $j^n_{(2)}$. Using this connection, we prove the following theorem.

THEOREM 4.12. *Suppose that $\theta^n(Z, Q)$ is the theta-series* (1.13) *for the matrix $Q \in \mathbf{A}^+_m$ with $m = 2k + 1$ odd, q is the level of Q, $M = \begin{pmatrix} A & B \\ C & D \end{pmatrix} \in \Gamma^n_0(q)$, and $j^n_{(2)}(M, Z)$ is the automorphy factor* (4.37). *Then the theta-series satisfies the functional equation*

(4.41) $$\theta^n(M\langle Z\rangle, Q) = \chi_Q(\det D)j^n_{(2)}(M, Z)^m\theta^n(Z, Q),$$

where

(4.42) $$\chi_Q(d) = \left(\frac{2\det Q}{|d|}\right)$$

is a Dirichlet character modulo q, and $(\frac{\cdot}{\cdot})$ is the Jacobi symbol.

PROOF. Using the definition (1.13) of theta-series, we easily see that

(4.43) $$\theta^n(Z, Q)\theta^n(Z, (2)) = \theta^n(Z, Q_1) \quad \text{with } Q_1 = Q \oplus (2).$$

According to Proposition 4.11, the level of Q_1 is also q. Since $Q_1 \in \mathbf{A}^+_{m+1}$ and $m + 1$ is even, it follows from (3.30) and Theorem 4.10 that

(4.44) $$\theta^n(M\langle Z\rangle, Q_1) = \chi_{Q_1}(\det D)\det(CZ + D)^{(m+1)/2}\theta^n(Z, Q_1).$$

On the other hand, the functional equation (3.30) implies that

(4.45) $$\theta^n(M\langle Z\rangle, Q) = \chi(M)j^n_{(2)}(M, Z)^m\theta^n(Z, Q).$$

If we now multiply this equation and the equation (4.36) and take (4.43) into account, we find that the last equation is preserved if Q and m are replaced by Q_1 and $m + 1$. Since all of our theta-series are nonzero functions, it follows from (4.45) for Q_1 and from (4.44) that

$$(4.46) \qquad \chi(M) = \chi_{Q_1}(\det D) \left(\frac{\det(CZ + D)}{j^n_{(2)}(M, Z)^2} \right)^{(m+1)/2}.$$

Furthermore, if we square the equality (4.36) and let $Q = (2)$ in (4.43) and (4.44), we obtain

$$(4.47) \qquad j^n_{(2)}(M, Z)^2 = \chi_{Q_2}(\det D) \det(CZ + D) \quad \text{with } Q_2 = \begin{pmatrix} 2 & 0 \\ 0 & 2 \end{pmatrix}.$$

Hence, by (4.46) and the definition of the characters of quadratic forms in (4.32), (4.33), and (4.42), we conclude that

$$\chi(M) = \chi_{Q_1}(\det D)\chi_{Q_2}(\det D)^{(m+1)/2} = \chi_Q(\det D). \qquad \square$$

Although the automorphy factor $j^n_{(2)}$ is simpler than the automorphy factor j^n_Q for an arbitrary matrix Q of odd order, nevertheless it has a rather complicated structure. In certain cases, however, it is possible to express $j^n_{(2)}$ in terms of $j^1_{(2)}$, and hence, because of (4.37)–(4.40), in terms of the one-dimensional Gauss sums

$$(4.48) \qquad G_d(c) = \sum_{r \in \mathbf{Z}/d\mathbf{Z}} \exp\left(2\pi i \frac{c}{d} r^2 \right),$$

which are the subject of the next two lemmas.

LEMMA 4.13. *Suppose that* $c, d \in \mathbf{Z}$, d *is a positive odd number,* $(c, d) = 1$, *and* $\left(\frac{\cdot}{\cdot}\right)$ *is the Jacobi symbol. Then the Gauss sum satisfies the relation*

$$G_d(c) = \left(\frac{c}{d} \right) G_d(1).$$

PROOF. If $d = p$ is an odd prime, then the lemma follows from (4.28). Suppose that $d = p^n$ with $n > 1$. If we set $r = r_1 + p^{n-1}r_2$ in (4.48), where r_1 runs through $\mathbf{Z}/p^{n-1}\mathbf{Z}$ and r_2 runs through $\mathbf{Z}/p\mathbf{Z}$, we find that

$$G_{p^n}(c) = pG_{p^{n-2}}(c),$$

and the proof of the lemma for $d = p^n$ can be obtained from this relation by induction on n. Now suppose that $d = d_1 \cdot d_2$, where d_1 is prime to d_2, and suppose that b_1 and b_2 are integers such that $b_1 d_1 + b_2 d_2 = 1$. In (4.48) let $r = d_2 r_1 + d_1 r_2$, where r_i runs through $\mathbf{Z}/d_i\mathbf{Z}$, and replace c by $c(b_1 d_1 + b_2 d_2)$. We then find that the Gauss sum satisfies the relation

$$(4.49) \qquad G_d(c) = G_{d_1}(cb_2)G_{d_2}(cb_1).$$

We assume, by induction, that the lemma holds for d_1 and d_2, and we prove that it holds for d. From (4.49) we have

$$G_d(c) = \left(\frac{c}{d_1}\right)\left(\frac{b_2}{d_1}\right)G_{d_1}(1)\left(\frac{c}{d_2}\right)\left(\frac{b_1}{d_2}\right)G_{d_2}(1)$$

$$= \left(\frac{c}{d_1}\right)\left(\frac{c}{d_2}\right)G_{d_1}(b_2)G_{d_2}(b_1) = \left(\frac{c}{d}\right)G_d(1). \qquad \square$$

LEMMA 4.14. *If d is a positive odd integer, then*

$$(4.50) \qquad\qquad G_d(1)\varepsilon_d \cdot d^{1/2},$$

where the square root is positive and ε_d is the function (4.39).

PROOF. We compute the value of the theta-function $\theta^1(z;0,0)$ (see (3.2)) at the point $z = 2c/d + i\lambda$, where $\lambda > 0$, c and $d \neq 0$ are integers. Let $d_1 = d$ if d is odd and $d_1 = d/2$ if d is even. In the definition of $\theta^1(z;0,0)$ in (3.2) we divide the summation into two parts: we set $N = r + d_1 m$, where r runs through the set of residues modulo d_1 and m runs through all integers. Then, after some simple transformations, we obtain the identity

$$(4.51) \qquad \theta^1(z;0,0) = \sum_r e\left\{2\left(\frac{c}{d}+\frac{i\lambda}{2}\right)r^2\right\}\theta^1(i\lambda d_1^2; i\lambda d_1 r, 0).$$

Let $c = 1$, and let $d > 0$ be an odd number. Then (4.51) and the inversion formula for the theta-function (3.5) imply that

$$(4.52) \qquad \lim_{\lambda\to+0}\lambda^{1/2}\theta^1(z;0,0) = \sum_{r\in\mathbf{Z}/d\mathbf{Z}}e\left\{2\frac{c}{d}r^2\right\}d^{-1}\lim_{\lambda\to+0}\theta^1\left(-\frac{1}{i\lambda d^2};0,-i\lambda dr\right)$$

$$= d^{-1}G_d(1).$$

We now compute the same limit in another way. Since $(-iz)^{-1/2} \to e\{\tfrac{1}{4}\}(d/2)^{1/2}$ as $\lambda \to +0$, it follows from (3.5) that the first limit in (4.52) is equal to

$$(4.53) \qquad\qquad e\left\{\frac{1}{4}\right\}\left(\frac{d}{2}\right)^{1/2}\lim_{\lambda\to+0}\lambda^{1/2}\theta^1(-z^{-1};0,0).$$

If we observe that $-z^{-1} = 2(-d)/4 + i\lambda_1$, where $\lambda_1 = [(2/d)(2/d + i\lambda)]^{-1}$ and apply (4.51) and (3.5), we find that the limit in (4.53) is equal to

$$\sum_{r\in\mathbf{Z}/2\mathbf{Z}}e\left\{\frac{-d}{2}r^2\right\}\lim_{\lambda\to+0}\frac{\lambda^{1/2}}{(4\lambda_1)^{1/2}}\lim_{\lambda\to+0}\theta^1\left(-\frac{1}{4i\lambda_1},0,-2i\lambda_1 r\right)$$

$$= d^{-1}\sum_{r\in\mathbf{Z}/2\mathbf{Z}}e\left\{\frac{-d}{2}r^2\right\}.$$

This, along with (4.53) and (4.39), implies that the first limit in (4.52) is equal to $\varepsilon_d \cdot d^{-1/2}$; in view of (4.52), we hence obtain (4.50). $\qquad \square$

From Lemmas 4.13 and 4.14 we obtain the following proposition.

PROPOSITION 4.15. *In the case $n = 1$ the automorphy factor* (4.37) *is given by the explicit formula*

$$(4.54) \qquad j_{(2)}^1(M, z) = \varepsilon_d^{-1}\left(\frac{c}{d}\right)(cz + d)^{1/2} \quad \text{for } M = \begin{pmatrix} a & b \\ c & d \end{pmatrix} \in \Gamma_0^1(4),$$

where ε_d is the function (4.39), $(c/d) = (c/|d|)$ *is the Jacobi symbol, and the square root is determined by the condition* (4.7).

PROOF. By (4.16), for odd $d > 0$ we have $G(b/d, (2)) = G(-c/d, (2))$ and $(-1/d) = \varepsilon_d^2$. Thus, in the case $d > 0$ the formula (4.54) follows from (4.37), (4.38), and Lemmas 4.13 and 4.14. On the other hand, if $d < 0$, then to prove (4.54) it suffices to replace $-c/d$ by $c/(-d)$ in the Gauss sum. $\square$

To conclude this section, we give a simple but important property of the multiplier $\chi_{(2)}^n$ in (4.38). From (4.47) and (4.37) it follows that $[\chi_{(2)}^n(M)]^2 = \chi_{Q_2}(\det D)$, and hence, by (4.32) and (4.33), we find that

$$(4.55) \qquad\qquad [\chi_{(2)}^n(M)]^4 = 1 \quad \text{for } M \in \Gamma_0^n(4).$$

CHAPTER 2

Modular Forms

The development in Chapter 1 of the analytic and group-theoretic properties of theta-series of integral quadratic forms provides the basis for the axiomatic definition and the study of all functions with similar properties. The resulting class of modular forms, although not generally exhausted by the set of theta-series, share many of their analytic and structural features. Moreover, the invariant definition of modular forms enables one to find and prove these properties much more easily.

§1. Fundamental domains for subgroups of the modular group

The functional equations for theta-series that were proved in the previous chapter show that all of the values of a theta-series at points of a fixed *orbit*

$$(1.1) \qquad K\langle Z \rangle = \{ M\langle Z \rangle; M \in K \} \quad (Z \in \mathbf{H}_n)$$

of the corresponding subgroup K of the modular group are determined if we know the value at one of the points. Thus, a theta-series is uniquely determined by its restriction to any subset of the upper half-plane which intersects with all of the orbits of K. In this section we shall construct a *fundamental domain* in $\mathbf{H}_n$ for an arbitrary subgroup K of finite index in Γ^n, i.e., we shall give a set of representatives of the orbits (1.1) that deserves to be called a "domain".

1. The modular triangle. For brevity we shall call the imaginary part y of a complex number $z = x + iy \in \mathbf{H}_1$ the *height* of z, denoted $h(z)$. By Lemma 2.8 of Chapter 1 (or a direct computation) we see that

$$h(M\langle z \rangle) = \frac{h(z)}{|cz + d|^2}, \quad \text{where } M = \begin{pmatrix} a & b \\ c & d \end{pmatrix} \in \Gamma^1.$$

Since the inequality

$$|cz + d|^2 = (cz + d)^2 + (cy)^2 < 1$$

has only finitely many solutions in integers c and d for any fixed $z = x + iy \in \mathbf{H}_1$, it follows that there are only finitely many values of h on the orbit $\Gamma^1\langle z \rangle$ that are greater than $h(z)$. Consequently, every orbit of Γ^1 in $\mathbf{H}_1$ has points of maximal height, and these points are characterized by the inequality $|cz + d| \geqslant 1$, which must hold for any pair of integers c, d that form the second row of a matrix $M = \begin{pmatrix} a & b \\ c & d \end{pmatrix} \in \Gamma^1$, i.e., for any pair of relatively prime integers. The transformation $z \to \begin{pmatrix} 1 & b \\ 0 & 1 \end{pmatrix} \langle z \rangle = z + b = (x + b) + iy$, where $b \in \mathbf{Z}$, does not affect the height of z. Here b can always be chosen so that $|x + b| \leqslant 1/2$. We thus see that every orbit of Γ^1 in $\mathbf{H}_1$ has a point in the set

$$
(1.2) \qquad \begin{aligned} D_1' = \{z = x + iy \in \mathbf{H}_1; |x| \leqslant 1/2, |cz + d| \geqslant 1, \\ \text{if } c, d \in \mathbf{Z} \text{ and } (c, d) = 1\}. \end{aligned}
$$

We now show that the set D_1' is actually given by a finite number of inequalities. Let

$$
(1.3) \qquad D_1 = \{z = x + iy \in \mathbf{H}_1; |x| \leqslant 1/2, |z|^2 = x^2 + y^2 \geqslant 1\}.
$$

Since $1, 0 \in \mathbf{Z}$ and $(1, 0) = 1$, it follows that $D_1' \subset D_1$. Conversely, if $z = x + iy \in D_1$, $c, d \in \mathbf{Z}$, and $(c, d) = 1$, then

$$
|cz + d|^2 = c^2(x^2 + y^2) + 2cdx + d^2 \geqslant c^2 - |cd| + d^2 \geqslant 1,
$$

so that $z \in D_1'$. Thus, $D_1 = D_1'$, and every orbit of the modular group Γ^1 in the upper half-plane $\mathbf{H}_1$ intersects the set D_1. This set may be regarded as a "triangle" (the *modular triangle*) with vertices at ρ, ρ^2, and $i\infty$ (see Figure 1).

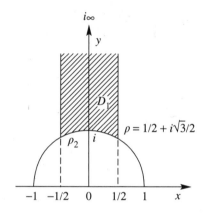

FIGURE 1

We now show that D_1 is a fundamental domain for Γ^1 on $\mathbf{H}_1$. More precisely, we have the following theorem.

THEOREM 1.1. (1) *For every point $z \in \mathbf{H}_1$ there exists a matrix $M \in \Gamma^1$ such that $M\langle z \rangle \in D_1$.*

(2) *If $z = x + iy$ and z' are two distinct points in D_1 that lie in the same orbit of Γ^1, then either $x = \pm 1/2$ and $z' = z \mp 1$, or else $|z| = 1$ and $z' = -z^{-1}$. In particular, no two interior points of D_1 lie in the same orbit of Γ^1.*

PROOF. The first part of the theorem was proved above. Suppose that $z = x + iy \in D_1$, $M = \begin{pmatrix} a & b \\ c & d \end{pmatrix} \in \Gamma^1$, $z' = x' + iy' = M\langle z \rangle \in D_1$, and $z \neq z'$. Since $D_1 = D_1'$, we have $h(z) = h(M\langle z \rangle)$, and hence $|cz + d| = 1$.

If $c = 0$, then $d = \pm 1$. Then $M = \begin{pmatrix} \pm 1 & b \\ 0 & \pm 1 \end{pmatrix}$ and $z' = z \pm b$. Since $-1/2 \leqslant x, x' \leqslant 1/2$ and $x' = x \pm b$, it follows that $x = \pm 1/2$, $b = \mp 1$, $x' = \mp 1/2$.

If $d = 0$, then $c = \pm 1$, and $|z| = 1$. Then $M = \begin{pmatrix} a & \mp 1 \\ \pm 1 & 1 \end{pmatrix}$, and $z' = \pm a - z^{-1}$.
Since we have $-z^{-1} \in D_1$ and $|-z^{-1}| = 1$, it follows that $a = 0$, except in the cases $z = \rho$ or ρ^2 and $a = -1$ or 1, respectively. But in those cases $z' = z$, contradicting our assumption.

Finally, suppose that $c \neq 0$ and $d \neq 0$. Then the inequalities

$$1 = |cz + d|^2 = c^2(x^2 + y^2) + 2cdx + d^2 \geqslant c^2 - |cd| + d^2 \geqslant 1$$

imply that $x^2 + y^2 = 1$, $x = \pm 1/2$, $c = \pm 1$, $d = \pm 1$, and the product cd has the opposite sign of x. Thus, $(c, d) = \pm(1, 1)$ and $z = \rho^2$, or else $(c, d) = \pm(1, -1)$ and $z = \rho$. In the first case, up to a sign the matrix M is equal to

$$\begin{pmatrix} a & a-1 \\ 1 & 1 \end{pmatrix} \quad \text{and} \quad z' = \frac{a\rho^2 + (a-1)}{\rho^2 + 1} = a - \frac{1}{\rho^2 + 1} = a + \rho^2,$$

so that (because $z \neq z'$) we have $a = 1$ and $z' = 1 + \rho^2 = \rho = -1/\rho^2$. The second case is similar. $\qquad\square$

PROBLEM 1.2. Show that the stabilizer

$$S(z) = \{M \in \Gamma^1; M\langle z \rangle = z\}$$

of the point $z \in \mathbf{H}_1$ in the group Γ^1 is $\{\pm E_2\}$ if z does not belong to either of the two orbits $\Gamma^1\langle i \rangle$, $\Gamma^1\langle \rho \rangle$. Show that

$$S(i) = \left\{ \pm E_2, \pm \begin{pmatrix} 0 & 1 \\ -1 & 0 \end{pmatrix} \right\},$$

$$S(\rho) = \left\{ \pm E_2, \pm \begin{pmatrix} -1 & 1 \\ -1 & 0 \end{pmatrix}, \pm \begin{pmatrix} 0 & -1 \\ 1 & -1 \end{pmatrix} \right\}.$$

PROBLEM 1.3. Two binary quadratic forms

$$Q(x, y) = ax^2 + bxy + cy^2 \quad \text{and} \quad Q_1(x_1, y_1) = a_1 x_1^2 + b_1 x_1 y_1 + c_1 y_1^2$$

are said to be equivalent (over $\mathbf{Z}$) if

$$(1.4) \qquad Q_1(x_1, y_1) = Q(\alpha x_1 + \beta y_1, \gamma x_1 + \delta y_1), \quad \text{where} \begin{pmatrix} \alpha & \beta \\ \gamma & \delta \end{pmatrix} \in \Gamma^1.$$

Show that any real positive definite form Q is equivalent to a form Q_1 with $|b_1| \leqslant a_1 \leqslant c_1$. Further show that in the interior of the region defined by these inequalities in the space of coefficients of binary quadratic forms, there are no two distinct points that correspond to equivalent forms.

[**Hint:** Let ω and ω_1 be the roots of the quadratic equations $Q(t, 1) = 0$ and $Q_1(t, 1) = 0$, respectively, that belong to $\mathbf{H}_1$. Show that (1.4) is equivalent to the conditions $b_1^2 - 4a_1 c_1 = b^2 - 4ac$ and $\omega_1 = M^{-1}\langle \omega \rangle$ with $M = \begin{pmatrix} \alpha & \beta \\ \gamma & \delta \end{pmatrix}$, and use Theorem 1.1.]

PROBLEM 1.4. Show that there are only finitely many equivalence classes of positive definite integral binary quadratic forms $Q = ax^2 + bxy + cy^2$ with fixed discriminant $b^2 - 4ac < 0$.

2. The Minkowski reduction domain. The construction of a fundamental domain for the modular group Γ^n for $n > 1$ is based on the same idea as in the case $n = 1$ above. Again, every orbit of Γ^n in $\mathbf{H}_n$ has points $Z = X + iY$ of maximal height $h(Z) = \det Y$, and in the set of such points we make a further reduction by means of transformations in Γ^n that do not affect the height. But whereas all such transformations were of the form $z \to z + b$ ($b \in \mathbf{Z}$) in the case $n = 1$, when $n > 1$, in addition to the analogous transformations $Z \to Z + B$ ($B \in S_n$), the transformations $Z \to Z[V]$ ($V \in GL_n(\mathbf{Z})$) also have this property. Minkowski reduction theory is devoted to the construction and the study of the properties of a fundamental domain for the group

$$(1.5) \qquad\qquad \Lambda = \Lambda^n = GL_n(\mathbf{Z})$$

of $n \times n$ unimodular matrices acting on the set

$$(1.6) \qquad\qquad P = P_n = \{ Y \in S_n(\mathbf{R}); Y > 0 \}$$

of matrices of real positive definite quadratic forms in n variables, where the action is given by:

$$\Lambda \ni U : Y \to Y[U] = {}^tUYU.$$

We shall let $u_1, \ldots, u_n$ denote the columns of $U \in \Lambda^n$, so that $U = (u_1, \ldots, u_n)$. In order to choose a "reduced" representative $Y[U]$ in the orbit

$$(1.7) \qquad\qquad \{ Y \}_\Lambda = \{ Y[U]; U \in \Lambda \}$$

of a point $Y \in P$, we construct the matrix $U \in \Lambda$ column by column, starting from certain minimality conditions. We let Λ_k^n denote the set of integer matrices made up from the first k columns of the matrices in Λ:

$$(1.8) \qquad\qquad \Lambda_k^n = \{ (u_1, \ldots, u_k); (u_1, \ldots, u_k, *, \ldots, *) \in \Lambda \}.$$

In other words, Λ_k^n is the set of $n \times k$ integer matrices which can be completed to an $n \times n$ unimodular matrix. Starting with a fixed matrix $Y \in P$, we choose $u_1 \in \Lambda_1^n$ in such a way that the value $Y[u_1]$ is minimal in the set Λ_1^n. This can be done, because Λ_1^n consists of integer vectors and $Y > 0$. After choosing u_1, we choose u_2 so that $(u_1, u_2) \in \Lambda_2^n$ and the value $Y[u_2]$ is minimal. Possibly replacing u_2 by $-u_2$, without loss of generality we may assume that ${}^tu_1 Y u_2 \geqslant 0$. Continuing this process, at the kth step we find a column u_k for which $(u_1, \ldots, u_k) \in \Lambda_k^n$, $Y[u_k]$ is minimal, and ${}^tu_{k-1} Y u_k \geqslant 0$. After n steps we have a matrix $U = (u_1, \ldots, u_n) \in \Lambda_n^n = \Lambda$ and a matrix $T = (t_{\alpha\beta}) = Y[U] \in \{ Y \}_\Lambda$ which we call *reduced*.

We now explain what the reduced property of a matrix means in terms of the entries of the matrix.

LEMMA 1.5. *Let $r \geqslant 1$ and $l \in M_{r,1}$. Then $l \in \Lambda_1^r$ if and only if the components of the vector are relatively prime.*

PROOF. Necessity is obvious. Conversely, if the components of l are relatively prime, then, by Lemma 3.8 of Chapter 1, there exists $V \in \Lambda^r$ such that

$$Vl = {}^t(1, 0, \ldots, 0), \quad \text{and hence } l = V^{-1} \cdot {}^t(1, 0, \ldots, 0)$$

and l coincides with the first column of the matrix V^{-1}. $\square$

LEMMA 1.6. *Let $U, U' \in \Lambda^n$. Then the first r columns of U' coincide with the first r columns of U if and only if*

$$U' = U \begin{pmatrix} E_r & B \\ 0 & D \end{pmatrix}, \qquad \text{where } D \in \Lambda^{n-r}, \quad B \in M_{r,n-r}.$$

PROOF. The direct implication is obvious. To prove the converse, we let $u_1, \ldots, u_n$ denote the columns of U, and we suppose that the first r columns of U' are $u_1, \ldots, u_r$. We set

$$U'U^{-1} = V = \begin{pmatrix} A & B \\ C & D \end{pmatrix} \in \Lambda^n,$$

where $A = (a_{\alpha\beta})$, B, $C = (c_{\alpha\beta})$, and D are $(r \times r)$-, $(r \times (n-r))$-, and $((n-r) \times (n-r))$-matrices, respectively. Since the β-column of the matrix $U' = U \begin{pmatrix} A & B \\ C & D \end{pmatrix}$ is equal to u_β for $1 \leqslant \beta \leqslant r$, it follows that

$$\sum_{\alpha=1}^{r} a_{\alpha\beta} u_\alpha + \sum_{\alpha=1}^{n-r} c_{\alpha\beta} u_{r+\alpha} = u_\beta,$$

which, by the linear independence of the columns $u_1, \ldots, u_n$, implies that $a_{\alpha\beta} = 1$ for $\alpha = \beta$, $a_{\alpha\beta} = 0$ for $\alpha \neq \beta$, and $c_{\alpha\beta} = 0$. Thus, $A = E_r$ and $C = 0$; hence $D \in \Lambda^{n-r}$. $\qquad\square$

Let $U = (u_1, \ldots, u_n) \in \Lambda^n$ and $1 \leqslant k \leqslant n$. By Lemma 1.6, the set of kth columns of all of the matrices $U' \in \Lambda^n$ with first $k-1$ columns $u_1, \ldots, u_{k-1}$ coincides with the set of columns of the form

$$Ul, \qquad \text{where } l = \begin{pmatrix} l_1 \\ \vdots \\ l_n \end{pmatrix} \in M_{n,1}$$

and $l_k, \ldots, l_n$ are the components of the first column of some matrix $D \in \Lambda^{n-k+1}$. By Lemma 1.5, the latter condition means that $l_k, \ldots, l_n$ are relatively prime. We thus find that, if $U = (u_1, \ldots, u_n) \in \Lambda^n$ and $1 \leqslant k \leqslant n$, then

(1.9) $$\{u \in M_{n,1}; (u_1, \ldots, u_{k-1}, u) \in \Lambda_k^n\} = UL_{k,n},$$

where $L_{k,n}$ is the set of columns in $M_{n,1}$ whose last $(n-k+1)$ components are relatively prime.

From the definition and the relations (1.9) it follows that $T = (t_{\alpha\beta}) = Y[U]$ is a reduced matrix if and only if

$$Y[Ul] \geqslant Y[u_k] \quad \text{for all } l \in L_{k,n} \text{ and } l \leqslant k \leqslant n$$

and

$${}^t u_{k-1} Y u_k \geqslant 0 \quad \text{for } 1 < k \leqslant n,$$

where $U = (u_1, \ldots, u_n) \in \Lambda$. Since $Y[U] = T$, $Y[u_k] = t_{kk}$, and ${}^t u_{k-1} Y u_k = t_{k-1,k}$, these conditions mean precisely that T belongs to the *Minkowski reduction domain*

(1.10) $$F_n = \{T = (t_{\alpha\beta}) \in P_n; t_{kk} \leqslant T[l],$$
$$\text{if } l \in L_{k,n} \ (l \leqslant k \leqslant n), \text{ and } t_{k-1,k} \geqslant 0 \ (1 < k \leqslant n)\}.$$

THEOREM 1.7. *In every orbit $\{Y\}_\Lambda$ of the group Λ^n in P_n there exists at least one point—and no more than finitely many points—belonging to the reduction domain F_n. If T and T' are two interior points of F_n with $T' = T[U]$, where $U \in \Lambda^n$, then $U = \pm E_n$. In particular, any two interior points of F_n are in different orbits of Λ^n.*

PROOF. The above discussion shows that for every matrix $Y \in P_n$ there exists $U \in \Lambda^n$ such that $Y[U] \in F_n$, and each column of this matrix U can be chosen in only finitely many ways.

Let $e_1, \ldots, e_n$ denote the columns of the identity matrix E_n. We set

$$F_n^0 = \{T = (t_{\alpha\beta}) \in P_n; \ t_{kk} < T[l], \text{ if } l \in L_{k,n},$$
$$l \neq \pm e_k \ (1 \leqslant k \leqslant n), \text{ and } t_{k-1,k} > 0 \ (1 < k \leqslant n)\}.$$

Clearly $F_n^0 \subset F_n$, and every interior point of F_n is contained in F_n^0. If $T = (t_{\alpha\beta})$, $T' = (t'_{\alpha\beta}) \in F_n^0$, and $T' = T[U]$, where $U = (u_1, \ldots, u_n) \in \Lambda^n$, then

$$t_{kk} = t'_{kk} = T[u_k] \qquad (1 \leqslant k \leqslant n).$$

Since $u_1 \in L_{1,n}$, this equality and the definition of F_n^0 imply that $u_1 = \pm e_1$. Then obviously $u_2 \in L_{2,n}$, and we find that $u_2 = \pm e_2$. Continuing in this way, we obtain $u_k = \pm e_k$ for all $1 \leqslant k \leqslant n$. Furthermore, from the conditions

$$t_{k-1,k} > 0, \quad t'_{k-1,k} = {}^t u_{k-1} T u_k > 0 \qquad (1 < k \leqslant n)$$

it follows that either $u_1 = e_1, \ldots, u_n = e_n$, or else $u_1 = -e_1, \ldots, u_n = -e_n$. Thus, $U = \pm E_n$ and $T' = T$.   □

The inequalities that determine the reduction domain imply a series of useful inequalities for the entries in a reduced matrix $T = (t_{\alpha\beta})$. In the first place, since $t_{kk} \leqslant T[e_{k+1}] = t_{k+1,k+1}$ $(1 \leqslant k \leqslant n)$, it follows that

$$(1.11) \qquad\qquad t_{11} \leqslant t_{22} \leqslant \cdots \leqslant t_{nn}.$$

In addition, since $t_n \leqslant T[e_k \pm e_l] = t_{kk} + 2t_{kl} + t_{ll}$ for $1 \leqslant k < l \leqslant n$, it follows that

$$(1.12) \qquad\qquad |2t_{kl}| \leqslant t_{kk}, \quad \text{if } k \neq l.$$

Finally, we have the following important theorem.

THEOREM 1.8. *If $T = (t_{\alpha\beta}) \in F_n$, then*

$$(1.13) \qquad\qquad t_{11} t_{22} \cdots t_{nn} \leqslant c_n \det T,$$

where c_n depends only on n.

PROOF. For $\alpha = 1, \ldots, n$ we determine the nonnegative integer $\mu_\alpha = \mu_\alpha(t)$ by the following conditions:

(1) The columns of integers m satisfying $T[m] \leqslant \mu_\alpha$ include at least α linearly independent columns.

(2) The maximum number of linearly independent columns of integers m satisfying the inequality $T[m] < \mu_\alpha$ is at most $\alpha - 1$.

The numbers $\mu_1, \ldots, \mu_n$ are called the *successive minima of the matrix* $T > 0$. It is clear that $\mu_1 \leqslant \mu_2 \leqslant \cdots \leqslant \mu_n$, and there exist linearly independent columns $m_1, \ldots, m_n$ such that $T[m_\alpha] = \mu_\alpha$.

We prove the theorem in three stages.

LEMMA 1.9. *Let* $T = (t_{\alpha\beta}) \in F_n$, *and let* $\mu_1, \ldots, \mu_n$ *be the successive minima of* T. *Then*

$$(1.14) \qquad t_{\alpha\alpha} \leqslant c(\alpha)\mu_\alpha \qquad (1 \leqslant \alpha \leqslant n),$$

where $c(\alpha)$ *depends only on* α.

PROOF OF THE LEMMA. As before, let $m_1, \ldots, m_n$ be linearly independent columns such that $T[m_\alpha] = \mu_\alpha$, and let $e_1, \ldots, e_n$ be the columns of E_n. If α is fixed, then at least one of the columns $m_1, \ldots, m_\alpha$ is not a linear combination of the columns $e_1, \ldots, e_{\alpha-1}$. Suppose that m_k is such a column. Then there exists a column e_α^* such that $(e_1, \ldots, e_{\alpha-1}, e_\alpha^*) \in \Lambda_\alpha^n$ (see (1.8)) and

$$m_k = h_1 e_1 + \cdots + h_{\alpha-1} e_{\alpha-1} + s e_\alpha^*,$$

where $s > 0$ and $h_1, \ldots, h_{\alpha-1}$ are integers. If we replace e_α^* by $e_\alpha + b_1 e_1 + \cdots + b_{\alpha-1} e_{\alpha-1}$ for suitable integers $b_1, \ldots, b_{\alpha-1}$, without loss of generality we may assume that $|h_\beta| \leqslant s/2$ $(1 \leqslant \beta < \alpha)$. Since

$$e_\alpha^* = (1/s)m_k - (n_1/s)e_1 - \cdots - (h_{\alpha-1}/s)e_{\alpha-1},$$

it follows from the triangle inequality for the norm $\|x\| = (T[x])^{1/2}$ in the space $M_{n,1}(\mathbf{R})$ that

$$(1.15) \qquad \|e_\alpha^*\| \leqslant \frac{1}{s}\|m_k\| + \sum_{\beta<\alpha} \frac{h_\beta}{s}\|e_\beta\| \leqslant \|m_k\| + \frac{1}{2}\sum_{\beta<\alpha}\|e_\beta\|.$$

Since (1.4) obviously holds for $\alpha = 1$ with $c(1) = 1$, we can proceed to prove (1.14) by induction on α. If the inequality holds for all $\beta < \alpha$, then

$$T[e_\beta] = t_{\beta\beta} \leqslant c(\beta)\mu_\beta \leqslant c(\beta)\mu_\alpha.$$

Since $k \leqslant \alpha$, we have $T[m_k] = \mu_k \leqslant \mu_\alpha$. From this and (1.15) we obtain $T[e_\alpha^*] \leqslant c(\alpha)\mu_\alpha$, where

$$c(\alpha) = \left(1 + \frac{1}{2}\sum_{\beta<\alpha} c(\beta)^{1/2}\right)^2.$$

Since $(e_1, \ldots, e_{\alpha-1}, e_\alpha^*) \in \Lambda_\alpha^n$ and T is a reduced matrix, it follows that $t_{\alpha\alpha} \leqslant T[e_\alpha^*]$. □

LEMMA 1.10. *Let* $T \in P_n$, *and let* μ_1 *be the first minimum of* T. *Then*

$$\mu_1 \leqslant \gamma_n(\det T)^{1/n},$$

where γ_n *depends only on* n.

PROOF. We regard the set of columns $M_{n,1}(\mathbf{R})$ as an n-dimensional real space with the usual coordinates. For any $\mu > 0$ the set

$$\{X \in M_{n,1}(\mathbf{R}); T[X] \leqslant \mu\}$$

is obviously a centrally symmetric convex set centered at the origin, and its volume v is equal to $s_n \mu^{n/2}(\det T)^{-1/2}$, where s_n is the volume of the unit sphere in n-dimensional space. By Minkowski's theorem on convex solids, this set contains a point other than the origin with integer coordinates, provided that $v > 2^n$, i.e., $\mu > 4s_n^{-2/n}(\det T)^{1/n}$. Thus,

$$\mu_1 = \inf_{m \in M_{n,1}, m \neq 0} T[m] \leqslant 4s_n^{-2/n}(\det T)^{1/n}. \qquad \square$$

LEMMA 1.11. *Let $T \in P_n$, and let $\mu_1, \ldots, \mu_n$ be the successive minima of T. Then*

$$\mu_1 \cdots \mu_n \leqslant (\gamma_n)^n \det T,$$

where γ_n is the same constant as in Lemma 1.10.

PROOF. As before, let $m_1, \ldots, m_n$ be linearly independent columns of integers such that $T[m_\alpha] = \mu_\alpha$ $(1 \leqslant \alpha \leqslant n)$. Then the matrix $M = (m_1, \ldots, m_n)$ is nonsingular, and by Theorem 1.5 of Appendix 1, the matrix $T[M]$ can be represented in the form $T[M] = {}^t L \cdot L$, where $L = (l_{\alpha\beta})$, $l_{\alpha\beta} = 0$ for $\alpha > \beta$. We set

$$Q = D[LM^{-1}], \quad \text{where } D = \mathrm{diag}(\mu_1^{-1}, \ldots, \mu_n^{-1}),$$

and we show that $Q[m] \geqslant 1$ for nonzero $m \in M_{n,1}$.

In fact, let $m = Mh$, where ${}^t h = (h_1, \ldots, h_n)$, and let α be the greatest index for which $h_\alpha \neq 0$. Then m is a linear combination of the columns $m_1, \ldots, m_\alpha$ with coefficients $h_1, \ldots, h_\alpha$, and it is not a linear combination of the columns $m_1, \ldots, m_{\alpha-1}$. From the definition of the minimum μ_α it now follows that $T[m] \geqslant \mu_\alpha$. Hence, taking into account that the components $(Lh)_\beta$ of the column Lh are zero if $\beta > \alpha$, we obtain

$$Q[m] = D[Lh] = \sum_{\beta=1}^{\alpha} \mu_\beta^{-1}(Lh)_\beta^2 \geqslant \mu_\alpha^{-1} \sum_{\beta=1}^{\alpha} (Lh)_\beta^2$$

$$= \mu_\alpha^{-1} E_n[Lh] = \mu_\alpha^{-1} T[m] \geqslant 1.$$

From this inequality and Lemma 1.10 it follows that

$$\gamma_n(\det Q)^{1/n} = \gamma_n((\mu_1 \cdots \mu_n)^{-1} \det T)^{1/n} \geqslant 1. \qquad \square$$

Returning to the proof of Theorem 1.8, we see that the inequality (1.13) follows from Lemma 1.9 and Lemma 1.11. The theorem is proved. $\qquad \square$

COROLLARY. *Suppose that $T = (t_{\alpha\beta}) \in F_n$. Then*

$$(1.16) \qquad n^{1-n}c_n^{-1}T_0 \leqslant T \leqslant nT_0, \quad \text{where } T_0 = \mathrm{diag}(t_{11}, \ldots, t_{nn})$$

and c_n is the same constant as in Theorem 1.8.

PROOF. Let $\rho_1, \ldots, \rho_n$ denote the eigenvalues of the matrix $T[T_0^{-1/2}]$, where $T_0^{1/2} = \mathrm{diag}(t_{11}^{1/2}, \ldots, t_{nn}^{1/2})$. Then

$$\rho_1 + \cdots + \rho_n = \sigma(T[T_0^{-1/2}]) = n,$$

and by Theorem 1.8

$$\rho_1 \cdots \rho_n = \det T(t_{11} \cdots t_{nn})^{-1} \geqslant 1/c_n.$$

Thus, $n^{1-n}c_n^{-1} \leqslant \rho_\alpha \leqslant n$ for $\alpha = 1, \ldots, n$. If V is an orthogonal matrix such that

$$T[T_0^{-1/2}][V] = \mathrm{diag}(\rho_1, \ldots, \rho_n),$$

then we have

$$n^{1-n}c_n^{-1}E_n \leqslant T[T_0^{-1/2}][V] \leqslant nE_n,$$

which is equivalent to the inequalities (1.16).   □

The inequalities proved above imply the following

THEOREM 1.12. *The number of classes*

$$\{R\}_Z = \{R[U]; U \in \Lambda^n\}$$

of matrices $R \in A_n^+$ *of fixed determinant* $\det R = d$ *is finite.*

PROOF. By Theorem 1.7, every class contains a reduced representative $R' \in A_n^+ \cap F_n$. From (1.11)–(1.13) it follows that all of the entries in R' are bounded in absolute value by a constant that depends only on n and $\det R' = d$. But these entries are integers, and hence the number of such R' is finite.   □

COROLLARY. *The number of classes* $\{R\}_Z$ *of matrices* $R \in A_n^+$ *of fixed level* q *is finite.*

PROOF. If q is the level of $R \in A_n^+$, then the inclusion $qR^{-1} \in A_n^+$ implies that $\det R$ divides q^n. Hence, $\det R$ can only take finitely many values.   □

PROPOSITION 1.13. *Set*

$$c_n = \sup_{T=(t_{\alpha\beta}\in F_n)} t_{11}t_{22}\cdots t_{nn}(\det T)^{-1}.$$

Then $c_n \geqslant c_{n-1}$ *for every* $n > 1$.

PROOF. If $T' = (t_{\alpha\beta}) \in F_{n-1}$ and $t = t_{nn}$ is a real number not less than $t_{11}, \ldots, t_{n-1,n-1}$, then the matrix $T = \begin{pmatrix} T' & 0 \\ 0 & t \end{pmatrix}$ is contained in F_n. Namely, if $l = \begin{pmatrix} l' \\ l_{nn} \end{pmatrix} \in L_{k,n}$ and $l_{nn} \neq 0$, then $T[l] = T'[l'] + tl_{nn}^2 \geqslant t \geqslant t_{kk}$. If, on the other hand, $l_{nn} = 0$, then $k < n$ and $l' \in L_{k,n-1}$, so that $T[l] = T'[l'] \geqslant t_{kk}$. Thus,

$$c_n \geqslant \sup_{T=\begin{pmatrix} T' & 0 \\ 0 & t \end{pmatrix}\in F_n} t_{11}\cdots t_{n-1,n-1}t(t \det T')^{-1} = c_{n-1}.   □$$

PROBLEM 1.14. Show that

$$F_2 = \left\{ \begin{pmatrix} t_{11} & t_{12} \\ t_{12} & t_{22} \end{pmatrix} \in P_2; \quad 0 \leqslant 2t_{12} \leqslant t_{11} \leqslant t_{12} \right\}.$$

[**Hint:** See Problem 1.3.]

PROBLEM 1.15. Show that Lemma 1.10 for $n = 2$ holds with $\gamma_2 = 2/\sqrt{3}$. By considering the matrix $T = \begin{pmatrix} 1 & 1/2 \\ 1/2 & 1 \end{pmatrix}$, show that this value of γ_2 cannot be improved.

PROBLEM 1.16. Show that Theorem 1.8 for $n = 2$ holds with $c_2 = 4/3$, and that this value cannot be improved.

3. The fundamental domain for the Siegel modular group. Just as in the case of the classical modular group, the basic step in the construction of a fundamental domain for Γ^n on $\mathbf{H}_n$ is to choose in each orbit a representative Z that satisfies the inequalities $|\det(CZ + D)| \geqslant 1$ for every pair of $n \times n$-matrices (C, D) that occurs in a matrix $\begin{pmatrix} A & B \\ C & D \end{pmatrix} \in \Gamma^n$. The proof that such a choice is possible is based on an explicit description of all possible bottom rows of the matrices in Γ^n.

We shall examine pairs (C, D) of $n \times n$ integer matrices, where n is fixed. Such a pair is said to be *symmetric* if $C \cdot {}^t D = D \cdot {}^t C$. A pair is said to be *relatively prime* if, whenever GC and GD are integer matrices for an $n \times n$ rational matrix G, it follows that G itself is an integer matrix.

LEMMA 1.17. *Let (C, D) be a pair of $n \times n$ integer matrices. Then the following conditions are equivalent:*

(1) *there exist matrices A and B such that $M = \begin{pmatrix} A & B \\ C & D \end{pmatrix} \in \Gamma^n$;*

(2) *the pair (C, D) is symmetric and relatively prime.*

PROOF. If (C, D) satisfies (1), then the pair is symmetric by the second relation (2.8) of Chapter 1. Furthermore, if GC and GD are integer matrices, then the same relations imply integrality of the matrix

$$G = G \cdot {}^t(-B^t C + A^t D) = -(GC)\,{}^t B + (GD)\,{}^t A.$$

Now suppose that (C, D) satisfies (2). Note that the pair

$$(C', D') = (C, D)M'' = (CA'' + DC'', CB'' + DD''),$$

where $M'' = \begin{pmatrix} A'' & B'' \\ C'' & D'' \end{pmatrix} \in \Gamma^n$, then also satisfies (2). According to the conditions (2.8) of Chapter 1, the matrix

$$C'\,{}^t D' = CA''\,{}^t B''\,{}^t C + DC''\,{}^t B''\,{}^t C + CA''\,{}^t D''\,{}^t D + DC''\,{}^t D''\,{}^t D$$
$$= CA''\,{}^t B''\,{}^t C + DC''\,{}^t D''\,{}^t D + DC''\,{}^t B''\,{}^t C + CB''\,{}^t C''\,{}^t D + C\,{}^t D$$

is symmetric; in addition, it is clear that the pair (C', D') is relatively prime. We choose a matrix $M'' \in \Gamma^n$ such that $(C', D') = (E, 0)$. Let $\mathbf{t}$ be the first row of the matrix (C, D). Since (C, D) is a relatively prime pair, it follows that $\mathbf{t} \neq 0$. By Lemma 3.9 of Chapter 1, there exists a matrix $M_0 \in \Gamma^n$ such that $M_0 \cdot {}^t\mathbf{t} = (t, 0, \dots, 0)$, where $t \in \mathbf{N}$. Then

$$(C, D)\,{}^t M_0 = (C', D')$$

and

$$C' = \begin{pmatrix} t & 0 & \dots & 0 \\ c'_{21} & & & \\ \vdots & & C_1 & \\ c'_{n1} & & & \end{pmatrix}, \qquad D' = \begin{pmatrix} 0 & & \dots & 0 \\ d'_{21} & & & \\ \vdots & & D_1 & \\ d'_{n1} & & & \end{pmatrix}.$$

Since $C''D' = D''C'$, it follows that $c'_{21} = \cdots = c'_{n1} = d'_{21} = \cdots = d'_{n1} = 0$, and $C_1\,{}^t D_1$ is a symmetric matrix. Since the pair (C', D') is relatively prime, it follows that $t = 1$, (C_1, D_1) is a relatively prime pair, and our claim is proved for $n = 1$. If $n \geqslant 2$, then by induction we may assume that the assertion holds for the pair (C_1, D_1), i.e.,

$$(C_1, D_1)M_2 = (E_{n-1}, 0) \text{ for some } M_2 = \begin{pmatrix} A_2 & B_2 \\ C_2 & D_2 \end{pmatrix} \in \Gamma^{n-1}. \text{ The matrix } \widehat{M_2} \text{ with}$$

blocks $\begin{pmatrix} 1 & 0 \\ 0 & A_2 \end{pmatrix}$, $\begin{pmatrix} 0 & 0 \\ 0 & B_2 \end{pmatrix}$, $\begin{pmatrix} 0 & 0 \\ 0 & C_2 \end{pmatrix}$, $\begin{pmatrix} 1 & 0 \\ 0 & D_2 \end{pmatrix}$ belongs to the group Γ^n, and $(C, D)^t M_0 \widehat{M_2} = (E, 0)$, or

$$(C, D) = (E, 0)(M'')^{-1} = (0, E)J_n^{-1}(M'')^{-1}, \quad \text{where } M'' \in \Gamma^n.$$

Consequently, (C, D) satisfies (1) if we set $M = J_n^{-1}(M'')^{-1} \in \Gamma^n$.   □

We shall say that two symmetric and relatively prime pairs (C, D) and (C', D') are *equivalent* (or *belong to the same class*) if

(1.17) $$(C', D') = U(C, D) = (UC, UD), \quad \text{where } U \in \Lambda^n.$$

In this case obviously

(1.18) $$C \cdot {}^t D' = D \cdot {}^t C'.$$

Conversely, if (1.18) holds, and if M, M' are matrices in Γ^n with bottom rows (C, D) and (C', D'), respectively, then $M' = (M'M^{-1})M$ and $M'M^{-1} = \begin{pmatrix} U_1 & V \\ 0 & U \end{pmatrix} \in \Gamma^n$. Hence, $U \in \Gamma^n$, and the pairs satisfy (1.17). Thus, the conditions (1.17) and (1.18) are equivalent.

LEMMA 1.18. *Every symmetric and relatively prime pair* (C, D) *such that* rank $C = r$, *where* $0 \leqslant r \leqslant n$, *is equivalent to a pair of the form*

(1.19) $$\left(\begin{pmatrix} C_1 & 0 \\ 0 & 0 \end{pmatrix} {}^t U_1, \begin{pmatrix} D_1 & 0 \\ 0 & E_{n-r} \end{pmatrix} U_1^1 \right)$$

(*to the pair* $(0, E_n)$ *if* $r = 0$), *where* (C_1, D_1) *is a symmetric and relatively prime pair of* $r \times r$-*matrices*, rank $C_1 = r$, *and* $U_1 \in \Lambda^n$.

Two symmetric and relatively prime pairs of the form (1.19), *one of which corresponds to* C_1, D_1, U_1 *and the other of which corresponds to* C_2, D_2, U_2, *are equivalent if and only if*

(1.20) $$U_2 = U_1 \begin{pmatrix} V & B \\ 0 & V' \end{pmatrix}, \quad \text{where } V \in \Lambda^r,$$

and the pair $(C_2 {}^t V, D_2 V^{-1})$ *is equivalent to the pair* (C_1, D_1).

REMARK. If $U_1 = (Q_1, Q_1')$ and $U_2 = (Q_2, Q_2')$ are two matrices in the group Λ^n, where Q_1 and Q_2 are $n \times r$-blocks, then (1.20) is equivalent to the condition

(1.21) $$Q_2 = Q_1 V, \quad \text{where } V \in \Lambda^r.$$

Namely, (1.21) obviously follows from (1.20). Conversely, from (1.21) it follows that the matrix

$$U_1' = U_2 \begin{pmatrix} V & 0 \\ 0 & E_{n-r} \end{pmatrix}^{-1} \in \Lambda^n$$

has the same first r columns as U_1; then, by Lemma 1.6,

$$U_1' = U_1 \begin{pmatrix} E_r & B' \\ 0 & V' \end{pmatrix},$$

which implies (1.20).

PROOF OF THE LEMMA. If $r = 0$ or n, then the lemma is obvious. Suppose that $0 < r < n$. In this case the homogeneous system of linear equations $Cx = 0$, where ${}^t x = (x_1, \ldots, x_n)$, has a nonzero integer solution l, where we clearly may suppose that the components of the column l are relatively prime. Let V be a unimodular matrix with last column l, which exists by Lemma 1.5. Then the last column of the matrix CV consists of zeros. Repeating the same argument for the rows of CV, we find that there exist matrices $V, V_1 \in \Lambda^n$ such that $V_1 CV = \begin{pmatrix} C' & 0 \\ 0 & 0 \end{pmatrix}$, where C' is an $(n-1) \times (n-1)$-matrix. If $r = \operatorname{rank} C = \operatorname{rank} C' < n - 1$, then we similarly find $V_1', V' \in \Lambda^{n-1}$ such that the last row and last column of the matrix $V_1' C' V'$ consist of zeros. Then

$$\begin{pmatrix} V_1' & 0 \\ 0 & 1 \end{pmatrix} V_1 CV \begin{pmatrix} V' & 0 \\ 0 & 1 \end{pmatrix} = \begin{pmatrix} C'' & 0 \\ 0 & 0 \end{pmatrix},$$

where C'' is an $(n-2) \times (n-2)$-matrix. Continuing this process, we eventually obtain two unimodular matrices, which we shall denote U' and U_1^*, such that

$$U' CU_1^* = \begin{pmatrix} C_1 & 0 \\ 0 & 0 \end{pmatrix} \quad \text{or} \quad U' C = \begin{pmatrix} C_1 & 0 \\ 0 & 0 \end{pmatrix} {}^t U_1,$$

where C_1 is an $r \times r$-matrix of rank r. We set

$$U' D = \begin{pmatrix} D_1 & D_2 \\ D_3 & D_4 \end{pmatrix} U_1^{-1},$$

where D_1 is an $r \times r$-matrix, and the sizes of the other blocks are determined by the size of D_1. The pair $(U'C, U'D)$, and hence also the pair

$$\left(\begin{pmatrix} C_1 & 0 \\ 0 & 0 \end{pmatrix}, \begin{pmatrix} D_1 & D_2 \\ D_3 & D_4 \end{pmatrix} \right),$$

is clearly symmetric and relatively prime. From this it easily follows that (C_1, D_1) is a symmetric pair, $D_3 = 0$, and $D_4 \in \Lambda^{n-r}$. If we now set

$$U = \begin{pmatrix} E_r & -D_2 D_4^{-1} \\ 0 & D_4^{-1} \end{pmatrix} U' \in \Lambda^n,$$

we see that

$$UC = \begin{pmatrix} C_1 & 0 \\ 0 & 0 \end{pmatrix} {}^t U_1, \qquad UD = \begin{pmatrix} D_1 & 0 \\ 0 & E_{n-r} \end{pmatrix} U_1^{-1}.$$

This implies, in particular, that (C_1, D_1) is a relatively prime pair, and the first part of the lemma is proved.

Now suppose that we are given two symmetric and relatively prime pairs of the form (1.19), written in terms of the matrices C_1, D_1, U_1 and C_2, D_2, U_2, respectively. If they are equivalent, then, by (1.18), we have the equality

$$\begin{pmatrix} C_1 & 0 \\ 0 & 0 \end{pmatrix} {}^t U_1 U_2^* \begin{pmatrix} {}^t D_2 & 0 \\ 0 & E \end{pmatrix} = \begin{pmatrix} D_1 & 0 \\ 0 & E \end{pmatrix} U_1^{-1} U_2 \begin{pmatrix} {}^t C_2 & 0 \\ 0 & 0 \end{pmatrix}.$$

If we divide the matrices ${}^t U_1 U_2^*$ and $U_1^{-1} U_2$ into blocks of the corresponding sizes

$$ {}^t U_1 U_2^* = \begin{pmatrix} W & H' \\ H & W' \end{pmatrix}, \qquad U_1^{-1} U_2 = \begin{pmatrix} V & B \\ B' & V' \end{pmatrix},$$

then we can rewrite the last relation in the form

$$\begin{pmatrix} C_1 W \,{}^t D_2 & C_1 H' \\ 0 & 0 \end{pmatrix} = \begin{pmatrix} D_1 V \,{}^t C_2 & 0 \\ B' \,{}^t C_2 & 0 \end{pmatrix},$$

from which it follows that $C_1 W^t D_2 = D_1 V^t C_2$ and $B' = 0$, since C_2 is a nonsingular matrix. In particular, $U_2 = U_1 \begin{pmatrix} V & B \\ 0 & V' \end{pmatrix}$. Furthermore, since ${}^t U_1 U_2^* = (U_1^{-1} U_2)^*$, it follows that $W = V^*$. Thus,

$$C_1 \,{}^t(D_2 V^{-1}) = D_1 \,{}^t(C_2 \,{}^t V),$$

so that the pair $(C_2^t V, D_2 V^{-1})$ is equivalent to the pair (C_1, D_1). The converse assertion is obvious. $\square$

We now examine the orbits (1.1) of the Siegel modular group $K = \Gamma^n$ on $\mathbf{H}_n$. By the *height of a point* $Z = X + iY \in \mathbf{H}_n$, denoted $h(Z)$, we mean the determinant $\det Y$. By Lemma 2.8 of Chapter 1 we have

$$(1.22) \qquad h(M\langle Z \rangle) = |\det(CZ + D)|^{-2} h(Z), \quad \text{if } M = \begin{pmatrix} A & B \\ C & D \end{pmatrix} \in \Gamma^n.$$

LEMMA 1.19. *Every orbit of Γ^n on $\mathbf{H}_n$ contains points Z of maximal height, i.e., points with the property*

$$(1.23) \qquad\qquad |\det(CZ + D)| \geqslant 1$$

for any symmetric and relatively prime pair (C, D).

PROOF. If $h(Z) \geqslant h(M\langle Z \rangle)$ for all $M \in \Gamma^n$, then from (1.22) it follows that (1.23) holds for any pair (C, D) that gives the bottom rows of a matrix in Γ^n, i.e. (by Lemma 1.17), for any symmetric and relatively prime pair. Conversely, if all of the inequalities (1.23) hold, then $h(M\langle Z \rangle) \leqslant h(Z)$ for all $M \in \Gamma^n$. Thus, it remains to prove that every orbit has a point of maximal height. This, in turn, follows if we show that for every fixed $Z \in \mathbf{H}_n$ the inequality

$$(1.24) \qquad\qquad |\det(CZ + D)| < 1$$

has only finitely many solutions in nonequivalent symmetric and relatively prime pairs (C, D). Namely, in that case the function $h(M\langle Z \rangle)$ takes only finitely many values greater than $h(Z)$ on the orbit of Z.

Suppose that rank $C = r$. By Lemma 1.18, we may assume that the pair (C, D) has the form (1.19). Hence, setting $U_1 = (Q, Q')$, where Q is an $n \times r$-block, we obtain

$$\begin{aligned} CZ + D &= \left[\begin{pmatrix} C_1 & 0 \\ 0 & 0 \end{pmatrix} \begin{pmatrix} {}^t Q \\ {}^t Q' \end{pmatrix} Z(Q, Q') + \begin{pmatrix} D_1 & 0 \\ 0 & E \end{pmatrix} \right] U_1^{-1} \\ &= \begin{pmatrix} C_1 Z[Q] + D_1 & C_1 \,{}^t Q Z Q' \\ 0 & E \end{pmatrix} U_1^{-1}. \end{aligned}$$

Thus, $|\det(CZ + D)| = |\det C_1| \cdot |\det(Z[Q] + P)|$, where $P = C_1^{-1} D_1$ is a rational symmetric $r \times r$-matrix. By Theorem 1.7, if we replace Q by QV for a suitable $V \in \Lambda^n$ (see Lemma 1.18 and the subsequent remark), we may assume that $Y[Q] \in F_r$. We note that the class of the pair (C_1, D_1) is uniquely determined by the symmetric matrix $P = C_1^{-1} \cdot D_1$. In fact, if $C_1^{-1} \cdot D_1 = C_2^{-1} \cdot D_2$ for another symmetric and relatively prime pair of $r \times r$-matrices (C_2, D_2), and if $\det C_2 \neq 0$, then $C_1^{-1} \cdot D_1 = {}^t D_2 D_2^*$, and hence

$C_1 \cdot {}^t D_2 = D_1 \cdot {}^t C_2$, so that our two pairs satisfy the condition for equivalence in the form (1.18). We set $T = Y[Q]$ and $S = X[Q] + P$. Since $T > 0$ and S is symmetric, there exists a real $r \times r$-matrix F such that $T[F] = E_r$ and $S[F] = H = \text{diag}(h_1, \ldots, h_r)$. Since $(\det F)^{-2} = \det T$, we have

$$|\det(Z[Q] + P)| = |\det(S + iT)| = |\det(H + iE)[F^{-1}]|$$

$$= \left| \det T \cdot \prod_{\alpha=1}^{r} (h_\alpha + i) \right|.$$

Thus, (1.24) is equivalent to the inequality

$$(1.25) \qquad\qquad |\det C_1| \det T \prod_{\alpha=1}^{r} (1 + h_\alpha^2)^{1/2} < 1.$$

From this inequality it follows that $\det T < 1$. Let $q_1, \ldots, q_r$ denote the columns of the matrix Q. Since $T = ({}^t q_\alpha Y q_\beta)$ is a reduced matrix, it follows by Theorem 1.8 that

$$\prod_{\alpha=1}^{r} Y[q_\alpha] \leqslant c_r \det T < c_r.$$

On the other hand, if λ is the smallest eigenvalue of the matrix Y, then $Y[q_\alpha] \geqslant \lambda \, {}^t q_\alpha \cdot q_\alpha \geqslant \lambda$. These inequalities imply that $Y[q_\alpha] < \lambda^{1-r} c_r$ $(1 \leqslant \alpha \leqslant r)$, and so all of the q_α belong to a certain finite set of integer vectors. In particular, there are only finitely many matrices Q that are not connected by relations of the form (1.21) and have the property that a pair of the form (1.19) satisfies (1.24). Furthermore, $\det T = \det Y[Q]$ takes only finitely many values, and hence the inequality (1.25) implies that the numbers $|\det C_1|, h_1, \ldots, h_r$ are bounded from above. In addition, since $T = F^* F^{-1}$, it follows that all of the entries in the matrix F^{-1} are bounded from above. Consequently, all of the entries in $S = H[F^{-1}]$—and hence all of the entries in $P = S - X[Q]$—are bounded from above. We conclude that P is a rational matrix with bounded entries, all of whose denominators are also bounded, since they are divisors of a finite number of values of $\det C_1$. There are only finitely many such P, and hence only finitely many nonequivalent pairs (C_1, D_1). Applying the second part of Lemma 1.18, we complete the proof of the lemma. $\qquad\qquad\qquad\qquad\qquad\qquad\qquad\qquad \square$

THEOREM 1.20. *Let D_n be the subset of the upper half-plane $\mathbf{H}_n$ that consists of all matrices $Z = X + iY \in \mathbf{H}_n$ that satisfy the conditions:*

(1) $|\det(CZ + D)| \geqslant 1$ for all symmetric and relatively prime pairs of $n \times n$-matrices (C, D);

(2) $Y \in F_n$, where F_n is the reduction domain (1.10);

(3) $X \in X_n = \{X = (x_{\alpha\beta}) \in S_n(\mathbf{R}); |x_{\alpha\beta}| \leqslant 1/2 \, (1 \leqslant \alpha, \beta \leqslant n)\}$.
Then D_n intersects with every orbit of Γ^n on $\mathbf{H}_n$. If Z and Z' are two interior points of D_n and $Z' = M\langle Z \rangle$ with $M \in \Gamma^n$, then $M = \pm E_{2n}$. In particular, all of the interior points of D_n lie in distinct orbits of Γ^n.

PROOF. We consider the orbit $\Gamma^n \langle Z'' \rangle$ of an arbitrary point $Z'' \in \mathbf{H}_n$. By Lemma 1.19, there exists a point $Z' \in \Gamma^n \langle Z'' \rangle$ having maximal height, and this point satisfies all of the inequalities in (1). Any transformation of the form

$$Z' \to \begin{pmatrix} {}^t V & S V^{-1} \\ 0 & V^{-1} \end{pmatrix} \langle Z' \rangle = X'[V] + S + iY'[V],$$

where $V \in \Lambda^n$ and $S \in S_n$, belongs to Γ^n and does not change the height of Z'. By Theorem 1.7, there exists a matrix $V \in \Lambda^n$ such that $Y = Y'[V] \in F_n$. There also obviously exists a symmetric integer matrix S such that $X = X'[V] + S \in X_n$. Then

$$Z = X + iY \in \Gamma^n \langle Z'' \rangle \cap D_n.$$

Now suppose that Z and $Z' \in D_n$, $Z' = M\langle Z \rangle$, where $M = \begin{pmatrix} A & B \\ C & D \end{pmatrix} \in \Gamma^n$. Then $h(Z) = h(Z')$ by Lemma 1.19, and from (1.22) it follows that $|\det(CZ+D)| = 1$. Similarly, because $Z = M^{-1}\langle Z' \rangle$, we have $|\det(-{}^t CZ' + {}^t A)| = 1$ (see (2.9) of Chapter 1). If $C \neq 0$, then these equations are nontrivial, and consequently the points Z and Z' lie on the boundary of D_n. If $C = 0$, then M can obviously be written in the form $M = \begin{pmatrix} {}^t V & SV^{-1} \\ 0 & V^{-1} \end{pmatrix}$, where $V \in \Lambda^n$, $S \in S_n$. Then $Z' = X' + iY' = X[V] + S + iY[V]$, where $X + iY = Z$. In particular, $Y' = Y[V]$. Since $Y, Y' \in F_n$, it follows by Theorem 1.7 that either Y and Y' are boundary points of F_n, or else $V = \pm E_n$. In the latter case $X' = X + S$, and hence $S = 0$ if X and X' do not lie on the boundary of F_n. We conclude that $M = \pm E_{2n}$ if Z and Z' are not boundary points of D_n. $\square$

THEOREM 1.21. *Any matrix* $Z = X + iY \in D_n$ *satisfies the inequalities*

(1.26) $$Y \geqslant c_n' E_n, \quad \text{where } c_n' = \frac{n^{1-n}\sqrt{3}}{2c_n},$$

and c_n *is the constant in Theorem* 1.8. *In particular,*

(1.27) $$\sigma(Y^{-1}) \leqslant \sigma_n, \quad \text{where } \sigma_n = \frac{2n^n \cdot c_n}{\sqrt{3}}.$$

PROOF. We set $Z = (z_{\alpha\beta})$, $z_{\alpha\beta} = x_{\alpha\beta} + iy_{\alpha\beta}$. From the inequality $|\det(CZ+D)| \geqslant 1$ for the symmetric and relatively prime pair

$$(C, D) = \left(\begin{pmatrix} 1 & 0 \\ 0 & 0 \end{pmatrix}, \begin{pmatrix} 0 & 0 \\ 0 & E_{n-1} \end{pmatrix} \right)$$

we obtain: $|z_{11}| = (x_{11}^2 + y_{11}^2)^{1/2} \geqslant 1$. Since $|x_{11}| \leqslant 1/2$, this implies that $y_{11}^2 \geqslant 1 - 1/4$, i.e., $y_{11} \geqslant \sqrt{3}/2$. The last inequality and (1.11) imply that $y_{\alpha\alpha} \geqslant \sqrt{3}/2$ for $\alpha = 1, \ldots, n$. These inequalities, along with (1.16), prove (1.26). From (1.26) it follows that the smallest eigenvalue of Y is greater than or equal to c'. Hence, the largest eigenvalue of Y^{-1} is less than or equal to $1/c'$, and $\sigma(Y^{-1}) \leqslant n/c'$. $\square$

From Theorem 1.21 it follows that the fundamental domain D_n is closed in the space of all complex symmetric matrices. Siegel proved that D_n is a connected domain bounded by a finite number of algebraic hypersurfaces.

4. Subgroups of finite index. Let K be an arbitrary subgroup of finite index in the modular group Γ^n. We set

(1.28) $$K' = (-E_{2n})K \cup K.$$

Clearly, K' is also a subgroup of Γ^n. We let $M_1, \ldots, M_\mu$ denote a complete set of left coset representatives for Γ^n modulo K', so that

$$(1.29) \qquad \Gamma^n = \bigcup_{\alpha=1}^{\mu} K' M_\alpha \quad \text{and} \quad K' M_\alpha \cap K' M_\beta = \varnothing, \quad \text{if } \alpha \neq \beta,$$

and we set

$$(1.30) \qquad\qquad\qquad D_K = \bigcup_{\alpha=1}^{\mu} M_\alpha \langle D_n \rangle,$$

where D_n is a fundamental domain for Γ^n. We then have the following theorem.

THEOREM 1.22. *The K-orbit $K\langle Z \rangle$ of every point $Z \in \mathbf{H}_n$ intersects with the set D_K. If Z and Z' are two interior points of D_K and $Z' = g\langle Z \rangle$, where $g \in K$, then $g = \pm E_{2n}$. In particular, all of the interior points of D_K lie on different orbits of K.*

PROOF. Since all of the orbits of K and K' coincide and $D_K = D_{K'}$, it follows that, replacing K by K', we may suppose that $K = K'$. By Theorem 1.20, there exists a matrix $M \in \Gamma^n$ such that $M\langle Z \rangle \in D_n$. Let KM_α be the left coset in (1.29) that contains the matrix M^{-1}. Then, writing M^{-1} in the form $g^{-1} M_\alpha$, where $g \in K$, we obtain $Z \in M^{-1}\langle D_n \rangle = g^{-1} M_\alpha \langle D_n \rangle$, and hence $gZ \in M_\alpha \langle D_n \rangle \subset D_K$.

Suppose that Z and $Z' = g\langle Z \rangle$ are interior points of D_K. By replacing Z, if necessary, by a sufficiently nearby point, we may assume that Z is an interior point of one of the sets in the decomposition (1.30), say, the set $M_\alpha \langle D_n \rangle$. Then $Z = M_\alpha \langle Z_1 \rangle$, where Z_1 is an interior point of D_n. Similarly, $g\langle Z \rangle = M_\beta \langle Z_2 \rangle$, where $Z_2 \in D_n$. Since

$$M_\beta^{-1} g M_\alpha \langle Z_1 \rangle = M_\beta^{-1} g \langle Z \rangle = Z_2,$$

it follows by Theorem 1.20 that $M_\beta^{-1} g M_\alpha = \pm E_{2n}$, and this obviously implies that $\alpha = \beta$ and $g = \pm E_{2n}$. $\qquad\square$

From Theorem 1.22 it follows that

$$(1.31) \qquad\qquad\qquad \mathbf{H}_n = \bigcup_{g \in K'/\{\pm E_{2n}\}} g\langle D_K \rangle,$$

and the intersection of any pair of subsets on the right in this decomposition does not contain any interior points of the subsets.

PROPOSITION 1.23. (1) *Let $K \subset \Gamma^n$ be a subgroup of finite index, and let D_K be a fundamental domain for K in $\mathbf{H}_n$. Then the volume*

$$v(K) = v(D_K) = \int_{D_K} d^*Z,$$

where d^*Z is the invariant volume element defined in Proposition 2.9 of Chapter 1, is finite and does not depend on the choice of fundamental domain.

(2) *Let $K_1 \subset K$ be subgroups of finite index in Γ^n. Then in the notation (1.28)*

$$(1.32) \qquad\qquad\qquad v(K_1) = [K': K_1'] v(K).$$

PROOF. From Proposition 2.9 of Chapter 1 it follows that $v(D_K)$ does not depend on the choice of D_K. If we choose D_K in the form (1.30), we have

$$v(D_K) = \sum_{\alpha=1}^{\mu} \int_{M_\alpha\langle D_n\rangle} d^*Z = \sum_{\alpha=1}^{\mu} \int_{D_n} d^*M_\alpha\langle Z\rangle$$

$$= \sum_{\alpha=1}^{\mu} \int_{D_n} d^*Z = [\Gamma^n : K']v(D_n),$$

from which (1.32) follows. To prove finiteness of the volume it suffices to treat the case $v(D_n)$. From Theorems 1.21 and 1.8 and inequalities (1.11) and (1.12) we obtain

$$v(D_n) \leqslant \int_{\{Y\in F_n:\ Y\geqslant c_n'E\}} (\det Y)^{-(n+1)}\,dY$$

$$\leqslant c \int_{\substack{c_n'\leqslant y_{11}\leqslant\cdots\leqslant y_{nn}\\ |y_{\alpha\beta}|\leqslant y_{\alpha\alpha}/2\ (\alpha\neq\beta)}} (y_{11}\cdots y_{nn})^{-(n+1)} \prod_{1\leqslant\alpha\leqslant\beta\leqslant n} dy_{\alpha\beta}$$

$$\leqslant c \int_{y_{11},\ldots,y_{nn}\geqslant c_n'} (y_{11}\cdots y_{nn})^{-(n+1)} \prod_{\alpha=1}^{n} y_{\alpha\alpha}^{n-\alpha} \prod_{\alpha=1}^{n} dy_{\alpha\alpha}$$

$$= c \prod_{\alpha=1}^{n} \int_{c_n'}^{\infty} y_{\alpha\alpha}^{-(\alpha+1)}\,dy_{\alpha\alpha} < \infty,$$

where c denotes suitable constants. □

PROBLEM 1.24. Sketch connected fundamental domains for $\Gamma_0^1(2)$ and $\Gamma^1(2)$.

PROBLEM 1.25. Compute $v(\Gamma^1)$.

§2. Definition of modular forms

1. Congruence subgroups of the modular group. The *principal congruence subgroup of level q* in the Siegel modular group is the group

(2.1) $$\Gamma^n(q) = \{M \in \Gamma^n; M \equiv E_{2n}(\operatorname{mod} q)\};$$

this is a normal subgroup of finite index in Γ^n, since it is the kernel of the homomorphism from Γ^n to the finite group $GL_{2n}(\mathbf{Z}/q\mathbf{Z})$ that is defined by reduction modulo q. If for some q a subgroup K satisfies

(2.2) $$\Gamma^n(q) \subset K \subset \Gamma^n,$$

then it is called a *congruence subgroup*. It is clear that

(2.3) $$\mu(K) = [\Gamma^n : K] < \infty.$$

PROBLEM 2.1. Show that

$$\mu(\Gamma^1(q)) = q^3 \prod_{p|q}(1 - p^{-2}),$$

where p runs through all prime divisors of q.

2. Modular forms of integer weight. Let K be a congruence subgroup of the Siegel modular group Γ^n, and let χ be a *finite character* of K, i.e., a homomorphism from K to a finite group of roots of unity. A function F on the Siegel upper half-plane $\mathbf{H}_n$ is said to be a *modular form of degree n, weight k* (where k is an integer), *and character χ for the group K* if it satisfies the following three conditions:

(1) F is a holomorphic function in $\langle n \rangle$ complex variables on all of $\mathbf{H}_n$;

(2) for every $M = \begin{pmatrix} A & B \\ C & D \end{pmatrix} \in K$, F satisfies the functional equation

(2.4) $$\det(CZ + D)^{-k} F(M\langle Z \rangle) = \chi(M)F(Z);$$

(3) if $n = 1$, then for any matrix $\begin{pmatrix} a & b \\ c & d \end{pmatrix} \in \Gamma^1$ the function

$$(cz + d)^{-k} F((az + b)(cz + d)^{-1})$$

is bounded on any region $\mathbf{H}_1(\varepsilon) \subset \mathbf{H}_1$, where $\varepsilon > 0$ (see (1.11) of Chapter 1).

The set $\mathfrak{M}_k(K, \chi)$ of all modular forms of degree n, weight k, and character χ for the group K is obviously a vector space over the field of complex numbers. We let $\mathfrak{M}_k^n$ denote the space $\mathfrak{M}_k(\Gamma^n, 1)$.

3. Definition of modular forms of half-integer weight. Let K be a congruence subgroup of the group $\Gamma_0^n(4)$, let χ be a finite character of K, and let k be an odd integer. A function on $\mathbf{H}_n$ is said to be a *modular form of degree n, weight $k/2$, and character χ for the group K* if the following conditions are fulfilled:

(1) F is a holomorphic function in $\langle n \rangle$ complex variables on all of $\mathbf{H}_n$;

(2) for every $M = \begin{pmatrix} A & B \\ C & D \end{pmatrix} \in K$, F satisfies the functional equation

(2.5) $$j_{(2)}^n(M, Z)^{-k} F(M\langle Z \rangle) = \chi(M)F(Z),$$

where $j_{(2)}^n(M, Z)$ is the automorphy factor (4.37) of Chapter 1;

(3) if $n = 1$, then for any matrix $\begin{pmatrix} a & b \\ c & d \end{pmatrix} \in \Gamma^1$ the function

$$(cz + d)^{-k/2} F((az + b)(cz + d)^{-1})$$

is bounded on any region $\mathbf{H}_1(\varepsilon) \subset \mathbf{H}_1$, where $\varepsilon > 0$.

The set $\mathfrak{M}_{k/2}(K, \chi)$ of all modular forms of degree n, weight $k/2$, and character χ for the group K is a vector space over $\mathbf{C}$.

4. Theta-series as modular forms. Let $Q \in \mathbf{A}_m$ be the matrix of a positive definite integral quadratic form in m variables, let q be the level of Q, and let $n \in \mathbf{N}$. Then Theorems 1.4 and 3.13, along with Theorem 4.10 if m is even or Theorem 4.12 and Proposition 4.11 if m is odd (see Chapter 1), show that the theta-series $\theta^n(Z, Q)$ of degree n for Q satisfies conditions (1) and (2) in the definition of a modular form of weight $m/2$ and character χ_Q^n for the group $\Gamma_0^n(q)$. Furthermore, from Proposition 3.14 and Theorem 3.6 of Chapter 1 it is easy to see that any function of the form

$$(2.6) \qquad \det(CZ + D)^{-m/2}\theta^n(M\langle Z\rangle, Q), \quad \text{where } M = \begin{pmatrix} & \\ C & D \end{pmatrix} \in \Gamma^n,$$

is a linear combination with constant coefficients of a finite number of generalized theta-series $\theta^n(Z, Q|T)$. Since each of these series is absolutely and uniformly convergent on any $\mathbf{H}_n(\varepsilon)$ for $\varepsilon > 0$ (see (1.11) of Chapter 1), it follows that the function (2.6) has this property. Thus, we have

THEOREM 2.2. *The theta-series* $\theta^n(Z, Q)$ *of degree* n *for matrices* $Q \in \mathbf{A}_m$ *are modular forms of weight* $m/2$ *and character* χ_Q^n *for the group* $\Gamma_0^n(q)$:

$$(2.7) \qquad\qquad \theta^n(Z, Q) \in \mathfrak{M}_{m/2}(\Gamma_0^n(q), \chi_Q^n),$$

where q *is the level of* Q *and* χ_Q^n *is the character in Theorem* 4.10 *of Chapter* 1 *if* m *is even or in Theorem* 4.12 *of Chapter* 1 *if* m *is odd.*

This theorem reduces the study of the theta-series of quadratic forms to the investigation of modular forms. The latter study often turns out to be simpler, because of the invariant definition of such forms.

PROBLEM 2.3. Let $n, m \in \mathbf{N}$, $Q \in \mathbf{A}_m^+$, and $T \in T^n(Q)$. Show that the generalized theta-series $\theta^n(Z, Q|T)$ (see (3.32) of Chapter 1) is a modular form of weight $m/2$ and trivial character for the group $\Gamma^n(q)$, where q is the level of Q.

[**Hint:** See Problem 3.17 of Chapter 1.]

§3. Fourier expansions

1. Modular forms for triangular subgroups. Some of the important properties of modular forms actually depend only on their analyticity and their invariance under certain subgroups of infinite index in the modular group. It is convenient to examine the corresponding function spaces on an abstract level, because this will be useful later when we construct a theory of Hecke operators and when we analyze the connections between modular forms for different congruence subgroups.

We introduce the triangular subgroup of the modular group:

$$(3.1) \qquad\qquad \Gamma_0^n = \left\{ \begin{pmatrix} A & B \\ 0 & D \end{pmatrix} \in \Gamma^n \right\}.$$

Let $T = T^n$ be a subgroup of Γ_0^n, and let χ be a finite character of T. We say that a function F on $\mathbf{H}_n$ is a *modular form of character* χ *for the group* T if it satisfies the following three conditions:

(1) F is a holomorphic function in $\langle n\rangle$ complex variables on all of $\mathbf{H}_n$;

(2) for every matrix $M \in T$ the function F satisfies the functional equation

$$(3.2) \qquad\qquad F(M\langle Z\rangle) = \chi(M)F(Z);$$

(3) if $n = 1$, then F is bounded in any region $\mathbf{H}_1(\varepsilon)$, where $\varepsilon > 0$.

The space of all such functions will be denoted $\mathfrak{M}(T, \chi)$. If K is a congruence subgroup of Γ^n, $K \supset \Gamma^n(q)$, and χ is a finite character of K, then from the above definitions it follows that

$$(3.3) \qquad\qquad \mathfrak{M}_w(K, \chi) \subset \mathfrak{M}(T_q^n, \chi),$$

where

$$(3.4) \qquad T_q^n = \left\{ \begin{pmatrix} A & B \\ 0 & D \end{pmatrix} \in \Gamma^n(q); \det D = 1 \right\}$$

is a subgroup of finite index in Γ_0^n, and $w = k$ or $k/2$ is an integer or half-integer. Since $\chi^m = 1$ for some natural number m (the smallest such m is called the *order* of χ), it follows that $\chi(T(mqS)) = 1$ for any matrix $S \in S_n$. In addition, $T(mqS) \in T_q^n$.

2. The Koecher effect. We now prove the following fact.

THEOREM 3.1. *Every modular form $F \in \mathbf{M}(T, \chi)$, where T is a subgroup of finite index in Γ_0^n, $n \geqslant 1$, and χ is a finite character of T, has a series expansion of the form*

$$(3.5) \qquad F(Z) = \sum_{R \in \mathbf{A}_n} f(R) e\{q^{-1} R Z\} \quad (Z \in \mathbf{H}_n),$$

where $\mathbf{A}_n$ is the set (1.7) of Chapter 1, $e\{\cdots\}$ is the function (3.23) of Chapter 1, and $q = q(T, \chi)$ is the smallest natural number such that

$$(3.6) \qquad T(qS) \in T \quad and \quad \chi(T(qS)) = 1 \quad for \ any \ S \in S_n.$$

The series (3.5) is absolutely convergent on all of $\mathbf{H}_n$, and it converges uniformly on every $\mathbf{H}_n(\varepsilon)$, where $\varepsilon > 0$. In particular, the function $F(Z)$ is bounded on each $\mathbf{H}_n(\varepsilon)$.

For every matrix V such that

$$(3.7) \qquad\qquad\qquad U(^tV) \in T,$$

the coefficients $f(R)$ in (3.5) satisfy the relation

$$(3.8) \qquad\qquad\qquad f(R[V]) = \chi(U(^tV)) f(R).$$

We call (3.5) the *Fourier expansion of the form F*, and we call the numbers $f(R)$ its *Fourier coefficients*.

PROOF. In the case of matrices of the form (3.6), the functional equation (3.2) becomes

$$F(Z + qS) = F(Z) \quad (Z = (z_{\alpha\beta}) \in \mathbf{H}_n),$$

which holds for any symmetric $n \times n$ integer matrix. This means that F is a periodic function with period q in each variable $z_{\alpha\beta} = z_{\beta\alpha}$. Since it is a regular analytic function, F then has a Fourier expansion of the form

$$F(Z) = \sum_{\substack{r_{\alpha\beta} = -\infty \\ 1 \leqslant \alpha \leqslant \beta \leqslant n}}^{+\infty} g((r_{\alpha\beta}), Y) \exp\left(\frac{2\pi i}{q} \sum_{1 \leqslant \alpha \leqslant \beta \leqslant n} r_{\alpha\beta} x_{\alpha\beta} \right),$$

where $Z = X + iY$ and $X = (x_{\alpha\beta})$. This expansion can be differentiated term by term any number of times with respect to any of its variables. Since $2 \sum_{1 \leqslant \alpha \leqslant \beta \leqslant n} r_{\alpha\beta} x_{\alpha\beta} = \sigma(RX)$, where R is the matrix with entries $2r_{\alpha\alpha}$ and $r_{\alpha\beta}$ for $\alpha \neq \beta$, the above expansion can be rewritten in the form

$$F(Z) = \sum_R f(R, Y) e\{q^{-1} R Z\},$$

where $f(R, Y) = g((r_{\alpha\beta}), Y) e\{-iq^{-1} R Y\}$, and R runs through the set $\mathbf{E}_n$ of all matrices in S_n with even main diagonal. Since $F(Z)$ is holomorphic in each of the variables $z_{\alpha\beta}$, using term-by-term differentiation and uniqueness of the Fourier expansion we see that the Cauchy–Riemann equations

$$\frac{\partial F}{\partial \bar{z}_{\alpha\beta}} = 0, \quad \text{where} \quad \frac{\partial}{\partial \bar{z}_{\alpha\beta}} = \frac{1}{2}\left(\frac{\partial}{\partial x_{\alpha\beta}} + i\frac{\partial}{\partial y_{\alpha\beta}}\right),$$

lead to the equations

$$\frac{\partial f(R, Y)}{\partial \bar{z}_{\alpha\beta}} = \frac{i}{2}\frac{\partial f(R, Y)}{\partial y_{\alpha\beta}} = 0 \quad (1 \leqslant \alpha \leqslant \beta \leqslant n),$$

which show that the coefficients $f(R, Y)$ do not depend on Y. We thus obtain the expansion

(3.9)
$$F(Z) = \sum_R f(R)e\{q^{-1}RZ\}$$

with constant coefficients $f(R)$, where R runs through the same set as above.

The expansion (3.9) may be regarded as the Laurent series for the analytic function F in the variables $t_{\alpha\beta} = \exp(2\pi i z_{\alpha\beta}/q)$ ($1 \leqslant \alpha \leqslant \beta \leqslant n$). Consequently, the series (3.9) converges absolutely on all of $\mathbf{H}_n$.

We now substitute the expansion (3.9) for F in the functional equation (3.2) for a matrix of the form (3.7). If we replace R by $R[V]$ and equate coefficients, we then obtain (3.8).

To complete the proof of the proposition it remains to verify that $f(R) = 0$ if $R \notin \mathbf{A}_n$, and that the series converges uniformly on $\mathbf{H}_n(\varepsilon)$.

We first consider the case $n = 1$. In this case the expansion (3.9) takes the form

$$F(z) = \sum_{r=-\infty}^{+\infty} f(2r)\exp\left(\frac{2\pi i}{q}rz\right) = \sum_{r=-\infty}^{+\infty} f(2r)t^r$$

$$\left(t = \exp\left(\frac{2\pi i}{q}z\right)\right).$$

If we regard this as the Laurent expansion of an analytic function in t in the region $|t| = \exp(-2\pi/qy) < 1$ (where $z = x + iy \in \mathbf{H}_1$), and if we take into account that, by condition (3) in the definition of a modular form for T, the function F is bounded for $|t| < \exp(-2\pi\varepsilon/q)$, where $\varepsilon > 0$, then we see that F as a function of t is holomorphic in the interior of the unit disc including its center. Thus, $f(2r) = 0$ for $r < 0$, and our series converges uniformly on any $\mathbf{H}_1(\varepsilon)$, where $\varepsilon > 0$.

Now let $n \geqslant 2$. Using the absolute convergence of (3.9) and the relations (3.8), we can rewrite the expansion (3.9) in the form

$$F(Z) = \sum_{\{R\}_{T,\chi}} f(R)\varepsilon(Z, \{R\}_{T,\chi}),$$

where the sum is taken over a complete set of representatives of the classes

(3.10)
$$\{R\}_{T,\chi} = \{R[V]; U({}^tV) \in T, \chi(U({}^tV)) = 1\}$$

of matrices $R \in \mathbf{E}_n$, and

(3.11)
$$\varepsilon(Z, \{R\}_{t,\chi}) = \sum_{R' \in \{R\}_{T,\chi}} e\{q^{-1}R'Z\}.$$

If $f(R) \neq 0$, then the series $f(R)\varepsilon(Z, \{R\}_{T,\chi})$ converges for all $Z \in \mathbf{H}_n$, since it is a partial sum of the absolutely convergent series for F. In particular, in this case the following series converges:

$$\varepsilon(iE_n, \{R\}_{T,\chi}) = \sum_{R' \in \{R\}_{T,\chi}} \exp\left(-\frac{2\pi}{q}\sigma(R')\right).$$

Since the traces $\sigma(R')$ of the matrices R' are integers, from the convergence of the last series it follows that the inequality $\sigma(R') < 0$ can hold for at most a finite number of different matrices in $\{R\}_{T,\chi}$.

We show that for any symmetric $n \times n$ integer matrix R, $n \geqslant 2$, with even diagonal, the function $\sigma(R')$ takes infinitely many negative values on the class $\{R\}_{T,\chi}$ if R is not semidefinite. If $R \notin \mathbf{A}_n$, then there obviously exists a column vector h of n integers such that $R[h] < 0$. We set $V_s = E_n + sH$, where $H = (t_1 h, \ldots, t_n h) \in M_n$ and $s, t_1, \ldots, t_n$ are integers. Since the matrix sH has rank 0 or 1, we clearly have

$$\det V_s = 1 + \sigma(sH) = 1 + s(t_1 h_1 + \cdots + t_n h_n),$$

where $h_1, \ldots, h_n$ are the coordinates of h. Since $n \geqslant 2$, the integers $t_1, \ldots, t_n$ can be chosen so that $t_1 h_1 + \cdots + t_n h_n = 0$ and $t_1^2 + \cdots + t_n^2 > 0$. Then $V_s \in SL_n(\mathbf{Z})$ and $H^2 = (h_\alpha i_\beta (t_1 h_1 + \cdots + t_n h_n)) = 0$, from which it follows that, in particular, $V_s = V_1^s$. Since the index $[\Gamma_0^n : T]$ and the character χ are finite, it follows that $U({}^t V_r) = U({}^t V_1)^r$ lies in T for some $r \in \mathbf{N}$, and also $\chi(U({}^t V_r)) = 1$; this implies that for any integer l, $R_l = R[V_{rl}]$ is contained in $\{R\}_{T,\chi}$ and

$$\sigma(R_l) = \sigma(R) + 2rl\sigma(RH) + r^2 l^2 R[h](t_1^2 + \cdots + t_n).$$

Since the last expression is a quadratic trinomial in l with negative coefficient of l^2, it takes negative values of arbitrarily large absolute value for suitable integers l.

From what we have proved it follows that the coefficient $f(R)$ in (3.9) vanishes if $R \notin \mathbf{A}_n$. This proves the existence of the expansion (3.5).

Finally, suppose that $Z = X + iY \in \mathbf{H}_n(\varepsilon)$ for some $\varepsilon > 0$. Then, by the inequality (1.6) of Appendix 1, we find that for any $R \in \mathbf{A}_n$

$$(3.12) \qquad\qquad \sigma(YR) \geqslant \varepsilon\sigma(R).$$

Since the series (3.5) converges absolutely at $Z_0 = (i\varepsilon/2)E_n \in \mathbf{H}_n$, it follows that for any matrix $R \in \mathbf{A}_n$ we have the inequality $|f(R)e\{q^{-1}Z_0 R\}| \leqslant c$, and hence

$$|f(R)| \leqslant c_1 \exp\left(\frac{\pi\varepsilon}{q}\sigma(R)\right),$$

where c_1 depends on ε. From these inequalities and (3.12) we obtain

$$\sum_{R \in \mathbf{A}_n} |f(r)e\{q^{-1}RZ\}| \leqslant c_1 \sum_{R \in \mathbf{A}_n} \exp(-\pi\varepsilon\sigma(R)/2q).$$

Now if $K = (R_{\alpha\beta}) \in \mathbf{A}_n$ and $\sigma(R) \leqslant N$, then from the inequality (1.5) of Appendix 1 we obtain $|R_{\alpha\beta}| \leqslant N$. Thus, the number of different matrices $R \in \mathbf{A}_n$ with $\sigma(R) \leqslant N$ is no greater than

$$(3.13) \qquad\qquad (N/2 + 1)^n (2N + 1)^{\langle n-1 \rangle}$$

(note that the $R_{\alpha\alpha}$ are nonnegative even integers), and

$$\sum_{R \in \mathbf{A}_n} |f(R)e\{q^{-1}RZ\}| \leqslant c_1 \sum_{N=0}^{\infty} (N/2 + 1)^n (2N + 1)^{\langle n-1 \rangle} \exp(-\pi\varepsilon N/2q).$$

Since the latter series converges, it follows that the series (3.5) converges uniformly on $\mathbf{H}_n(\varepsilon)$. $\qquad\square$

PROBLEM 3.2. Suppose that $n, q \in \mathbf{N}$, χ_0 is a finite character of the group Λ^n, and F is a series of the form (3.5) with Fourier coefficients satisfying the condition $f(R[V]) = \chi_0(V)f(R)$ for any matrix $V \equiv E_n \pmod{q}$ in Λ^n. Show that any such series F having the analytic properties listed in Theorem 3.1 determines a modular form in $\mathfrak{M}(T_q^n, \chi)$, where χ is given by the relation $\chi(M) = \chi_0({}^t D)$ for any $M = \begin{pmatrix} * & * \\ 0 & D \end{pmatrix} \in T_q^n$.

PROBLEM 3.3. Suppose that $T = T_q^n$ and $\{R\}_T$ is the set (3.10) with $\chi = 1$. Show that any series $\varepsilon(Z, \{R\}_T)$ of the form (3.11) with $R \in \mathbf{A}_n$ is a modular form of trivial character for the group T.

3. Fourier expansions of modular forms. The inclusions (3.3) show that Theorem 3.1 can be applied, in particular, to modular forms for congruence subgroups of the modular group. Thus, every such form has a Fourier expansion with the properties described above. However, both in the development of the theory of modular forms and in applications of the theory it turns out that one also needs to consider analogous expansions of functions obtained from modular forms by means of certain standard transformations. In addition, one wants to have bounds on the Fourier coefficients for all such expansions.

In the case of modular forms of integer weight k, the transformations we are referring to can be expressed in terms of the elementary transformations that take a function F on $\mathbf{H}_n$ to the function

$$(3.14) \qquad F|_k M = \det(CZ + D)^{-k} F(M\langle Z \rangle), \quad \text{where } M = \begin{pmatrix} A & B \\ C & D \end{pmatrix} \in S_R^n.$$

By Lemma 4.2 of Chapter 1, the function $F|_k M$ is also a function on $\mathbf{H}_n$, and it is analytic on all of $\mathbf{H}_n$ if F is analytic there. According to the same lemma, the expression $\det(CZ + D)^k$ is an automorphy factor of the group S_R^n. By condition (3) in Lemma 4.1 of Chapter 1, we then have the relation

$$(3.15) \qquad F|_k M_1 |_k M_2 = F|_k M_1 M_2 \quad \text{for } M_1, M_2 \in S_R^n.$$

Using the transformations (3.14), we can write the second condition in the definition of a modular form of integer weight k in the form

$$(3.16) \qquad F|_k M = \chi(M)F, \quad \text{if } F \in \mathfrak{M}_k(K, \chi) \text{ and } M \in K.$$

If we want to define analogous transformations for half-integer weight $k/2$ and require that they satisfy the property (3.15), then instead of S_R^n we must consider a *covering group* of S_R^n, denoted $\mathfrak{G}$. By definition, $\mathfrak{G}$ consists of all pairs (M, φ), where $M = \begin{pmatrix} * & * \\ C & D \end{pmatrix} \in S_R^n$ and φ is any holomorphic function $\varphi = \varphi(Z)$ on $\mathbf{H}_n$ such that

$$(3.17) \qquad \varphi(Z)^2 = t \cdot \det(CZ + D),$$

where t is an arbitrary complex number in the set

$$(3.18) \qquad\qquad \mathbf{C}_1 = \{z \in \mathbf{C}; |z| = 1\},$$

and the multiplication law on $\mathfrak{G}$ is given by the formula

$$(3.19) \qquad (M_1, \varphi_1) \cdot (M_2, \varphi_2) = (M_1 M_2, \varphi_1(M_2\langle Z \rangle)\varphi_2(Z)).$$

It is not hard to check that $\mathfrak{G}$ is a group: this follows immediately from the definition of the group operation and the basic property of the automorphy factor $\det(CZ + D)$. The groups S_R^n and $\mathfrak{G}$ are related by the epimorphism

$$(3.20) \qquad\qquad \mathfrak{G} \xrightarrow{P} S_R^n : (M, \varphi) \xrightarrow{P} M,$$

whose kernel is contained in the center of the group $\mathfrak{G}$ and is obviously isomorphic to the multiplicative group $\mathbf{C}_1$.

We are now ready to define the transformations in the case of half-integer weight $k/2$ that are analogous to the transformations (3.14). We set

$$(3.21) \qquad F|_{k/2}\widehat{M} = \varphi(Z)^{-k} F(M, \langle Z \rangle), \quad \text{where } \widehat{M} = (M, \varphi) \in \mathfrak{G};$$

this, along with (3.19), implies the relation

$$(3.22) \qquad F|_{k/2}\widehat{M}_1|_{k/2}\widehat{M}_2 = F|_{k/2}\widehat{M}_1\widehat{M}_2 \quad \text{for } \widehat{M}_1, \widehat{M}_2 \in \mathfrak{G}.$$

Using the automorphy factor $j_{(2)}^n$ of Chapter 1, we define the imbedding

$$(3.23) \qquad\qquad \Gamma_0^n(4) \xrightarrow{j^n} : M \to \widehat{M} = (M, j_{(2)}^n(M, Z)).$$

From (4.36) in Chapter 1 it follows that j^n is a group homomorphism. If we agree to let M denote the element $j^n(M) \in \mathfrak{G}$ for every $M \in \Gamma_0^n(4)$, then condition (2) of the definition of a modular form of half-integer weight $k/2$ can be written in the form

$$(3.24) \qquad F|_{k/2}\widehat{M} = \chi(M)F, \quad \text{if } F \in \mathfrak{M}_{k/2}(K, \chi) \text{ and } M \in K.$$

Suppose that K is a congruence subgroup of Γ^n, $K \supset \Gamma^n(q)$, and M is any matrix in the group $S^n = S_{\mathbf{Q}}^n$. Let

$$(3.25) \qquad\qquad K_M = M^{-1}KM \cap \Gamma,$$

where from now on Γ denotes either Γ^n or $\Gamma_0^n(4)$, depending on whether we are considering forms of integer or half-integer weight; in the latter case, we always assume that $K \subset \Gamma_0^n(4)$. We show that K_M is also a congruence subgroup of Γ^n. To prove this, we write M in the form $M = tM_0$, where M_0 is an integer matrix, we set $q_0 = r(M_0)$, and we verify that $\Gamma^n(qq_0) \subset K_M$. In fact, if $L \in \Gamma^n(qq_0)$, then $L \in \Gamma$ and $MLM^{-1} \in \Gamma^n(q) \subset K$, since

$$ML(q_0 M^{-1}) \equiv M(q_0 M^{-1}) \equiv q_0 E_{2n} (\text{mod } qq_0).$$

LEMMA 3.4. *Let K be a congruence subgroup of $\Gamma_0^n(4)$, and let $M \in S^n$. Let the map*

$$(3.26) \qquad t_M : K_M = M^{-1}KM \cap \Gamma_0^n(4) \to \mathbf{C}_1$$

be defined for any $M_0 \in K_M$ by the equality

$$(3.27) \qquad \widehat{MM_0M}^{-1} = \widehat{M}\,\widehat{M_0}\widehat{M}^{-1}(E_{2n}, t_M(M_0)),$$

where $\widehat{M} = (M, \varphi)$ is any P-preimage of M in $\mathfrak{G}$ and $\widehat{L} = j_{(2)}^n(L)$ for all $L \in \Gamma_0^n(4)$. Then this map does not depend on the choice of $\widehat{M}$, it is a character of the group K_M, and, in addition, $t_M^4 = 1$.

PROOF. Since the P-images of the elements on the left and right sides of (3.27) are the same, it follows that they differ from one another by a factor in the kernel of P. It is easy to see that this kernel consists of elements of the center of $\mathfrak{G}$ of the form (E_{2n}, t), where $t \in \mathbf{C}_1$. We thus find that the equality (3.27) uniquely determines a number $t_M(M_0) \in \mathbf{C}_1$.

We now show that t_M is a homomorphism. For any matrices $M_1, M_2 \in K_M$ we have

$$\begin{aligned}
\widehat{MM_1M_2M}^{-1} &= (\widehat{MM_1M}^{-1})(\widehat{MM_2M}^{-1}) \\
&= \widehat{M}\widehat{M_1}\widehat{M}^{-1}(E_{2n}, t_M(M_1)) \cdot \widehat{M}\widehat{M_2}\widehat{M}^{-1}(E_{2n}, t_M(M_2)) \\
&= \widehat{M}\widehat{M_1M_2}\widehat{M}^{-1}(E_{2n}, t_M(M_1)t_M(M_2)),
\end{aligned}$$

which, together with the relation (3.27) for the matrix $M_0 = M_1M_2$, implies that $t_M(M_1M_2) = t_M(M_1)t_M(M_2)$.

Finally, if we multiply the elements on the right in (3.27) using (3.19) and recall the definition (3.23) of the homomorphism j^n, we obtain the relation
$$(3.28) \qquad j_{(2)}^n(MM_0M^{-1}, Z) = \varphi(M_0M^{-1}\langle Z \rangle)j_{(2)}^n(M_0, M^{-1}\langle Z \rangle)\varphi(M^{-1}\langle Z \rangle)t_M(M_0),$$

since

$$(3.29) \qquad (M, \varphi)^{-1} = (M^{-1}, \varphi(M^{-1}\langle Z \rangle)^{-1}).$$

Squaring both sides of (3.28) and using Lemma 4.2 of Chapter 1, formula (4.47) of Chapter 1, and the definition (3.17), we find that $t_M(M_0)$ satisfies the relation

$$(3.30) \qquad t_M(M_0)^2 = \chi_{Q_2}^n(MM_0M^{-1})\chi_{Q_2}^n(M_0),$$

where $\chi_{Q_2}^n$ is the character (4.31) of Chapter 1 for the matrix $Q_2 = 2E_2$. Thus, $[t_M(M_0)]^4 = 1$. $\qquad\square$

Given an arbitrary character χ of a congruence subgroup K of Γ^n and $M \in S_Q^n$, we define the character

$$(3.31) \qquad \chi_M(W) = \chi(MWM^{-1}) \quad (W \in K_M)$$

of the group K_M, and, if $K \subset \Gamma_0^n(4)$ and k is any integer, then we also define the character

$$(3.32) \qquad \chi_{M,k}(W) = \chi_M(W)t_M(W)^k \quad (W \in K_M)$$

of the group K_M, where we naturally take $\Gamma = \Gamma_0^n(4)$ in the definition (3.25). From (3.31), (3.32), and Lemma 3.4 it follows that the characters χ_M and $\chi_{M,k}$ are finite if χ is a finite character.

THEOREM 3.5. *Let $F \in \mathfrak{M}_w(K, \chi)$ be a modular form of degree $n \geqslant 1$, of integer or half-integer weight w ($w = k$ or $k/2$), and of character χ (where χ has order m) for the congruence subgroup K of Γ^n, $K \supset \Gamma^n(q_1)$; and let $K \subset \Gamma_0^n(4)$ if $w = k/2$. Then for every matrix $M \in \Gamma^n$ one has the expansion*

$$(3.33) \qquad F|_w \xi = \sum_{R \in \mathbf{A}_n} f_\xi(R) e\{q^{-1} R Z\},$$

where $\xi = M$ and $q = q_1 m$ if $w = k$, while if $w = k/2$, then $\xi = \widehat{M}$ is any P-preimage of M in the symplectic covering group $\mathfrak{G}$ and $q = 4 q_1 m$. Each of these series converges absolutely on all of $\mathbf{H}_n$ and uniformly in $\mathbf{H}_n(\varepsilon)$ for any $\varepsilon > 0$. In particular, each function $F|_w \xi$ is bounded on any of the sets $\mathbf{H}_n(\varepsilon)$.

The Fourier coefficients in the expansion (3.33) satisfy the relations

$$(3.34) \qquad f_\xi(R[V]) = \chi'(U({}^tV)) f_\xi(R) \quad (R \in \mathbf{A}_n),$$

where $\chi' = \chi_M$ or $\chi_{M,k}$ for $w = k$ or $k/2$, respectively, $V \in \Lambda^n$, and $V \equiv E_n \pmod{q_1}$. If $w \geqslant 0$, then

$$(3.35) \qquad |f_\xi(R)| \leqslant \gamma_F (\det R)^w \quad (R \in \mathbf{A}_n^+),$$

where γ_F depends only on F.

REMARK. We shall soon see that $\mathfrak{M}_w(K, \chi) = \{0\}$ if $w < 0$. So there is no loss of generality in the condition $w \geqslant 0$ in (3.35).

PROOF. Since obviously $F \in \mathfrak{M} = \mathfrak{M}_w(\Gamma^n(q_1), \chi)$, we can start by replacing $\mathfrak{M}_w(K, \chi)$ by $\mathfrak{M}$.

We first consider the case $w = k$. As we already noted, the function $F|_k M$ has the same analytic properties as F. If $M_0 \in \Gamma^n(q_1)$, then $M M_0 M^{-1} \in \Gamma^n(q_1)$, and hence

$$(3.36) \qquad F|_k M|_k M_0 = F|_k M M_0 M^{-1}|_k M = \chi_M(M_0) F|_k M.$$

Furthermore, any function $F|_k M|_k M'$, where $M' \in \Gamma^n$, is also of the form $F|_k M_1$, where $M_1 \in \Gamma^n$, and so is bounded wherever the latter type of function is bounded. From this and the inclusions (3.3) it follows that $F|_k M \in \mathfrak{M}(T_{q_1}^n, \chi_M)$. If we now apply Theorem 3.1 with $q = q_1 m$ to the function $F|_k M$, we obtain all of the statements in our theorem, except for (3.35), in the case $w = k$.

In the case $w = k/2$, the same parts of the theorem follow if we use the above argument, (3.21)–(3.24), Lemma 3.4, and the relations

$$
\begin{aligned}
(3.37) \qquad F|_{k/2} \widehat{M}|_{k/2} \widehat{M_0} &= F|_{k/2} \widehat{M} \widehat{M_0} \widehat{M}^{-1}|_{k/2} \widehat{M} \\
&= F|_{k/2} \widehat{M M_0 M^{-1}}|_{k/2} (E_{2n}, t_M(M_0)^{-1})|_{k/2} \widehat{M} \\
&= \chi_M(M_0) t_M(M_0)^k F|_{k/2} \widehat{M} = \chi_{M,k}(M_0) F|_{k/2} \widehat{M}.
\end{aligned}
$$

To prove (3.35) for the Fourier coefficients $f_M(R)$, we consider the function

$$(3.38) \qquad G = G_F = |F|_h M_1| + \cdots + |F|_k M_\mu|,$$

where $\{M_\alpha\}$ is a complete set of representatives of $\Gamma^n(q_1)\backslash\Gamma^n$. From (3.15) and (3.16) it follows that G does not depend on the choice of representatives; since the set $\{M_\alpha \cdot M\}$, where $M \in \Gamma^n$, is also a set of representatives, we have

$$(3.39) \qquad |G|_k M| = \sum_{\alpha=1}^{\mu} |F|_k M_\alpha|_k M| = \sum_{\alpha=1}^{\mu} |F|_k M_\alpha M| = G.$$

In addition, because the functions $F|_k M_\alpha$ are bounded on $\mathbf{H}_n(\varepsilon)$ for $\varepsilon > 0$, it follows from Theorem 1.21 that any of these functions—and hence also G—are bounded on the fundamental domain D_n for the group Γ^n.

LEMMA 3.6. *Suppose that the nonnegative real-valued function G on $\mathbf{H}_n$ satisfies the functional equation* (3.39) *for any $M \in \Gamma^n$ and is bounded on D_n. Then the following bound holds uniformly in $X \in S_n(\mathbf{R})$ and $R \in \mathbf{A}_n^+$:*

$$G(X + iR^{-1}) \leqslant \gamma(\det R)^k,$$

where γ depends only on G.

PROOF OF THE LEMMA. Let $Z = X + iY \in \mathbf{H}_n$. Then, since D_n is a fundamental domain for Γ^n on $\mathbf{H}_n$, there exists a matrix $M = \begin{pmatrix} * & * \\ C & D \end{pmatrix} \in \Gamma^n$ such that $Z_0 = M\langle Z\rangle \in D_n$. If $\det C \neq 0$, then $|\det C| \geqslant 1$, and by (3.39) and the boundedness of G on D_n we obtain

$$(3.40) \qquad \begin{aligned} G(Z) &= |\det(CZ + D)|^{-k} G(M\langle Z\rangle) \\ &\leqslant |\det C|^{-k} |\det(X + C^{-1}D + iY)|^{-k} \delta_G \leqslant \delta_G (\det Y)^{-k} \end{aligned}$$

(see the inequality (1.10) in Appendix 1). If $\det C = 0$, we replace Z by the point

$$Z_1 = X_1 + iY_1 = M_1\langle Z\rangle, \quad \text{where } M_1 = \begin{pmatrix} 0 & E_n \\ -E_n & 0 \end{pmatrix}, \quad S \in S_n.$$

Then

$$Z_0 = M\langle Z\rangle = MM_1^{-1}\langle Z_1\rangle \quad \text{and} \quad MM_1^{-1} = \begin{pmatrix} AS + B & -A \\ CS + D & -D \end{pmatrix}.$$

We show that there exists a symmetric $n \times n$ integer matrix S such that

$$(3.41) \qquad\qquad \det(CS + D) \neq 0.$$

Let r be the rank of C. From Lemmas 1.17 and 1.18 we see that in this case the matrices C and D can be represented in the form

$$C = U \begin{pmatrix} C_1 & 0 \\ 0 & 0 \end{pmatrix} {}^t U_1, \qquad D = U \begin{pmatrix} D_1 & 0 \\ 0 & E_{n-r} \end{pmatrix} U_1^{-1},$$

where $U, U_1 \in \Lambda^n$, (C_1, D_1) is a symmetric and relatively prime pair of $r \times r$-matrices, and $\det C_1 \neq 0$. Then for t a sufficiently large integer the matrix $S = \begin{pmatrix} tE_r & 0 \\ 0 & E_{n-r} \end{pmatrix} [U_1^{-1}]$ satisfies the inequality (3.41). By (3.40) and Lemma 2.8 of Chapter 1, for any such matrix S we have

$$G(Z_1) \leqslant \delta_G (\det Y_1)^{-k} = \delta_G (\det Y)^{-k} |\det(-Z + S)|^{2k}.$$

Hence,

$$(3.42) \qquad G(Z) = |\det(-Z + S)|^{-k} G(M_1\langle Z\rangle) \leqslant \delta_G (\det Y)^{-k} |\det(-Z + S)|^k.$$

The expression $\det(CS + D)$ is a polynomial in the entries $s_{\alpha\beta}$ of the matrix S of degree at most two in each variable $s_{\alpha\beta}$. Since this polynomial takes nonzero values, using induction on n it is easy to see that it is nonzero for certain integer values of the $s_{\alpha\beta}$ satisfying the inequalities $-2 < s_{\alpha\beta} - x_{\alpha\beta} < 2$ $(\alpha, \beta = 1, \ldots, n)$, where $x_{\alpha\beta}$ are the entries in the real part $X = (x_{\alpha\beta})$ of the matrix Z. Supposing that these inequalities hold, we see that $|\det(-Z + S)|^2 = |\det(S - X + iY)|^2$ is a polynomial of degree $2n$ with bounded coefficients in the entries $y_{\alpha\beta}$ of the matrix Y. Since $Y > 0$, it follows by inequality (1.5) of Appendix 1 that $|y_{\alpha\beta}| \leqslant \sigma(Y)$. Hence,

$$|\det(-Z + S)| \leqslant \delta_n (1 + \sigma(Y))^n,$$

where δ_n depends only on n. From this bound and (3.42) we obtain

$$G(Z) \leqslant \delta'(1 + \sigma(Y))^{nk} (\det Y)^{-k}.$$

Now let $Y = R^{-1}$, where $R \in \mathbf{A}_n^+$. The matrix R can be written in the form $R = R_0[U^{-1}]$, where R_0 is Minkowski reduced and $U \in \Lambda^n$. Then, using (3.39) for $M = \begin{pmatrix} U & 0 \\ 0 & U^* \end{pmatrix} \in \Gamma^n$ and the last inequality, we obtain

$$G(X + iR^{-1}) = G(U(U^{-1}XU^* + R_0^{-1})\,{}^tU)$$
$$= G(U^{-1}XU^* + R_0^{-1}) \leqslant \delta'(1 + \sigma(R_0^{-1}))^{nk} (\det R_0)^k.$$

Let $R_{\alpha\alpha}$ denote the matrix obtained from R_0 by crossing out the αth row and column, and let r_α denote the αth diagonal entry of R_0. Since $R_{\alpha\alpha} > 0$ and R_0 is Minkowski reduced, we can apply the inequality (1.8) of Appendix 1 to the matrices $R_{\alpha\alpha}$ and use Theorem 1.8 to obtain

$$\sigma(R_0^{-1}) = \sum_{\alpha=1}^{n} \frac{\det R_{\alpha\alpha}}{\det R_0} \leqslant \sum_{\alpha=1}^{n} \frac{r_\alpha^{-1} r_1 \cdots r_n}{c_n r_1 \cdots r_n} = c_n^{-1} \sum_{\alpha=1}^{n} r_\alpha^{-1} \leqslant c_n^{-1} n.$$

Thus, we finally have

$$G(X + iR^{-1}) \leqslant \gamma (\det R_0)^k = \gamma (\det R)^k,$$

where $\gamma = \delta'(1 + nc_n^{-1})^{nk}$. $\qquad\qquad\qquad\qquad\qquad\qquad\qquad\qquad\qquad\square$

We return to the proof of the bound (3.35) on the coefficients $f_M(R)$. Since $F|_k M$ is obviously equal to one of the functions $F|_k M_\alpha$ in (3.38), if we apply Lemma 3.6 to $G = G_F$ we obtain

$$(3.43) \qquad |(F|_k M)(Z)| \leqslant \gamma (\det R)^k, \qquad \text{where } Z = (x_{\alpha\beta}) + iR^{-1},$$

and hence

$$(3.44) \qquad |f_M(R)| = \left| q^{-\langle n\rangle} \int \cdots \int_{\substack{0 \leqslant x_{\alpha\beta} \leqslant q \\ (1 \leqslant \alpha \leqslant \beta \leqslant n)}} (F|_k M)(Z) e\{-q^{-1}RZ\} \prod_{1 \leqslant \alpha \leqslant \beta \leqslant n} dx_{\alpha\beta} \right|$$

$$\leqslant q^{-\langle n\rangle} \gamma (\det R)^k \exp(\pi n q^{-1}) q^{\langle n\rangle} = \gamma_F (\det R)^k.$$

Before examining the case $w = k/2$, we prove the following lemma.

LEMMA 3.7. *Let $F_i \in \mathfrak{M}_{k_i/2}(K_i, \chi_i)$, where $i = 1, 2$, the k_i are odd integers, and the K_i are congruence subgroups in $\Gamma_0^n(4)$. Then the product $F = F_1 F_2$ is a modular form of integer weight $k = (k_1 + k_2)/2$ belonging to $\mathfrak{M}_k(K, \chi)$, where $K = K_1 \cap K_2$ is a congruence subgroup in $\Gamma_0^n(4)$, $\chi = \chi_1 \chi_2 (\chi_{Q_2}^n)^k$, and $\chi_{Q_2}^n$ is the character* (4.31) *of Chapter 1 for the matrix $Q_2 = 2E_2$.*

PROOF. Since K_i contains a principal congruence subgroup $\Gamma^n(q_i)$, it follows that $K \supset \Gamma^n(q_1 q_2)$. We also obviously have $K \subset \Gamma_0^n(4)$. Next, according to (3.24) we can write $F_i|_{k_i/2}\widehat{M} = \chi_i(M)F_i$ for any matrix $M \in K$. If we multiply these equalities together for $i = 1$ and 2, then from (4.47) of Chapter 1 and the definition of modular forms we obtain all of the claims in the lemma. $\qquad\square$

To prove (3.35) for a modular form F of half-integer weight $w = k/2$, instead of F we consider its square F^2, which, by Lemma 3.7, is a modular form of integer weight k for the same group K as F. For F^2 we can use the bound (3.43):

$$(3.45) \qquad |(F^2|_k M)(Z)| \leqslant \gamma (\det R)^k, \qquad \text{where } Z = X + iR^{-1}.$$

On the other hand, the definitions of the transformations (3.14) and (3.21) imply that

$$(3.46) \qquad |F|_{k/2}\widehat{W}|^2 = |F^2|_k W| \qquad \text{for } \widehat{W} \in \mathfrak{G},$$

where $W = P(\widehat{W})$. From this and (3.45) we find that

$$(3.47) \qquad |(F|_{k/2}\widehat{M})(Z)| \leqslant \gamma (\det R)^{k/2}, \qquad \text{where } Z = X + iR^{-1},$$

$P(\widehat{M}) \in \Gamma^n$, and γ is a new constant that depends only on F. We now use the expansion (3.33) for the function $F|_{k/2}\widehat{M}$ and the bound (3.47). By analogy with (3.44), we obtain the bound (3.35) for the coefficients $f_{\widehat{M}}(R)$ with $w = k/2$. $\qquad\square$

At the beginning of the proof of Theorem 3.5 we saw that the operators $|_w \xi$, where $\xi = M$ or $\widehat{M}$ and $M \in \Gamma^n$, map modular forms to modular forms. We now show that this is also a property of the analogous operators for any rational matrix (or matrix proportional to a rational one).

PROPOSITION 3.8. *Let $F \in \mathfrak{M}_w(K, \chi)$ be a modular form of integer or half-integer weight w (where $w = k$ or $k/2$) and finite character χ for the congruence subgroup K of Γ^n; if $w = k/2$, then also $K \subset \Gamma_0^n(4)$. Further let $\xi = M$ if $w = k$ and $\xi = \widehat{M} \in \mathfrak{G}$ if $w = k/2$, where M is any matrix in S_Q^n, and $P(\widehat{M}) = M$. Then*

$$F|_w \xi \in \mathfrak{M}_w(K_M, \chi'),$$

where K_M is the congruence subgroup (3.25) *and χ' is the finite character χ_M if $w = k$ and $\chi_{M,k}$ if $w = k/2$.*

PROOF. By Lemma 4.2 of Chapter 1 and the definition of the transformations (3.14) and (3.21), the function $G = F|_w \xi$ is regular on all of $\mathbf{H}_n$. Furthermore, the functional equations (3.16) and (3.24) for G follow from the relations (3.36) and (3.37), where we take the matrix M_0 from the group K_M. Thus, to show that G is a modular form it remains to verify part three of the definition.

Suppose that $n = 1$ and $w = k$. Then every function $F|_k W'$ for $W' \in \Gamma^1$ is bounded on $\mathbf{H}_1(\varepsilon) \subset \mathbf{H}_1$ for any $\varepsilon > 0$, and, by (3.15), we have $G|_k W = F|_k MW$ for any $W \in \Gamma^1$. By Proposition 3.7 of Chapter 1, we can write $MW = W'M'$, where $W' \in \Gamma^1$, $M' = \begin{pmatrix} a & b \\ 0 & d \end{pmatrix}$ and $ad = r(M) > 0$; hence,

$$(3.48) \qquad G|_k W = F|_k W'|_k M' = d^{-k}(F|_k W')((az + b)/d),$$

and boundedness of this function on $\mathbf{H}_1(\varepsilon)$ follows from the boundedness of $F|_k W'$ on $\mathbf{H}_1(a\varepsilon/d)$.

Finally, if $n = 1$ and $w = k/2$, then the function $F|_{k/2}\widehat{W'}$, where $P(\widehat{W'}) \in \Gamma^1$, is also bounded on $\mathbf{H}_1(\varepsilon)$ for any $\varepsilon > 0$. By analogy with (3.48), the function $G = F|_{k/2}\widehat{M}$ satisfies the relation

$$G|_{k/2}\widehat{W} = t|d|^{-k/2}(F|_{k/2}\widehat{W'})((az + b)/d),$$

where $P(\widehat{W}) \in \Gamma^1$ and t is a complex number in $\mathbf{C}_1$. This implies that the function $G|_{k/2}\widehat{W}$ is bounded on $\mathbf{H}_1(\varepsilon)$. $\qquad\square$

PROBLEM 3.9. Prove that if $K = \Gamma^n(4q)$ and $M \in \Gamma^n$, then the homomorphism t_M in Lemma 3.4 satisfies the condition $t_M^2 = 1$. If $M \in \Gamma_0^n(4)$, then show that $t_M = 1$.

PROBLEM 3.10. Let K be a congruence subgroup of $\Gamma_0^n(4)$, and let M be any matrix in $S_\mathbf{Q}^n$. Show that the characters t_M and $t_{M^{-1}}$ are related as follows: $t_M(W)^{-1} = t_{M^{-1}}(MWM^{-1})$ for any $W \in K_M$.

4. The Siegel operator. In this subsection we shall establish connections between modular forms of degree n and degree $n - 1$. These connections come from the properties of the Fourier expansion of a modular form that were described in Theorems 3.1 and 3.5. Let $\mathfrak{F}_q^n$ denote the set of all Fourier series of the form (3.5) that converge absolutely and uniformly on $\mathbf{H}_n(\varepsilon)$ for $\varepsilon > 0$. Let $F \in \mathfrak{F}_q^n$. If $Z \in \mathbf{H}_{n-1}(\varepsilon)$ and $\lambda > \varepsilon$, then obviously

$$Z_\lambda = \begin{pmatrix} Z & 0 \\ 0 & i\lambda \end{pmatrix} \in \mathbf{H}_n(\varepsilon).$$

Because of the uniform convergence of the series (3.5) for F, we have

$$\lim_{\lambda \to +\infty} F(Z_\lambda) = \sum_{R \in \mathbf{A}_n} f(R) \lim_{\lambda \to +\infty} e\{q^{-1}RZ_\lambda\}.$$

If

$$R = \begin{pmatrix} R' & * \\ * & 2r_{nn} \end{pmatrix},$$

then $\sigma(RZ_\lambda) = \sigma(R'Z) + 2r_{nn}\lambda i$, and hence

$$(3.49) \qquad \lim_{\lambda \to +\infty} e\{q^{-1}RZ_\lambda\} = \begin{cases} e\{q^{-1}R'Z\}, & \text{if } r_{nn} = 0, \\ 0, & \text{if } r_{nn} > 0. \end{cases}$$

Since $R \geqslant 0$, the equality $r_{nn} = 0$ implies that $r_{1n} = r_{n1} = \cdots = r_{n-1,n} = r_{n,n-1} = 0$, i.e., $R = \begin{pmatrix} R' & 0 \\ 0 & 0 \end{pmatrix}$. Thus, for $Z \in \mathbf{H}_{n-1}$ we have

$$(3.50) \qquad (F|\Phi)(Z) = \lim_{\lambda \to +\infty} F(Z_\lambda) = \sum_{R' \in \mathbf{A}_{n-1}} f\left(\begin{pmatrix} R' & 0 \\ 0 & 0 \end{pmatrix}\right) e\{q^{-1}R'Z\}.$$

Since this last series is a partial sum for the expansion (3.5) of F, it converges absolutely and uniformly on $\mathbf{H}_{n-1}(\varepsilon)$. Thus, $F|\Phi \in \mathfrak{F}_q^{n-1}$. If $n = 1$, we set

$$(3.51) \qquad F|\Phi = \lim_{\lambda \to +\infty} F(i\lambda).$$

As before, the limit exists and is equal to the constant term of the Fourier expansion of F. Setting $\mathfrak{F}_q^0 = \mathbf{C}$, for all $n, q \geqslant 1$ we obtain the linear operator

$$(3.52) \qquad \Phi : \mathfrak{F}_q^n \to \mathfrak{F}_q^{n-1},$$

which is called the *Siegel operator*.

The Siegel operator has an especially simple action on the theta-series of positive quadratic forms.

PROPOSITION 3.11. *The image of the generalized theta-series* $\theta^n(W, Q|T)$*, where* $W \in \mathbf{H}_n$*,* $Q \in \mathbf{A}_m^+$*,* $T \in T^n(Q)$*,* $m, n \geqslant 1$*, under the action of the Siegel operator* Φ *is equal to* $\theta^{n-1}(Z, Q|T')$ *or* 0*, depending on whether* $t \equiv 0$ *or* $\not\equiv 0 \pmod{q}$*, respectively, where* q *is the level of* Q*,* t *is the last column of the matrix* $T = (T', t)$*, and* $Z \in \mathbf{H}_{n-1}$ *(we set* $\theta^0 = 1$*). In particular,* Φ *takes* $\theta^n(W, Q)$ *to* $\theta^{n-1}(Z, Q)$*.*

PROOF. Since the series defining the function $\theta^n(W, Q|T)$ converges uniformly on $\mathbf{H}_n(\varepsilon)$ for $\varepsilon > 0$, it follows that

$$\lim_{\lambda \to +\infty} \theta^n(Z_\lambda, Q|T) = \sum_{M \in M_{m,n}} \lim_{\lambda \to +\infty} e\left\{ Q[M + q^{-1}T] \begin{pmatrix} Z & 0 \\ 0 & i\lambda \end{pmatrix} \right\}.$$

If $M = (M', m')$ and m' is the last column of the matrix M, then the entry in the lower-right corner of the matrix $Q[M + q^{-1}T]$ is obviously equal to $Q[m' + q^{-1}t]$. Using (3.49) and the positivity of Q, we see that the limit of the corresponding term in the sum on the right is equal to $e\{Q[M' + q^{-1}T']Z\}$ or 0, depending on whether $m' + q^{-1}t = 0$ or $\neq 0$, respectively. $\qquad \square$

We now consider the action of the Siegel operator on modular forms for congruence subgroups of the modular group. For $n > 1$ we define the monomorphism

$$(3.53) \qquad \Gamma^{n-1} \overset{\varphi}{\to} \Gamma^n : M' = \begin{pmatrix} A' & B' \\ C' & D' \end{pmatrix} \to M = \begin{pmatrix} A & B \\ C & D \end{pmatrix},$$

where

$$A = \begin{pmatrix} A' & 0 \\ 0 & 1 \end{pmatrix}, \quad B = \begin{pmatrix} B' & 0 \\ 0 & 0 \end{pmatrix}, \quad C = \begin{pmatrix} C' & 0 \\ 0 & 0 \end{pmatrix}, \quad D = \begin{pmatrix} D' & 0 \\ 0 & 1 \end{pmatrix},$$

and we let

$$(3.54) \qquad \Gamma^{n,n-1} = \varphi(\Gamma^{n-1}).$$

For an arbitrary subgroup $K = K^n \subset \Gamma^n$ we set

$$(3.55) \qquad K^{n,n-1} = K \cap \Gamma^{n,n-1}, \qquad K^{[n-1]} = \varphi^{-1}(K^{n,n-1}).$$

If K is a congruence subgroup of Γ^n, $K \supset \Gamma^n(q)$, and χ is a finite character of the group K, then, since $K^{n,n-1} \supset \Gamma^n(q) \cap \Gamma^{n,n-1}$, we have

$$\Gamma^{n-1}(q) = \varphi^{-1}(\Gamma^{n,n-1}(q)) \subset K^{[n-1]},$$

so that $K^{[n-1]}$ is a congruence subgroup of Γ^{n-1}, and

$$(3.56) \qquad \chi^{[n-1]}(M') = \chi(\varphi(M')) \quad (M' \in K^{[n-1]})$$

is a finite character of this group. Now let $F \in \mathfrak{M}_w(K, \chi)$ be a modular form of integer or half-integer weight w (where $w = k$ or $w = k/2$). If $Z \in \mathbf{H}_{n-1}$, $M' \in \Gamma^{n-1}$, and $M = \varphi(M')$, then for $\lambda > 0$ we have

$$(3.57) \qquad M\langle Z_\lambda \rangle = M'\langle Z \rangle_\lambda, \quad \text{where } Z_\lambda = \begin{pmatrix} Z & 0 \\ 0 & i\lambda \end{pmatrix} \in \mathbf{H}_n;$$

$$(3.58) \qquad \det(CZ_\lambda + D) = \det(C'Z + D'),$$

where $M = \begin{pmatrix} A & B \\ C & D \end{pmatrix}$ and $M' = \begin{pmatrix} A' & B' \\ C' & D' \end{pmatrix}$. Thus,

$$(3.59) \qquad (F|\Phi)|_k M' = (F|_k M|)|\Phi.$$

In particular, if $M' \in K^{[n-1]}$, then

$$(3.60) \qquad (F|\Phi)|_k M' = \chi(M)F|\Phi = \chi^{[n-1]}(M')F|\Phi.$$

We next suppose that $w = k/2$ and $K \subset \Gamma_0^n(4)$. In the earlier notation, if we use the functional equation (4.36) of Chapter 1, the relation (3.57), and Proposition 3.11, we obtain

$$\lim_{\lambda \to +\infty} j_{(2)}^n(M, Z_\lambda) = \lim_{\lambda \to +\infty} \frac{\theta^n(M\langle Z_\lambda \rangle, (2))}{\theta^n(Z_\lambda, (2))}$$
$$= \frac{\theta^{n-1}(M'\langle Z \rangle, (2))}{\theta^{n-1}(Z, (2))} = j_{(2)}^{n-1}(M', Z).$$

But in view of (4.37) of Chapter 1 and (3.58), the function $j_{(2)}^n(M, Z_\lambda)$ does not depend on λ; hence,

$$(3.61) \qquad j_{(2)}^n(M, Z_\lambda) = j_{(2)}^{n-1}(M', Z),$$

where $M' \in \Gamma_0^{n-1}(4)$, $M = \varphi(M')$, and $Z_\lambda = \begin{pmatrix} Z & 0 \\ 0 & i\lambda \end{pmatrix} \in \mathbf{H}_n$.

We now consider the more general case when $M' \in \Gamma^{n-1}$ and $\widehat{M'} = (M', \varphi')$, $\widehat{M} = (M, \varphi)$ are arbitrary P-preimages of M' and M in $\mathfrak{G}$. From the definition (3.17) of $\mathfrak{G}$ and the equality (3.58) we see that

$$(3.62) \qquad \varphi(Z_\lambda) = t\varphi'(Z), \quad \text{where } t \in \mathbf{C}_1.$$

Hence, using the definition (3.21) of the transformation $|_{k/2}\widehat{M'}$, by analogy with (3.59) we obtain the relation

$$(3.63) \qquad (F|\Phi)|_{k/2}\widehat{M'} = t^k(F|_{k/2}\widehat{M})|\Phi.$$

If $M' \in \Gamma_0^{n-1}(4)$, then (3.61) and the definition (3.23) of $\widehat{M'}$ and $\widehat{M}$ imply that in this case $t = 1$. We thus have

$$(3.64) \qquad (F|\Phi)|_{k/2}\widehat{M'} = (F|_{k/2}\widehat{M})|\Phi \quad \text{for } M' \in \Gamma_0^{n-1}(4),$$

which implies, in particular, that for $F \in \mathfrak{M}_{k/2}(K, \chi)$ and $M \in K^{[n-1]}$

$$(3.65) \qquad (F|\Phi)|_{k/2}\widehat{M'} = \chi(M)F|\Phi = \chi^{[n-1]}(M')F|\Phi.$$

Thus, by (3.60) and (3.65), the function $F|\Phi$ satisfies the functional equations for a modular form in $\mathfrak{M}_w(K^{[n-1]}, \chi^{[n-1]})$ with finite character $\chi^{[n-1]}$. In addition, from (3.59) and (3.63) it follows that for any $M' \in \Gamma^{n-1}$ the functions $(F|\Phi)|_k M'$ and $(F|\Phi)|_{k/2} \widehat{M}'$ are bounded on $\mathbf{H}_{n-1}(\varepsilon)$, provided that the functions $F|_k M$ and $F|_{k/2} M'$, where $M \in \Gamma^n$, are bounded on $\mathbf{H}_n(\varepsilon)$. Finally, if $K \supset \Gamma^n(q_1)$ and the character χ has order m, then by (3.33) we have $F|\Phi \in \mathfrak{F}_q^{n-1}$, where $q = 4q_1 m$, and hence this function is analytic on $\mathbf{H}_{n-1}$. We have thereby proved

PROPOSITION 3.12. *Suppose that K is a congruence subgroup of the modular group Γ^n, χ is a finite character of K, and w is an integer or half-integer ($w = k$ or $w = k/2$). Set $\mathfrak{M}_w(K^{[0]}, \chi^{[0]}) = \mathbf{C}$. Then the Siegel operator Φ gives a linear map*

$$(3.66) \qquad \Phi: \mathfrak{M}_w(K, \chi) \to \mathfrak{M}_w(K^{[n-1]}, \chi^{[n-1]}),$$

where $K \subset \Gamma_0^n(4)$ if $w = k/2$.

5. Cusp-forms. A modular form $F \in \mathfrak{M}_w(K, \chi)$ is called a *cusp-form* if all of the Siegel operators Φ_ξ, where $\xi = M \in \Gamma^n$ if $w = k$ and $\xi = \widehat{M} \in \mathfrak{G}$ with $P(\widehat{M}) \in \Gamma^n$ if $w = k/2$, map the form to zero:

$$(3.67) \qquad F|\Phi_\xi = (F|_w \xi)|\Phi = 0.$$

The condition (3.67) means that F approaches zero as the argument makes certain "rational" approaches to the boundary of the upper half-plane $\mathbf{H}_n$, i.e., in some sense F is small near the boundary. This circumstance makes it possible to substantially strengthen the bounds (3.35) on the Fourier coefficients of the functions $F|_w \xi$ for such F.

THEOREM 3.13. *Let $F \in \mathfrak{M}_w(K, \chi)$ be a cusp-form of degree $n \geqslant 1$, integer or half-integer weight w (where $w = k$ or $w = k/2$), and finite character χ of order m for a congruence subgroup K in Γ^n, where $K \supset \Gamma^n(q_1)$ and if $w = k/2$, then $K \subset \Gamma_0^n(4)$. Then, in the notation of Proposition 3.8:*
(1) for any matrix $M \in \Gamma^n$ the Fourier expansion (3.33) of the function $F|_w \xi \in \mathfrak{M}_w(K_M, \chi')$ has the form

$$(3.68) \qquad F|_w \xi = \sum_{R \in \mathbf{A}_n^+} f_\xi(R) e\{q^{-1} R Z\},$$

where $q = q_1 m$ if $w = k$ and $4q_1 m$ if $w = k/2$, i.e., only positive definite matrices appear in the expansion;
(2) if $w \geqslant 0$, then the functions $F|_w \xi$ ($M \in \Gamma^n$) and its Fourier coefficients satisfy the bounds

$$(3.69) \qquad |(F|_w \xi)(X + iY)| \leqslant \delta'_F (\det Y)^{-w/2} \quad \text{for } X + iY \in \mathbf{H}_n,$$

$$(3.70) \qquad |f_\xi(R)| \leqslant \delta_F (\det R)^{w/2} \quad \text{for } R \in \mathbf{A}_n^+,$$

where δ'_F and δ_F depend only on F;
(3) the function $F|_w \xi$ is a cusp-form for every $M \in S_{\mathbf{Q}}^n$.

We first prove a lemma.

LEMMA 3.14. *Let* $R \in \mathbf{A}_m$, $m > 1$, *and* $\det R = 0$. *Then there exists a matrix* $V \in SL_m(\mathbf{Z})$ *such that*

$$(3.71) \qquad\qquad R[V] = \begin{pmatrix} R' & 0 \\ 0 & 0 \end{pmatrix}, \qquad \text{where } R' \in \mathbf{A}_{m-1}.$$

PROOF. Since R is a singular integer matrix, there exists a nonzero m-dimensional integer column-vector v such that $Rv = 0$. Without loss of generality we may assume that the coordinates of v are relatively prime. Then by Lemma 1.5 there exists an $m \times (m-1)$ integer matrix V' such that $V = (V', v) \in SL_m(\mathbf{Z})$. This matrix obviously satisfies the lemma. $\qquad\square$

PROOF OF THE THEOREM. Let $w = k$. We shall show that the coefficients $f_M(R)$ in the expansion (3.33) of a cusp-form are zero for matrices with $\det R = 0$. If $n = 1$, then this follows immediately from (3.67), since in this case $(F|_k M)|\Phi = f_M(0)$. Suppose that $n > 1$ and $V \in SL_n(\mathbf{Z})$ satisfies (3.71) for the matrix $R = R_0$. Then $M_0 = U(V^*) \in \Gamma^n$, and

$$F|_k M|_k M_0 = \sum_{R \in \mathbf{A}_n} f_M(R[V^{-1}]) e\{q^{-1}RZ\}.$$

On the other hand, the Fourier coefficients of the function $F|_k M|_k M_0 = F|_k M M_0$ are $f_{MM_0}(R)$, so that

$$f_M(R[V^{-1}]) = f_{MM_0}(R) \quad (R \in \mathbf{A}_n).$$

If we set $R = R_0[V]$ here, we obtain

$$f_M(R_0) = f_{MM_0}(R_0[V]) = f_{MM_0}\left(\begin{pmatrix} R' & 0 \\ 0 & 0 \end{pmatrix} \right) = 0,$$

since $F|\Phi_{MM_0} = 0$, and the last expression is one of the Fourier coefficients of this function (see (3.50)). If we use (3.33) with $w = k/2$, then the above argument goes through for modular forms of half-integer weight as well.

LEMMA 3.15. *Suppose that a series of the form*

$$Q(Y) = \sum_{R \in \mathbf{A}_n^+} q(R)\exp(-\xi\sigma(RY)),$$

where all of the coefficients $q(R)$ *are nonnegative and* $\xi > 0$, *converges for all* $Y \in P_n$. *Then the following bounds hold for any matrix* Y *that belongs to the Minkowski reduction domain* F_n *and satisfies the inequality* $Y \geqslant \varepsilon E_n$, *where* $\varepsilon > 0$:

$$Q(Y) \leqslant \delta_1 \exp(-\delta_2 \sigma(Y)),$$
$$Q(Y) \leqslant \delta_1 \exp(-\delta_2 n (\det Y)^{1/n}),$$

where δ_1 *and* δ_2 *are positive constants, the first of which depends only on* Q *and* ε *and the second of which depends only on* n *and* ξ.

PROOF OF THE LEMMA. Since $Y \in F_n$, it follows from (1.16) and also (1.6) of Appendix 1 that

$$\sigma(RY) \geqslant b_n \sigma(R \operatorname{diag}(y_{11}, \ldots, y_{nn}))$$

$$= b_n \sum_{\alpha=1}^{n} 2r_{\alpha\alpha} y_{\alpha\alpha} \geqslant 2b_n \sigma(Y) \quad (R \in \mathbf{A}_n^+),$$

where $b_n = n^{1-n} c_n^{-1}$ depends only on n and $2r_{\alpha\alpha} \geqslant 2$ (i.e., the diagonal entries in R). On the other hand, since $Y \geqslant \varepsilon E_n$, it follows that $\sigma(RY) \geqslant \varepsilon \sigma(R)$ for $R \geqslant 0$. From these inequalities we obtain $\sigma(RY) \geqslant b_n \sigma(Y) + \varepsilon/2 \sigma(R)$, and hence

$$Q(Y) \leqslant \exp(-\xi b_n \sigma(Y)) \sum_{R \in \mathbf{A}_n^+} q(R) \exp(-\xi \varepsilon \sigma(R)/2)$$

$$= \exp(-\xi b_n \sigma(Y)) Q((\varepsilon/2) E_n),$$

proving the first inequality. The second inequality follows from the first, since, if we use the inequality between the arithmetic and geometric means and the inequality (1.8) of Appendix 1, we obtain

$$\sigma(Y)/n \geqslant (y_{11} \cdots y_{nn})^{1/n} \geqslant (\det Y)^{1/n} \quad \text{for } Y \in P_n. \qquad \square$$

We now return to the proof of Theorem 3.13 for $w = k$. By analogy with the function (3.38) we consider the function

$$(3.72) \qquad G = G_F = \sum_{M \in K \backslash \Gamma^n} |F|_k M|.$$

From (3.15) and (3.16) it follows that G does not depend on the choice of left coset representatives of Γ^n modulo K. Consequently, for any matrix $M' = \begin{pmatrix} * & * \\ C & D \end{pmatrix} \in \Gamma^n$ we have

$$(3.73) \quad |\det(CZ + D)|^{-k} G(M'\langle Z \rangle) = \sum_M |F|_k M|_k M'| = \sum_M |F|_k MM'| = G(Z).$$

From these relations and Lemma 2.8 of Chapter 1 it follows that the function

$$\Psi_F(Z) = \Psi_F(X + iY) = (\det Y)^{k/2} G_F(Z)$$

on $\mathbf{H}_n$ is invariant relative to all transformations in Γ^n:

$$\Psi_F(M'\langle Z \rangle) = \Psi_F(Z) \quad \text{for } M' \in \Gamma^n.$$

Hence, any value it takes on $\mathbf{H}_n$ is already taken on the fundamental domain D_n of the group Γ^n. However, if $Z = X + iY \in D_n$, then, by the definition of D_n and the inequality (1.26), Y satisfies the conditions of Lemma 3.15 for suitable $\varepsilon > 0$. If we apply the first inequality in Lemma 3.15 to each of the functions

$$\sum_{R \in \mathbf{A}_n^+} |f_M(R)| \exp(-\pi \sigma(RY)/q) \geqslant |F|_k M|$$

we obtain

$$\Psi_F(X + iY) \leqslant \delta_3 (\det Y)^{k/2} \exp(-\delta_2 \sigma(Y)),$$

where δ_2 and δ_3 are positive constants depending only on F. If $y_{\alpha\alpha}$ are the diagonal entries in Y, then, by (1.8) of Appendix 1 (recall that $k \geqslant 0$), the last expression is no greater than

$$\delta_3 \prod_{\alpha=1}^{n} y_{\alpha\alpha}^{k/2} \exp(-\delta_2 y_{\alpha\alpha}),$$

and consequently it is bounded for all $Y > 0$. Thus, the function Ψ_F is bounded on D_n, and hence on all of $\mathbf{H}_n$: $\Psi_F(Z) \leqslant \delta'_F$ for $Z \in \mathbf{H}_n$. Since for an arbitrary matrix $M \in \Gamma^n$ the function $|F|_k M|$ is obviously equal to one of the terms in (3.72), the last bound will give us (3.69) with $w = k$:

$$\begin{aligned}(3.74) \qquad (F|_k M)(X + iY) &\leqslant G_F(X + iY) \\ &= (\det Y)^{-k/2} \Psi_F(X + iY) \leqslant \delta'_F (\det Y)^{-k/2}.\end{aligned}$$

If we now substitute this bound in the integral in (3.44), we obtain the bound (3.70) for the coefficients $f_M(R)$. To prove (3.69) and (3.70) for $w = k/2$ one can repeat the proof of (3.35) in Theorem 3.5, i.e., instead of the modular form F of half-integer weight $k/2$ one considers the modular form F^2 of integer weight k. Then F^2 satisfies (3.74), and from that, along with (3.46), one obtains

$$|(F|_{k/2}\widehat{M})(X + iY)|^2 = |(F^2|_k M)(X + iY)| \leqslant \delta'_F (\det Y)^{-k/2},$$

i.e., (3.69) with $w = k/2$. If we substitute this bound in the integral in (3.44), we obtain (3.70) for the coefficients $f_{\widehat{M}}(R)$.

Finally, to prove the last part of the theorem, by Proposition 3.8 we must show that

$$(3.75) \qquad\qquad (F|_w \xi|_w \xi')|\Phi = 0,$$

where $\xi' = M' \in \Gamma^n$ if $w = k$ and $\xi' = \widehat{M'} \in \mathfrak{G}$, $P(\widehat{M'}) = M'$ if $w = k/2$. By Proposition 3.7 of Chapter 1, the matrix MM' can be written in the form $M_1 M_0$, where $M_0 = \begin{pmatrix} A_0 & B_0 \\ 0 & D_0 \end{pmatrix}, M_1 \in \Gamma^n$. Let ξ_1 and ξ_2 be defined in the same manner as ξ; in addition, if $w = k/2$, then let $\xi_0 = \widehat{M_0} \in \mathfrak{G}$ be chosen so that we have $\widehat{M}\widehat{M'} = \widehat{M_1}\widehat{M_2}$ (this is always possible). Then, by the first part of the theorem, we have the expansion

$$F|_w \xi_1 = \sum_{R \in \mathbf{A}_n^+} f_{\xi_1}(R) e\{q^{-1} RZ\},$$

and hence

$$\begin{aligned}F|_w \xi\xi' = F|_w \xi_1 \xi_0 &= F|_w \xi_1|_w \xi_0 \\ &= t(\det D_0)^{-w} \sum_{R \in \mathbf{A}_n^+} f_{\xi_1}(R) e\{q^{-1} RB_0 D_0^{-1}\} e\{q^{-1} D_0^{-1} RA_0 Z\},\end{aligned}$$

where t is a complex number in $\mathbf{C}_1$. Since the matrix $q^{-1} D_0^{-1} RA_0$ is positive definite, from (3.49) and the definition of the Siegel operator it follows that $F|_w \xi\xi'|\Phi = 0$. This, along with the relation $F|_w \xi\xi' = F|_w \xi|_w \xi'$, implies (3.75). $\square$

We let $\mathfrak{N}_w(K, \chi)$ denote the space of all cusp-forms of weight w and character χ for the group K.

§4. Spaces of modular forms

In §2, starting from certain properties of theta-series, we defined modular forms for congruence subgroups of the modular group. Generally speaking, there are more modular forms than theta-series; nevertheless, the conditions in the definition of a given type of modular form turn out to be so rigid that they are satisfied by only a finite number of linearly independent functions.

1. Zeros of modular forms for Γ^1. By the *order at a point* $p \in \mathbf{H}_1$ of a nonzero holomorphic function F on $\mathbf{H}_1$ we mean the integer $n = v_p(F)$ for which the function $F(z)/(z-p)^n$ is holomorphic and nonzero at p. If $F \in \mathfrak{M}_k = \mathfrak{M}_k(\Gamma^1, 1)$ is a modular form of weight k for Γ^1, then the functional equation

$$(4.1) \qquad (cz+d)^{-k}F(M\langle z\rangle) = F(z) \quad \text{for } M = \begin{pmatrix} a & b \\ c & d \end{pmatrix} \in \Gamma^1$$

implies that $v_p(F) = v_{M\langle p\rangle}(F)$ for $M \in \Gamma^1$; thus, the order $v_p(F)$ depends only on the Γ^1-orbit $\Gamma^1\langle p\rangle$ of p. In addition, by Theorem 3.1, in this case F has a series expansion of the form

$$(4.2) \qquad F(z) = \sum_{r=0}^{\infty} f(2r)\exp(2\pi i r z),$$

which converges uniformly on $\mathbf{H}_1(\varepsilon)$ for any $\varepsilon > 0$. This implies that F may be regarded as a function of the variable $q = \exp(2\pi i z)$:

$$(4.3) \qquad F(z) = \widetilde{F}(q) = \sum_{r=0}^{\infty} f(2r)q^r,$$

and, as a function of q, it is holomorphic in the open unit disc $|q| < 1$, including the center $q = 0 = \lim_{z \to i\infty} \exp(2\pi i z)$. The order of $\widetilde{F}(q)$ at the point $q = 0$ is called the order of F at the point $i\infty$; it is denoted $v_{i\infty}(F)$. In other words, $v_{i\infty}(F) = n$ if $f(0) = f(2) = \cdots = f(2(n-1)) = 0$ but $f(2n) \neq 0$ in the expansion (4.2).

PROPOSITION 4.1. *Any nonzero modular form F of weight k and trivial character for the modular group Γ^1 vanishes on only a finite number of Γ^1-orbits in $\mathbf{H}_1$. If $p_1, \ldots, p_m$ are a set of representatives of these orbits, then*

$$(4.4) \qquad v_{i\infty}(F) + \sum_{\alpha=1}^{m} e(p_\alpha)^{-1} v_{p_\alpha}(F) = k/12,$$

where $e(p) = 2$ if p belongs to the orbit of the point i, $e(p) = 3$ if p belongs to the orbit of the point $\rho = (1 + i\sqrt{3})/2$, and $e(p) = 1$ otherwise.

PROOF. By Theorem 1.1, every Γ^1-orbit in $\mathbf{H}_1$ intersects with the fundamental domain D_1. Thus, to prove the first part of the proposition it suffices to verify that F vanishes at only finitely many points $p \in D_1$. Since the function $\widetilde{F}(q)$ (see (4.3)) is holomorphic at $q = 0$ and is not identically zero, it must be nonzero in some region of the form $0 < |q| < \varepsilon$, where $\varepsilon < 1$. This implies that $F(x+iy) \neq 0$ if $y > (\ln \varepsilon^{-1})/2\pi$. But the subset of D_1 consisting of all points $x + iy$ for which $y \leqslant (\ln \varepsilon^{-1})/2\pi$ is compact, and hence can contain only finitely many zeros of the holomorphic function F.

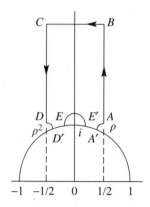

FIGURE 2

In proving (4.4) we may assume that $p_1, \ldots, p_m \in D_1$. We first suppose that the boundary of D_1 does not contain any zeros of F, except possibly for i, ρ, ρ^2. Then one can draw the contour L_r shown in Figure 2, where DD', EE', and $A'A$ are arcs of small circles all of radius r centered at ρ^2, i, and ρ, respectively, such that L_r contains all of the zeros $p_1, \ldots, p_m$ that are distinct from ρ, ρ^2, and i. Since all of the interior points of D_1 lie in different Γ^1-orbits, it follows that there are no other zeros of F inside L_r, and so, by the residue theorem, we have

$$(4.5) \qquad \sum_{p_\alpha \neq \rho, \rho^2, i} v_{p_\alpha}(F) = \frac{1}{2\pi i} \int_{L_r} \frac{dF}{F} = \lim_{r \to 0} \frac{1}{2\pi i} \int_{L_r} \frac{dF}{F},$$

where the integral is taken counterclockwise around L_r.

The integral over L_r can be computed by dividing L_r into pieces with endpoints at $A, B, C, D, D', E, E', A'$ (see Figure 2). First of all, since $F(z-1) = F(z)$ and the transformation $z \to z - 1$ takes the segment AB to the segment DC, we have

$$\frac{1}{2\pi i} \int_{AB} \frac{dF}{F} + \frac{1}{2\pi i} \int_{CD} \frac{dF}{F} = 0.$$

Next, since the map $z \to q$ takes the segment BC to a (clockwise) circle R centered at $q = 0$ that does not contain any zeros of F—with the possible exception of a zero of order $v_{i\infty}(F)$ at $q = 0$—it follows that

$$\frac{1}{2\pi i} \int_{BC} \frac{dF}{F} = \frac{1}{2\pi i} \int_R \frac{d\widetilde{F}}{\widetilde{F}} = -v_{i\infty}(F).$$

The integral of $\frac{1}{2\pi i}\frac{dF}{F}$ over the entire circle containing the arc DD' (taken in the same direction as the arc) is equal to $-v_{\rho^2}(F) = -v_\rho(F)$ for small r. Since the angle between the radii from ρ^2 to D and from ρ^2 to D' is obviously $2\pi/6$, we have

$$\lim_{r \to 0} \frac{1}{2\pi i} \int_{DD'} \frac{dF}{F} = -\frac{1}{6} v_\rho(F).$$

Similarly,

$$\lim_{r \to 0} \frac{1}{2\pi i} \int_{EE'} \frac{dF}{F} = -\frac{1}{2} v_i(F)$$

and

$$\lim_{r \to 0} \frac{1}{2\pi i} \int_{A'A} \frac{dF}{F} = -\frac{1}{6} v_\rho(F).$$

Finally, the transformation $z \to z^{-1}$ takes the arc $A'E'$ to the arc $D'E$; and the relation $F(-1/z) = z^k F(z)$ implies that

$$\frac{dF(-1/z)}{F(-1/z)} = \frac{k\,dz}{z} + \frac{dF}{F},$$

hence,

$$\frac{1}{2\pi i} \int_{D'E} \frac{dF}{F} + \frac{1}{2\pi i} \int_{E'A'} \frac{dF}{F} = \frac{1}{2\pi i} \int_{A'E'} \frac{dF(-1/z)}{F(-1/z)} + \frac{1}{2\pi i} \int_{E'A'} \frac{dF}{F}$$

$$= \frac{k}{2\pi i} \int_{A'E'} \frac{dz}{z} + \frac{1}{2\pi i} \left(\int_{A'E'} \frac{dF}{F} + \int_{E'A'} \frac{dF}{F} \right) = \frac{k}{2\pi i} \int_{A'E'} \frac{dz}{z}.$$

From this it follows that as $r \to 0$ our sum approaches the limit

$$\frac{k}{2\pi i} \int_{(\rho,i)} \frac{dz}{z} = \frac{k}{12},$$

since the length of the arc from ρ to i is $2\pi/12$. If we substitute these expressions into (4.5), we obtain (4.4).

If F has zeros other than ρ, ρ^2, or i on the boundary of D_1, then the same argument goes through if we deform L_r in such a way that its interior contains only one from each pair of zeros lying in the same orbit. For example, if we have a pair of zeros λ and $\lambda + 1$ on the lines $x = \pm 1/2$ and another pair β and $-1/\beta$ on the circle $|z| = 1$, then we draw the contour shown in Figure 3, where the small circular arcs have the same radius and are centered at the points indicated. □

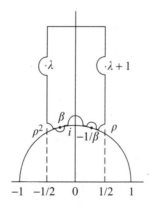

FIGURE 3

PROBLEM 4.2. Prove that
(1) $\dim \mathfrak{M}_k(\Gamma^1, 1) = 0$ if k is negative, if k is odd, or if $k = 2$;
(2) $\dim \mathfrak{M}_k(\Gamma^1, 1) \leqslant 1$ if $k = 0, 4, 6, 8, 10$;
(3) $\dim \mathfrak{M}_k(\Gamma^1, 1) \leqslant \left[\frac{k}{12}\right] + 1$ if $k \geqslant 0$.

2. Modular forms with zero initial Fourier coefficients. Proposition 4.1 implies that
for any nonzero $F \in \mathfrak{M}_k(\Gamma^1, 1)$ we have

$$(4.6) \qquad\qquad\qquad v_{i\infty}(F) \leqslant \frac{k}{12}.$$

This means that F must be identically zero if sufficiently many of its initial Fourier
coefficients are zero. The analogous fact holds for all modular forms for congruence
subgroups of the Siegel modular group, and it is this fact that implies finite dimension-
ality of the space of such forms.

THEOREM 4.3. *Suppose that K is a congruence subgroup of the modular group Γ^n,
$K \supset \Gamma^n(q_1)$, χ is a finite character of K of order m, and w is an integer or half-integer
($w = k$ or $w = k/2$), where $K \subset \Gamma_0^n(4)$ if $w = k/2$. Then a modular form*

$$F(Z) = \sum_{R \in \mathbf{A}_n} F(R)e\{q^{-1}RZ\} \in \mathfrak{M}_w(K, \chi),$$

*where $q = q_1 m$ or $4q_1 m$ for $w = k$ or $k/2$, respectively, is identically zero if its coefficients
satisfy the condition*

$$(4.7) \qquad\qquad f(R) = 0, \quad if\ \sigma(R) \leqslant \frac{w_1}{2\pi}\sigma_n\mu(K)q,$$

*where $w_1 = k$ or $(k+1)/2$ for $w = k$ or $k/2$, respectively, $\sigma_n = 2n^n c_n\sqrt{3}$, c_n is the
constant in Theorem 1.8, and $\mu(K)$ is the index of the subgroup K in Γ^n.*

PROOF. Supposing that the theorem has been proven for modular forms of integer
weight, we show that this implies the theorem for modular forms of half-integer
weight. Thus, let F be a modular form of weight $w = k/2$ that satisfies (4.7). By Lemma 3.7
and Theorem 2.2, the product $G(Z) = F(Z) \cdot \theta^n(Z, (2))$ is a modular form of integer
weight $w_1 = (k+1)/2$ and character $\chi(\chi_{Q_2}^n)^k$ (of order m' dividing $2m$) for the
same group $K \supset \Gamma^n(q_1)$ as F. By the definition (1.13) of Chapter 1, the theta-series
$\theta^n(Z, (2))$, like F, has an expansion with Fourier coefficients $f_{(2)}(R) = r((2), q^{-1}R)$.
Hence, G has Fourier coefficients of the form

$$g(R) = \sum_{R_i \in \mathbf{A}_n, R_1+R_2=R} f(R_1)f_{(2)}(R_2).$$

Let $R \in \mathbf{A}_n$ and $\sigma(R) \leqslant \gamma = (w_1/2\pi)\sigma_n\mu(K)(4q_1 m)$. If $R = R_1 + R_2$, then $\sigma(R_1) \leqslant$
$\sigma(R) \leqslant \gamma$, since any matrix in $\mathbf{A}_n$ has nonnegative trace. From this and the assumption
(4.7) it follows that $f(R_1) = 0$, and hence $g(R) = 0$ if $\sigma(R) \leqslant \gamma$. Since $q_1 m' \leqslant 4q_1 m$,
all of the conditions of the theorem hold for the modular form G. Hence $G = 0$, and
therefore $F = 0$, since $\theta^n(Z, (2)) \not\equiv 0$.

The proof of the theorem for $w = k$ will be divided into three stages. First, using
(4.6), we examine the case $K = \Gamma^1$ and $\chi = 1$; we then use induction on n to prove
the theorem for $K = \Gamma^n$ and $\chi = 1$; finally, we deduce the general case from the case
$K = \Gamma^n$ and $\chi = 1$.

Suppose that $K = \Gamma^1$ and $\chi = 1$. If the Fourier coefficients $f(2r)$ of a modular form $F \in \mathfrak{M}_k(\Gamma^1, 1)$ satisfy (4.7), then

$$v_{i\infty}(F) > \frac{k}{4\pi}\sigma_1 \geqslant \frac{k}{2\pi\sqrt{3}} \geqslant \frac{k}{12},$$

since $c_1 \geqslant 1$ and $\sigma_1 \geqslant 2/\sqrt{3}$. This contradicts (4.6). Hence, F must be identically zero.

We now prove the theorem for $K = \Gamma^n$ and $\chi = 1$ by induction on n. To be definite, we shall take the constant c_n in (4.7) to be the constant in Proposition 1.13. The case $n = 1$ has already been considered. Suppose that $n > 1$, and the theorem has already been proved for $K = \Gamma^{n-1}$ and $\chi = 1$. Let F be a modular form in $\mathfrak{M}_k(\Gamma^n, 1)$ that satisfies the conditions of the theorem. If we apply Proposition 3.12 with $K = \Gamma^n$ and $M = E_n$ to the function F, we conclude that the function

$$F' = F|\Phi = \sum_{R' \in \mathbf{A}_{n-1}} f\left(\begin{pmatrix} R' & 0 \\ 0 & 0 \end{pmatrix}\right)e\{R'Z'\} \quad (Z' \in \mathbf{H}_{n-1}),$$

where Φ is the Siegel operator (3.50), is contained in $\mathfrak{M}_k(\Gamma^{n-1}, 1)$. If $\sigma(R') \leqslant (k/2\pi)\sigma_{n-1}$, then

$$\sigma\left(\begin{pmatrix} R' & 0 \\ 0 & 0 \end{pmatrix}\right) = \sigma(R') \leqslant \frac{k}{2\pi}\sigma_n,$$

since $c_{n-1} \leqslant c_n$ by Proposition 1.13, and hence $\sigma_{n-1} \leqslant \sigma_n$. Then, using the assumption on F, we have

$$f'(R') = f\left(\begin{pmatrix} R' & 0 \\ 0 & 0 \end{pmatrix}\right) = 0.$$

By the induction assumption, F' is identically zero. This means that F is a cusp-form. Then, by Theorem 3.13, F has a Fourier expansion of the form

$$F = \sum_{R \in \mathbf{A}_n^+} f(R)e\{RZ\},$$

which, by Theorem 1.21 and the second inequality in Lemma 3.15 for the function

$$\sum_{R \in \mathbf{A}_n^+} |f(R)|\exp(-\pi\sigma(RY)) \geqslant |F|$$

implies the following bound for all $Z = X + iY$ in the fundamental domain D_n of Γ^n:

$$|F(Z)| \leqslant \delta_1\exp(-\delta_1'(\det Y)^{1/n}),$$

where δ_1 and δ_1' are positive constants. From this bound it follows that the function $G(Z) = (\det Y)^{k/2}|F(Z)|$ approaches zero as $\det Y \to +\infty$ and $Z = X + iY$ remains in D_n. On the other hand, from the definition of D_n and Theorem 1.21 it follows that any subset of D_n of the form $\{X + iY \in D_n; \det Y \leqslant \delta\}$ with $\delta > 0$ is closed and bounded, and hence compact. Thus, the function $G(Z)$ attains its maximum μ on D_n at some finite point $Z_0 = X_0 + iY_0 \in D_n$. Next, from Lemma 2.8 of Chapter 1 and the definition of a modular form it follows that for any $M = \begin{pmatrix} A & B \\ C & D \end{pmatrix} \in \Gamma^n$ the function G satisfies the relation

$$G(M\langle Z \rangle) = (|\det(CZ + D)|^{-2}\det Y)^{k/2}|\det(CZ + D)^k F(Z)|$$
$$= (\det Y)^{k/2}|F(Z)| = G(Z)$$

and so is constant on every Γ^n-orbit in $\mathbf{H}_n$. According to Theorem 1.20, the set D_n intersects with each Γ^n-orbit in $\mathbf{H}_n$; thus, the maximum μ of G on D_n is also its maximum on $\mathbf{H}_n$: $G(Z) \leqslant G(Z_0) = \mu$ for all $z \in \mathbf{H}_n$. We introduce the complex parameter $t = u + iv$, set $Z_t = Z_0 + tE_n$, and consider the function

$$g(t) = F(Z_t)\exp(-i\lambda\sigma(Z_1)),$$

where λ is determined from the condition $n\lambda/\pi = 1 + \left[\frac{k}{2\pi}\sigma_n\right]$ (here $[\cdots]$ denotes the greatest integer function). If we substitute the Fourier expansion for F, we obtain the expansion

$$\begin{aligned}
g(t) &= \sum_{R \in A_n^+} f(R)e\{RZ_t\}\exp(-i\lambda\sigma(Z_t)) \\
&= \sum_{R \in A_n^+} f(R)e\{RZ_0\}\exp(-i\lambda\sigma(Z_0))q^{\sigma(R)-\lambda n/\pi} = \tilde{g}(q),
\end{aligned}$$

where $q = \exp(\pi i t)$. The assumptions of the theorem imply that $f(R) = 0$ if $\sigma(R) - \lambda n/\pi < 0$. Hence, the series for $\tilde{g}(q)$ does not contain negative powers of q. If $\varepsilon > 0$ is small enough so that $Z_t \in \mathbf{H}_n$ for $v \geqslant -\varepsilon$, then the series for $g(t)$ converges absolutely and uniformly in the half-plane $v \geqslant -\varepsilon$. Then $\tilde{g}(q)$ is a holomorphic function in the disc $|q| \leqslant \exp(-\pi\varepsilon) = \rho$. Since $\rho > 1$, it follows from the maximum principle that there exists a point $q_0 = \exp(\pi i t_0)$ such that

$$|q_0| = \rho \quad \text{and} \quad |\tilde{g}(1)| \leqslant |\tilde{g}(q_0)|.$$

If we return to the variable t and recall the definition of g, we can rewrite the last inequality in the form

$$|F(Z_0)|\exp(\lambda\sigma(Y_0)) \leqslant |F(Z_{t_0})|\exp(\lambda\sigma(Y_0))\exp(\lambda n v_0),$$

where $t_0 = u_0 + iv_0$; hence,

$$(\det Y_0)^{-k/2}G(Z_0) \leqslant (\det Y_{t_0})^{-k/2}G(Z_{t_0})\exp(\lambda n v_0).$$

Since $G(Z_0) = \mu$ and $G(Z_{t_0}) \leqslant \mu$, this inequality implies that

$$(4.8) \qquad \mu \leqslant \mu(\det Y_0)^{k/2}(\det Y_{t_0})^{-k/2}\exp(\lambda n v_0) = \mu\varphi(v_0),$$

where $\varphi(v) = \det(E_n + vY_0^{-1})^{-k/2}\exp(\lambda n v)$. Clearly, $\varphi(0) = 1$. We show that φ has positive derivative φ' at the point $v = 0$. In fact,

$$\varphi(v) = \prod_{i=1}^{n}(1 + v\lambda_i)^{-k/2}\exp(\lambda n v),$$

where $\lambda_1, \ldots, \lambda_n$ are the eigenvalues of the matrix Y_0^{-1}. Hence,

$$\begin{aligned}
\varphi'(0) &= \lambda n - k(\lambda_1 + \cdots + \lambda_n)/2 = \lambda n - k\sigma(Y_0^{-1})/2 \\
&\geqslant \lambda n - k\sigma_n/2 = \pi(1 + [k\sigma_n/2\pi] - k\sigma_n/2\pi) > 0,
\end{aligned}$$

since $X_0 + iY_0 \in D_n$ and, by Theorem 1.21, $\sigma(Y_0^{-1}) \leqslant \sigma_n$. This implies that $\varphi(v_0) = \varphi(-\varepsilon) < 1$ if ε is sufficiently small. This, along with (4.8), proves that $\mu = 0$. Consequently, the function F is identically zero, and the theorem is proved in the case $K = \Gamma^n$ and $\chi = 1$.

Finally, we consider the general case. Suppose that $F \in \mathfrak{M}_k(K, \chi)$, $M \in \Gamma^n$, and $F|_k M$ is defined by (3.14). From (3.15) and (3.16) it follows that any function of the form $(F|_k M)^m$ depends only on the left coset KM of the group K. Thus, the function

$$(4.9) \qquad G(Z) = \prod_{\alpha=1}^{\mu} (F|_k M_\alpha)^m,$$

where $M_1, \ldots, M_\mu$ is a complete set of representatives of $K \backslash \Gamma^n$, does not depend on the choice of representatives. Since for any $M \in \Gamma^n$ the set $M_1 M, \ldots, M_\mu M$ is also a set of representatives of $K \backslash \Gamma^n$, if we again use (3.15) we find that $G|_{km\mu} M = G$ for $M \in \Gamma^n$ (compare with (3.73)).

On the other hand, G is obviously a holomorphic function on $\mathbf{H}_n$, and, by Theorem 3.5, it is bounded on $\mathbf{H}_n(\varepsilon)$ for any $\varepsilon > 0$. Thus, $G \in \mathfrak{M}_{km\mu}(\Gamma^n, 1)$. Letting $f_\alpha(R)$ denote the Fourier coefficients (3.33) of the function $F|_k M_\alpha$, we easily see that the Fourier coefficients $g(R)$ of G are given by the formula

$$g(R) = \sum_{\substack{R_{\alpha\beta} \in \mathbf{A}_n \\ \sum_{\alpha,\beta} R_{\alpha\beta} = qR}} \prod_{\alpha,\beta} f_\alpha(R_{\alpha\beta}),$$

where $1 \leqslant \alpha \leqslant \mu$, $1 \leqslant \beta \leqslant m$. We suppose that F satisfies (4.7), and we let $R \in \mathbf{A}_n$, $\sigma(R) \leqslant \frac{km\mu}{2\pi} \sigma_n$. Since any matrix in $\mathbf{A}_n$ has nonnegative trace, it follows from the last inequality that the following holds for any partition $qR = \sum_{\alpha,\beta} R_{\alpha\beta}$, where $R_{\alpha\beta} \in \mathbf{A}_n$, and for any $\alpha = 1, \ldots, \mu$:

$$\sigma\left(\sum_{\beta=1}^{m} R_{\alpha\beta}\right) = \sum_{\beta=1}^{m} \sigma(R_{\alpha\beta}) \leqslant \frac{km\mu}{2\pi} \sigma_n q,$$

and this implies that for any α there exists a β such that

$$\sigma(R_{\alpha\beta}) \leqslant \frac{k}{2\pi} \sigma_n \mu q.$$

To be definite, we suppose that $F|_k M_1 = F$, and hence $f_1 = f$. We then see that in the expression $g(R)$ every term contains a factor of the form $f_1(R_{1\beta}) = f(R_{1\beta})$, which is equal to zero, because F satisfies (4.7). Consequently, $g(R) = 0$; and, by what was proved above, G and so also F are identically zero. $\qquad \square$

PROBLEM 4.4. Prove that a function $F \in \mathfrak{M}_w(K, \chi)$, where K, χ, and w are as in Theorem 4.3, is identically zero if for every $M \in \Gamma^n$ the coefficients $f_\xi(R)$ in (3.33), where $\xi = M$ for $w = k$ and $\xi = \widehat{M} \in \mathfrak{G}$, $P(\widehat{M}) = M$ for $w = k/2$, are equal to zero for all $R \in \mathbf{A}_n$ satisfying the condition $\sigma(R) \leqslant w_1 \sigma_n q / 2\pi$.

3. Finite dimensionality of the spaces of modular forms.

THEOREM 4.5. *Suppose that K is a congruence subgroup of the modular group Γ^n, $K \supset \Gamma^n(q_1)$, χ is a character of K of order m, w is an integer or half-integer ($w = k$ or $w = k/2$), and $K \subset \Gamma_0^n(4)$ if $w = k/2$. Then the $\mathbf{C}$-dimension of the space of modular forms of weight w and character χ for K satisfies the following relations:*

$$\dim \mathfrak{M}_w(K, \chi) \leqslant d_n (w_1 \mu(K) q)^{\langle n \rangle}, \qquad \text{if } w > 0,$$

where $w_1 = k$ and $q = q_1 m$ if $w = k$, $w_1 = (k+1)/2$ and $q = 4q_1 m$ if $w = k/2$, d_n depends only on n, and $\mu(K)$ is the index of K in Γ^n;

$$\dim \mathfrak{M}_0(K, \chi) = \begin{cases} 1, & \text{if } \chi = 1, \\ 0, & \text{if } \chi \neq 1; \end{cases}$$

$$\dim {}_w(K, \chi) = 0, \quad \text{if } w < 0.$$

PROOF. The case $w > 0$. If we use the bound (3.13) for the number of matrices $R \in \mathbf{A}_n$ with $\sigma(R) \leqslant N$, we see that the number of different $R \in \mathbf{A}_n$ satisfying (4.7) is at most $d = d_n(w_1 \mu(K)q)^{\langle n \rangle}$, where d_n depends only on n. Then any $d + 1$ functions in $\mathfrak{M}_w(K, \chi)$ are linearly dependent, since one can always find complex numbers not all zero such that the corresponding linear combination of the functions satisfies (4.7), and so is equal to zero.

The case $w = k = 0$. In this case Theorem 4.3 shows that $F = 0$ if $f(0) = 0$. Since obviously $1 \in \mathfrak{M}_0 = \mathfrak{M}_0(K, 1)$, it follows that for any form $F \in \mathfrak{M}_0$ we have $F - f(0) \times 1 \in \mathfrak{M}_0$. Hence, $F - f(0) = 0$ and $F = f(0)$. If $F \in \mathfrak{M}_0(K, \chi)$ and $\chi(M) \neq 1$ for some $M \in K$, then from the functional equation for the function F and matrix M and the uniqueness of Fourier expansions it follows that $f(0) = 0$. Hence, $F = 0$.

The case $w = k < 0$. We first prove the theorem for $K = \Gamma^n$ and $\chi = 1$ by induction on n. If $n = 1$, the result follows from Proposition 4.1, since the left side of (4.4) is nonnegative. Now suppose that $n > 1$, and we have already shown that $\mathfrak{M}_k^{n-1} = \{0\}$. If $F \in \mathfrak{M}_k^n$, then $F|\Phi \in \mathfrak{M}_k^{n-1}$ by Proposition 3.12, and hence $F|\Phi = 0$ and F is a cusp-form. Let G be an arbitrary nonzero modular form of positive integer weight l and character χ_1 for some congruence subgroup K_1 of Γ^n (for example, a theta-series). Then obviously $F^l G^{|k|} \in \mathfrak{M}_0(K', \chi_1^{|k|})$, and the constant coefficient in the Fourier expansion of this function is equal to zero. By what was proved before, the function is zero, and hence $F = 0$.

The general case when $w = k < 0$ reduces to the case $K = \Gamma^n$ and $\chi = 1$ by the same method as in the proof of Theorem 4.3. That is, if $F \in \mathfrak{M}_k(K, \chi)$, then the function G in (4.9) belongs to $\mathfrak{M}_{km\mu(K)}$. Hence $G = 0$, and then $F = 0$.

If $w = k/2 < 0$, then, by Lemma 3.7, the function F^2 is a modular form of weight $2w = k < 0$, and the proof reduces to the previous case. $\square$

PROBLEM 4.6. Prove that
(1) $\dim \mathfrak{M}_k(\Gamma^2, 1) = 0$ for $k = 1, 2, 3$;
(2) $\dim \mathfrak{M}_4(\Gamma^2, 1) \leqslant 1$.

[**Hint:** Use Theorem 4.3 and Problem 1.16; verify that the space $\mathfrak{M}_k(\Gamma^2, 1)$ has no nonzero cusp-forms for $k = 1, 2, 3, 4$; and then use Problem 4.2.]

PROBLEM 4.7. Prove that $\dim \mathfrak{M}_k(\Gamma^n, 1) = 0$ if nk is odd.

§5. Scalar product and orthogonal decomposition

In the space of modular forms one can introduce an invariant scalar product that makes it possible, in particular, to find a natural complement to the space of cusp-forms.

1. The scalar product. Given any two functions F and G on $\mathbf{H}_n$, we consider the differential form

$$(5.1) \qquad \omega_w(F, G) = F(Z)\overline{G(Z)}h(Z)^w d^*Z,$$

where w is an integer or half-integer ($w = k$ or $w = k/2$), $h(Z) = \det Y$ is the height of $Z = X + iY$, and d^*Z is the invariant volume element in Proposition 2.9 of Chapter 1.

LEMMA 5.1. *For an arbitrary matrix* $M = \begin{pmatrix} A & B \\ C & D \end{pmatrix}$ *in* $S_{\mathbf{R}}^n$ *one has the transformation formulas*

$$(5.2) \qquad Y' = r(M) \cdot {}^t(C\overline{Z} + D)^{-1} Y (CZ + D)^{-1},$$

where $Z = X + iY \in \mathbf{H}_n$, $M\langle Z\rangle = X' + iY'$; *in particular,*

$$(5.3) \qquad h(M\langle Z\rangle) = \det Y' = r(M)^n |\det(CZ + D)|^{-2} h(Z);$$

$$(5.4) \qquad d^*M\langle Z\rangle = d^*Z;$$

$$(5.5) \qquad \omega_w(F, G)(M\langle Z\rangle) = r(M)^{nw} \omega_w(F|_w\xi, G|_w\xi)(Z),$$

where, as before, $\xi = M$ *for* $w = k$ *and* $\xi = \widehat{M} \in \mathfrak{G}$, $P(\widehat{M}) = M$ *for* $w = k/2$, *and* $|_w$ *is the operator* (3.15) *or* (3.21).

PROOF. The formulas (5.2) and (5.3) follow from Lemma 3.8 of Chapter 1, if we note that $M_1 = r(M)^{-1/2}M \in \mathrm{Sp}_n(\mathbf{R})$. Since $d^*\lambda Z = d^*Z$ for $\lambda > 0$, (5.4) follows from Proposition 2.9 of Chapter 1. The relation (5.5) is a direct consequence of (5.3), (5.4); moreover, from (5.1) and (3.17) it follows that the right side of (5.5) does not depend on the choice of $\widehat{M} \in \mathfrak{G}$. □

Now suppose that $F, G \in \mathfrak{M}_w(K, \chi)$, $M \in K$, and $\widehat{M}$ is as in (3.23). Then by (5.5) we have

$$\omega_w(F, G)(M\langle Z\rangle) = \omega_w(F|_w\xi, G|_w\xi)(Z)$$
$$= \omega_w(\chi, (M)F, \chi(M)G)(Z) = \omega_w(F, G)(Z),$$

i.e., the differential form (5.1) is invariant under the group K. This implies that the integral

$$(5.6) \qquad \int_{D_K} \omega_w(F, G)(Z),$$

where D_K is a fundamental domain for K in $\mathbf{H}_n$, does not depend on the choice of D_K, provided that the integral is absolutely convergent.

LEMMA 5.2. *If at least one of the two forms* $F, G \in \mathfrak{M}_w(K, \chi)$ *is a cusp-form, then the integral* (5.6) *is absolutely convergent.*

PROOF. Since D_K is a finite union of sets of the form $M\langle D_n\rangle$, where $M \in \Gamma^n$ and D_n is the fundamental domain for Γ^n described in Theorem 1.20, to prove the lemma it suffices to verify absolute convergence of an integral of the form

$$\int_{M\langle D_n\rangle} \omega_w(F, G)(Z) = \int_{D_n} \omega_w(F|_w\xi, G|_w\xi)(Z),$$

where ξ has the same meaning as in Lemma 5.1. To be definite, suppose that F is a cusp-form. According to Theorems 3.13 and 3.5, we have the Fourier expansions

$$F|_w\xi = \sum_{R \in \mathbf{A}_n^+} f_\xi(R)e\{q^{-1}RZ\},$$

$$G|_w\xi = \sum_{R \in \mathbf{A}_n} g_\xi(R)e\{q^{-1}RZ\}.$$

Since both expansions converge absolutely on $\mathbf{H}_n$, it follows that

$$|(F|_w\xi)(Z)\overline{(G|_w\xi)}(Z)| \leqslant \sum_{R \in \mathbf{A}_n^+} t(R)\exp(-\pi q^{-1}\sigma(RY)),$$

where

$$t(R) = \sum_{R_1+R_2=R} |f_\xi(R_1)g_\xi(R_2)|$$

is a finite sum and the last series converges on $\mathbf{H}_n$. Then, by Theorem 1.21 and the first inequality in Lemma 3.15, we obtain the inequality

$$|(F|_w\xi)(Z)\overline{(G|_w\xi)}(Z)| \leqslant \delta_1 \exp(-\delta_2\sigma(Y)) \quad (Z = X + iY \in D_n),$$

where δ_1 and δ_2 are positive constants. Hence, the lemma follows if we verify convergence of the integrals

$$\int_{D_n} \exp(-\delta\sigma(Y))(\det Y)^w d^*Z$$

$$= \int_{D_n} \exp(-\delta\sigma(Y))(\det Y)^{w-n-1} \prod_{1\leqslant\alpha\leqslant\beta\leqslant n} dx_{\alpha\beta}\, dy_{\alpha\beta},$$

where $\delta > 0$. If $(x_{\alpha\beta}) + i(y_{\alpha\beta}) \in D_n$, then the definition of D_n and the inequalities (1.12) imply that $|x_{\alpha\beta}| \leqslant 1/2$, $|y_{\alpha\beta}| \leqslant y_{\alpha\alpha}/2$ $(\alpha \neq \beta)$. In addition, at the beginning of the proof of Theorem 1.21 it was shown that in this case $y_{\alpha\alpha} \geqslant \sqrt{3}/2$. Thus, applying the inequality (1.13) if $w < n + 1$ and the inequality (1.8) of Appendix 1 if $w \geqslant n + 1$, we see that the last integral is majorized by an integral of the form

$$\int_{\substack{|x_{\alpha\beta}|\leqslant 1/2 \\ y_{\alpha\alpha}\sqrt{3}/2, |y_{\alpha\beta}|\leqslant y_{\alpha\alpha}/2}} \prod_{\alpha=1}^{n} y_{\alpha\alpha}^{w-n-1}\exp(-\delta y_{\alpha\alpha}) \prod_{1\leqslant\alpha\leqslant\beta\leqslant n} dx_{\alpha\beta}\, dy_{\alpha\beta}$$

$$= c \prod_{\alpha=1}^{n} \int_{\sqrt{3}/2}^{\infty} y_{\alpha\alpha}^{w-n-1+n-\alpha}\exp(-\delta y_{\alpha\alpha})\, dy_{\alpha\alpha} < \infty. \qquad \square$$

We are now ready to prove the following fact.

THEOREM 5.3. *Suppose that K is a congruence subgroup of Γ^n, χ is a finite character of K, w is an integer or half-integer ($w = k$ or $w = k/2$), and $K \subset \Gamma_0^n(4)$ if $w = k/2$. If at least one of the modular forms $F, G \in \mathfrak{M}_w(K, \chi)$ is a cusp-form, then define their scalar product by the formula*

$$(5.7) \qquad (F, G) = \mu(K')^{-1} \int\limits_{D_K} \omega_w(F, G)(Z),$$

where $K' = K \cup (-E_{2n})K$, $\mu(K') = [\Gamma^n : K']$, and D_K is a fundamental domain for K in $\mathbf{H}_n$. This scalar product then has the following properties:

(1) (F, G) converges absolutely and does not depend on the choice of fundamental domain D_K;

(2) (F, G) does not depend on the choice of group K such that $F, G \in \mathfrak{M}_w(K, \chi)$;

(3) (F, G) is a positive definite nondegenerate hermitian scalar product;

(4) if $M \in S_\mathbf{Q}^n$, then

$$(5.8) \qquad (F|_w \xi, G|_w \xi) = r(M)^{-nw}(F, G),$$

where the functions $F|_w \xi$ and $G|_w \xi$ are regarded as elements of $\mathfrak{M}_w(K_M, \chi')$ (see Proposition 3.8 and Theorem 3.13(3)).

PROOF. Property (1) has already been proved, and (3) follows immediately from the definitions.

We prove (2). If $F, G \in \mathfrak{M}_w(K_1, \chi_1)$, then, replacing K_1 by $K_1 \cap K$, we may assume that $K_1 \subset K$. Let

$$\Gamma_n = \bigcup_\alpha K' M_\alpha, \qquad K' = \bigcup_\beta K_1 N_\beta,$$

where $k_1' = k_1 \cup (-e_{2n})k_1$, be partitions into left cosets. Then

$$\Gamma^n = \bigcup_{\alpha, \beta} K_1' N_\beta M_\alpha$$

is also a partition into disjoint left cosets. By Theorem 1.22, we can take $D_{K_1} = \bigcup_{\alpha, \beta} N_\beta M_\alpha \langle D_n \rangle$; hence,

$$\mu(K_1')^{-1} \int\limits_{D_{K_1}} \omega_w(F, G)(Z) = \mu(K_1')^{-1} \sum_{\alpha, \beta} \int\limits_{N_\beta \langle M_\alpha \langle D_n \rangle \rangle} \omega_w(F, G)(Z)$$

$$= \frac{[K' : K_1']}{[\Gamma^n : K_1']} \sum \int\limits_{M_\alpha \langle D_n \rangle} \omega_w(F, G)(Z)$$

$$= \mu(K')^{-1} \int\limits_{D_K} \omega_w(F, G)(Z),$$

where for the second equality we used the invariance of the differential form ω_w under the group K'. This proves property (2).

We now prove (5.8). Since K and K_M are both congruence subgroups of Γ^n, their intersection $K_{(M)} = K \cap K_M$ is also a congruence subgroup, and so has finite index in

Γ^n. Then if we use (5.5) and property (2), we obtain

$$(F|_w\xi, G|_w\xi) = \mu(K'_{(M)})^{-1} \int_D \omega_w(F|_w\xi, G|_w\xi)(Z)$$

$$= r(M)^{-nw}\mu(K'_{(M)})^{-1} \int_D \omega_w(F, G)(M, \langle Z \rangle)$$

$$= r(M)^{-nw}\mu(K'_{(M)})^{-1} \int_{M\langle D \rangle} \omega_w(F, G)(Z),$$

where $D = D_{K_{(M)}}$. It is easy to see that the set $M\langle D \rangle$ is a fundamental domain for the group $MK_{(M)}M^{-1} = MKM^{-1} \cap K = K_{(M^{-1})}$. Thus, again using property (2), we can rewrite the last expression in the form

$$r(M)^{-nw}\mu(K'_{(M)})^{-1}\mu(K'_{(M^{-1})})(F, G),$$

and it remains for us to verify that $\mu(K'_{(M)}) = \mu(K'_{(M^{-1})})$. Since we obviously have

$$[\Gamma_{(M^{-1})} : K'_{(M^{-1})}] = [M\Gamma_{(M)}M^{-1} : MK'_{(M)}M^{-1}] = [\Gamma_{(M)} : K'_{(M)}],$$

where $\Gamma = \Gamma^n$, we can limit ourselves to the case $K = \Gamma$. For future reference we shall prove a more general fact. $\square$

LEMMA 5.4. *Let G be a congruence subgroup of Γ^n. Then for every matrix $M \in S^n_{\mathbf{Q}}$ the group $G_{(M)} = G \cap M^{-1}GM$ is a congruence subgroup of Γ^n, and one has*

$$(5.9) \qquad\qquad\qquad [G : G_{(M)}] = [G : G_{(M^{-1})}].$$

PROOF. The first part follows from the relation $G_{(M)} = G \cap G_M$, where G_M is defined as in (3.25). Now let D be a fundamental domain for $G_{(M)}$ in $\mathbf{H}_n$. Since $G_{(M^{-1})} = MG_{(M)}M^{-1}$, it follows that $M\langle D \rangle$ is a fundamental domain for $G_{(M^{-1})}$. Then from Proposition 1.23(2) we have the relations

$$v(D) = [G : G_{(M)}]v(D_G), \qquad v(M\langle D \rangle) = [G : G_{(M^{-1})}]v(D_G),$$

where D_G is a fundamental domain for G. On the other hand, by (5.4) we have $v(D) = v(M\langle D \rangle)$. But this, along with the above relations, implies (5.9). $\square$

2. The orthogonal complement. We now define the subspace $\mathfrak{E}_w(K, \chi)$ of the space $\mathfrak{M}_w(K, \chi)$ of modular forms of integer or half-integer weight w ($w = k$ or $w = k/2$) and character χ for the congruence subgroup K of Γ^n, where $K \subset \Gamma^n_0(4)$ if $w = k/2$. This subspace is the set of all forms that are orthogonal to the subspace of cusp-forms with respect to the scalar product (5.7):

$$(5.10) \qquad \mathfrak{E}_w(K, \chi) = \{F \in \mathfrak{M}_w(K, \chi); (F, G) = 0 \text{ for all } G \in \mathfrak{N}_w(K, \chi)\}.$$

PROPOSITION 5.5. *The space of all modular forms splits into the direct sum*

$$\mathfrak{M}_w(K, \chi) = \mathfrak{E}_w(K, \chi) \oplus \mathfrak{N}_w(K, \chi) \tag{5.11}$$

of orthogonal subspaces with respect to the scalar product (5.7). *In addition, for any matrix* $M \in S_{\mathbf{Q}}^n$ *the map* (*see Proposition* 3.8)

$$\mathfrak{M}_w(K, \chi) \xrightarrow{\ |_w\xi\ } \mathfrak{M}_w(K_M, \chi'): F \to F|_w\xi$$

is an isomorphic imbedding; here cusp-forms go to cusp-forms, and if $F, G \in \mathfrak{M}_w(K, \chi)$ *with at least one of these forms a cusp-form, then*

$$(F, G) = r(M)^{nw}(F|_w\xi, G|_w\xi).$$

PROOF. The decomposition (5.11) follows from Theorem 5.3(3) and standard linear algebra. Next, since $(F|_w\xi)|_w\xi^{-1} = F$ by (3.15) and (3.22), it follows from Proposition 3.8 that the map $|_w\xi$ is an isomorphic imbedding. The remaining claims in the theorem follow from Theorem 3.13(3) and (5.8). □

From (5.11) it follows that any modular form $F \in \mathfrak{M}_w(K, \chi)$ can be uniquely represented in the form

$$F = F_1 + F_2, \quad \text{where } F_1 \in \mathfrak{E}_w(K, \chi), F_2 \in \mathfrak{N}_w(K, \chi).$$

Equating Fourier coefficients, we obtain

$$f(R) = f_1(R) + f_2(R) \quad (R \in \mathbf{A}_n), \tag{5.12}$$

where $f(R)$, $f_1(R)$, and $f_2(R)$ are the Fourier coefficients of the functions F, F_1, and F_2, respectively. The Fourier coefficients $f_1(R)$ of $F_1 \in \mathfrak{E}_w(K, \chi)$ can sometimes be computed in explicit form. On the other hand, the Fourier coefficients $f_2(R)$ of the cusp-form F_2 are relatively small, by (3.70). Starting from these considerations, in many cases one can prove that as $\det R \to +\infty$ the decomposition (5.12) gives an asymptotic formula for the function $f(R)$ with principal term $f_1(R)$.

CHAPTER 3

Hecke Rings

One of the most fruitful ideas in the theory of modular forms—the notion of a Hecke operator—is based on a procedure for taking the average of a function over suitable double cosets of subgroups of the modular group. Chapter 4 is devoted to the theory and application of Hecke operators. The properties of Hecke operators are to a large extent a reflection of the connections that exist between the corresponding double cosets. The present chapter examines these connections.

§1. Abstract Hecke rings

1. Averaging over double cosets. As in §4.1 of Chapter 1, let S be a multiplicative semigroup that acts on a set $H: h \to g(h)$ $(h \in H, g \in S)$ as a subsemigroup of the group of all 1-to-1 maps from H onto itself. Let φ be an automorphy factor of the semigroup S on H with values in a group T, and let V be a left T-module. A function $F: H \to V$ is called an *automorphic form* of weight φ for a subgroup $\Gamma \subset S$ if for any $\gamma \in \Gamma$ it satisfies the functional equation

$$(1.1) \qquad (F|_\varphi \gamma)(h) = \varphi(\gamma, h)^{-1} F(\gamma(h)) = F(h).$$

It is clear that the set of such functions forms an abelian group, which we shall denote $\mathfrak{M} = \mathfrak{M}_\varphi(\Gamma)$.

If $F \in \mathfrak{M}$ and $g \in S$, then the function $F'(h) = (F|_\varphi g)(h)$ does not generally lie in $\mathfrak{M}$. Moreover, there might be infinitely many pairwise distinct functions of the form

$$(1.2) \qquad F'|_\varphi \gamma = F|_\varphi g|_\varphi \gamma = F|_\varphi g\gamma \quad (\gamma \in \Gamma).$$

However, it often turns out that there are only finitely many distinct functions among the functions (1.2). In that case, if we sum these functions, we might again obtain a function in $\mathfrak{M}$. A typical situation of this sort occurs if the double coset $\Gamma g\Gamma$ contains only finitely many left cosets modulo Γ:

$$(1.3) \qquad |\Gamma \setminus \Gamma g\Gamma| < \infty.$$

Namely, each product $g\gamma$ $(\gamma \in \Gamma)$ is contained in the double coset $\Gamma g\Gamma$; if $g\gamma$ lies in a fixed left coset $\Gamma g' \subset \Gamma g\Gamma$, then $g\gamma = \gamma'g'$, where $\gamma' \in \Gamma$, and, by Lemma 4.1(3) of Chapter 1 and the equality (1.1) above, we find that

$$F'|_\varphi \gamma = F|_\varphi g\gamma = F|_\varphi \gamma'|_\varphi g' = F|_\varphi g',$$

so that any function (1.2) is equal to one of the functions $F|_\varphi g_1, \ldots, F_\varphi g_\mu$, where $g_1, \ldots, g_\mu$ are a complete set of representatives of the left cosets modulo Γ contained in $\Gamma g\Gamma$.

We consider the following average of the function F over the double coset $\Gamma g \Gamma$ (or, if we want, the average of the function F' over the group Γ):

$$(1.4) \qquad F|(g) = F|_\varphi(g) = \sum_{g_i \in \Gamma \setminus \Gamma g \Gamma} F|_\varphi g_i.$$

LEMMA 1.1. *Suppose that* $F \in \mathfrak{M} = \mathfrak{M}_\varphi(\Gamma)$, *and the double coset* $\Gamma g \Gamma$, *where* $g \in S$, *satisfies the condition* (1.3). *Then the function* $F|(g)$ *does not depend on the choice of representatives* $g_1, \ldots, g_\mu$ *of* $\Gamma \setminus \Gamma g \Gamma$, *and it is an automorphic form of weight* φ *for* Γ.

PROOF. If $g_i' = \gamma_i g_i$ $(i = 1, \ldots, \mu)$, where $\gamma_i \in \Gamma$, form another set of representatives, then, by Lemma 4.1(3) of Chapter 1 and the definition of an automorphic form, we have

$$\sum_i F|_\varphi \gamma_i g_i = \sum_i F|_\varphi \gamma_i|_\varphi g_i = \sum_i F|_\varphi g_i.$$

Let $\gamma \in \Gamma$. Since the set $g_1 \gamma, \ldots, g_\mu \gamma$ is also obviously a set of representatives of $\Gamma \setminus \Gamma g \Gamma$, it follows that

$$(F|_\varphi(g))|_\varphi \gamma = \sum_i F|_\varphi g_i \gamma = F|_\varphi(g),$$

so that $F|_\varphi(g) \in \mathfrak{M}$. $\square$

The operators

$$(1.5) \qquad |_\varphi(g) \colon \mathfrak{M}_\varphi(\Gamma) \to \mathfrak{M}_\varphi(\Gamma)$$

are called *Hecke operators on the space* $\mathfrak{M}_\varphi(\Gamma)$.

2. Hecke rings. In order to study the connections between the Hecke operators corresponding to different double cosets, we first examine the connections between the double cosets themselves, where we suppose that the double cosets satisfy (1.3).

LEMMA 1.2. *Suppose that* G *is an arbitrary group,* Γ *is a subgroup of* G, $g \in G$, *and*

$$(1.6) \qquad \Gamma = \bigcup_{\gamma_i \in \Gamma_{(g)} \setminus \Gamma} \Gamma_{(g)} \gamma_i, \qquad \text{where } \Gamma_{(g)} = \Gamma \cap g^{-1} \Gamma g,$$

is the partition of Γ *into a disjoint union of left cosets of the subgroup* $\Gamma_{(g)}$. *Then*

$$(1.7) \qquad \Gamma g \Gamma = \bigcup_{\gamma_i \in \Gamma_{(g)} \setminus \Gamma} \Gamma g \gamma_i$$

and the left cosets in this union are pairwise disjoint. In particular,

$$(1.8) \qquad \mu(g) = \mu_\Gamma(g) = |\Gamma \setminus \Gamma g \Gamma| = [\Gamma : \Gamma_{(g)}].$$

PROOF. The right hand side of (1.7) is clearly contained in the left hand side. Conversely, suppose that $g' = \gamma g \delta$, where $\gamma, \delta \in \Gamma$. By (1.6), the element δ lies in one of the left cosets $\Gamma_{(g)} \gamma_i$, i.e., $\delta = \alpha \gamma_i$, where $\alpha \in \Gamma$ and $g \alpha g^{-1} \in \Gamma$. Then $g' = \gamma g \alpha g^{-1} g \gamma_i \in \Gamma g \gamma_i$. If $\Gamma g \gamma_i$ and $\Gamma g \gamma_j$ intersect, then for some $\gamma, \delta \in \Gamma$ we have the equality $\gamma g \gamma_i = \delta g \gamma_j$, and hence $g^{-1} \delta^{-1} \gamma g \gamma_i = \gamma_j$ and $\Gamma_{(g)} \gamma_i = \Gamma_{(g)} \gamma_j$. $\square$

Thus, if g is an invertible element, the condition (1.3) holds if and only if $\Gamma \cap g^{-1}\Gamma g$ has finite index in Γ.

Two subgroups Γ_1 and Γ_2 of a group G are said to be *commensurable* if their intersection $\Gamma_1 \cap \Gamma_2$ has finite index both in Γ_1 and in Γ_2; in this case we write $\Gamma_1 \sim \Gamma_2$.

LEMMA 1.3. *The commensurability relation is transitive on the set of subgroups of a group G.*

PROOF. Suppose that $\Gamma_1 \sim \Gamma_2$ and $\Gamma_2 \sim \Gamma_3$. If we take left cosets modulo $\Gamma_2 \cap \Gamma_3$, the imbedding $\Gamma_1 \cap \Gamma_2 \subset \Gamma_2$ gives the imbedding

$$\Gamma_1 \cap \Gamma_2 \cap \Gamma_3 \setminus \Gamma_1 \cap \Gamma_2 \subset \Gamma_2 \cap \Gamma_3 \setminus \Gamma_2,$$

so that

$$[\Gamma_1 \cap \Gamma_2 : \Gamma_1 \cap \Gamma_2 \cap \Gamma_3] \leqslant [\Gamma_2 : \Gamma_2 \cap \Gamma_3] < \infty,$$

and

$$[\Gamma_1 : \Gamma_1 \cap \Gamma_3] \leqslant [\Gamma_1 : \Gamma_1 \cap \Gamma_2 \cap \Gamma_3] = [\Gamma_1 : \Gamma_1 \cap \Gamma_2][\Gamma_1 \cap \Gamma_2 : \Gamma_1 \cap \Gamma_2 \cap \Gamma_3] < \infty.$$

Similarly, $[\Gamma_2 \cap \Gamma_3 : \Gamma_1 \cap \Gamma_2 \cap \Gamma_3] < \infty$ and $[\Gamma_3 : \Gamma_1 \cap \Gamma_3] < \infty$. Thus, $\Gamma_1 \sim \Gamma_3$. □

LEMMA 1.4. *Let G be a group, and let Γ be a subgroup. Then the set*

$$\widetilde{\Gamma} = \{g \in G; g^{-1}\Gamma g \sim \Gamma\}$$

is a group.

PROOF. If Γ_1 and Γ_2 are two commensurable subgroups of G and $g \in G$, then clearly the subgroups $g^{-1}\Gamma_1 g$ and $g^{-1}\Gamma_2 g$ are also commensurable. Thus, if $g \in \widetilde{\Gamma}$, then $\Gamma \sim g^{-1}\Gamma g$, and hence $g\Gamma g^{-1} \sim g(g^{-1}\Gamma g)g^{-1} = \Gamma$, and $g^{-1} \in \widetilde{\Gamma}$. Next, if $g_1, g_2 \in \widetilde{\Gamma}$, then $g_1^{-1}\Gamma g_1 \sim \Gamma$, and hence $g_2^{-1}g_1^{-1}\Gamma g_1 g_2 \sim g_2^{-1}\Gamma g_2 \sim \Gamma$; then, by Lemma 1.3, $g_1 g_2 \in \widetilde{\Gamma}$. □

The group $\widetilde{\Gamma}$ is called the *commensurator of the subgroup Γ in G,* and its elements are called Γ-*rational elements of G.*

Let Γ be a subgroup of G, and let S be a multiplicatively closed subset of G. We call (Γ, S) a *Hecke pair* if

$$(1.9) \qquad\qquad \Gamma \subset S \subset \widetilde{\Gamma},$$

where $\widetilde{\Gamma}$ is the commensurator of Γ in G. To each Hecke pair (Γ, S) we associate the free **Z**-module $L = L(\Gamma, S)$ whose generators over **Z** are the symbols (Γg) $(g \in S)$, one for each left coset Γg. The elements of S act as linear transformations of the module L according to the rule

$$S \ni g : t = \sum_i a_i(\Gamma g_i) \to tg = \sum_i a_i(\Gamma g_i g).$$

We consider the submodule $D = D(\Gamma, S)$ of L that consists of all Γ-invariant elements:

$$D(\Gamma, S) = \{t \in L(\Gamma, S); t\gamma = t \text{ for all } \gamma \in \Gamma\}.$$

If $t = \sum_i a_i(\Gamma g_i)$ and $t' = \sum_j b_j(\Gamma h_j)$ are two elements of D, then the element

$$(1.10) \qquad\qquad t \cdot t' = \sum_{i,j} a_i b_j(\Gamma g_i h_j) \in L$$

does not depend on the choice of left coset representatives, and it also belongs to the module D. Namely, $t \cdot t'$ obviously does not depend on the choice of representatives g_i. Let $\gamma_j h_j$, where $\gamma_j \in \Gamma$, be different representatives of the left cosets Γh_j. Since, by assumption, $t\gamma_j = t$ for each j, it follows that

$$\sum_{i,j} a_i b_j (\Gamma g_i \gamma_j h_j) = \sum_j b_j (t\gamma_j) h_j = t \cdot t',$$

and the element (1.10) does not depend on the choice of representatives h_j. Finally, if $\gamma \in \Gamma$, then $(t \cdot t')\gamma = t(t'\gamma) = t \cdot t'$, so that $t \cdot t' \in D$. Since the multiplication map $(t, t') \to t \cdot t'$ on elements of D is obviously bilinear and associative, it follows that D becomes an associative ring, called the *Hecke ring of the pair* (Γ, S).

If (Γ, S) is a Hecke pair, then, by Lemma 1.2, the double coset $\Gamma g \Gamma$ of any $g \in S$ is a finite union of disjoint left cosets of Γ:

$$\Gamma g \Gamma = \bigcup_{i=1}^{\mu} \Gamma g_1.$$

If $\gamma \in \Gamma$, then the set $g_1\gamma, \ldots, g_\mu\gamma$ obviously is also a full set of representatives of the distinct left cosets $\Gamma \setminus \Gamma g \Gamma$. Thus, the elements

$$(1.11) \qquad\qquad (g) = (g)_\Gamma = \sum_{g_i \in \Gamma \setminus \Gamma g \Gamma} (\Gamma g_i)$$

of the module $L(\Gamma, S)$ satisfy the condition $(g)\gamma = (g)$ for $\gamma \in \Gamma$, and hence belong to the Hecke ring $D = D(\Gamma, S)$. From the definition of multiplication in D it follows that the element

$$(e) = (e)_\Gamma = (\Gamma),$$

where e is the identity of the group G, is the unit of the ring D.

LEMMA 1.5. *The elements* (1.11) *corresponding to the distinct* Γ-*double cosets of* S *form a* **Z**-*basis of the module* $D(\Gamma, S)$. *The product of elements of the form* (1.1) *in the ring* D *can be calculated from the following formulas.*

Let $g, g' \in S$ *and let*

$$\Gamma g \Gamma = \bigcup_{i=1}^{\mu(g)} \Gamma g_i, \quad \Gamma g' \Gamma = \bigcup_{j=1}^{\mu(g')} \Gamma g_i'$$

be the decomposition of the double cosets into distinct left cosets. Then

$$(g)(g') = \sum_{\Gamma h \Gamma \in \Gamma g \Gamma g' \Gamma} c(g, g'; h)(h),$$

where h *runs through a set of representatives of the* Γ-*double cosets contained in the set* $\Gamma g \Gamma g' \Gamma$, *and for each* h *the coefficient* $c(g, g'; h)$ *is equal to the number of pairs* (g_i, g_j') *such that* $g_i g_j' \in \Gamma h$. *The coefficients* $c(g, g'; h)$ *can also be expressed in the form*

$$c(g, g'; h) = \nu(g, g'; h)\mu(g')\mu(h)^{-1},$$

where $\nu(g, g'; h)$ *is the number of elements* g_i *such that* $g_i g' \in \Gamma h \Gamma$, *and* $\mu(g')$, $\mu(h)$ *are the indices* (1.8).

PROOF. Let D' denote the submodule of D consisting of all finite linear combinations of elements of the form (1.11) with coefficients in $\mathbf{Z}$. Any nonzero element $t \in D$ can be written in the form

$$(1.12) \qquad t = \sum_{i=1}^{\mu} a_i (\Gamma g_i),$$

where all of the coefficients a_i are nonzero and all of the left cosets Γg_i are pairwise distinct. We then call $\mu = \mu(t)$ the *length* of t, and we prove that $t \in D'$ by induction on μ. If $\mu = 1$, then $t = a(\Gamma g)$. Since $t \in D$, it follows that $t\gamma = t$ for all $\gamma \in \Gamma$, i.e., $a(\Gamma g\gamma) = a(\Gamma g)$ for $\gamma \in \Gamma$, and hence $\Gamma g = \Gamma g \Gamma$ and $t = a(g) \in D'$. Now suppose that $\mu > 1$, and we have already verified that all elements of D of length less than μ are contained in D'. Let t be an element of D of the form (1.12) that has length μ. Since $t\gamma = t$ for $\gamma \in \Gamma$, it follows that, if the left coset (Γg_i) appears in (1.12), then all of the left cosets $(\Gamma g_i \gamma)$ for $\gamma \in \Gamma$ appear in (1.12) with the same coefficient. By Lemma 1.2, every left coset in the double coset $\Gamma g_i \Gamma$ can be written in the form $\Gamma g_i \gamma$ for some $\gamma \in \Gamma$. Thus, all left cosets in the decomposition of the double coset $\Gamma g_i \Gamma$ appear in (1.12) with coefficient a_i. Hence, the length of the element $t - a_i(g_i)$ is less than μ. By the induction assumption, $t - a_i(g_i) \in D'$, and so $t \in D'$. The first part of the lemma is proved.

By definition, we have $(g) = \sum_i (\Gamma g_i)$, $(g') = \sum_j (\Gamma g'_j)$, and

$$(g)(g') = \sum_{i,j} (\Gamma g_i g'_j).$$

Since all of the products $g_i g'_j$ obviously lie in the set $\Gamma g \Gamma g' \Gamma$, it follows from what was just proved that the product $(g)(g')$ can also be written in the form

$$\sum_{\Gamma h \Gamma \in \Gamma g \Gamma g' \Gamma} c(g, g'; h)(h) = \sum_{\Gamma h \Gamma \in \Gamma g \Gamma g' \Gamma} c(g, g'; h) \sum_{h_k \in \Gamma \backslash \Gamma h \Gamma} (\Gamma h_k)$$

with certain coefficients $c(g, g'; h)$. If we equate coefficients of (Γh) in these two left coset decompositions of the product $(g)(g')$, we find that $c(g, g'; h)$ is equal to the number of pairs (g_i, g'_j) such that $\Gamma g_i g'_j = \Gamma h$, i.e., $g_i g'_j \in \Gamma h$. From what we proved before it follows that $c(g, g'; h)$ depends only on the double cosets of g, g', and h. If we sum the numbers $c(g, g'; h)$ over all left cosets in $\Gamma h \Gamma$, we find that $\mu(h) c(g, g'; h)$ is equal to the number of pairs (g_i, g'_j) such that $g_i g'_j \in \Gamma h \Gamma$. Taking the set of $g' \gamma_j$ with $\gamma_j \in \Gamma$ (see Lemma 1.2) as our set of representatives g'_j, we see that the last number is equal to the product of $\mu(g')$ with the number of elements g_i for which $g_i g' \in \Gamma h \Gamma$. $\square$

Let G and $\widehat{G}$ be multiplicative groups, let Γ be a subgroup of G, and let

$$(1.13) \qquad P: \widehat{G} \to G, \quad \delta: \Gamma \to \widehat{G}$$

be a group epimorphism and a group monomorphism that satisfy the conditions: $P(\delta(\gamma)) = \gamma$ for any $\gamma \in \Gamma$, and $\operatorname{Ker} P$ is contained in the center of $\widehat{G}$. Note that we already encountered these conditions in Chapter 2 when we studied modular forms of half-integer weight. For any Hecke pair (Γ, S) for G we define a new pair

$$(1.14) \qquad (\widehat{\Gamma}, \widehat{S}), \quad \text{where } \widehat{\Gamma} = \delta(\Gamma) \text{ and } \widehat{S} = P^{-1}(S),$$

and we examine the conditions under which $(\widehat{\Gamma}, \widehat{S})$ is a Hecke pair for $\widehat{G}$.

To do this, we consider the map

$$(1.15) \qquad \rho = \rho_g : \Gamma_{(g)} = \Gamma \cap g^{-1}\Gamma g \to \widehat{G},$$

where g is an arbitrary element of G, that is defined for $\gamma \in \Gamma_{(g)}$ by the equality

$$(1.16) \qquad \widehat{(g\gamma g^{-1})} = \xi \widehat{\gamma} \xi^{-1} \rho(\gamma),$$

where $\widehat{\alpha} = \delta(\alpha)$ for $\alpha \in \Gamma$ and $\xi \in \widehat{G}$ is any P-preimage of g. Since Ker P is contained in the center of $\widehat{G}$, it is clear that $\rho(\gamma)$ does not depend on the choice of ξ; moreover, $\rho(\gamma)$ belongs to the center of $\widehat{G}$.

LEMMA 1.6. *The map* ρ_g $(g \in G)$ *is a homomorphism, and for any* $\gamma_1, \gamma_2 \in \Gamma$ *it satisfies the relation*

$$(1.17) \qquad \rho_g(\gamma) = \rho_{\gamma_1 g \gamma_2}(\gamma_2^{-1}\gamma\gamma_2) \quad \text{for } \gamma \in \Gamma_{(g)}.$$

PROOF. The proof of the first part of the lemma is similar to the proof of Lemma 3.4 of Chapter 2. As for (1.17), we first note that the right side is in fact well defined, because, by (1.6), $\gamma' = \gamma_2^{-1}\gamma\gamma_2 \in \Gamma_{(g')}$ for $g' = \gamma_1 g \gamma_2$ and $\gamma \in \Gamma_{(g)}$. We now use the definition of the map $\rho_{g'}$ and choose $\widehat{g}' \in \widehat{G}$ to be the product $\widehat{\gamma}_1 \widehat{g} \widehat{\gamma}_2$. We then have

$$g'\widehat{\gamma'(g')}^{-1} = \widehat{g}'\widehat{\gamma}'(\widehat{g}')^{-1}\rho_{g'}(\gamma') = \widehat{\gamma}_1(\widehat{g\gamma g}^{-1}\rho_{g'}(\gamma'))\widehat{\gamma}_1^{-1},$$

since $\widehat{\gamma}' = \widehat{\gamma}_2^{-1}\widehat{\gamma\gamma}_2$, and the element $\rho_{g'}(\gamma')$ is in the center of the group $\widehat{G}$. On the other hand, using the definition of the map ρ_g, we have

$$g'\widehat{\gamma'(g')}^{-1} = \widehat{\gamma}_1(\widehat{g\gamma g}^{-1})\widehat{\gamma}_1^{-1} = \widehat{\gamma}_1(\widehat{g\gamma g}^{-1}\rho_g(\gamma))\widehat{\gamma}_1^{-1},$$

which implies that $\rho_g(\gamma) = \rho_{g'}(\gamma')$. $\qquad\qquad\square$

We shall call ρ the *lifting homomorphism from G to $\widehat{G}$*. Using this homomorphism, we can easily formulate a condition under which $(\widehat{\Gamma}, \widehat{S})$ is a Hecke pair.

LEMMA 1.7. *Let* (Γ, S) *be a Hecke pair for the group* G, *which is related to the group* $\widehat{G}$ *by the relations* (1.13). *If the kernel of the homomorphism* $\rho = \rho_g$ *has finite index in* Γ *for every* $g \in S$, *then* $(\widehat{\Gamma}, \widehat{S})$ *is a Hecke pair.*

PROOF. We first prove the equality

$$(1.18) \qquad \delta(\text{Ker } \rho) = \widehat{\Gamma}_{(\xi)} = \widehat{\Gamma} \cap \xi^{-1}\widehat{\Gamma}\xi, \quad \text{where } \xi \in \widehat{G}, P(\xi) = g.$$

If $\gamma \in \text{Ker } \rho$, then by (1.16) we have

$$(1.19) \qquad \widehat{g\gamma g^{-1}} = \xi\widehat{\gamma}\xi^{-1}.$$

Since $\gamma \in \Gamma_{(g)}$, it follows that $g\gamma g^{-1} \in \Gamma$, and hence the right side of (1.19) is contained in the group $\widehat{\Gamma} = \delta(\Gamma)$. Using this and the fact that $\widehat{\gamma} \in \widehat{\Gamma}$, we see that $\delta(\text{Ker } \rho) \subset \widehat{\Gamma}_{(\xi)}$. We now prove the reverse inclusion. Let $\widehat{\gamma} \in \widehat{\Gamma}_{(\xi)}$, i.e., $\widehat{\gamma} \in \widehat{\Gamma}$ and $\widehat{\gamma} = \xi^{-1}\widehat{\gamma}_1\xi$, where $\widehat{\gamma}_1 \in \widehat{\Gamma}$. Then $\gamma \in \Gamma_{(g)}$, since $\gamma = g^{-1}\gamma_1 g$, and by (1.16) we have $\widehat{g\gamma g}^{-1} = \xi\widehat{\gamma}\xi^{-1}\rho(\gamma) = \widehat{\gamma}_1\rho(\gamma)$. If we note that $\widehat{g\gamma g}^{-1} = \widehat{\gamma}_1$, we find that $\rho(\gamma) = 1$, and so $\widehat{\gamma} \in \delta(\text{Ker } \rho)$. This proves (1.18).

We now prove that $(\widehat{\Gamma}, \widehat{S})$ is a Hecke pair. Since $\widehat{\Gamma} \subset \widehat{S}$, by (1.19) it suffices to verify that

(1.20) $$[\widehat{\Gamma} : \widehat{\Gamma}_{(\xi)}] < \infty \quad \text{and} \quad [\xi^{-1}\widehat{\Gamma}\xi : \widehat{\Gamma}_{(\xi)}] < \infty.$$

By the hypothesis of the lemma, $\delta \colon \Gamma \to \widehat{\Gamma}$ is an isomorphism of groups. Hence, using (1.18), we have

$$[\widehat{\Gamma} : \widehat{\Gamma}_{(\xi)}] = [\widehat{\Gamma} : \delta(\operatorname{Ker} \rho)] = [\Gamma : \operatorname{Ker} \rho] < \infty.$$

Finiteness of the second index in (1.20) follows from the equalities

$$[\xi^{-1}\widehat{\Gamma}\xi : \widehat{\Gamma}_{(\xi)}] = [\widehat{\Gamma} : \xi^{-1}\widehat{\Gamma}_{(\xi)}\xi] = [\widehat{\Gamma} : \widehat{\Gamma}_{(\xi^{-1})}]$$

and from the previous argument. $\square$

If (Γ, S) is a Hecke pair that satisfies the conditions in Lemma 1.7, we let

(1.21) $$\widehat{D}(\Gamma, S) = D(\widehat{\Gamma}, \widehat{S})$$

denote the Hecke ring of the pair (1.14), and we call this ring the *Hecke ring obtained by lifting of the ring $D(\Gamma, S)$*.

In order to clarify how the Hecke ring $\widehat{D}(\Gamma, S)$ differs from the original ring $D(\Gamma, S)$, we look at the relation between the partition of the double coset $\widehat{\Gamma}\xi\widehat{\Gamma}$ into $\widehat{\Gamma}$-left cosets and the partition of the double coset $\Gamma g \Gamma$, where $g = P(\xi) \in S$, into Γ-left cosets. Suppose we are given the partitions

(1.22) $$\Gamma = \bigcup_i \Gamma_{(g)}\gamma_i \quad \text{and} \quad \Gamma_{(g)} = \bigcup_j (\operatorname{Ker} \rho_g)\beta_j.$$

Since δ gives an imbedding of Γ in $\widehat{G}$, it follows from (1.22) and (1.18) that we have the partition

(1.23) $$\widehat{\Gamma} = \bigcup_{i,j} \widehat{\Gamma}_{(\xi)}\widehat{\beta}_j\widehat{\gamma}_i,$$

which, in conjunction with Lemma 1.2, gives us the corresponding partition of the double coset $\widehat{\Gamma}\xi\widehat{\Gamma}$:

(1.24) $$\widehat{\Gamma}\xi\widehat{\Gamma} = \bigcup_{i,j} \widehat{\Gamma}\xi\widehat{\beta}_j\widehat{\gamma}_i.$$

On the other hand, from (1.22) and Lemma 1.2 we also have the partition

(1.25) $$\Gamma g \Gamma = \bigcup_i \Gamma g \gamma_i,$$

which, when compared to (1.24), shows the similarities and differences between the partitions of the double cosets $\widehat{\Gamma}\xi\widehat{\Gamma}$ and $\Gamma g \Gamma$, and the role played by the lifting homomorphism ρ. Using (1.24) and (1.25), we obtain the following result.

LEMMA 1.8. *The equality* $\operatorname{Ker} \rho_g = \Gamma_{(g)}$ *holds if and only if the map*

$$P \colon \widehat{\Gamma}\xi\widehat{\Gamma} \to \Gamma g \Gamma, \quad \text{where } g = P(\xi),$$

is a one-to-one correspondence.

3. The imbedding ε. We now examine the connection between the Hecke rings corresponding to different Hecke pairs for the same group G. Let (Γ, S) and (Γ_0, S_0) be two Hecke pairs. Suppose that the following conditions hold:

$$(1.26) \qquad \Gamma_0 \subset \Gamma, \quad S \subset \Gamma S_0, \quad \text{and} \quad \Gamma \cap S_0 \cdot S_0^{-1} \subset \Gamma_0.$$

According to the second of these conditions, every left coset Γg, where $g \in S$, contains an element $g_0 \in S_0$. If we now set $\varepsilon((\Gamma g)) = (\Gamma_0 g_0)$, then, by the third condition in (1.26), we see that $(\Gamma_0 g_0) \in L(\Gamma_0, S_0)$ does not depend on the choice of g_0. The first condition in (1.26) shows that the map ε takes distinct Γ-left cosets to distinct Γ_0-left cosets. Thus, if we extend ε by **Z**-linearity onto all of $L(\Gamma, S)$, we obtain an imbedding of this module into $L(\Gamma_0, S_0)$.

PROPOSITION 1.9. *Suppose that the Hecke pairs (Γ, S) and (Γ_0, S_0) satisfy (1.26). Then the restriction of ε to the Hecke ring $D(\Gamma, S)$ is a monomorphism from this ring to the Hecke ring $D(\Gamma_0, S_0)$:*

$$(1.27) \qquad \varepsilon \colon D(\Gamma, S) \to D(\Gamma_0, S_0).$$

If, in addition,

$$(1.28) \qquad S_0 \subset S \quad \text{and} \quad \mu_\Gamma(g) = \mu_{\Gamma_0}(g) \quad \text{for all } g \in S_0,$$

where μ denotes the index (1.8), then the map (1.27) is an isomorphism of rings.

PROOF. The first part follows directly from the definitions and the assumption that $\Gamma_0 \subset \Gamma$. To prove the second part, by Lemma 1.5 it suffices to verify that under our assumptions

$$(1.29) \qquad \varepsilon(g)_\Gamma = (g)_{\Gamma_0} \quad \text{for } g \in S_0.$$

Let $\gamma_1, \ldots, \gamma_\mu$, where $\mu = \mu_{\Gamma_0}(g)$, be a set of left coset representatives of Γ_0 modulo $\Gamma_0 \cap g^{-1} \Gamma_0 g$. Then, by the definition of $(g)_{\Gamma_0}$ and Lemma 1.2, we have

$$(g)_{\Gamma_0} = \sum_{i=1}^{\mu} (\Gamma_0 g \gamma_i).$$

On the other hand, the elements $g\gamma_1, \ldots, g\gamma_\mu$ all lie in $\Gamma g \Gamma$ and belong to different left Γ-cosets, since if we had $g\gamma_i = \delta g \gamma_j$ with $\delta \in \Gamma$ it would follow that $\delta \in \Gamma \cap S_0 \cdot S_0^{-1} \subset \Gamma_0$, and hence $i = j$ and $\delta = e$. By (1.28) and Lemma 1.2, the number of elements is equal to the number of left Γ-cosets in $\Gamma g \Gamma$. Hence,

$$(g)_\Gamma = \sum_{i=1}^{\mu} (\Gamma g \gamma_i).$$

Then, by the definition of ε, we have $\varepsilon(g)_\Gamma = (g)_{\Gamma_0}$. $\qquad \square$

Again suppose that the Hecke pairs (Γ, S) and (Γ_0, S_0) are related as in (1.26), and let γ be an arbitrary element of Γ. We further suppose that $S_0 \subset S$, and we consider the commutative diagram

$$(1.30) \qquad
\begin{array}{ccc}
L(\Gamma, S) & \xrightarrow{\ \varepsilon\ } & L(\Gamma_0, S_0) \\
\downarrow{\scriptstyle \gamma} & & \downarrow{\scriptstyle \gamma} \\
L(\Gamma, S) & \xrightarrow{\ \varepsilon\ } & L(\Gamma_0, S_0),
\end{array}$$

where the vertical arrows denote the **Z**-linear homomorphisms that take $(\Gamma g) \in L(\Gamma, S)$ and $(\Gamma_0 g_0) \in L(\Gamma_0, S_0)$, respectively, to

$$(\Gamma g) \cdot \gamma = (\Gamma g \gamma) \quad \text{and} \quad (\Gamma_0 g_0) \cdot \gamma = (\Gamma_0 g_0'),$$

where g_0' is any element of $S_0 \cap \Gamma g_0 \gamma$. From the inclusions $S_0 \subset S$ and $\Gamma \subset S$ and the second property in (1.26) we find that $g_0 \gamma \in S$, and this product can be written in the form $\gamma' g_0'$ with $\gamma' \in \Gamma$ and $g_0' \in S_0$. From this, together with the third property in (1.26), it follows that $g_0' \in S_0 \cap \Gamma g_0 \gamma$, and the left coset $\Gamma_0 g_0'$ does not depend on the choice of g_0'.

LEMMA 1.10. *Suppose that the Hecke pairs (Γ, S) and (Γ_0, S_0) satisfy (1.26), and $S_0 \subset S$. Then the map ε in the diagram (1.30) is an isomorphism between $L(\Gamma, S)$ and $L(\Gamma_0, S_0)$, and the ε-image of the Hecke ring $D(\Gamma, S)$ coincides with the set of $t \in L(\Gamma_0, S_0)$ such that $t \cdot \gamma = t$ for all $\gamma \in \Gamma$.*

PROOF. From the inclusion $S_0 \subset S$ it follows that ε is an epimorphism. Since ε is also an imbedding, it must in fact be an isomorphism. The second part of the lemma follows from the commutativity of the diagram (1.30) and the definition of the Hecke ring $D(\Gamma, S)$.   □

4. The anti-isomorphism j.

PROPOSITION 1.11. *Let (Γ, S) be a Hecke pair for the group G. Then the pair (Γ, S^{-1}), where $S^{-1} = \{g^{-1}; g \in S\}$, is also a Hecke pair, and the **Z**-linear map of Hecke rings*

$$(1.31) \qquad\qquad j \colon D(\Gamma, S) \to D(\Gamma, S^{-1}),$$

which is defined on elements of the form (1.11) by setting

$$j((g)_\Gamma) = (g^{-1})_\Gamma \quad (g \in S),$$

is an anti-isomorphism of rings. In particular, if S is a group, then j is an anti-automorphism of the Hecke ring $D(\Gamma, S)$.

We first prove a lemma.

LEMMA 1.12. *Let Γ be a subgroup of G, and let $\widetilde{\Gamma}$ be the commensurator of Γ in G. Then the map*

$$g \to \lambda(g) = \mu(g)\mu(g^{-1})^{-1}, \quad g \in \widetilde{\Gamma},$$

where $\mu(h) = [\Gamma : \Gamma_{(h)}]$, is a homomorphism from $\widetilde{\Gamma}$ to the multiplicative group of rational numbers.

PROOF. We let X denote the set of all subgroups of G that are commensurable with Γ. If $\Gamma_1, \Gamma_2 \in X$, then there exists $\Gamma' \in X$ having finite index both in Γ_1 and in Γ_2 (for example, by Lemma 1.3 we can take $\Gamma' = \Gamma_1 \cap \Gamma_2$). We then set

$$\lambda(\Gamma_1 / \Gamma_2) = [\Gamma_1 : \Gamma'][\Gamma_2 : \Gamma']^{-1}.$$

It is easy to see that this number does not depend on the choice of Γ', and it satisfies the relations

$$(1.32) \qquad \begin{aligned} &\lambda(\Gamma_1/\Gamma_2)\lambda(\Gamma_2/\Gamma_3) = \lambda(\Gamma_1/\Gamma_3) \quad (\Gamma_i \in X), \\ &\lambda(g^{-1}\Gamma_1 g / g^{-1}\Gamma_2 g) = \lambda(\Gamma_1/\Gamma_2) \quad (\Gamma_i \in X, g \in G). \end{aligned}$$

Let $g \in \widetilde{\Gamma}$ and $\Gamma' \in X$; we set $\lambda'(g) = \lambda(\Gamma'/g^{-1}\Gamma'g)$. It is easily verified that $\lambda'(g)$ does not depend on the choice of $\Gamma' \in X$. Using (1.32), we find that, on the one hand,

$$\lambda'(g_1 \cdot g_2) = \lambda(\Gamma'/g_1^{-1}\Gamma'g_1)\lambda(g_1^{-1}\Gamma'g_1/g_2^{-1}(g_1^{-1}\Gamma'g_1)g_2) = \lambda'(g_1)\lambda'(g_2)$$

for any $g_1, g_2 \in \widetilde{\Gamma}$, and, on the other hand, for any $g \in \widetilde{\Gamma}$

$$\lambda'(g) = \lambda(\Gamma/\Gamma_{(g)})\lambda(\Gamma_{(g)}/g^{-1}\Gamma g) = \mu(g)\lambda(\Gamma_{(g^{-1})}/\Gamma) = \mu(g)\mu(g^{-1})^{-1} = \lambda(g). \quad \square$$

PROOF OF PROPOSITION 1.11. Since the elements of the form (1.11) form a **Z**-basis of the module $D(\Gamma, S)$, it suffices to prove that for any $g_1, g_2 \in S$ one has

$$j((g_1)(g_2)) = j((g_2))j((g_1)) = (g_2^{-1})(g_1^{-1}).$$

By Lemma 1.5 and the definition of j, this relation will be proved if we show that for any $h \in \Gamma g_1 \Gamma g_2 \Gamma$

$$(1.33) \qquad \nu(g_1, g_2; h)\mu(g_2)\mu(h)^{-1} = \nu(g_2^{-1}, g_1^{-1}; h^{-1})\mu(g_1^{-1})\mu(h^{-1})^{-1}.$$

Since the map $g \to \lambda(g)$ in Lemma 1.12 is trivial on the group Γ, Lemma 1.12 implies that

$$\mu(h)\mu(h^{-1})^{-1} = \lambda(h) = \lambda(g_1)\lambda(g_2) = \mu(g_1)\mu(g_1^{-1})^{-1}\mu(g_2)\mu(g_2^{-1})^{-1}.$$

From this it follows that the equality (1.33) is equivalent to the relation

$$(1.34) \qquad \nu(g_1, g_2; h)\mu(g_2^{-1}) = \nu(g_2^{-1}, g_1^{-1}; h^{-1})\mu(g_1).$$

From the definition of $\mu(g_1, g_2; h)$ and the decomposition (1.7) it follows that $\nu(g_1, g_2; h)$ is equal to the number of elements in the set

$$\{\gamma \in \Gamma_{(g_1)} \setminus \Gamma; g_1\gamma g_2 \in \Gamma h \Gamma\}.$$

It is easy to see that for $\gamma \in \Gamma$ the double coset $\Gamma g_1 \gamma g_2 \Gamma$ depends only on $\Gamma_{(g_1)} \gamma \Gamma_{(g_2^{-1})}$, and for $\gamma \in \Gamma$ and $t \in \Gamma_{(g_2^{-1})}$ we have $\Gamma_{(g_1)}\gamma t = \Gamma_{(g_1)}\gamma$ if and only if $t \in \gamma^{-1}\Gamma_{(g_1)}\gamma$. Thus, $\nu(g_1, g_2; h)$ can be written in the form

$$(1.35) \qquad \sum_{\substack{\gamma \in \Gamma_{(g_1)} \setminus \Gamma/\Gamma_{(g_2^{-1})} \\ g_1\gamma g_2 \in \Gamma h \Gamma}} [\Gamma_{(g_2^{-1})} : (\Gamma_{(g_2^{-1})} \cap \gamma^{-1}\Gamma_{(g_1)}\gamma)].$$

Similarly,

$$\nu(g_2^{-1}, g_1^{-1}; h^{-1}) = \sum_{\gamma} [\Gamma_{(g_1)} : (\Gamma_{(g_1)} \cap \gamma\Gamma_{(g_2^{-1})}\gamma^{-1})].$$

Using the notation $\lambda(\Gamma_1/\Gamma_2)$ (see the proof of Lemma 1.12) and the properties (1.32) of this symbol, we obtain

$$v(g_2^{-1}, g_1^{-1}; h^{-1}) = \sum_\gamma \lambda(\Gamma_{(g_1)}/\Gamma_{(g_1)} \cap \gamma\Gamma_{(g_2^{-1})}\gamma^{-1})$$

$$= \sum_\gamma \lambda(\gamma^{-1}\Gamma_{(g_1)}\gamma/(\gamma^{-1}\Gamma_{(g_1)}\gamma \cap \Gamma_{(g_2^{-1})}))$$

$$= \sum_\gamma \lambda(\gamma^{-1}\Gamma_{(g_1)}\gamma/\Gamma_{(g_1)})\lambda(\Gamma_{(g_1)}/\Gamma)$$

$$\times \lambda(\Gamma/\Gamma_{(g_2^{-1})})\lambda(\Gamma_{(g_2^{-1})}/\Gamma_{(g_2^{-1})} \cap \gamma^{-1}\Gamma_{(g_1)}\gamma)$$

$$= \sum_\gamma \lambda(\gamma)\mu(g_1)^{-1}\mu(g_2^{-1})\lambda(\Gamma_{(g_2^{-1})}/\Gamma_{(g_2^{-1})} \cap \gamma^{-1}\Gamma_{(g_1)}\gamma)$$

$$= \mu(g_1)^{-1}\mu(g_2^{-1})v(g_2, g_2; h),$$

where all of the summations are over the same γ as in (1.35). $\qquad\square$

The next lemma shows that the anti-isomorphism j is compatible with the monomorphism ε that was defined in the previous subsection.

LEMMA 1.13. *Let* (Γ, S) *and* (Γ_0, S_0) *be two Hecke pairs satisfying* (1.26). *Suppose that the Hecke pairs* (Γ, S^{-1}) *and* (Γ_0, S_0^{-1}) *also satisfy* (1.26). *Then the following diagram is commutative*:

$$
\begin{array}{ccc}
L(\Gamma, S) & \xrightarrow{\varepsilon} & L(\Gamma_0, S_0) \\
\downarrow{j} & & \downarrow{j} \\
L(\Gamma, S) & \xrightarrow{\varepsilon} & L(\Gamma_0, S_0^{-1})
\end{array}
$$

PROOF. From the definition of ε it follows that $\varepsilon(g)_\Gamma = \sum_i (g_i)_{\Gamma_0}$, where the summation is over the double Γ_0-cosets in $\Gamma g \Gamma \cap S_0$. The map $g_i \to g_i^{-1}$ gives a one-to-one correspondence between the double Γ_0-cosets in $\Gamma g \Gamma \cap S_0$ and in $\Gamma g^{-1}\Gamma \cap S_0^{-1}$. We thus have

$$j\varepsilon(g)_\Gamma = \sum_i (g_i^{-1})_{\Gamma_0} = \varepsilon(g^{-1})_\Gamma = \varepsilon j(g)_\Gamma.$$

The lemma follows from this relation and Lemma 1.5. $\qquad\square$

5. Representations in spaces of automorphic forms. We return to the situation described at the beginning of the section: a semigroup S acts on a set H, φ is an automorphy factor of S on H with values in a group T, V is a left T-module, and $\mathfrak{M} = \mathfrak{M}_\varphi(\Gamma)$ is the additive group of all Γ-automorphic forms of weight φ with values in V, where $\Gamma \subset S$ is a subgroup. We further suppose that (Γ, S) is a Hecke pair, and $D = D(\Gamma, S)$ is its Hecke ring. If $\sum_i a_i(\Gamma g_i) \in D$ and $F \in \mathfrak{M}$, then, in the notation of Lemma 4.1(3) of Chapter 1, we have

(1.36) $$F|t = F|_\varphi t = \sum_i a_i F|_\varphi g_i.$$

Using property (3) of Lemma 4.1 of Chapter 1 and (1.1), we see that $F|_{\gamma_i g_i} = F|g_i$, where $\gamma_i \in \Gamma$, and hence the function (1.36) does not depend on the choice of left coset representatives in t. Furthermore, if $\gamma \in \Gamma$, then $(F|t)|\gamma = F|t\gamma = F|t$, so that the function $F|t$ also belongs to $\mathfrak{M}$. Thus, to every $t \in D$ is associated a linear operator

$$|t = |_\varphi t : \mathfrak{M}_\varphi(\Gamma) \to \mathfrak{M}_\varphi(\Gamma),$$

which coincides with the operator (1.4) in the case $t = (g)_\Gamma$. Like the operator (1.4), this operator is called a *Hecke operator*. By Lemma 1.5, any such operator is a **Z**-linear combination of the operators (1.4).

PROPOSITION 1.14. *The map*

$$t \to |_\varphi t \quad (t \in D(\Gamma, S))$$

is a homomorphism from the Hecke ring $D(\Gamma, S)$ to the endomorphism ring of the **Z**-*module $\mathfrak{M}_\varphi(\Gamma)$.*

PROOF. The map is obviously linear. Hence, it suffices to prove that a product of elements of the Hecke ring is taken to the corresponding product of operators. If $F \in \mathfrak{M}_\varphi(\Gamma)$ and $t = \sum_i a_i(\Gamma g_i)$, $t' = \sum_j b_j(\Gamma h_j) \in D(\Gamma, S)$, then, by Lemma 4.1(3) of Chapter 1, we have

$$F|tt' = \sum_{i,j} a_i b_j F|g_i h_j = \sum_j b_j \left(\sum_i a_i F|g_i \right)|h_j = F|t|t'. \qquad \square$$

From Proposition 1.14 it follows that any algebraic relation between elements of the Hecke ring is also valid for the corresponding Hecke operators.

6. Hecke algebras over a commutative ring. Let (Γ, S) be a Hecke pair, and let **A** be an arbitrary commutative ring with unit. Just as in subsection 2 for the case of **Z**, we can define the free **A**-module $L_A = L_A(\Gamma, S)$ whose generators over **A** are the symbols (Γg) $(g \in S)$, one for each left coset $\Gamma g \subset S$, and we can define the submodule $D_A = D_A(\Gamma, S)$ consisting of all Γ-invariant elements. Again, the multiplication

$$\left(\sum_i a_i(\Gamma g_i) \right) \left(\sum_j b_j(\Gamma h_j) \right) = \sum_{i,j} a_i b_j(\Gamma(g_i h_j))$$

of elements in D_A does not depend on the choice of left coset representatives, and it makes D_A into an associative ring with unit, called the *Hecke algebra of the pair* (Γ, S) *over* **A**. All of the results of subsections 2–4 carry over without any change to the Hecke algebras $D_A(\Gamma, S)$. The results of subsections 1 and 5 concerning representations of Hecke rings also carry over (along with their proofs), if we suppose that V is also an **A**-module and the actions on V of the group T and the ring **A** commute with one another. Clearly, if $\mathbf{Z} \subset \mathbf{A}$, then

(1.37) $$D_A(\Gamma, S) = D(\Gamma, S) \otimes_{\mathbf{Z}} \mathbf{A}$$

(tensor product of rings).

PROBLEM 1.15. Let Γ be a normal subgroup of finite index in the group S. Show that (Γ, S) is a Hecke pair, and the Hecke algebra $D_A(\Gamma, S)$ of the Hecke pair (Γ, S) over the ring **A** is isomorphic to the group-ring of the quotient group $\Gamma \setminus S$ over **A**.

PROBLEM 1.16. Let (Γ, S) be a Hecke pair. Show that the **A**-linear map N from the Hecke algebra $D_{\mathbf{A}}(\Gamma, S)$ to **A** that is defined on elements of the form (1.11) by setting $N((g)) = \mu_\Gamma(g)e_{\mathbf{A}}$, where $\mu_\Gamma(g)$ is the index (1.8) and $e_{\mathbf{A}}$ is the unit of the ring **A**, is a ring homomorphism.

§2. Hecke rings for the general linear group

Our ultimate goal is to study the Hecke rings of the symplectic group and their representations in spaces of modular forms. To a large extent we do this by reducing the questions to the case of Hecke rings for the general linear group.

1. Global rings. We consider the groups

$$(2.1) \qquad \Lambda = \Lambda^n = GL_n(\mathbf{Z}) \quad \text{and} \quad G = G^n = GL_n(\mathbf{Q}),$$

where $n \geqslant 1$ and **Q** is the field of rational numbers.

LEMMA 2.1. *The commensurator of the subgroup Λ in the group G is G itself. In particular, (Λ, G) is a Hecke pair.*

PROOF. If in every matrix of Λ we replace the entries by their residue classes modulo q, where q is a positive integer, we obtain a homomorphism of the group Λ to the finite group $GL_n(\mathbf{Z}/q\mathbf{Z})$. The kernel of this homomorphism is the principal congruence subgroup of level q

$$(2.2) \qquad \Lambda(q) = \Lambda^n(q) = \{\lambda \in \Lambda; \lambda \equiv E_n(\mathrm{mod}\, q)\},$$

which is a normal subgroup of finite index in Λ.

Let $g = tg_0$, where $t \in \mathbf{Q}$ and g_0 is an integer matrix. It is easy to verify that $g\Lambda(d)g^{-1} \subset \Lambda$ and $g^{-1}\Lambda(d)g \subset \Lambda$, where $d = |\det g_0|$. This means that $\Lambda_{(g)} = \Lambda \cap g^{-1}\Lambda g$ contains the subgroup $\Lambda(d)$, and so it has finite index in Λ. The intersection $\Lambda_{(g^{-1})}$ also has finite index in Λ; hence, the group $g^{-1}\Lambda_{(g^{-1})}g = \Lambda_{(g)}$ has finite index in the group $g^{-1}\Lambda g$. All of this implies that $g^{-1}\Lambda g$ is commensurable with Λ for any element $g \in G$. $\qquad\square$

The main object of study in this section will be the Hecke ring

$$(2.3) \qquad H = H^n = D_{\mathbf{Q}}(\Lambda^n, G^n) = D(\Lambda^n, G^n) \otimes_{\mathbf{Z}} \mathbf{Q}$$

of the Hecke pair (Λ^n, G^n) over the field **Q** and over subrings of **Q**. By Lemma 1.5, we can take as a **Q**-basis of H the elements $(g) = (g)_\Lambda$ of the form (1.11), one for each double Λ-coset $\Lambda g\Lambda$ of the group G. In order to visualize this set of double cosets, we prove that there is a special type of diagonal representative in each double coset.

LEMMA 2.2. *Every double coset $\Lambda g\Lambda$, where $\Lambda = GL_n(\mathbf{Z})$ and $g \in G^n = GL_n(\mathbf{Q})$, has one and only one representative of the form*

$$(2.4) \qquad \mathrm{ed}(g) = \mathrm{diag}(d_1, \ldots, d_n), \quad \text{where } d_i > 0, d_i | d_{i+1}.$$

PROOF. Let d_1 be the greatest common divisor of the entries of the matrix g, i.e., d_1 is the positive rational number such that $g = d_1 g_1$, where g_1 is an integer matrix with relatively prime entries. Using induction on the minimum $\delta = \delta(g_1)$ of the greatest common divisors of the columns of these matrices g_1, we prove that the double coset $\Lambda g_1\Lambda$ of such a matrix contains a representative of the form $\begin{pmatrix} 1 & 0 \\ 0 & g_2 \end{pmatrix}$, where g_2 is an

$(n - 1) \times (n - 1)$ integer matrix. First suppose that $\delta = 1$, i.e., g_1 contains a column with relatively prime entries. If we multiply g_1 on the right by a suitable permutation matrix—an operation which does not cause us to leave the double coset—we may assume that the first column of g_1 has relatively prime entries. By Lemma 3.8 of Chapter 1, we can then multiply g_1 on the left by a suitable matrix in Λ that reduces g_1 to the form

$$\begin{pmatrix} 1 & l \\ 0 & g_2 \end{pmatrix} = \begin{pmatrix} 1 & 0 \\ 0 & g_2 \end{pmatrix} \begin{pmatrix} 1 & l \\ 0 & E_{n-1} \end{pmatrix}, \qquad \text{where } l \in M_{1,n-1},$$

and this proves the claim in the case $\delta = 1$. Now suppose that $\delta > 1$, and the claim has already been proved for all g_1 with $\delta(g_1) < \delta$. If we are given a matrix g_1 with $\delta(g_1) = \delta$, by permuting the columns we may assume that the greatest common divisor of the entries in the first column is equal to δ. Again using Lemma 3.8 of Chapter 1, we multiply g_1 on the left by a suitable matrix in Λ so as to reduce it to the form

$$\begin{pmatrix} \delta & l \\ 0 & h \end{pmatrix} = \begin{pmatrix} \delta & l + \delta l' \\ 0 & h \end{pmatrix} \begin{pmatrix} 1 & -l' \\ 0 & E_{n-1} \end{pmatrix},$$

where $l \in M_{1,n-1}$, l' is an arbitrary $(n - 1)$-dimensional integer row, and h is an $(n - 1) \times (n - 1)$ integer matrix. The row l' can be chosen in such a way that all of the entries in the row $l'' = l + \delta l'$ are positive and do not exceed δ. Then the matrix $g_1' = \begin{pmatrix} \delta & l'' \\ 0 & h \end{pmatrix} \in \Lambda g_1 \Lambda$ satisfies the inequality $\delta(g_1') < \delta$ (the inequality $\delta(g_1') \leqslant \delta$ is obvious; equality here would mean that all of the columns of g_1' were divisible by δ, but that contradicts the assumption that the entries are relatively prime). By the induction assumption, our claim holds for g_1', and hence it holds for g_1.

Returning to the proof of the lemma, we see that the double coset $\Lambda g \Lambda = d_1 \Lambda g_1 \Lambda$ contains a representative of the form $g' = \begin{pmatrix} d_1 & 0 \\ 0 & d_1 g_2 \end{pmatrix}$, where d_1 is the greatest common divisor of the entries of g and g_2 is an $(n - 1) \times (n - 1)$ integer matrix. If g_2 varies within the double coset $\Lambda^{n-1} g_2 \Lambda^{n-1}$, then clearly g' remains in the double coset $\Lambda g \Lambda$. Hence, we can continue this diagonalization process, applying it to the matrix $d_1 g_2$, and so on. We finally obtain a representative of the form (2.4).

If $D = \mathrm{diag}(d_1, \ldots, d_n)$ and $D' = \mathrm{diag}(d_1', \ldots, d_n')$ are two diagonal matrices of the form (2.4) and $D' = \lambda_1 D \lambda_2$, where $\lambda_i \in \Lambda$, then it follows from the Cauchy–Binet formula that all of the $r \times r$-minors of D', where $1 \leqslant r \leqslant n$—and, in particular, $d_1' \cdots d_r'$—are divisible by $d_1 \cdots d_r$. Similarly, $d_1 \cdots d_r$ is divisible by $d_1' \cdots d_r'$. Since the diagonal entries are positive, this implies that $d_1' \cdots d_r' = d_1 \cdots d_r$ $(1 \leqslant r \leqslant n)$, and so $D' = D$. $\square$

We shall call $\mathrm{ed}(g) = \mathrm{diag}(d_1, \ldots, d_n)$ the *matrix of elementary divisors of the matrix* g, and we shall call the numbers $d_r = d_r(g)$ the *elementary divisors of* g. Using an argument similar to the proof of uniqueness in the previous lemma, we see that the product $d_1 \cdots d_r$ is equal to the greatest common divisor of the $r \times r$-minors of g. In particular,

$$(2.5) \qquad\qquad d_1(g) \cdots d_n(g) = |\det g|.$$

We now turn our attention to the multiplicative properties of the Hecke rings H^n.

THEOREM 2.3. *The Hecke ring H^n, $n \geqslant 1$, is commutative.*

PROOF. According to Lemma 1.5, it suffices to prove that

$$(g)(g') = (g')(g), \quad \text{where } g, g' \in G^n.$$

Since every double coset $\Lambda g \Lambda$ ($g \in G^n$) contains a diagonal matrix, the coset is taken to itself by the transpose: ${}^t(\Lambda g \Lambda) = \Lambda \, {}^t g \Lambda = \Lambda g \Lambda$. This implies that the number of Λ-left cosets in $\Lambda g \Lambda$ is equal to the number of right cosets; but since each of the left cosets intersects with every right coset, there exists a set of representatives $g_1, \ldots, g_\mu$ of the left cosets which is also a set of representatives of the right cosets. It is easy to see that the set ${}^t g_1, \ldots, {}^t g_\mu$ has the same property. Let $g'_1, \ldots, g'_\nu$ be a similar set of representatives for the double coset $\Lambda g' \Lambda$. Then Lemma 1.5 implies that

$$(g)(g') = \sum_{\Lambda h \Lambda \in \Lambda g \Lambda g' \Lambda} v'(g, g'; h) \mu(h)^{-1}(h),$$

where $v'(g, g'; h) = v(g, g'; h)\mu(g')$ is equal to the number of pairs (i, j) such that $g_i \cdot g'_j \in \Lambda h \Lambda$. If we replace the representatives $\{g_i\}$ and $\{g'_j\}$ by $\{{}^t g_i\}$ and $\{{}^t g'_j\}$, we see that $v'(g, g'; h)$ is also equal to the number of pairs (i, j) such that

$${}^t g_i \cdot {}^t g'_j = {}^t(g'_j g_i) \in \Lambda h \Lambda = {}^t(\Lambda h \Lambda), \quad \text{i.e., } g'_j g_i \in \Lambda h \Lambda.$$

Thus, $v'(g, g'; h) = v'(g', g; h)$, and

$$(g)(g') = \sum_{\Lambda h \Lambda \in \Lambda g \Lambda g' \Lambda} v'(g', g; h) \mu(h)^{-1}(h) = (g')(g),$$

since the set $\Lambda g \Lambda g' \Lambda$, as a finite union of double Λ-cosets, coincides with its transpose ${}^t(\Lambda g \Lambda g' \Lambda) = \Lambda \, {}^t g' \Lambda \, {}^t g \Lambda = \Lambda g' \Lambda g \Lambda$. □

The rule for multiplying double cosets becomes especially simple when one of the cosets is proportional to the identity coset. In this case the definition of multiplication in the Hecke ring immediately implies

LEMMA 2.4. *The following relation holds in the Hecke ring H^n for any $g \in G^n$ and $r \in \mathbf{Q}^*$:*

$$(2.6) \qquad\qquad (rE_n)(g) = (g)(rE_n) = (rg).$$

PROPOSITION 2.5. *Let $g, g' \in G^n$. Suppose the ratios $d_n(g)/d_1(g)$ and $d_n(g')/d_1(g')$, where d_r denotes the rth elementary divisor, are relatively prime. Then the following equality holds in the Hecke ring H^n:*

$$(2.7) \qquad\qquad (g)(g') = (gg').$$

PROOF. Lemma 2.4 obviously implies that it suffices to prove the proposition in the case when $d_1(g) = d_1(g') = 1$. Hence, we suppose that g and g' are integer matrices, and the numbers $d = |\det g|$ and $d' = |\det g'|$ are relatively prime (see (2.5) and (2.6)). From Lemma 1.5 it follows that

$$(g)(g') = \sum_{\Lambda h \Lambda \in \Lambda g \Lambda g' \Lambda} c(g, g'; h)(h),$$

where $c(g, g'; h)$ are positive integers that depend only on the double Λ-coset of g, g', and h. Since $\Lambda gg'\Lambda \subset \Lambda g\Lambda g'\Lambda$, it follows that $c(g, g'; gg') \geqslant 1$, and the last relation can be rewritten in the form

$$(2.8) \qquad (g)(g') = (gg') + \sum_{\Lambda h\Lambda \in \Lambda g\Lambda g'\Lambda} c'(g, g'; h)(h),$$

where $c'(g, g'; h)$ are nonnegative integers that depend only on the double cosets of g, g', and h. For $m \geqslant 1$ we let

$$(2.9) \qquad\qquad ED(m) = ED_n(m)$$

denote the set of all integer matrices of the form (2.4) having determinant m. By Lemma 2.2, we may assume that the matrices g, g', and h in (2.8) belong to the sets $ED(d)$, $ED(d')$, and $ED(dd')$, respectively, since every h in (2.8) is an integer matrix with $|\det h| = dd'$. Since the $ED_n(m)$ are obviously finite sets, we can define the following elements of H^n:

$$(2.10) \qquad\qquad t(m) = t_n(m) = \sum_{g \in ED_n(m)} (g).$$

Summing (2.8) over all $g \in ED(d)$ and $g' \in ED(d')$, we obtain the relation

$$(2.11) \qquad t(d)t(d') = \sum_{g,g'}(gg') + \sum_{h \in ED(dd')} \left(\sum_{g,g'} c'(g, g'; h) \right)(h).$$

It is easy to see that for d prime to d' the map $(g, g') \rightarrow gg'$ gives a bijection of $ED(d) \times ED(d')$ with $ED(dd')$. Hence, the first sum on the right in (2.11) is equal to $t(dd')$. If we prove that $t(d)t(d') = t(dd')$ for d prime to d', then it will follow from (2.11) that the double sum on the right is equal to zero; hence, all of the coefficients $c'(g, g'; h)$, since they are nonnegative, must equal zero. Thus, (2.8) would turn into (2.7), and the proposition would be proved. In other words, to prove the proposition it suffices to prove the following lemma, which is also of independent interest. $\qquad \square$

LEMMA 2.6. *Let $d \geqslant 1$. Then the set*

$$(2.12) \qquad\qquad M_n(\pm d) = \{ M \in M_n; \det M = \pm d \}$$

is the union of finitely many left cosets modulo the group $\Lambda = \Lambda^n$, and the element $t(d) = t_n(d)$ of the Hecke ring H^n can be written as a sum of the form

$$(2.13) \qquad\qquad t(d) = \sum_{g \in \Lambda \backslash M_n(\pm d)} (\Lambda g).$$

If d and d' are relatively prime and the matrices g and g' run through sets of representatives of the left cosets $\Lambda \backslash M_n(\pm d)$ and $\Lambda \backslash M_n(\pm d')$, respectively, then the product gg' runs through a set of representatives of the left cosets $\Lambda \backslash M_n(\pm dd')$. In particular,

$$(2.14) \qquad\qquad t(d)t(d') = t(dd'), \quad if (d, d') = 1.$$

PROOF. Lemma 2.2 implies that the set $M_n(\pm d)$ is the union of finitely many double Λ-cosets, and each of these double cosets has exactly one representative in the set $ED(d)$. The first part of the lemma now follows from Lemma 2.1 and the definition of $t(d)$ and (g).

Suppose that d and d' are relatively prime, and $\{g_1, \ldots, g_\mu\}$ and $\{g'_1, \ldots, g'_\mu\}$ are fixed sets of representatives of the left cosets $\Lambda \backslash M_n(\pm d)$ and $\Lambda \backslash M_n(\pm d')$, respectively. Each product $g_i g'_j$ is clearly contained in $M_n(\pm dd')$. Suppose that two such products belong to the same left Λ-coset:

$$\lambda g_i g'_i = g_{i_1} g'_{j_1}, \quad \text{where } \lambda \in \Lambda.$$

We set $h = g_{i_1}^{-1} \lambda g_i = g'_{j_1}(g'_j)^{-1}$. Then $dh = dg_{i_1}^{-1}\lambda g_i$ and $d'h = g'_{j_1}d'(g'_j)^{-1}$ are integer matrices; since d is prime to d', it follows that h is an integer matrix. Furthermore, $\det h = \pm 1$. Thus, $h \in \Lambda$, so that $g'_{j_1} = hg'_j \in \Lambda g'_j$, and hence $j_1 = j$. Then also $\lambda g_i = g_{i_1}$, and so $i_1 = i$. We have thus proved that the products $g_i g'_j$ belong to distinct left cosets of $M_n(\pm dd')$ modulo the group Λ. Now suppose that Λg, where $g \in M_n(\pm dd')$, is an arbitrary left coset of $M_n(\pm dd')$ modulo Λ. Then Lemma 2.2 implies that g can be written in the form $g = vw$, where $v \in M_n(\pm d)$, $w \in M_n(\pm d')$. The element w lies in some left coset $\Lambda g'_j$, i.e., $w = \lambda_1 g'_j$, where $\lambda_1 \in \Lambda$, and $v\lambda_1$ lies in some left coset Λg_i, i.e., $v\lambda_1 = \lambda g_i$, where $\lambda \in \Lambda$. Consequently, $g = \lambda g_i g'_j$, and the left coset Λg contains the product $g_i g'_j$. The second part of the lemma is proved.

The relation (2.14) follows from what has already been proved and from the definition of multiplication in Hecke rings. This proves Lemma 2.6, and hence also Proposition 2.5. $\qquad\square$

The next lemma turns out to be useful for explicitly computing the left coset decomposition of elements of the Hecke ring H^n.

LEMMA 2.7. *Every left coset Λg, where g is an integer matrix in G^n, contains one and only one reduced representative $C = (c_{ij})$ with*

$$(2.15) \qquad 0 \leqslant c_{ij} < c_{jj}, \quad c_{ji} = 0 \quad \text{for } 1 \leqslant i < j \leqslant n.$$

PROOF. We use induction on n to prove the existence of a reduced representative. The assertion is obvious if $n = 1$. Suppose that $n > 1$, and the claim has already been proved for $n - 1$. If we apply Lemma 3.8 of Chapter 1 to the first column of g, we find that Λg contains a representative of the form $g' = \begin{pmatrix} c_{11} & * \\ 0 & g_1 \end{pmatrix}$, where $c_{11} > 0$ and g_1 is an integer matrix in G^{n-1}. By the induction assumption, we have $g_1 = \varepsilon g_0$, where $\varepsilon \in \Lambda^{n-1}$ and g_0 is a reduced matrix. It is then easy to see that there exists a row $l \in M_{1,n-1}$ such that $\begin{pmatrix} 1 & l \\ 0 & E_{n-1} \end{pmatrix} \begin{pmatrix} 1 & 0 \\ 0 & \varepsilon^{-1} \end{pmatrix} g' \in \Lambda g$ is a reduced matrix. Finally, uniqueness of the reduced representative easily follows using proof by contradiction. $\square$

The study of the global Hecke ring H reduces to the study of its local subrings H_p, where p runs through all prime numbers. Let p be a prime. We set

$$(2.16) \qquad G_p = G_p^n = GL_n(\mathbf{Z}[p^{-1}]),$$

where

$$(2.17) \qquad \mathbf{Z}[p^{-1}] = \{ap^b \in \mathbf{Q}; a, b \in \mathbf{Z}\}$$

is the ring of rational numbers that are integral outside p. Since $\Lambda \subset G_p \subset G$, we can consider the Hecke ring

$$(2.18) \qquad H_p = H_p^n = D_{\mathbf{Q}}(\Lambda^n, G_p^n),$$

and this ring can be regarded as a subring of the Hecke ring H. The subrings $H_p \subset H$ for prime p are called the *local Hecke rings of the group* G.

THEOREM 2.8. *The Hecke ring H^n is generated by the local subrings H_p^n as p runs through the prime numbers.*

PROOF. Given a nonzero rational number r and a prime p, we let $v_p(r)$ denote the exponent of p that occurs in the prime factorization of r. If $g \in G^n$, we define the matrix $\mathrm{ed}_p(g)$ of *elementary p-divisors of g* by setting

$$(2.19) \qquad \mathrm{ed}_p(g) = \mathrm{diag}(p^{\alpha_1}, \ldots, p^{\alpha_n}) \quad \text{with } \alpha_i = v_p(d_i(g)),$$

where $d_i(g)$ are the elementary divisors of g. The numbers p^{α_i} are called the *elementary p-divisors of g*. For fixed g, the matrices $\mathrm{ed}_p(g)$ are clearly equal to the identity matrix for all but finitely many p; furthermore,

$$(2.20) \qquad \mathrm{ed}(g) = \prod_p \mathrm{ed}_p(g) \quad (g \in G),$$

where the product is taken over all prime numbers. Since $g \in \Lambda \mathrm{ed}(g)\Lambda$, this product formula implies that g can be written in the form

$$(2.21) \qquad g = \prod_p g_p, \quad \text{where } \mathrm{ed}(g_p) = \mathrm{ed}_p(g)$$

and $g_p = E_n$ for all but finitely many p. It now follows from Proposition 2.5 that we have the expansion $(g) = \prod_p(g_p)$ for the corresponding double cosets. Since $(g_p) \in H_p$, and H consists of finite linear combinations of elements of the form (g) $(g \in G)$, we conclude that every element in H is a finite sum of finite products of elements of the subrings H_p. $\qquad \square$

PROBLEM 2.9. Let $g, g' \in G^n$. Suppose that the numbers $d_n(g)/d_1(g)$ and $d_n(g')/d_1(g')$ are relatively prime. Show that $d_i(gg') = d_i(g)d_i(g')$ for $i = 1, \ldots, n$.

PROBLEM 2.10. Show that the set of reduced integer matrices $C \in M_n$ with $\det C = d$, where $d \in \mathbf{N}$, can be taken as a set of representatives of the left cosets $\Lambda \setminus M_n(\pm d)$. Conclude from this that the zeta-function of the ring H^n, which is defined as the Dirichlet series

$$Z(s, N) = \sum_{m=1}^{\infty} \frac{N(t(m))}{m^s},$$

where $N \colon H^n \to \mathbf{Q}$ is the homomorphism in Problem 1.16 and $\mathrm{Re}\, s > n$, is equal to the product $\zeta(s)\zeta(s-1) \cdots \zeta(s - n + 1)$ of Riemann zeta-functions.

PROBLEM 2.11. Prove the following identities for formal Dirichlet series with coefficients in the ring H^n:

$$Z(s) = \sum_{m=1}^{\infty} \frac{t(m)}{m^s} = \prod_p Z_p(p^{-s}),$$

where p runs through the prime numbers, and

$$Z_p(v) = \sum_{\delta=0}^{\infty} t(p^\delta)v^\delta;$$

$$Z(s_1,\ldots,s_n) = \sum_{\mathrm{diag}(d_1,\ldots,d_n)} \frac{(\mathrm{diag}(d_1,\ldots,d_n))_\Lambda}{d_1^{s_1}\cdots d_n^{s_n}} = \prod_p Z_p(p^{-s_1},\ldots,p^{-s_n}),$$

where $\mathrm{diag}(d_1,\ldots,d_n)$ runs through the set $\bigcup_{m=1}^{\infty} ED(m)$, p runs through the prime numbers, and

$$Z_p(v_1,\ldots,v_n) = \sum_{0\leqslant\delta_1\leqslant\cdots\leqslant\delta_n} (\mathrm{diag}(p^{\delta_1},\ldots,p^{\delta_n}))_\Lambda v_1^{\delta_1}\cdots v_n^{\delta_n}.$$

PROBLEM 2.12. Using the coset representatives in Problem 2.10, prove that in the case $n = 2$ the following relation holds for every prime p:

$$t(p)t(p^\delta) = t(p^{\delta+1}) + p(pE_2)_\Lambda t(p^{\delta-1}) \quad (\delta \geqslant 1).$$

From this derive the following identities in the ring of formal power series over H_p^2:

$$Z_p(v) = (1 - t(p)v + p(pE_2)v^2)^{-1},$$
$$Z_p(v_1,v_2) = (1 - (pE_2)v_2^2)(1 - (pE_2)v_1v_2)^{-1}(1 - t(p)v_2 + p(pE_2)v_2^2)^{-1}.$$

PROBLEM 2.13. Show that in the ring H^2 we have the relations

$$t(m)t(m_1) = \sum_{d|m,m_1} d(dE_2)t(mm_1/d^2) \quad (m, m_1 \in \mathbf{N}),$$

where d runs through the positive common divisors of m and m_1.

PROBLEM 2.14. Show that G_p^n coincides with the subset of G^n consisting of all $g \in G^n$ all of whose elementary divisors $d_1(g),\ldots,d_n(g)$ are powers of p.

2. Local rings. In this subsection we study the structure of the local Hecke rings H_p^n defined by (2.18). We first note that the investigation of H_p^n reduces to that of its integral subring

(2.22) $$\underline{H}_p^n = D_{\mathbf{Q}}(\Lambda^n, G_p^n \cap M_n),$$

which consists of all linear combinations of elements (g) where $g \in G_p^n$ is an integer matrix.

LEMMA 2.15. *The element*

(2.23) $$\pi(p) = \pi_n(p) = (pE_n)_\Lambda \in H_p^n$$

is invertible in the ring H_p^n, and $\pi(p)^{-1} = (p^{-1}E_n)_\Lambda$. The ring H_p^n is generated by the subring $\underline{H}_p^n$ and the element $\pi(p)^{-1}$.

PROOF. The lemma follows from Lemma 2.4 and the definitions. $\square$

For brevity we denote
$$(\mathrm{diag}(p^{\delta_1}, \ldots, d^{\delta_n}))_\Lambda = t(p^{\delta_1}, \ldots, d^{\delta_n}).$$

In this notation the ring $\underline{H}^n_p$ consists of linear combinations of elements of the form $t(p^{\delta_1}, \ldots, p^{\delta_n})$, where $0 \leqslant \delta_1 \leqslant \cdots \leqslant \delta_n$. We say that such an element is *primitive* if $\delta_1 = 0$, and we say that it is *imprimitive* if $\delta_1 \geqslant 1$. An arbitrary element $t \in \underline{H}^n_p$ is said to be *primitive* (respectively, *imprimitive*) if it is a linear combination of primitive (resp. imprimitive) double cosets $t(p^{\delta_1}, \ldots, p^{\delta_n})$. Clearly, any $t \in \underline{H}^n_p$ can be uniquely written in the form $t = t^{\mathrm{pr}} + t^{\mathrm{im}}$, where t^{pr} is primitive and t^{im} is imprimitive. Lemma 2.4 implies that the set I of all imprimitive elements in $\underline{H}^n_p$ is the principal ideal generated by $\pi(p)$.

LEMMA 2.16. *Let* $\Psi \colon \underline{H}^n_p \to \underline{H}^{n-1}_p$ *be the* **Q***-linear map defined by setting*

$$\Psi(t(p^{\delta_1}, \ldots, d^{\delta_n})) = \begin{cases} t(p^{\delta_1}, \ldots, d^{\delta_n}), & \text{if } 0 = \delta_1 \leqslant \delta_2 \leqslant \cdots \leqslant \delta_n, \\ 0, & \text{if } 0 < \delta_1 \leqslant \delta_2 \leqslant \cdots \leqslant \delta_n. \end{cases}$$

Then Ψ *is an epimorphism of rings. The kernel of* Ψ *is the ideal* I *of imprimitive elements of* $\underline{H}^n_p$.

PROOF. From Lemmas 1.5 and 2.2 it follows that the map Ψ gives an isomorphism between the subspace of all primitive elements of $\underline{H}^n_p$ and the space $\underline{H}^{n-1}_p$. In particular, Ψ is an epimorphism. It is clear that the kernel of Ψ is I. Hence, it remains to prove that Ψ is a ring homomorphism. To do this it suffices to verify that

$$(2.24) \qquad \Psi(t(1, p^{\delta_2}, \ldots, d^{\delta_n}) t(1, p^{\delta'_2}, \ldots, d^{\delta'_n})) = t(p^{\delta_2}, \ldots, d^{\delta_n}) t(p^{\delta'_2}, \ldots, d^{\delta'_n}),$$

where $0 \leqslant \delta_2 \leqslant \cdots \leqslant \delta_n$, $0 \leqslant \delta'_2 \leqslant \cdots \leqslant \delta'_n$. We set $g_0 = \mathrm{diag}(p^{\delta_2}, \ldots, p^{\delta_n})$, $g'_0 = \mathrm{diag}(p^{\delta'_2}, \ldots, p^{\delta'_n})$, $g = \begin{pmatrix} 1 & 0 \\ 0 & g_0 \end{pmatrix}$, and $g' = \begin{pmatrix} 1 & 0 \\ 0 & g'_0 \end{pmatrix}$. By Lemma 1.5 we have

$$(g)(g') = \sum_{\Lambda h \Lambda \in \Lambda g \Lambda g' \Lambda} c(g, g'; h)(h).$$

From Lemma 1.5 and the relations (2.5) it follows that every element (h) in this expansion has the form $t(p^{\gamma_1}, \ldots, p^{\gamma_n})$, where $0 \leqslant \gamma_1 \leqslant \cdots \leqslant \gamma_n$ and $\gamma_1 + \cdots + \gamma_n = \delta_2 + \cdots + \delta_n + \delta'_2 + \cdots + \delta'_n = \delta$. Thus, by the definition of Ψ, we find that

$$\Psi((g)(g')) = \sum_{\substack{0 \leqslant \gamma_2 \leqslant \cdots \leqslant \gamma_n \\ \gamma_2 + \cdots + \gamma_n = \delta}} c(g, g'; \mathrm{diag}(1, p^{\gamma_2}, \ldots, p^{\gamma_n})) t(p^{\gamma_2}, \ldots, p^{\gamma_n}).$$

Similarly, in the ring $\underline{H}^{n-1}_p$ we have the relation

$$(g_0)(g'_0) = \sum_{\substack{0 \leqslant \gamma_2 \leqslant \cdots \leqslant \gamma_n \\ \gamma_2 + \cdots + \gamma_n = \delta}} c(g_0, g'_0; \mathrm{diag}(p^{\gamma_2}, \ldots, p^{\gamma_n})) t(p^{\gamma_2}, \ldots, p^{\gamma_n}).$$

These relations imply that to prove (2.24) it suffices to verify that

$$(2.25) \qquad\qquad c(g, g'; h) = c(g_0, g'_0; h_0),$$

if $h = \begin{pmatrix} 1 & 0 \\ 0 & h_0 \end{pmatrix}$ and $h_0 = \mathrm{diag}(p^{\gamma_2}, \ldots, p^{\gamma_n})$. Since $c(g, g'; h)$ depends only on

the double coset of h, we have $c(g, g'; h) = c(g, g'; h')$, where $h' = \begin{pmatrix} h_0 & 0 \\ 0 & 1 \end{pmatrix}$. By

Lemmas 1.5 and 2.7, the coefficient $c(g, g'; h')$ is equal to the number of pairs (C, C') of reduced matrices such that $C \in \Lambda g \Lambda$, $C' \in \Lambda g' \Lambda$, and $CC' = \lambda h'$, where $\lambda \in \Lambda$. Since $\lambda = CC'(h')^{-1}$, it follows that λ must be a triangular matrix. In particular, it can

be written in the form $\lambda = \begin{pmatrix} \lambda_0 & l \\ 0 & \lambda_{nn} \end{pmatrix}$, where $\lambda_0 \in \Lambda^{n-1}$ and $\lambda_{nn} = \pm 1$. If we similarly

divide C and C' into blocks, we see that these matrices can be written in the form

$C = \begin{pmatrix} C_0 & v \\ 0 & c_{nn} \end{pmatrix}$, $C' = \begin{pmatrix} C_0' & v' \\ 0 & c_{nn}' \end{pmatrix}$, where C_0 and C_0' are reduced $(n-1) \times (n-1)$-

matrices, $c_{nn}, c_{nn}' > 0$, and the entries in the columns v and v' are nonnegative and less than c_{nn} and c_{nn}', respectively. From the relation $CC' = \lambda h'$ it now follows that $c_{nn} c_{nn}' = \lambda_{nn}$ and $C_0 C_0' = \lambda_0 h_0$. The first equality gives us $c_{nn} = c_{nn}' = 1$, and so

$v = v' = 0$. Thus, $C = \begin{pmatrix} C_0 & 0 \\ 0 & 1 \end{pmatrix}$ and $C' = \begin{pmatrix} C_0' & 0 \\ 0 & 1 \end{pmatrix}$, where C_0 and C_0' are reduced

$(n-1) \times (n-1)$-matrices satisfying the relation $C_0 C_0' = \lambda_0 h_0$ with $\lambda_0 \in \Lambda^{n-1}$. Clearly $C_0 \in \Lambda^{n-1} g_0 \Lambda^{n-1}$ and $C_0' \in \Lambda^{n-1} g_0' \Lambda^{n-1}$. Conversely, any two matrices C and C' with these properties are reduced, belong to the double cosets $\Lambda g \Lambda$ and $\Lambda g' \Lambda$, respectively, and satisfy the relation $CC' = \lambda h'$ with $\lambda \in \Lambda$. Hence, $c(g, g'; h') = c(g_0, g_0'; h_0)$. $\square$

We can now completely determine the structure of the rings $\underline{H}_p^n$ and H_p^n.

THEOREM 2.17. *Let $n \geqslant 1$, and let p be a prime number. Then:*
(1) *the ring $\underline{H}_p^n$ is generated over $\mathbf{Q}$ by the elements*

$$(2.26) \qquad \pi_i(p) = \pi_i^n(p) = (\mathrm{diag}(\underbrace{1, \ldots, 1}_{n-i}, \underbrace{p, \ldots, p}_{i})) \quad (1 \leqslant i \leqslant n);$$

(2) *the Hecke ring H_p^n is generated over $\mathbf{Q}$ by the elements $\pi_1(p), \ldots, \pi_{n-1}(p)$ and $\pi_n(p)^{\pm 1}$;*
(3) *the elements $\pi_1(p), \ldots, \pi_n(p)$ are algebraically independent over $\mathbf{Q}$.*

REMARK. We are identifying $\mathbf{Q}$ with the subring $\mathbf{Q}(E_n)_\Lambda \subset \underline{H}_p^n$.

PROOF. To prove the first part it suffices to verify that every element $t(p^{\delta_1}, \ldots, p^{\delta_n})$, where $1 \leqslant \delta_1 \leqslant \cdots \leqslant \delta_n$, is a polynomial in $\pi_1(p), \ldots, \pi_n(p)$ with rational coefficients. We prove this by induction on n, and for each fixed $n > 1$ by induction on $N = \delta_1 + \cdots + \delta_n$. If $n = 1$, the claim is obvious, since $t(p^\delta) = t(p)^\delta = \pi_1^1(p)^\delta$. Suppose that $n > 1$, and the claim has been verified for smaller orders. If $N = \delta_1 + \cdots + \delta_n = 1$, then $t(p^{\delta_1}, \ldots, p^{\delta_n}) = \pi_1(p)$. Suppose that for all $t(p^{\delta_1'}, \ldots, p^{\delta_n'})$ with $\delta_1' + \cdots + \delta_n' < N$ it is already known that they are polynomials in the elements (2.26), and let $0 \leqslant \delta_1 \leqslant \cdots \leqslant \delta_n$ and $\delta_1 + \cdots + \delta_n = N$. If $\delta_1 \geqslant 1$, then by Lemma 2.4 we have

$$t(p^{\delta_1}, \ldots, p^{\delta_n}) = \pi_n(p)^{\delta_1} t(1, p^{\delta_2 - \delta_1}, \ldots, p^{\delta_n - \delta_1}).$$

Since $(\delta_2 - \delta_1) + \cdots + (\delta_n - \delta_1) < N$, it follows that the element $t(1, p^{\delta_2 - \delta_1}, \ldots, p^{\delta_n - \delta_1})$ is a polynomial in $\pi_1(p), \ldots, \pi_n(p)$, and hence so is the element $t(p^{\delta_1}, \ldots, p^{\delta_n})$. Suppose that $\delta_1 = 0$. By the first induction assumption, the element

$$\Psi(t(1, p^{\delta_2}, \ldots, p^{\delta_n})) = t(p^{\delta_2}, \ldots, p^{\delta_n}),$$

where Ψ is the homomorphism in Lemma 2.16, is a polynomial in the $\pi_i^{n-1}(p)$:

$$t(p^{\delta_2}, \ldots, p^{\delta_n}) = F(\pi_1^{n-1}(p), \ldots, \pi_{n-1}^{n-1}(p)),$$

where

$$F(x_1, \ldots, x_{n-1}) = \sum_{\alpha=(\alpha_1,\ldots,\alpha_{n-1})} a_\alpha x_1^{\alpha_1} \cdots x_{n-1}^{\alpha_{n-1}}.$$

Since each element (h) in the expansion of the product $(\pi_1^{n-1}(p))^{\alpha_1} \cdots (\pi_{n-1}^{n-1}(p))^{\alpha_{n-1}}$ obviously satisfies the relation $|\det h| = p^{|\alpha|}$ with $|\alpha| = \alpha_1 + 2\alpha_2 + \cdots + (n-1)\alpha_{n-1}$, it follows that, after combining similar terms, we may assume that the only nonzero coefficients a_α in F are those for which $|\alpha| = \delta_2 + \cdots + \delta_n = N$. Since $\Psi(\pi_i^n(p)) = \pi_i^{n-1}(p)$ $(1 \leqslant i \leqslant n-1)$, it follows that Ψ takes the element

$$t = t(1, p^{\delta_2}, \ldots, p^{\delta_n}) - \sum_{|\alpha|=N} a_\alpha(\pi_1^n(p))^{\alpha_1} \cdots (\pi_{n-1}^n(p))^{\alpha_{n-1}}$$

to zero. Thus, t is imprimitive. On the other hand, from the form of t it follows that t is a linear combination of elements $t(p^{\delta_1'}, \ldots, p^{\delta_n'})$ with $\delta_1' + \cdots + \delta_n' = N$. Hence, $t = \pi_n(p)t'$, where t' is a linear combination of elements $t(p^{\gamma_1}, \ldots, p^{\gamma_n})$ with $\gamma_1 + \cdots + \gamma_n < N$. By the second induction assumption, t' is a polynomial in $\pi_1^n(p), \ldots, \pi_n^n(p)$, and so the same is true of t. Part (1) of the theorem is proved.

Part (2) follows from part (1) and Lemma 2.15.

We prove the third part by induction on n. If $n = 1$, it is obvious, since the elements $\pi_1^1(p)^\delta = (p^\delta)$, $\delta = 0, 1, 2, \ldots$, correspond to pairwise distinct double cosets modulo $\Lambda^1 = \{\pm 1\}$, and so they are linearly independent. Suppose that $n > 1$, and the claim has been verified for all $n' < n$. Suppose that the elements $\pi_1^n(p), \ldots, \pi_n^n(p)$ are algebraically dependent over $\mathbf{Q}$, and let $F(\pi_1^n(p), \ldots, \pi_n^n(p)) = 0$ be an algebraic relation in which the polynomial F has the smallest possible degree. If we now apply the homomorphism Ψ, we obtain

$$\begin{aligned} \Psi(F(\pi_1^n(p), \ldots, \pi_n^n(p))) &= F(\Psi(\pi_1^n(p)), \ldots, \Psi(\pi_n^n(p))) \\ &= F(\pi_1^{n-1}(p), \ldots, \pi_{n-1}^{n-1}(p), 0) = 0. \end{aligned}$$

By the induction assumption, this implies that $F = x_n F_1$, where $F_1 = F_1(x_1, \ldots, x_n)$ is also a polynomial. By Lemma 2.15, $\pi_n^n(p)$ is not a zero divisor. Hence, the relation $0 = F(\pi_1^n(p), \ldots, \pi_n^n(p)) = \pi_n^n(p)F_1(\pi_1^n(p), \ldots, \pi_n^n(p))$ implies that $F_1(\pi_1^n(p), \ldots, \pi_n^n(p)) = 0$, and this contradicts our assumption that F has minimal degree. $\square$

We conclude this subsection by proving some technical facts about the generators $\pi_i(p)$ and their pairwise products which will be needed later.

LEMMA 2.18. *Let $n \geqslant 1$, and let p be a prime number. Then:*
(1) *For $0 \leqslant i \leqslant n$ the subset of reduced matrices*

(2.27)
$$\{C = (c_{\alpha\beta}) \in M_n; \det C = p^i, c_{\alpha\alpha} = 1, \text{ or } p,$$
$$c_{\alpha\beta} = 0, \text{ if } \alpha > \beta, \text{ or } \alpha < \beta \text{ and } c_{\alpha\alpha} = p\}$$

is a complete set of representatives of the distinct left cosets of $\Lambda D_i \Lambda$ modulo $\Lambda = \Lambda^n$; the number of matrices in this subset is equal to

$$\mu_\Lambda(D_i) = \frac{\varphi_n}{\varphi_i \varphi_{n-1}},$$

where

$$(2.28) \qquad D_i = D_i^n(p) = \begin{pmatrix} E_{n-i} & 0 \\ 0 & pE_i \end{pmatrix}, \quad \varphi_r = \varphi_r(p),$$

$$(2.29) \qquad \varphi_r(x) = \prod_{i=1}^{r}(x^i - 1) \quad \text{for } r \geqslant 1 \text{ and } \varphi_0(x) = 1.$$

(2) *The double coset expansion of the product in the Hecke ring H_p^n of the two elements $\pi_i = \pi_i^n(p)$ and $\pi_j = \pi_j^n(p)$, where $1 \leqslant i, j \leqslant n$, has the form*

$$(2.30) \qquad \pi_i \pi_j = \sum_{\substack{0 \leqslant a \leqslant n-j \\ 0 \leqslant b \leqslant j \\ a+b=i}} \frac{\varphi_{a+j-b}}{\varphi_a \varphi_{j-b}} \pi_{a+j-b,b},$$

where

$$\pi_{\alpha\beta} = \pi_{\alpha\beta}^n(p) = (D_{\alpha\beta})_\Lambda$$

$$(2.31) \qquad and \ D_{\alpha\beta} = D_{\alpha\beta}^n(p) = \begin{pmatrix} E_{n-\alpha-\beta} & 0 & 0 \\ 0 & pE_\alpha & 0 \\ 0 & 0 & p^2 E_\beta \end{pmatrix}.$$

The number of Λ-left cosets in the double coset $\Lambda D_{\alpha\beta}\Lambda$ is given by the formula

$$(2.32) \qquad \mu_\Lambda(D_{\alpha\beta}) = p^{\beta(n-\alpha-\beta)} \frac{\varphi_n}{\varphi_{n-\alpha-\beta}\varphi_\alpha\varphi_\beta}.$$

PROOF. Every matrix in (2.27) can be written in the form $\mathrm{diag}(p^{\delta_1}, \ldots, p^{\delta_n})C'$, where $\delta_n = 0, 1, \delta_1 + \cdots + \delta_n = i$, and $C' \in \Lambda$. Thus, the matrix lies in $\Lambda D_i \Lambda$. By Lemma 2.7, all of these matrices belong to different Λ-left cosets. Now let C be any matrix of the form (2.15) that lies in $\Lambda D_i \Lambda$. Its diagonal entries $c_{\alpha\alpha}$ are positive integers, and their product is p^i. Hence, $c_{\alpha\alpha} = p^{\delta_\alpha}$, where $\delta_\alpha \geqslant 0, \delta_1 + \cdots + \delta_n = i$. From the integrality of the matrix $p \cdot C^{-1}$ it follows that each δ_α is either 0 or 1. Suppose that $c_{\alpha\beta} \neq 0$, where $1 \leqslant \alpha < \beta \leqslant n$. Then $\delta_\beta = 1$ and $c_{\alpha\beta}$ is not divisible by p. We let $\gamma_1, \ldots, \gamma_{n-i}$ denote the indices γ for which $\delta_\gamma = 0$. If $\delta_\alpha = 1$, then the γ_1th, $\ldots, \gamma_{n-i}$th, and βth columns of C are linearly independent modulo p; but the rank of C modulo p is obviously equal to $n - i$. Thus, $\delta_\alpha = 0$, and C lies in the set (2.27).

It is easy to see that the number of elements of the set (2.27) with fixed diagonal and with $\delta_\alpha = 1$ precisely when $\alpha = \alpha_1, \alpha_2, \ldots, \alpha_i$, where $1 \leqslant \alpha_1 < \alpha_2 < \cdots < \alpha_i \leqslant n$, is equal to $p^{\alpha_1 - 1} \cdots p^{\alpha_i - i}$; this implies that the number of elements in the set (2.27) is

$$p^{-\langle i \rangle} \sum_{1 \leqslant \alpha_1 < \cdots < \alpha_i \leqslant n} p^{\alpha_1 + \cdots + \alpha_i},$$

and to prove (2.28) it suffices to verify the following identity:

$$(2.33) \qquad \sum_{1 \leqslant \alpha_1 < \cdots < \alpha_i \leqslant n} x^{\alpha_1 + \cdots + \alpha_i} = x^{\langle i \rangle} \frac{\varphi_n(x)}{\varphi_i(x)\varphi_{n-i}(x)} \quad (1 \leqslant i \leqslant n),$$

where $\varphi_r(x)$ is the function (2.29). This identity can be obtained by equating coefficients of t^i on both sides of the identity

$$(2.34) \qquad \prod_{\alpha=1}^{n}(1+tx^\alpha) = \sum_{i=0}^{n} t^i \frac{x^{\langle i \rangle}\varphi_n(x)}{\varphi_i(x)\varphi_{n-i}(x)},$$

an identity which the reader can easily prove by induction on n, in a manner analogous to the standard proof of Newton's binomial expansion.

To prove (2.30) we use induction on n. If $n = 1$, we must prove the formula $\pi_1^1 \pi_1^1 = \pi_{0,1}^1$, which is obvious. Now let $n > 1$, and suppose that the formula has been proved for smaller orders. By Lemma 1.5 and the first part of Lemma 2.18, we have the expansion

$$\pi_i \pi_j = \sum_{\Lambda g \Lambda \in G_p} c_{ij}(g)(g)_\Lambda,$$

where

$$c_{ij}(g) = \nu(D_i, D_j; g)\mu_\Lambda(D_j)\mu_\Lambda(g)^{-1}$$

and $\nu(D_i, D_j; g)$ is the number of matrices C in the set (2.27) such that $CD_j \in \Lambda g \Lambda$. This number is easy to compute. If the diagonal entries in C are $p^{\delta_1}, \ldots, p^{\delta_n}$, then, as noted before, $C = \operatorname{diag}(p^{\delta_1}, \ldots, p^{\delta_n})C_1$, where C_1 is an upper-triangular matrix in Λ. Then

$$CD_j = \operatorname{diag}(p^{\delta_1}, \ldots, p^{\delta_n})D_j D_j^{-1} C_1 D_j \in \Lambda \operatorname{diag}(p^{\delta_1}, \ldots, p^{\delta_n})D_j \Lambda,$$

where $D_j = D_j^n(p)$, since obviously $D_j^{-1}C_1 D_j \in \Lambda$. Let $\alpha_1 < \cdots < \alpha_i$ be the indices α for which $\delta_\alpha = 1$. If a of the numbers $\alpha_1, \ldots, \alpha_i$ do not exceed $n-j$ and $b = i - a$ of these numbers are greater than $n-j$, then clearly

$$\operatorname{diag}(p^{\delta_1}, \ldots, p^{\delta_n})D_j \in \Lambda D_{a+j-b,b}\Lambda,$$

where $D_{\alpha\beta} = D_{\alpha\beta}^n(p)$. As noted before, the number of matrices C with fixed $\alpha_1, \ldots, \alpha_i$ is equal to $p^{\alpha_1-1} \cdots p^{\alpha_i-i}$. Thus, the number of matrices C in the set (2.27) such that $CD_j \in \Lambda D_{a+j-b,b}\Lambda$ for fixed a and b satisfying the inequalities $0 \leqslant a \leqslant n-j$, $0 \leqslant b \leqslant j$, and $a+b = i$, is equal to

$$\sum_{1 \leqslant \alpha_1 < \cdots < \alpha_a \leqslant n-j < \alpha_{a+1} < \cdots < \alpha_i \leqslant n} p^{\alpha_1+\cdots+\alpha_i - \langle i \rangle}$$

$$= p^{-\langle i \rangle} \sum_{\substack{1 \leqslant \alpha_1 < \cdots < \alpha_a \leqslant n-j \\ 1 \leqslant \beta_1 < \cdots < \beta_b \leqslant j}} p^{\alpha_1+\cdots+\alpha_a+(\beta_1+(n-j))+\cdots+(\beta_b+(n-j))}$$

$$= p^{b(n-j)-\langle i \rangle} \frac{p^{\langle a \rangle}\varphi_{n-j}}{\varphi_a\varphi_{n-j-a}} \cdot \frac{p^{\langle a \rangle}\varphi_j}{\varphi_b\varphi_{j-b}} = p^{b(n-j-a)} \frac{\varphi_{n-j}\varphi_j}{\varphi_a\varphi_{n-j-a}\varphi_b\varphi_{j-b}}$$

(we have used the identities (2.33)). These arguments, along with (2.28), imply that

$$(2.35) \qquad \pi_i \pi_j = \sum_{\substack{0 \leqslant a \leqslant n-j \\ 0 \leqslant b \leqslant j \\ a+b=i}} c_{ij}(D_{a+j-b,b})\pi_{a+j-b,b},$$

where

$$c_{ij}(D_{a+j-b,b}) = p^{b(n-j-a)} \frac{\varphi_n}{\varphi_a\varphi_{n-j-a}\varphi_b\varphi_{j-b}}\mu_\Lambda(D_{a+j-b,b})^{-1}.$$

In the case when $i = n$ or $j = n$, (2.30) follows from the definitions and Lemma 2.4. Hence, we may assume that $1 \leqslant i, j < n$. If we now apply the homomorphism Ψ of Lemma 2.16 to the left and right sides of (2.35), we find that

$$\pi_i^{n-1}(p)\pi_j^{(n-1)}(p) = \sum_{\substack{0 \leqslant a \leqslant n-j \\ 0 \leqslant b \leqslant j \\ a+b=i}} c_{ij}(D_{a+j-b,b})\pi_{a+j-b,b}^{n-1}(p).$$

On the other hand, by the induction assumption we have

$$\pi_i^{n-1}(p)\pi_j^{(n-1)}(p) = \sum_{\substack{0 \leqslant a \leqslant n-1-j \\ 0 \leqslant b \leqslant j \\ a+b=i}} \frac{\varphi_{a+j-b}}{\varphi_a \varphi_{j-b}} \pi_{a+j-b,b}^{n-1}(p).$$

Since the double cosets in this expansion are linearly independent, we obtain the relation

$$c_{ij}(D_{a+j-b,b}) = \frac{\varphi_{a+j-b}}{\varphi_a \varphi_{j-b}},$$

where $a + b = i, 0 \leqslant a \leqslant n - 1 - j, 0 \leqslant b \leqslant j$. The same formula can be obtained in the case $a = n - j$ if we use the original formula for $c_{ij}(D_{2n-i-j,i+j-n})$ and (2.28) and take into account that

$$\mu_\Lambda(D_{2n-i-j,i+j-n}) = \mu_\Lambda(D_{i+j-n}^n(p)) = \frac{\varphi_n}{\varphi_{i+j-n}\varphi_{2n-i-j}}.$$

This proves (2.30). Comparing the expressions for the coefficients $c_{ij}(D_{a+j-b,b})$, we find that

$$\mu_\Lambda(D_{a+j-b,b}) = p^{b(n-j-a)}\frac{\varphi_n}{\varphi_{n-j-a}\varphi_{a+j-b}\varphi_b},$$

from which (2.32) follows if we set $a + j - b = \alpha, b = \beta$. $\square$

PROBLEM 2.19. Let $\mathbf{Q}_p$ be the field of p-adic numbers, and let $\mathbf{Z}_p$ be the ring of p-adic integers. Then $G = GL_n(\mathbf{Q}_p)$ is a locally compact group in the p-adic topology, and $\Gamma = GL_n(\mathbf{Z}_p)$ is a (maximal) compact subgroup. Let D_K, where K is a subring of $\mathbf{C}$, denote the K-module consisting of all continuous functions $f : G \to K$ with compact support which satisfy the condition $f(\gamma_1 g \gamma_2) = f(g)$ for any $\gamma_1, \gamma_2 \in \Gamma$ and $g \in G$. Fix the Haar measure μ on G for which $\mu(\Gamma) = 1$, and define the product $f * f_1$ of functions $f, f_1 \in D_K$ by the formula

$$(f * f_1)(h) = \int_G g(hg^{-1})f_1(g)\,d\mu(g) \quad (h \in G).$$

Show that the K-linear map from the Hecke ring $D_K(\Lambda^n, G_p^n)$ to D_K that associates to an element $(g)_\Lambda$ $(g \in G_p^n)$ the characteristic function of the double coset $\Gamma g \Gamma \subset G$, is an isomorphism of rings.

3. The spherical map. We have shown that every element in the local Hecke ring of the general linear group can be uniquely expressed as a polynomial in a finite number of generators. But often it is not so simple to find this polynomial if the element is given, say, as a linear combination of left cosets. In order to solve this problem, we define certain maps from the local Hecke rings to rings of symmetric polynomials that are analogous to the spherical functions in the representation theory of locally compact groups.

As in the previous subsection, we suppose that $n \geqslant 1$ and p is a prime number. We first define a linear map

$$\omega = \omega_p^n \colon L_{\mathbf{Q}}(\Lambda^n, G_p^n) \to \mathbf{Q}[x_1^{\pm 1}, \ldots, x_n^{\pm 1}]$$

from the $\mathbf{Q}$-vector space spanned by the distinct left cosets (Λg) of G_p^n modulo Λ^n to the subring of the field $\mathbf{Q}(x_1, \ldots, x_n)$ of rational functions in n variables that is generated over $\mathbf{Q}$ by $x_1^{\pm 1}, \ldots, x_n^{\pm 1}$. Lemma 2.7 implies that every left coset Λg $(g \in G_p^n)$ has a representative of the form

$$(2.36) \qquad \begin{pmatrix} p^{\delta_1} & c_{12} & \cdots & c_{1n} \\ 0 & p^{\delta_2} & \cdots & c_{2n} \\ \cdots & \cdots & \cdots & \cdots \\ 0 & 0 & \cdots & p^{\delta_n} \end{pmatrix}, \qquad \text{where } \delta_1, \ldots, \delta_n \in \mathbf{Z},$$

and the diagonal $(p^{\delta_1}, \ldots, p^{\delta_n})$ is uniquely determined by the left coset. We set

$$\omega((\Lambda g)) = \prod_{i=1}^{n} (x_i p^{-i})^{\delta_i}$$

and for an arbitrary element

$$t = \sum_j a_j (\Lambda g_j) \in L_{\mathbf{Q}}(\Lambda^n, G_p^n)$$

we define

$$\omega(t) = \sum_j a_j \omega((\Lambda g_j)).$$

We call ω the *spherical map*. We would like to describe the image of the Hecke rings under ω. Recall that an element of the field $\mathbf{Q}(x_1, \ldots, x_n)$ is said to be symmetric if it does not change under any permutation of the variables $x_1, \ldots, x_n$.

THEOREM 2.20. *The restriction of the map* $\omega = \omega_p^n$ *to the Hecke ring* $H_p^n = D_{\mathbf{Q}}(\Lambda^n, G_p^n) \subset L_{\mathbf{Q}}(\Lambda^n, G_p^n)$ *is an isomorphism of this ring with the ring* $\mathbf{Q}[x_1^{\pm 1}, \ldots, x_n^{\pm 1}]_s$ *of all symmetric elements of* $\mathbf{Q}[x_1^{\pm 1}, \ldots, x_n^{\pm 1}]$. *The image of the integral subring* $\underline{H}_p^n$ *under the map* ω *is the ring* $\mathbf{Q}[x_1, \ldots, x_n]_s$ *of all symmetric polynomials in* $x_1, \ldots, x_n$ *over* $\mathbf{Q}$.

We first prove a lemma.

LEMMA 2.21. *The images of the elements* (2.26) *under the map* $\omega = \omega_p^n$ *are given by the formulas*

$$\omega(\pi_i^n(p)) = p^{-\langle i \rangle} s_i(x_1, \ldots, x_n) \quad (1 \leqslant i \leqslant n),$$

where

$$s_i(x_1, \ldots, x_n) = \sum_{1 \leqslant \alpha_1 < \cdots < \alpha_i \leqslant n} x_{\alpha_1} \cdots x_{\alpha_i}$$

is the i th elementary symmetric polynomial.

PROOF. In the expansion of $\pi_i^n(p)$ into left cosets we take the set (2.27) as the representatives of the different left cosets, and we use the fact that the number of elements in this set for which $\delta_\alpha = 1$ precisely when $\alpha = \alpha_1, \alpha_2, \ldots, \alpha_i$, where $1 \leqslant \alpha_1 < \cdots < \alpha_i \leqslant n$, is equal to $p^{\alpha_1 - 1} \cdots p^{\alpha_i - i}$. By the definition of ω we then have

$$\omega(\pi_i^n(p)) = \sum_{1 \leqslant \alpha_1 < \cdots < \alpha_i \leqslant n} p^{(\alpha_1 - 1) + \cdots + (\alpha_i - i)} (x_{\alpha_1} p^{-\alpha_1}) \cdots (x_{\alpha_i} p^{-\alpha_i})$$

$$= p^{-\langle i \rangle} s_i(x_1, \ldots, x_n). \qquad \square$$

PROOF OF THEOREM 2.20. We first show that the restriction of ω to H_p is a ring homomorphism. In fact, if $t = \sum_i a_i(\Lambda g_i)$, $t' = \sum_j b_j(\Lambda g_j')$ are two elements of H_p, then it follows from Lemma 2.7 that all of the left coset representatives g_i, g_j' can be assumed to have been chosen in the form (2.36). Then all of the representatives $g_i g_j'$ in the expansion of the product

$$tt' = \sum_{i,j} a_i b_j(\Lambda g_i g_j'),$$

have the same form, and the diagonal entries in the matrices $g_i g_j'$ are equal to the products of the corresponding diagonal entries in g_i and g_j'. Thus, from the definition of ω we have $\omega((\Lambda g_i g_j')) = \omega((\Lambda g_i))\omega((\Lambda g_j'))$. Using this and the linearity of ω, we obtain the relation $\omega(tt') = \omega(t)\omega(t')$.

Clearly $\omega((E_n)) = 1$, where $(E_n) = (E_n)_\Lambda$ is the unit of the ring H_p. From Theorem 2.17 it then follows that the ω-image of $\underline{H}_p$ consists of all polynomials over $\mathbf{Q}$ in the elements $\omega(\pi_i(p))$ $(1 \leqslant i \leqslant n)$, and the ω-image of H_p is generated by $\omega(\underline{H}_p)$ and the element $\omega(\pi_n(p)^{-1}) = \omega(\pi_n(p))^{-1}$. But then Lemma 2.21 and the fundamental theorem on symmetric polynomials imply that $\omega(\underline{H}_p)$ coincides with the ring of symmetric polynomials in n variables over $\mathbf{Q}$, and the ring $\omega(H_p)$ is generated over $\omega(\underline{H}_p)$ by the element $(x_1 \cdots x_n)^{-1}$, and so obviously coincides with $\mathbf{Q}[x_1^{\pm 1}, \ldots, x_n^{\pm 1}]_s$.

Finally, if $\omega(t) = 0$ for some $t \in H_p$, then by Lemma 2.15 we can write $t = \pi_n(p)^\delta t_1$, where $\delta \in \mathbf{Z}$ and $t_1 \in \underline{H}_p$. Then, since $\omega(\pi_n(p)^{\pm 1}) \neq 0$, we have $\omega(t_1) = 0$. By Theorem 2.17, $t_1 = \Phi(\pi_1(p), \ldots, \pi_n(p))$, where $\Phi(x_1, \ldots, x_n)$ is a polynomial with rational coefficients. But by Lemma 2.21, the equality $\omega(t_1) = 0$ means that

$$\Phi(p^{-1} s_1(x_1, \ldots, x_n), \ldots, p^{-\langle n \rangle} s_n(x_1, \ldots, x_n)) = 0,$$

and because of the algebraic independence of elementary symmetric polynomials, this implies that $\Phi(x_1, \ldots, x_n) = 0$. Hence, $t_1 = 0$ and $t = 0$. $\qquad \square$

Theorem 2.20 reduces the computation of the product of elements of a local Hecke ring to the computation of the product of the corresponding symmetric polynomials. The problem of expressing elements of the local Hecke rings in terms of the generators $\pi_i(p)$ reduces to the problem of expressing symmetric polynomials in terms of the elementary symmetric polynomials.

We illustrate the usefulness of the spherical map by discussing the problem of summing the formal generating series for elements of the form $t(p^\delta)$, where p is a prime number. Note that when $n > 2$ these elements do not have a simple multiplication table (the case $n = 1$ is trivial, and the case $n = 2$ is treated in Problem 2.12).

PROPOSITION 2.22. *In the notation* (2.10) *and* (2.26), *the following identity holds in the ring of formal power series over* $\underline{H}_p^n$:

$$\sum_{\delta=0}^{\infty} t(p^\delta) v^\delta = \left(\sum_{i=0}^{n} (-1)^i p^{\langle i-1 \rangle} \pi_i(p) v^i \right)^{-1}.$$

PROOF. From the definition of $t(p^\delta)$ and Lemma 2.7 it follows that

$$t(p^\delta) = \sum_{\substack{\delta_i \geqslant 0 \\ \delta_1+\cdots+\delta_n=\delta \\ 0 \leqslant c_{ij} < p^{\delta_j}}} \left(\Lambda \begin{pmatrix} p^{\delta_1} & c_{12} & \cdots & c_{1n} \\ 0 & p^{\delta_2} & \cdots & c_{2n} \\ \cdots & \cdots & \cdots & \cdots \\ 0 & 0 & \cdots & p^{\delta_n} \end{pmatrix} \right).$$

By the definition of the map $\omega = \omega_p^n$ we then have

(2.37)
$$\omega(t(p^\delta)) = \sum_{\substack{\delta_i \geqslant 0 \\ \delta_1+\cdots+\delta_n=\delta}} p^{\delta_2} p^{2\delta_3} \cdots p^{(n-1)\delta_n} (x_1 p^{-1})^{\delta_1} \cdots (x_n p^{-n})^{\delta_n}$$
$$= \sum_{\substack{\delta_i \geqslant 0 \\ \delta_1+\cdots+\delta_n=\delta}} (x_1 p^{-1})^{\delta_1} \cdots (x_n p^{-1})^{\delta_n},$$

so that
(2.38)
$$\sum_{\delta=0}^{\infty} \omega(t(p^\delta)) v^\delta = \sum_{\delta_1,\ldots,\delta_n=0}^{\infty} (x_1 p^{-1})^{\delta_1} \cdots (x_n p^{-1})^{\delta_n} v^{\delta_1} \cdots v^{\delta_n}$$
$$= \left(\prod_{i=1}^{n} (1 - p^{-1} x_i v) \right)^{-1} = \left(\sum_{i=0}^{n} (-1)^i p^{-i} s_i(x_1,\ldots,x_n) v^i \right)^{-1}$$
$$= \left(\sum_{i=0}^{n} (-1)^i p^{\langle i-1 \rangle} \omega(\pi_i(p)) v_i \right)^{-1},$$

where we used Lemma 2.21 for the last step. From the identity for formal power series over the ring of polynomials in $x_1, \ldots, x_n$ it follows that all of the coefficients in the formal series

$$\left(\sum_{\delta=0}^{\infty} t(p^\delta) v^\delta \right) \left(\sum_{i=0}^{n} (-1)^i p^{\langle i-1 \rangle} \pi_i(p) v_i \right),$$

except for the coefficient of v^0, belong to the kernel of ω, and hence are equal to zero. As for the constant term, it is obviously equal to the unit of the ring $\underline{H}_p$. $\qquad\square$

This proposition can be used to find explicit expressions for $t(p^\delta)$ in terms of the generators $\pi_i(p)$.

We conclude this section by describing the anti-automorphism j of the Hecke ring H_p^n (see §1.4) in terms of symmetric functions.

LEMMA 2.23. *Let $n \geqslant 1$, and let p be a prime number. Then the diagram*

$$
\begin{array}{ccc}
H_p^n & \xrightarrow{\ \omega\ } & \mathbf{Q}[x_1^{\pm 1}, \ldots, x_n^{\pm 1}] \\
\downarrow{\scriptstyle j} & & \downarrow{\scriptstyle w} \\
H_p^n & \xrightarrow{\ \omega\ } & \mathbf{Q}[x_1^{\pm 1}, \ldots, x_n^{\pm 1}]
\end{array}
$$

commutes, where ω is the spherical map, j is the anti-automorphism (1.31), *and $w = w_{n,p}$ denotes the $\mathbf{Q}$-linear ring homomorphism given by $w(x_i) = p^{n+1} x_i^{-1}$ $(1 \leqslant i \leqslant n)$ on the generators.*

PROOF. Since H_p^n is a commutative ring, all of the maps in this diagram are $\mathbf{Q}$-linear ring homomorphisms. Taking into account Theorem 2.17 and Lemma 2.4, it is therefore sufficient to verify that $\omega(j(\pi_i(p))) = w(\omega(\pi_i(p)))$ $(1 \leqslant i \leqslant n)$. We have

$$
\begin{aligned}
\omega(j(\pi_i(p))) &= \omega((p^{-1} D_{n-i})_\Lambda) = \omega(\pi_n(p)^{-1} \pi_{n-i}(p)) \\
&= p^{\langle n \rangle} (x_1 \cdots x_n)^{-1} p^{-\langle n-i \rangle} s_{n-i}(x_1 \cdots x_n) \\
&= p^{-\langle i \rangle} s_i(p^{n+1} x_1^{-1}, \ldots, p^{n+1} x_n^{-1}) = w(\omega(\pi_i(p))),
\end{aligned}
$$

where we have used Lemma 2.21 twice in the computation. $\qquad\square$

PROBLEM 2.24. Show that any $\mathbf{Q}$-linear ring homomorphism from H_p^n to $\mathbf{C}$ that takes the unit of H_p^n to 1 has the form

$$
t \to \lambda(t) = \omega(t)|_{x_1 = \lambda_1, \ldots, x_n = \lambda_n} \quad (t \in H_p^n),
$$

where ω is the spherical map and $\lambda_1, \ldots, \lambda_n$ are nonzero complex numbers that are determined except for their order by the homomorphism λ. The numbers $\lambda_1, \ldots, \lambda_n$ are called the *parameters of the homomorphism* λ. Show that for arbitrary nonzero $\lambda_1, \ldots, \lambda_n \in \mathbf{C}$ there exists a homomorphism $\lambda \colon H_p^n \to \mathbf{C}$ having parameters $\lambda_1, \ldots, \lambda_n$.

PROBLEM 2.25. Let $\lambda \colon H^n \to \mathbf{C}$ be a $\mathbf{Q}$-linear homomorphism. Show that the formal zeta-function of the ring H^n with character λ, i.e., the formal Dirichlet series

$$
Z(s, \lambda) = \sum_{d=1}^{\infty} \frac{\lambda(t(d))}{d^s},
$$

where $t(d)$ is the element (2.13), has a formal Euler product expansion of the form

$$
Z(s, \lambda) = \prod_p Z_p(s, \lambda_p),
$$

where p runs through all prime numbers, λ_p denotes the restriction of λ to H_p^n, and the local zeta-functions $Z_p(s, \lambda_p)$ have the form

$$
Z_p(s, \lambda_p) = \prod_{i=1}^{n} (1 - \lambda_i(p) p^{-s-1})^{-1},
$$

where $\lambda_1(p), \ldots, \lambda_n(p)$ are the parameters of the homomorphism λ_p.

PROBLEM 2.26. We return to the notation of Problem 2.19. Recall that a continuous function ω on the group $G = GL_n(\mathbf{Q}_p)$ with values in $\mathbf{C}$ is called a (zonal) spherical function if $\omega(\gamma_1 g \gamma_2) = \omega(g)$ for any $\gamma_1, \gamma_2 \in \Gamma = GL_n(\mathbf{Z}_p)$ and $g \in G$, $\omega(E_n) = 1$, and the map

$$f \to \widehat{\omega}(f) = \int_G f(g)\omega(g)\,d\mu(g)$$

is a ring homomorphism from $D_{\mathbf{C}}$ to $\mathbf{C}$. Show that every spherical function on G has the form

$$\omega_{(\lambda_1,\ldots,\lambda_n)}(g) = \int_\Gamma \varphi_{(\lambda_1,\ldots,\lambda_n)}(g\gamma)\,d\mu(\gamma),$$

where $\lambda_1,\ldots,\lambda_n$ are nonzero complex numbers and the function $\varphi = \varphi_{(\lambda_1,\ldots,\lambda_n)}$ is given by the conditions: $\varphi(\gamma g) = \varphi(g)$ for $\gamma \in \Gamma$ and $g \in G$, and

$$\varphi\left(\begin{pmatrix} p^{\delta_1} & & * \\ & \ddots & \\ 0 & & p^{\delta_n} \end{pmatrix}\right) = \prod_{i=1}^n (\lambda_i\, p^{-i})^{\delta_i}.$$

Show that the spherical functions $\omega_{(\lambda_1,\ldots,\lambda_n)}$ and $\omega_{(\lambda_1',\ldots,\lambda_n')}$ coincide if and only if the numbers $\lambda_1',\ldots,\lambda_n'$ are a permutation of the numbers $\lambda_1,\ldots,\lambda_n$.

[**Hint:** Use Problems 2.19 and 2.24, Theorem 2.20, and Lemma 2.21.]

§3. Hecke rings for the symplectic group

1. Global rings. In all of the Hecke pairs for subgroups of the symplectic group that we shall be considering, the second component is a subgroup of the group

$$S^n = S_{\mathbf{Q}}^n = G\mathrm{Sp}_n^+(\mathbf{Q}).$$

From the definition it easily follows that a rational $(2n \times 2n)$-matrix $M = \begin{pmatrix} A & B \\ C & D \end{pmatrix}$ lies in S^n if and only if its $(n \times n)$-blocks satisfy the relations

$$(3.1) \qquad {}^tAC = {}^tCA, \quad {}^tBD = {}^tDB, \quad {}^tAD - {}^tCB = r(M)E_n,$$

where $r(M) > 0$. Since the matrix ${}^tM = r(M)J_n M^{-1} J_n^{-1}$ lies in S^n whenever M does, the conditions (3.1) are equivalent to the conditions

$$(3.2) \qquad A\,{}^tB = B\,{}^tA, \quad C\,{}^tD = D\,{}^tC, \quad A\,{}^tD - B\,{}^tC = r(M)E_n.$$

LEMMA 3.1. *Let K be an arbitrary congruence subgroup of the modular group $\Gamma = \Gamma^n = \mathrm{Sp}_n(\mathbf{Z})$. Then the commensurator of K in the group S^n is all of S^n. In particular, (K, S^n) is a Hecke pair.*

PROOF. Let $M \in S^n$. According to §3.3 of Chapter 2, the intersection $M^{-1}KM \cap \Gamma$ is a congruence subgroup of Γ. The group $K_{(M)} = K \cap M^{-1}KM = \Gamma \cap K \cap M^{-1}KM$ is also a congruence subgroup, and so it has finite index in K. If we replace M by M^{-1}, we see that $K_{(M^{-1})}$ has finite index in K, and therefore $K_{(M)} = M^{-1}K_{(M^{-1})}M$ has finite index in $M^{-1}KM$. $\qquad\square$

Using Lemma 3.1 as a point of departure, one could determine the Hecke ring of the pair (K, S^n) and then consider its representations on spaces of modular forms for the group K. However, the structure of the Hecke rings that arise is in general unknown, and one does not yet have a concrete general theory of Hecke operators. Because our constructions are not meant as an end in themselves, but rather as a means for studying Diophantine problems in number theory, we shall simplify the situation by, in the first place, limiting ourselves to the types of congruence subgroups that arise in arithmetic, and, in the second place, considering certain subrings of the Hecke ring of the pair (K, S^n), rather than the entire Hecke ring.

We first prove an approximation lemma.

LEMMA 3.2. (1) *The natural homomorphisms* $\mod q$

$$SL_n(\mathbf{Z}) \to SL_n(\mathbf{Z}/q\mathbf{Z})$$

and

$$\Gamma^n = \mathrm{Sp}_n(\mathbf{Z}) \to \mathrm{Sp}_n(\mathbf{Z}/q\mathbf{Z}),$$

where $n, q \in \mathbf{N}$, *are epimorphisms.*

(2) *If* q *and* q_1 *are relatively prime, then*

$$\Gamma^n(q)\Gamma^n(q_1) = \Gamma^n.$$

PROOF. We prove the first part separately for each of the two groups using induction on n. For SL_n it is obvious in the case $n = 1$. Suppose that $n > 1$, and the claim has been proved for SL_m with $m < n$. Let T be an $n \times n$ integer matrix such that $\det T \equiv 1 (\mod q)$. We can replace T by a matrix congruent to it modulo q that has the property that the entries in its first column are relatively prime. Then, by Lemma 3.8 of Chapter 1, there exists $V_1 \in SL_n(\mathbf{Z})$ such that $V_1 T = \begin{pmatrix} 1 & T_2 \\ 0 & T_4 \end{pmatrix}$. Since $\det T_4 \equiv \det T \equiv 1 (\mod q)$, it follows by the induction assumption that there exists $V_2 \in SL_{n-1}(\mathbf{Z})$ that is congruent to T_4 modulo q. Then $T \equiv V_1^{-1} \begin{pmatrix} 1 & T_2 \\ 0 & V_2 \end{pmatrix} (\mod q)$, and the last matrix obviously lies in $SL_n(\mathbf{Z})$. This proves the first part of the lemma for SL_n. Since $\mathrm{Sp}_1(\mathbf{Z}) = SL_2(\mathbf{Z})$, part one of the lemma holds for $\mathrm{Sp}_1(\mathbf{Z})$. Suppose that $n > 1$, and the claim has already been proved for Sp_m with $m < n$. If M is a $2n \times 2n$ integer matrix satisfying the congruence $J_n[M] \equiv J_n (\mod q)$, then we first replace the entries in the first column of M by suitable integers congruent to them modulo q so that they are relatively prime. Applying Lemma 3.9 of Chapter 1 to the first column of M, we find that there exists $g \in \Gamma^n$ such that the first column of the matrix $M' = gM$ is the same as the first column of the identity matrix E_{2n}. The reader can easily verify that it is always possible to choose matrices $V \in SL_n(\mathbf{Z})$ and $S = {}^t S \in M_n$ in such a way that the first row of the matrix

$$M'' = \begin{pmatrix} A & B \\ C & D \end{pmatrix} = M'g_1, \quad \text{where } g_1 = \begin{pmatrix} V & 0 \\ 0 & V^* \end{pmatrix} \begin{pmatrix} E & S \\ 0 & E \end{pmatrix} \in \Gamma^n,$$

is the same as the first row of E_{2n}. Thus,

$$A = \begin{pmatrix} 1 & 0 \\ 0 & A_0 \end{pmatrix}, \quad B = \begin{pmatrix} 0 & 0 \\ b_3 & B_0 \end{pmatrix}, \quad C = \begin{pmatrix} 0 & c_2 \\ 0 & C_0 \end{pmatrix}, \quad D = \begin{pmatrix} d_1 & d_2 \\ d_3 & D_0 \end{pmatrix},$$

where A_0, B_0, C_0, and D_0 are $(n-1) \times (n-1)$-blocks. Like M, the matrix M'' satisfies the congruence $J_n[M''] \equiv J_n (\mod q)$, which is equivalent to the following congruences for the blocks:

$$ {}^tAC \equiv {}^tCA, \quad {}^tBD \equiv {}^tDB, \quad {}^tAD - {}^tCB \equiv E_n (\mod q).$$

The first of these congruences shows that

$$c_2 \equiv 0 (\text{mod } q) \quad \text{and} \quad {}^t A_0 C_0 \equiv {}^t C_0 A_0 (\text{mod } q).$$

From the third congruence it follows that

$$d_1 \equiv 1, \quad d_2 \equiv 0, \quad \text{and} \quad {}^t A_0 D_0 - {}^t C_0 B_0 \equiv E_{n-1} (\text{mod } q),$$

and the second congruence implies that ${}^t B_0 D_0 \equiv {}^t D_0 B_0$. In particular, it follows from all of these congruences that the matrix $M_0 = \begin{pmatrix} A_0 & B_0 \\ C_0 & D_0 \end{pmatrix}$ satisfies the condition $J_{n-1}[M_0] \equiv J_{n-1} (\text{mod } q)$. Hence, by the induction assumption, there exists a matrix $g_0 \in \Gamma^{n-1}$ such that $g_0 M_0 \equiv E_{2n-2} (\text{mod } q)$. Then the above congruences and the symplectic condition modulo q imply that $\widetilde{g}_0 M'' \equiv E_{2n} (\text{mod } q)$, where $\widetilde{g}_0$ is a matrix in $\Gamma^{n,n-1}$ such that $\varphi(\widetilde{g}_0) = g_0$ (here φ is the map (3.53) of Chapter 2). Returning to our original matrix, we see that $M \equiv g^{-1}(\widetilde{g}_0)^{-1} g_1^{-1} (\text{mod } q)$. The first part of the lemma is proved.

It follows from the first part of the lemma that for $q \in \mathbf{N}$ the index $\mu(\Gamma^n(q))$ is equal to the order of the group $\mathrm{Sp}_n(\mathbf{Z}/q\mathbf{Z})$, i.e., it is equal to the number of solutions of a certain system of polynomial congruences modulo q having rational coefficients that do not depend on q. It follows from the Chinese Remainder Theorem that the number of solutions of such a system modulo a composite number of the form qq_1 with q prime to q_1 is equal to the product of the number of solutions modulo q and the number of solutions modulo q_1. In particular, the indices are multiplicative:

$$\mu(\Gamma^n(qq_1)) = \mu(\Gamma^n(q))\mu(\Gamma^n(q_1)), \quad \text{if } (q, q_1) = 1.$$

Furthermore, because q and q_1 are relatively prime, we have $\Gamma^n(q) \cap \Gamma^n(q_1) = \Gamma^n(qq_1)$, from which we conclude that the imbedding $\Gamma^n(q) \to \Gamma^n$ determines an imbedding of quotient groups

$$\Gamma^n(q)/\Gamma^n(qq_1) \to \Gamma^n/\Gamma^n(q_1).$$

On the other hand, since the indices are multiplicative, it follows that both of these quotient groups have the same order. Hence, this imbedding is a one-to-one correspondence. Consequently, for any $\gamma \in \Gamma^n$ there exists $\gamma_1 \in \Gamma^n(q_1)$ such that $\gamma\gamma_1 \in \Gamma^n(q)$, and so $\gamma \in \Gamma^n(q)\Gamma^n(q_1)$. $\qquad \square$

We now turn to the Hecke rings. For $n, q \in \mathbf{N}$ we define the group

$$(3.3) \qquad S(\Gamma^n(q)) = \left\{ M \in S^n \cap GL_{2n}(\mathbf{Z}_q); M \equiv \begin{pmatrix} E_n & 0 \\ 0 & r(M)E_n \end{pmatrix} (\text{mod } q) \right\},$$

where $\mathbf{Z}_q$ is the ring of q-integral rational numbers. We shall limit ourselves to groups $K, \Gamma^n(q) \subset K \subset \Gamma^n$, which satisfy the following q-symmetry condition:

$$(3.4) \qquad\qquad\qquad KS(\Gamma^n(q)) = S(\Gamma^n(q))K.$$

For such groups the set

$$(3.5) \qquad S(K) = S(K)_q = KS(\Gamma^n(q))K = KS(\Gamma^n(q)) = S(\Gamma^n(q))K$$

is a subgroup of the group S^n. From Lemma 3.1 it follows that $(K, S(K))$ is a Hecke pair. We let

$$(3.6) \qquad\qquad\qquad L(K) = D_{\mathbf{Q}}(K, S(K))$$

denote the Hecke ring of this pair over $\mathbf{Q}$.

THEOREM 3.3. *Let K and K_1 be two groups such that*

$$\Gamma^n(q) \subset K_1 \subset K \subset \Gamma^n,$$

where $n, q \in \mathbf{N}$. Suppose that K and K_1 each satisfies the q-symmetry condition (3.4). *Then:*

(1) $S(K) = KS(K_1) = S(K_1)K$;

(2) $\Gamma^n \cap S(K) = K$;

(3) *if $M, M' \in S(K_1)$ and $M' \in KMK$, then $M' \in \Gamma^n(q)MK_1$*;

(4) *if $M \in S(K_1)$ and $(M)_K = \sum_i (KM_i)$ is the expansion into left cosets, and if all $M_i \in S(K_1)$, then $(M)_{K_1} = \sum_i (K_1 M_i)$ and conversely, where the double cosets have the form* (1.11);

(5) *the linear maps $\varepsilon \colon L(K) \to L(K_1)$ and $\zeta \colon L(K_1) \to L(K)$ given by the conditions*

$$\varepsilon((M)_K) = (M')_{K_1} \quad (M \in S(K), M' \in KMK \cap S(K_1)),$$

$$\zeta((M')_{K_1}) = (M')_K \quad (M' \in S(K_1))$$

are mutually inverse isomorphisms of rings.

PROOF. By (3.5) we have

$$S(K) = KS(\Gamma^n(q)) = KK_1 S(\Gamma^n(q)) = KS(K_1).$$

Similarly, $S(K) = S(K_1)K$. Part (1) is proved.

If $g \in \Gamma^n \cap S(K_1)$, then $g = g_1 M$, where $g_1 \in K_1$ and $M \in S(\Gamma^n(q))$. Then $M = g_1^{-1}g \in \Gamma^n \cap S(\Gamma^n(q))$. The latter intersection is clearly $\Gamma^n(q)$. Thus, $g \in K_1 \Gamma^n(q) \subset K_1$. Part (2) is proved.

We now prove part (3). By assumption, $M' = \gamma M \gamma'$, where $\gamma, \gamma' \in K$. We choose an integer q_1 prime to q such that $q_1 M^{\pm 1}$ are integer matrices. By Lemma 3.2(2), the matrix γ can be written in the form $\gamma = \gamma_1 \gamma_2$, where $\gamma_1 \in \Gamma^n(q)$ and $\gamma_2 \in \Gamma^n(q_1^2)$. Then $M' = \gamma_1 M \gamma_3$, where $\gamma_3 = M^{-1}\gamma_2 M \gamma'$. Because of our choice of γ_2, the last matrix is an integer matrix, and hence $\gamma_3 \in \Gamma^n$. On the other hand, $\gamma_3 = M^{-1}\gamma_1^{-1}M' \in S(K_1)$. Thus, $\gamma_3 \in \Gamma^n \cap S(K_1) = K_1$. This proves part (3).

To prove part (4) we note that the first of the expansions implies that the left cosets $K_1 M_i$ are pairwise distinct, and their union contains the double coset $K_1 M K_1$. Hence, to prove the claim in one direction it suffices to verify that M_i lies in $K_1 M K_1$ for every i, and this follows from part (3). This result along with part (1) obviously implies the claim in the converse direction. Part (4) is proved.

Finally, from parts (1), (2), and (4) it follows that the Hecke pairs $(K, S(K))$ and $(K_1, S(K_1))$ satisfy the conditions (1.26) and (1.27). By Proposition 1.9, we then have the map $\varepsilon \colon D(K, S(K)) \to D(K_1, S(K_1))$, and this map is an isomorphism of rings. From part (4) and the definition of ε it follows that ε coincides with the first of the maps in part (5). The second map is clearly the inverse of the first, and so is also an isomorphism. $\qquad\square$

For subgroups $K \subset \Gamma^n$ containing $\Gamma^n(q)$ and satisfying the q-symmetry condition, the above theorem says that the Hecke rings $L(K)$ are canonically isomorphic. Hence, from an abstract point of view, when studying the Hecke rings one can start with any K with these properties. When we investigated the symplectic transformations of theta-series of quadratic forms in Chapter 1, we arrived at congruence subgroups of the form $\Gamma_0^n(q)$. We shall use the same congruence subgroups as the basis for our development of a theory of Hecke rings, because, in addition to their arithmetic origin, these groups have a number of technical advantages that make the calculations easier. Using Theorem 3.3, a reader who has need of results for the rings $L(K)$ for other K will easily be able to obtain them from the corresponding results for $L(\Gamma_0^n(q))$.

We start with some technical lemmas that give information on the left and double cosets of $\Gamma_0^n(q)$. We set

(3.7)
$$S^n(q, q_1) = \left\{ M = \begin{pmatrix} A & B \\ C & D \end{pmatrix} \in S^n \cap GL_{2n}(\mathbf{Z}_q); \ C \equiv 0 (\mathrm{mod}\, q_1) \right\},$$
$$S^n(q) = S^n(q, q),$$

where in the first case we assume that q_1 divides q. These are clearly subgroups of S^n.

LEMMA 3.4. *Every left coset $\Gamma_0^n(q_1)M$, where $M \in S^n(q, q_1)$ and $q_1|q$, contains a representative of the form $M_0 = \begin{pmatrix} A & B \\ 0 & C \end{pmatrix}$.*

PROOF. By Proposition 3.7 of Chapter 1, there exists $\gamma \in \Gamma^n$ such that $\gamma M = M_0$ has the required form. But obviously $\gamma = M_0 M^{-1} \in S^n(q, q_1) \cap \Gamma^n = \Gamma_0^n(q_1)$. □

LEMMA 3.5. *We have*

$$S^n(q, q_1) = \Gamma_0^n(q_1)S(\Gamma^n(q)) = S(\Gamma^n(q))\Gamma_0^n(q_1),$$

where $q_1|q$ and $S(\Gamma^n(q))$ is the set (3.3). In particular, the group $\Gamma_0^n(q_1)$ has the q-symmetry property for any q that is divisible by q_1.

PROOF. It suffices to prove, say, the first equality, since the second one can then be obtained by applying the group anti-automorphism $M \to M^{-1}$. Let $M \in S^n(q, q_1)$. By the previous lemma, we may assume that $M = M_0 = \begin{pmatrix} A & B \\ 0 & D \end{pmatrix}$. The matrix $M_1 = \begin{pmatrix} A_1 & B_1 \\ 0 & D_1 \end{pmatrix}$, where $A_1 = A^{-1}$, $B_1 = -A^{-1}BD^{-1}$, $D_1 = {}^tA$, is obviously q-integral, and modulo q it belongs to the group $\mathrm{Sp}_n(\mathbf{Z}/q\mathbf{Z})$. Then by Lemma 3.2(1) there exists $\gamma \in \Gamma^n$ such that $\gamma \equiv M_1 (\mathrm{mod}\, q)$. Then

$$\gamma \in \Gamma_0^n(q) \subset \Gamma_0^n(q), \quad \text{and} \quad \gamma M_0 \equiv \begin{pmatrix} E & 0 \\ 0 & r(M_0)E \end{pmatrix} (\mathrm{mod}\, q). \qquad \square$$

Lemma 3.5 implies that $S(\Gamma_0^n(q))_q = S^n(q)$, and the Hecke ring (3.6) for the group $K = \Gamma_0^n(q)$ has the form

(3.8)
$$L^n(q) = L(\Gamma_0^n(q)) = D_{\mathbf{Q}}(\Gamma_0^n(q), S^n(q)).$$

By Lemma 1.5, we can take elements of the form (1.11) as a $\mathbf{Q}$-basis of the ring $L^n(q)$, one element for each double coset $\Gamma_0^n(q)M\Gamma_0^n(q)$ of the group $S^n(q)$ modulo the subgroup $\Gamma_0^n(q)$.

LEMMA 3.6. *Every double coset $\Gamma_0^n(q)M\Gamma_0^n(q)$, where $M \in S^n(q)$, contains one and only one representative of the form*

(3.9)
$$\mathrm{sd}(M) = \mathrm{diag}(d_1, \ldots, d_n; e_1, \ldots, e_n),$$

where $d_i, e_j > 0$, $d_i|d_{i+1}$, $d_n|e_n$, $e_{i+1}|e_i$, $d_i e_i = r(M)$.

PROOF. We first treat the case $q = 1$ using induction on n. Since $\Gamma^1 = SL_2(\mathbf{Z})$ and $S^1(1)$ is the group of 2×2 rational matrices with positive determinant, the lemma in the case $n = 1$ follows from Lemma 2.2. Suppose that $n > 1$, and the lemma for $q = 1$ has been proved for $(2(n - 1) \times 2(n - 1))$-matrices. In the argument below we shall systematically make use of partitions of the standard blocks of a matrix $M = \begin{pmatrix} A & B \\ C & D \end{pmatrix} \in S^n$ into smaller blocks of fixed size. For example, $A = \begin{pmatrix} A_1 & A_2 \\ A_3 & A_4 \end{pmatrix}$, $B = \begin{pmatrix} B_1 & B_2 \\ B_3 & B_4 \end{pmatrix}$, and so on, where A_4, B_4, C_4, D_4 are $(n - 1) \times (n - 1)$-blocks, and the sizes of the other blocks are determined correspondingly. Let $M' \in S^n = S^n(1)$, and let d_1 be the greatest common divisor of the entries in M'. Then $M' = d_1 M$, where $M \in S^n$ is an integer matrix with relatively prime entries. We let $\delta = \delta(M)$ denote the minimum of the greatest common divisors of entries in the columns of M, and we prove by induction on δ that the double coset $\Gamma^n M \Gamma^n$ contains a representative $\begin{pmatrix} A & B \\ C & D \end{pmatrix}$ with blocks of the form

(3.10)

$$A = \begin{pmatrix} 1 & 0 \\ 0 & A_4 \end{pmatrix}, \quad B = \begin{pmatrix} 0 & 0 \\ 0 & B_4 \end{pmatrix}, \quad C = \begin{pmatrix} 0 & 0 \\ 0 & C_4 \end{pmatrix}, \quad D = \begin{pmatrix} r(M) & 0 \\ 0 & D_4 \end{pmatrix},$$

where $\begin{pmatrix} A_4 & B_4 \\ C_4 & D_4 \end{pmatrix} \in S^{n-1}$. First suppose that $\delta = 1$, and let i be the index of the first column of M whose entries are relatively prime. By replacing M by MJ_n if necessary, we may suppose that $i \leqslant n$. If we then replace M by $MU(V^*)$, where $V \in \Lambda^n$ is a suitable permutation matrix, we may assume that $i = 1$. We now apply Lemma 3.9 of Chapter 1 to the first column of M; we find that the left coset $\Gamma^n M$ contains a representative whose A_1-block is 1 and whose A_3-, C_1-, and C_3-blocks consist of zeros. After multiplying this matrix on the right by

$$U(V^*) = \begin{pmatrix} V & 0 \\ 0 & V^* \end{pmatrix}, \quad \text{where } V = \begin{pmatrix} 1 & -A_2 \\ 0 & E_{n-1} \end{pmatrix},$$

we may assume that its A_2-block also consists of zeros. If we multiply the last matrix on the right by the matrix

$$T(S) = \begin{pmatrix} E_n & S \\ 0 & E_n \end{pmatrix}, \quad \text{where } S = \begin{pmatrix} -B_1 & -B_2 \\ -{}^tB_2 & 0 \end{pmatrix},$$

we obtain a matrix with zero B_1- and B_2-blocks. Thus, in our double coset we have found a matrix with $A_1 = 1$, $A_2 = 0$, $B_1 = 0$, $B_2 = 0$, $A_3 = 0$, $C_1 = 0$, and $C_3 = 0$. From (3.1)–(3.2) it now follows that this matrix has the form (3.10). For example, the first relation in (3.1) shows that $C_2 = 0$ and ${}^tA_4C_4 = {}^tC_4A_4$, the first relation in (3.2) leads to the equalities $B_3 = 0$ and $A_4 {}^tB_4 = B_4 {}^tA_4$, and so on. Now suppose that $\delta > 1$, the claim has been proved for all integer matrices $M' \in S^n$ with relatively prime entries and with $\delta(M') < \delta$, and $\delta(M) = \delta$. Just as in the above discussion, in the double coset of M we can find a representative M_0 with $A_1 = \delta$, with zero A_3-, C_1-, and C_3-blocks, and with the property that all of the entries in the A_2-, B_1-, and B_2-blocks are between 1 and δ. Then $\delta(M_0) < \delta$. In fact, we obviously have $\delta(M_0) \leqslant \delta$. If $\delta(M_0)$ were equal to δ, then all of the entries of M_0—and hence all of the entries of M—would be divisible by δ, contradicting the assumption that its entries are relatively prime. By the induction assumption, the double coset $\Gamma^n M_0 \Gamma^n = \Gamma^n M \Gamma^n$ contains a representative of the form (3.10), and the proof of the claim is complete. Returning to the proof of the lemma, we see that the Γ^n-double coset of an arbitrary matrix $M \in S^n$ contains a representative $M_0 = \begin{pmatrix} A & B \\ C & D \end{pmatrix}$ with blocks of the form

$$A = \begin{pmatrix} d_1 & 0 \\ 0 & A' \end{pmatrix}, \quad B = \begin{pmatrix} 0 & 0 \\ 0 & B' \end{pmatrix}, \quad C = \begin{pmatrix} 0 & 0 \\ 0 & C' \end{pmatrix}, \quad D = \begin{pmatrix} e_1 & 0 \\ 0 & D' \end{pmatrix},$$

where $d_1, e_1 > 0$, $d_1 | e_1$, $d_1 e_1 = r(M)$, $M' = \begin{pmatrix} A' & B' \\ C' & D' \end{pmatrix} \in S^{n-1}$, and all of the entries in M' are divisible by d_1. By the induction assumption, there exist $\xi, \xi_1 \in \Gamma^{n-1}$ such that the matrix $\xi M' \xi_1$ has the form (3.9). Then the matrix $\tilde{\xi} M_0 \tilde{\xi}_1$, where for $\xi = \begin{pmatrix} \alpha & \beta \\ \gamma & \delta \end{pmatrix} \in \Gamma^{n-1}$ we set

$$\tilde{\xi} = \begin{pmatrix} 1 & 0 & 0 & 0 \\ 0 & \alpha & 0 & \beta \\ 0 & 0 & 1 & 0 \\ 0 & \gamma & 0 & \delta \end{pmatrix} \in \Gamma^n,$$

has the form (3.9). The uniqueness of the Γ^n-double coset representative of the form (3.9) follows from Lemma 2.2, since the numbers $d_1, \ldots, d_n, e_n, e_{n-1}, \ldots, e_1$ obviously are the elementary divisors of this matrix in the sense of §2.1. The lemma is proved in the case $q = 1$.

We now turn to the case of arbitrary $q \geqslant 1$. If $M \in S^n(q)$, then, by what was proved above, there exist $\gamma_1, \gamma_2 \in \Gamma^n$ such that $\gamma_1 M \gamma_2 = \mathrm{sd}(M)$. By Lemma 3.5, the groups $K = \Gamma^n$ and $K_1 = \Gamma_0^n(q)$ satisfy the conditions of Theorem 3.3. Since $\mathrm{sd}(M) \in S^n(q)$, it follows from part (3) of that theorem that $\mathrm{sd}(M) \in \Gamma_0^n(q) M \Gamma_0^n(q)$; the uniqueness of this element follows from its uniqueness in the larger double coset $\Gamma^n M \Gamma^n$. $\square$

We call a matrix $\mathrm{sd}(M)$ of the form (3.9) the *symplectic divisor matrix* of M, and we call $d_i = d_i(M)$ and $e_i = e_i(M)$ $(i = 1, \ldots, n)$ the *symplectic divisors* of M. Clearly, the numbers $d_1, \ldots, d_n, e_n, \ldots, e_1$ are the elementary divisors of M. From Lemmas 1.5 and 3.6 it follows that the elements of the form

(3.11) $\qquad T_q(d_1, \ldots, d_n, e_1, \ldots, e_n) = (\mathrm{diag}(d_1, \ldots, d_n, e_1, \ldots, e_n))_{\Gamma_0^n(q)}$

form a basis of the space $L^n(q)$ over $\mathbf{Q}$, where d_i, e_j are positive rational numbers that are q-integral, have q-integral inverses, and satisfy the conditions

(3.12) $\qquad d_i | d_{i_1}, d_n | e_n, e_{i+1} | e_i, d_1 e_1 = \cdots = d_n e_n = r.$

We now turn to the multiplicative properties of the ring $L^n(q)$.

THEOREM 3.7. *For $n, q \in \mathbf{N}$ the ring $L^n(q)$ is commutative.*

PROOF. The map $M \to r(M^{-1})M$ is obviously an automorphism of the group $S^n(q)$ that does not affect the elements of $\Gamma = \Gamma_0^n(q)$. Hence, the $\mathbf{Q}$-linear map from $L^n(q)$ to itself that is given by the condition $(M)_\Gamma \to (r(M^{-1})M)_\Gamma$ for $M \in S^n(q)$

is an automorphism of the ring $L^n(q)$. It now follows from Proposition 1.11 that the map

$$(3.13) \qquad X = \sum_i a_i (M_i)_\Gamma \to X^* = \sum_i a_i (r(M_i) M_i^{-1})_\Gamma$$

is an anti-automorphism of the ring $L^n(q)$. In particular, $(XY)^* = Y^* X^*$ for $X, Y \in L^n(q)$. On the other hand, if $M_0 = \mathrm{sd}(M) = \mathrm{diag}(d_1, \ldots, d_n, e_1, \ldots, e_n)$ and $M \in S^n(q)$, then by Lemma 3.6 we have $(M)_\Gamma = (M_0)_\Gamma$, and hence

$$(M)_\Gamma^* = (M_0)_\Gamma^* = (\mathrm{diag}(e_1, \ldots, e_n, d_1, \ldots, d_n))_\Gamma = (J_n M_0 J_n^{-1})_\Gamma.$$

Since the Γ^n-double coset of the matrix $J_n M_0 J_n^{-1}$ coincides with $\Gamma^n M_0 \Gamma^n$, it follows that $\mathrm{sd}(J_n M_0 J_n^{-1}) = M_0$. Since the symplectic divisor matrix does not depend on q, by Lemma 3.6 we obtain $(J_n M_0 J_n^{-1})_\Gamma = (M_0)_\Gamma = (M)_\Gamma$. Thus,

$$(3.14) \qquad X^* = X \quad \text{for } X \in L^n(q),$$

and $XY = (XY)^* = Y^* X^* = YX$. $\qquad\qquad\qquad\qquad\qquad\qquad\qquad\qquad \square$

We now examine the products of concrete elements of $L^n(q)$.

LEMMA 3.8. *For any $M \in S^n(q)$ and $r \in \mathbf{Z}_q^*$ the following relations hold in the ring $L^n(q)$:*

$$(3.15) \qquad (rE)_\Gamma (M)_\Gamma = (M)_\Gamma (rE)_\Gamma = (rM)_\Gamma,$$

where $E = E_{2n}$ and $\Gamma = \Gamma_0^n(q)$.

PROOF. This lemma is proved in the same way as Lemma 2.4, but with Λ replaced by Γ and g replaced by M. $\qquad\qquad\qquad\qquad\qquad\qquad\qquad\qquad\qquad\qquad\qquad \square$

PROPOSITION 3.9. *Let $M, M' \in S^n(q)$. Suppose that the symplectic divisor ratios $e_1(M)/d_1(M)$ and $e_1(M')/d_1(M')$ are relatively prime. Then the relation*

$$(3.16) \qquad (M)_\Gamma (M')_\Gamma = (MM')_\Gamma,$$

where $\Gamma = \Gamma_0^n(q)$, holds in the Hecke ring $L^n(q)$.

PROOF. The proof of this proposition is similar to the proof of Proposition 2.5, with obvious modifications. So we shall be brief. From Lemma 3.8 and the definition of the symplectic divisors it follows that one need only prove the proposition in the case when M and M' are integer matrices and $r = r(M)$ and $r' = r(M')$ are relatively prime. In analogy with (2.8) we obtain

$$(3.17) \qquad (M)_\Gamma (M')_\Gamma = (MM')_\Gamma + \sum_{\Gamma H \Gamma \subset \Gamma M \Gamma M' \Gamma} a(M, M'; H)(H)_\Gamma,$$

where $a(M, M'; H)$ are nonnegative integers depending only on the Γ-double cosets of M, M', and H. For $m \in \mathbf{N}$ we let

$$(3.18) \qquad \begin{aligned} \mathrm{SD}(m) = \mathrm{SD}_n(m) = \{&\mathrm{diag}(d_1, \ldots, d_n, e_1, \ldots, e_n); \\ &d_i, e_j \in \mathbf{N}, d_i | d_{i+1}, d_n | e_n, e_{i+1} | e_i, d_i e_i = m\} \end{aligned}$$

denote the set of all integer matrices of the form (3.9) with $r(M) = m$. By Lemma 3.6, we may assume that $M \in \text{sd}(r)$, $M' \in \text{sd}(r')$, and $H \in \text{sd}(rr')$ for all H in (3.17). If m is prime to q, we can define the element

$$(3.19) \qquad T(m) = T^n(m) = \sum_{M \in \text{SD}_n(m)} (M)_\Gamma$$

of the ring $L^n(q)$. If m and m' are relatively prime and also prime to q, then, summing (3.17) over all $M \in \text{sd}(m)$ and $M' \in \text{sd}(m')$, we obtain

$$T(m)T(m') = T(mm') + \sum_{M \in \text{SD}_n(mm')} \left(\sum_{M,M'} a(M, M'; H) \right)(H)_\Gamma.$$

This implies that to prove (3.16) it suffices to show that

$$(3.20) \qquad T(m)T(m') = T(mm'), \quad \text{if } (m, m') = (mm', q) = 1,$$

where $T(l) = T^n(l)$, since in that case it will follow from the nonnegativity of $a(M, M'; H)$ that all of these coefficients are zero, and (3.17) will turn into (3.16). □

LEMMA 3.10. *Let* $m \in \mathbf{N}$, $(m, q) = 1$. *Then the set*

$$(3.21) \qquad \text{SM}(m) = \text{SM}_n(m, q) = \{M \in S^n(q) \cap M_{2n}; r(M) = m\}$$

is the union of finitely many left cosets of the group $\Gamma = \Gamma_0^n(q)$, *and the element* (3.19) *has an expansion of the form*

$$(3.22) \qquad T(m) = \sum_{M \in \Gamma \backslash \text{SM}(m)} (\Gamma M).$$

If m *and* m' *are relatively prime (and also prime to* q), *and if the matrices* M *and* M' *run through a set of representatives of the left cosets* $\Gamma \backslash \text{SM}(m)$ *and* $\Gamma \backslash \text{SM}(m')$, *respectively, then the product* MM' *runs through a complete set of representatives of the left cosets* $\Gamma \backslash \text{SM}(mm')$. *In particular, the relation* (3.20) *holds.*

PROOF. The first assertion follows from Lemmas 3.6 and 3.1.

If $\{M_1, \dots, M_\mu\}$ and $\{M_1', \dots, M_\nu'\}$ are fixed sets of representatives of the left cosets $\Gamma \backslash \text{SM}(m)$ and $\Gamma \backslash \text{SM}(m')$, respectively, then every product $M_i M_j'$ is obviously contained in $\text{SM}(mm')$. Suppose that two such products lie in the same Γ-left coset, say, $\gamma M_i M_j' = M_k M_l'$, where $\gamma \in \Gamma$. We set $H = M_k^{-1} \gamma M_i = M_l'(M_j')^{-1}$. Then mH and $m'H$ are integer matrices, and since $(m, m') = 1$ it follows that H is an integer matrix. On the other hand, $H \in S^n(q)$ and $r(H) = 1$. Thus, $H \in \Gamma^n \cap S^n(q) = \Gamma_0^n(q) = \Gamma$, so that $M_k \in \Gamma M_i$ and $M_l' \in \Gamma M_j'$. This means that $k = i$ and $l = j$. Thus, all of the products $M_i M_j'$ belong to different left cosets $\Gamma \backslash \text{SM}(mm')$. If ΓM_0 ($M_0 \in \text{SM}(mm')$) is an arbitrary left coset, then it follows from Lemma 3.6 that M_0 can be written in the form $M_0 = MM'$, where $M \in \text{SM}(m)$ and $M' \in \text{SM}(m')$. Then $M' = \gamma' M_j'$, where $\gamma' \in \Gamma$, and $M\gamma' = \gamma M_i$, where $\gamma \in \Gamma$; hence, $M_0 = \gamma M_i M_j'$, and the left coset ΓM_0 contains the product $M_i M_j'$. Lemma 3.10, and hence also Proposition 3.9, are proved. □

The next lemma is useful for explicitly computing the left coset expansions of elements of the Hecke ring $L^n(q)$.

LEMMA 3.11. *Every left coset ΓM, where $\Gamma = \Gamma_0^n(q)$ and $M \in S^n(q)$, contains one and only one representative of the form*

$$(3.23) \qquad \begin{pmatrix} A & B \\ 0 & D \end{pmatrix} = \begin{pmatrix} r(M)D^* & B \\ 0 & D \end{pmatrix},$$

where D belongs to a fixed Λ^n-left coset of $GL_n(\mathbf{Z}_q)$, and B belongs to a fixed residue class of the set

$$(3.24) \qquad B(D) = B(D)_\mathbf{Q} = \{B \in M_n(\mathbf{Q}); \ {}^tBD = {}^tDB\}$$

modulo D, where

$$(3.25) \qquad B \equiv B_1 (\operatorname{mod} D) \Leftrightarrow (B - B_1)D^{-1} \in M_n.$$

PROOF. The lemma follows directly from Lemma 3.4, the relations (3.1), and the definitions. $\qquad\square$

Just as in the case of the general linear group, the study of the global Hecke rings $L^n(q)$ reduces to the study of the local subrings. Let p be a prime number not dividing q. We set

$$(3.26) \qquad S_p(q) = S_p^n(q) = S^n(q) \cap GL_{2n}(\mathbf{Z}[p^{-1}]),$$

where $\mathbf{Z}[p^{-1}]$ is the ring (2.17). Since $\Gamma_0^n(q) \subset S_p^n(q) \subset S^n(q)$, it follows that $(\Gamma_0^n(q), S_p^n(q))$ is a Hecke pair, and the Hecke ring

$$(3.27) \qquad L_p(q) = L_p^n(q) = D_\mathbf{Q}(\Gamma_0^n(q), S_p^n(q))$$

may be regarded as a subring of the Hecke ring $L^n(q)$. The subrings $L_p^n(q) \subset L^n(q)$ as p runs through the primes not dividing q are called the *local subrings of $L^n(q)$*.

THEOREM 3.12. *For $n, q \in \mathbf{N}$ the Hecke ring $L^n(q)$ is generated by the local subrings $L_p^n(q)$ as p ranges over all primes not dividing q.*

PROOF. For $r \in \mathbf{Q}^*$ and p a prime number, as before we let $v_p(r)$ denote the power with which p occurs in the prime factorization of r. If $M \in S^n(q)$ and $p \nmid q$, we define the *symplectic p-divisor matrix of M* by setting

$$(3.28) \qquad \operatorname{sd}_p(M) = \operatorname{diag}(p^{v_p(d_1)}, \ldots, p^{v_p(d_n)}, p^{v_p(e_1)}, \ldots, p^{v_p(e_n)}),$$

where $d_i = d_i(M)$, $e_j = e_j(M)$ are the symplectic divisors of M. Clearly $\operatorname{sd}_p(M) \in S_p^n(q)$. If M is a fixed matrix, then obviously $\operatorname{sd}_p(M) = E_{2n}$ for almost all p, and

$$(3.29) \qquad \prod_{p \in \mathbf{P}_{(q)}} \operatorname{sd}_p(M) = \operatorname{sd}(M) \quad (M \in S^n(q)).$$

Since $M \in \Gamma_0^n(q)\operatorname{sd}(M)\Gamma_0^n(q)$, it follows from (3.29) that M can be represented in the form

$$(3.30) \qquad M = \prod_{p \in \mathbf{P}_{(q)}} M_p, \quad \text{where } M_p \in S_p^n(q), \operatorname{sd}_p(M_p) = \operatorname{sd}_p(M),$$

and $M_p = E_{2n}$ for all except finitely many p. From Proposition 3.9 it now follows that the corresponding double coset in the Hecke ring $L^n(q)$ has the expansion

$$(M)_\Gamma = \prod_{p \in \mathbf{P}_{(q)}} (M_p)_\Gamma,$$

where $\Gamma = \Gamma_0^n(q)$. Since $(M_p)_\Gamma \in L_p^n(q)$, and since $L^n(q)$ consists of finite linear combinations of $(M)_\Gamma$ (where $M \in S^n(q)$), this proves the theorem. $\square$

PROBLEM 3.13. Let $M, M' \in S^n(q)$. Suppose that the ratios $e_n(M)/d_1(M)$ and $e_n(M')/d_1(M')$ are relatively prime. Show that $\text{sd}(MM') = \text{sd}(M)\text{sd}(M')$.

PROBLEM 3.14. Show that the following set can be taken as a set of representatives of the left cosets of $\text{SM}_n(m, q)$ (where $(m, q) = 1$) modulo the group $\Gamma_0^n(q)$:

$$\left\{ \begin{pmatrix} A & B \\ 0 & D \end{pmatrix}; \quad D \in \Lambda^n \setminus GL_n(\mathbf{Q}) \cap M_n, d_n(D) | m, \right.$$

$$\left. B \in B(D) \cap M_n / \text{mod } D, A = mD^* \right\}.$$

PROBLEM 3.15. Let D be a nonsingular $n \times n$ integer matrix. Show that the number $\rho(D) = |B(D) \cap M_n / \text{mod } D|$ of left residue classes of the set $B(D) \cap M_n$ modulo D is finite and satisfies the relations

$$\rho(UDV) = \rho(D), \qquad \text{if } U, V \in \Lambda^n,$$

$$\rho(D) = d_1^n d_2^{n-1} \cdots d_n, \quad \text{if ed}(D) = \text{diag}(d_1, \ldots, d_n).$$

PROBLEM 3.16. Show that the zeta-function of the ring $L^n(q)$, which is defined as the Dirichlet series

$$Z_N(s) = \sum_{m \in \mathbf{N}_{(q)}} N(T(m)) m^{-s},$$

where $N: L^n(q) \to \mathbf{Q}$ is the homomorphism in Problem 1.16 and the real part of s is sufficiently large, converges and has an Euler product of the form

$$Z_N(s) = \prod_{p \in \mathbf{P}_{(q)}} \left(\sum_{\delta=0}^{\infty} N(T(p^\delta)) \right) p^{-\delta s}.$$

Show that

$$N(T(m)) = \sum_{\substack{d_i \in \mathbf{N} \\ d_1 | d_2 | \cdots | d_n | m}} N(t(d_1, \ldots, d_n)) d_1^n d_2^{n-1} \cdots d_n,$$

where $t(d_1, \ldots, d_n) = (\text{diag}(d_1, \ldots, d_n))_\Lambda \in H^n$ and the symbol N on the right denotes the corresponding homomorphism of the ring H^n.

[**Hint:** Use the two preceding problems to prove the last relation.]

PROBLEM 3.17. Show that for $L^2(q)$ one has

$$Z_N(s) = \zeta_q(s)\zeta_q(s-1)\zeta_q(s-2)\zeta_q(s-3)\zeta_q(2s-2)^{-1},$$

where

$$\zeta_q(s) = \sum_{m \in \mathbf{N}_{(q)}} m^{-s}.$$

[Hint: Use the previous problem and the following identity, which is a consequence of Problem 2.12:

$$\sum_{0 \leqslant \delta_1 \leqslant \delta_2} N(t(p^{\delta_1}, p^{\delta_2})) v_1^{\delta_1} v_2^{\delta_2} = \frac{1 - v_2^2}{(1 - v_1 v_2)(1 - (p+1)v_2 + p v_2^2)}.]$$

PROBLEM 3.18. Show that the Hecke pairs (Λ^2, G^2) and (Γ^1, S^1) satisfy the conditions (1.26) and (1.28), so that the map (1.27) gives a natural isomorphism between the rings H^2 and $L^1(1)$. From this and Problem 2.13 deduce that in the case $n = 1$ the elements (3.19) of the ring $L^1(q)$ can be multiplied by the rule

$$T(m)T(m_1) = \sum_{d \mid m, m_1} d(dE_2)_\Gamma T(mm_1/d^2),$$

where $(m, q) = (m_1, q) = 1$ and $\Gamma = \Gamma_0^1(q)$.

PROBLEM 3.19. Show that for $n = 1$ one has the formal identity

$$\sum_{m \in \mathbf{N}_{(q)}} T(m) m^{-s} \prod_{p \in \mathbf{P}_{(q)}} (1 - T(p)p^{-s} + (pE_2)_\Gamma p^{1-2s})^{-1}.$$

2. Local rings. In this subsection we study the structure of the local Hecke rings $L_p(q)$, where p is a prime not dividing q. We first note that this structure does not depend on q.

LEMMA 3.20. *Let p be a prime not dividing q. Then the restriction of the map $\varepsilon \colon D_\mathbf{Q}(\Gamma^n, S^n(q, 1)) \to L^n(q)$ to the subring*

$$L_p^n = L_p^n(1) = D_\mathbf{Q}(\Gamma^n, S_p^n(q))$$

is an isomorphism of this ring with the ring $L_p^n(q)$, and

(3.31) $$\varepsilon((M)_{\Gamma^n}) = (M)_{\Gamma_0^n(q)} \quad (M \in S_p^n(q)).$$

PROOF. The lemma follows from Theorem 3.3(5) and Lemma 3.5. □

This lemma enables us to restrict ourselves to the case $q = 1$ when proving structure theorems.

As in §2.2, it is convenient to reduce the study of $L_p^n(q)$ to the study of its *integral subring*

(3.32) $$\underline{L}_p^n(q) = D_\mathbf{Q}(\Gamma_0^n(q), S_p^n(q) \cap M_{2n}).$$

Lemma 3.20 implies that the ring $\underline{L}_p^n(q)$ is naturally isomorphic to the ring $\underline{L}_p^n = \underline{L}_p^n(1)$.

LEMMA 3.21. *The element*

(3.33) $$\Delta = \Delta_n(p) = (pE_{2n})_{\Gamma_0^n(q)}$$

of the Hecke ring $L_p^n(q)$, where $(p, q) = 1$, is invertible in $L_p^n(q)$, and $\Delta^{-1} = (p^{-1}E_{2n})$. The ring $L_p^n(q)$ is generated by Δ^{-1} and the subring $\underline{L}_p^n(q)$.

PROOF. The lemma follows from Lemma 3.8 and the definitions.   □

We set

(3.34) $T(p^{\delta_1}, \ldots, p^{\delta_n}, p^{\varepsilon_1}, \ldots, p^{\varepsilon_n}) = (\mathrm{diag}(p^{\delta_1}, \ldots, p^{\delta_n}, p^{\varepsilon_1}, \ldots, p^{\varepsilon_n}))_{\Gamma_0^n(q)},$

where $\delta_i, \varepsilon_j \in \mathbf{Z}$, $\delta_1 + \varepsilon_1 = \cdots = \delta_n + \varepsilon_n$. In this notation the ring $\underline{L}_p^n(q)$ consists of linear combinations of elements of the form (3.34), where

(3.35) $0 \leqslant \delta_1 \leqslant \cdots \leqslant \delta_n \leqslant \varepsilon_n \leqslant \cdots \leqslant \varepsilon_1.$

This follows from Lemmas 1.5 and 3.6. We say that such an element is *primitive* if $\delta_1 = 0$, and that it is *imprimitive* if $\delta_1 \geqslant 1$. An arbitrary element $T \in \underline{L}_p^n(q)$ is said to be *primitive* (or *imprimitive*) if it is a linear combination of primitive (resp. imprimitive) elements of the form (3.34)–(3.35). Clearly, any element T in $\underline{L}_p^n(q)$ can be uniquely represented in the form

(3.36) $T = T^{\mathrm{pr}} + T^{\mathrm{im}},$

where T^{pr} is primitive and T^{im} is imprimitive. Lemma 3.21 implies that the subset I of all imprimitive elements of $\underline{L}_p^n(q)$ is the principal ideal of this ring that is generated by the element (3.33):

(3.37) $I = \Delta \underline{L}_p^n(q).$

LEMMA 3.22. *Let $n > 1$. Let the $\mathbf{Q}$-linear map*

(3.38) $\Psi \colon \underline{L}_p^n(q) \to \underline{L}_p^{n-1}(q)$

be defined by giving its values on elements of the form (3.34)–(3.35) *as follows:*

$$\Psi(T(p^{\delta_1}, \ldots, p^{\varepsilon_n})) = \begin{cases} T(p^{\delta_2}, \ldots, p^{\delta_n}, p^{\varepsilon_2}, \ldots, p^{\varepsilon_n}), & \text{if } \delta_1 = 0, \\ 0, & \text{if } \delta_1 > 0. \end{cases}$$

Then Ψ is an epimorphism of rings, and its kernel is the ideal I of imprimitive elements of $\underline{L}_p^n(q)$.

PROOF. From Lemmas 1.5 and 3.6 and the definitions it follows that, as a map of vector spaces, Ψ is an epimorphism with kernel I. Hence, it remains to prove that Ψ is a ring homomorphism. This, in turn, will follow if we prove that the image of a product of primitive elements of the form (3.34)–(3.35) is equal to the product of the images. Let

$$M = \mathrm{diag}(p^{\delta_1}, \ldots, p^{\varepsilon_n}), \quad M' = \mathrm{diag}(p^{\delta_1'}, \ldots, p^{\varepsilon_n'}),$$

where the exponents satisfy the inequalities (3.35), $\delta_i + \varepsilon_i = \delta$, $\delta_i' + \varepsilon_i' = \delta'$. We suppose that $\delta_1 = \delta_1' = 0$, and we set

$$M_0 = \mathrm{diag}(p^{\delta_2}, \ldots, p^{\delta_n}, p^{\varepsilon_2}, \ldots, p^{\varepsilon_n}),$$
$$M_0' = \mathrm{diag}(p^{\delta_2'}, \ldots, p^{\delta_n'}, p^{\varepsilon_2'}, \ldots, p^{\varepsilon_n'}).$$

By Lemmas 1.5 and 3.6 and the definition of Ψ, we obtain

$$\Psi((M)_\Gamma (M')_\Gamma) = \sideset{}{'}\sum_{\substack{H = \mathrm{sd}H \\ r(H) = p^{\delta + \delta'}}} c(M, M'; H)(H_0)_{\Gamma'},$$

where the symbol $\sum'$ means that the H are primitive, $\Gamma = \Gamma_0^n(q)$, $\Gamma' = \Gamma_0^{n-1}(q)$, we set $H_0 = \operatorname{diag}(p^{\alpha_2}, \ldots, p^{\alpha_n}, p^{\beta_2}, \ldots, p^{\beta_n})$ for $H = \operatorname{diag}(p^{\alpha_1}, \ldots, p^{\alpha_n}, p^{\beta_1}, \ldots, p^{\beta_n})$, and $c(M, M'; H)$ is the number of pairs M_i, M'_j which belong to fixed sets of representatives of $\Gamma \setminus \Gamma M \Gamma$ and $\Gamma \setminus \Gamma M' \Gamma$, respectively, and which satisfy the relations

$$(3.39) \qquad M_i M'_j = \gamma H \quad \text{with } \gamma \in \Gamma.$$

Similarly—but now summing over integer matrices H_0—we have

$$(M_0)_{\Gamma'}(M'_0)_{\Gamma'} = \sum_{\substack{H_0 = \operatorname{sd} H_0 \\ r(H_0) = p^{\delta + \delta'}}} c(M_0, M'_0; H_0)(H_0)_{\Gamma'},$$

where $c(M_0, M'_0; H_0)$ is the number of pairs N_k, N'_l in $\Gamma' \setminus \Gamma' M_0 \Gamma'$ and $\Gamma' \setminus \Gamma' M'_0 \Gamma'$, respectively, which satisfy the relation

$$(3.40) \qquad N_k N'_l = \gamma' H_0 \quad \text{with } \gamma' \in \Gamma'.$$

Since the matrices H_0 in the above expansions obviously run through the same set, it follows that to prove that

$$\Psi((M)_\Gamma (M')_\Gamma) = (M_0)_{\Gamma'}(M'_0)_{\Gamma'} = \Psi((M)_\Gamma)\Psi((M')_\Gamma)$$

it suffices to verify that

$$(3.41) \qquad c(M, M'; H) = c(M_0, M'_0; H_0)$$

for primitive $n \times n$-matrices $H = \operatorname{sd}(H)$ with $r(H) = p^{\delta + \delta'}$. These coefficients depend only on the double cosets of the corresponding matrices. Hence, by Lemma 3.6, without loss of generality we may assume that

$$H = \operatorname{diag}(p^{\alpha_1}, \ldots, p^{\alpha_n}, p^{\beta_1}, \ldots, p^{\beta_n}),$$

where $\alpha_i + \beta_i = \delta + \delta'$, $\beta_n = 0$, and

$$H_0 = \operatorname{diag}(p^{\alpha_1}, \ldots, p^{\alpha_{n-1}}, p^{\beta_1}, \ldots, p^{\beta_{n-1}}).$$

By Lemmas 3.11 and 2.7, we may take

$$M_i = \begin{pmatrix} A_i & B_i \\ 0 & D_i \end{pmatrix}, \quad M'_j = \begin{pmatrix} A'_j & B'_j \\ 0 & D'_j \end{pmatrix},$$

where D_i, D'_j are matrices of the form (2.15), and B_i and B'_j are fixed modulo D_i and D'_j, respectively. It now follows from (3.39) that

$$\gamma = \begin{pmatrix} \gamma_{11} & \gamma_{12} \\ 0 & \gamma_{22} \end{pmatrix} \quad \text{and} \quad D_i D'_j = \gamma_{22} \operatorname{diag}(p^{\beta_1}, \ldots, p^{\beta_n}).$$

The last relation implies that $\gamma_{22} \in \Lambda^n$ is an upper-triangular matrix, and hence all of its diagonal entries are ± 1. But since D_i, D'_j are reduced matrices and $\beta_n = 0$, we conclude that

$$D_i = \begin{pmatrix} D_i^{(n-1)} & 0 \\ 0 & 1 \end{pmatrix}, \quad D'_j \begin{pmatrix} D_j'^{(n-1)} & 0 \\ 0 & 1 \end{pmatrix}, \quad \gamma_{22} = \begin{pmatrix} \gamma_{22}^{(n-1)} & 0 \\ 0 & 1 \end{pmatrix},$$

and

$$D_i^{(n-1)} D_j'^{(n-1)} = \gamma_{22}^{(n-1)} \operatorname{diag}(p^{\beta_1}, \ldots, p^{\beta_{n-1}}).$$

From these formulas it follows that

$$A_i = \begin{pmatrix} A_i^{(n-1)} & 0 \\ 0 & p^\delta \end{pmatrix}, \quad A_j' = \begin{pmatrix} A_j'^{(n-1)} & 0 \\ 0 & p^{\delta'} \end{pmatrix}, \quad \gamma_{11} = \begin{pmatrix} \gamma_{11}^{(n-1)} & 0 \\ 0 & 1 \end{pmatrix},$$

and

$$A_i^{(n-1)} A_j'^{(n-1)} = \gamma_{11}^{(n-1)} \mathrm{diag}(p^{\alpha_1}, \ldots, p^{\alpha_{n-1}}).$$

If we replace the matrices B_i and B_j' by $B_i + SD_i$ and $B_j' + S'D_j'$, respectively, where S and S' are suitably chosen symmetric integer matrices, and if we take into account the symmetry of the matrices ${}^tD_i B_i$ and ${}^tD_j' B_j'$, we may assume that

$$B_i = \begin{pmatrix} B_i^{(n-1)} & 0 \\ 0 & 0 \end{pmatrix}, \quad B_j' = \begin{pmatrix} B_j'^{(n-1)} & 0 \\ 0 & 0 \end{pmatrix},$$

so that

$$\gamma_{12} = \begin{pmatrix} \gamma_{12}^{(n-1)} & 0 \\ 0 & 0 \end{pmatrix},$$

and

$$A_i^{(n-1)} B_j'^{(n-1)} + B_i^{(n-1)} D_j'^{(n-1)} = \gamma_{12}^{(n-1)} \begin{pmatrix} p^{\beta_1} & & 0 \\ & \ddots & \\ 0 & & p^{\beta_{n-1}} \end{pmatrix}.$$

The above discussion and our assumptions imply that the matrices

$$\widetilde{A}_i = \begin{pmatrix} A_i^{(n-1)} & B_i^{(n-1)} \\ 0 & D_i^{(n-1)} \end{pmatrix} \quad \text{and} \quad \widetilde{A}_j' = \begin{pmatrix} A_j'^{(n-1)} & B_j'^{(n-1)} \\ 0 & D_j'^{(n-1)} \end{pmatrix}$$

belong to different Γ'-left cosets in $\Gamma' M_0 \Gamma'$ and $\Gamma' M_0' \Gamma'$, respectively, and they satisfy the relation

$$\widetilde{A}_i \widetilde{A}_j' = \widetilde{\gamma} H_0, \quad \text{where } \widetilde{\gamma} = \begin{pmatrix} \gamma_{11}^{(n-1)} & \gamma_{12}^{(n-1)} \\ 0 & \gamma_{22}^{(n-1)} \end{pmatrix} \in \Gamma'.$$

If we repeat the same argument in the reverse order, we see that, with a suitable choice of Γ'-left coset representatives, any pair N_k, N_l' that satisfies (3.40) can be obtained in the manner indicated. This proves (3.41), and hence the lemma. $\qquad \square$

We can now completely determine the structure of the rings $\underline{L}_p^n(q)$ and $L_p^n(q)$.

THEOREM 3.23. *Suppose that* $n, q \in \mathbf{N}$, *and* p *is a prime number not dividing* q. *Then:*

(1) *the ring* $\underline{L}_p^n(q)$ *is generated over* $\mathbf{Q}$ *by the elements*

$$(3.42) \qquad\qquad T(p) = T^n(p) = T(\underbrace{1, \ldots, 1}_{n}; \underbrace{p, \ldots, p}_{n})$$

and

$$T_i(p^2) = T_i^n(p^2) = T(\underbrace{1, \ldots, 1}_{n-i}, \underbrace{p, \ldots, p}_{n}, \underbrace{p^2, \ldots, p^2}_{n-i}, \underbrace{p, \ldots, p}_{i})$$

for $i = 1, \ldots, n$;

(2) *the ring* $L_p^n(q)$ *is generated over* $\mathbf{Q}$ *by the elements* (3.42) *and the element* $T_n(p^2)^{-1} = \Delta_n(p)^{-1}$;

(3) *the elements* (3.42) *are algebraically independent over* $\mathbf{Q}$.

PROOF. First let $n = 1$. Since

$$\Gamma^1 = SL_2(\mathbf{Z}) = \{M \in \Lambda^1; \det M > 0\} \quad \text{and} \quad S_p^1 = \{M \in G_p^1; \det M > 0\},$$

it follows from Proposition 1.9 that there exists an isomorphism ε from the Hecke ring H_p^1 to the ring L_p^1. According to (1.29), we have $\varepsilon(t(p^{\delta_1}, p^{\delta_2})) = T(p^{\delta_1}, p^{\delta_2})$ for $\delta_1, \delta_2 \in \mathbf{Z}$. From this and Theorem 2.17 it follows that Theorem 3.23 holds in the case $n = 1$ and $q = 1$. Lemma 3.20 then gives us the theorem in the case $n = 1$ and q arbitrary.

We now consider the case of arbitrary n. To prove part (1) it suffices to show that every element of the form (3.34), where the exponents satisfy (3.35), is a polynomial in the elements (3.42). We prove this by induction on n, and for fixed $n > 1$ we use induction on $\delta = \delta_i + \varepsilon_i$. We have already treated the case $n = 1$. Suppose that $n > 1$, and our claim has been proved for degrees less than n. If $\delta = 1$, then $T(p^{\delta_1}, \ldots, p^{\varepsilon_n}) = T(p)$. We now suppose that for all $T(p^{\delta_1'}, \ldots, p^{\varepsilon_n'})$ with $\delta' = \delta_i' + \varepsilon_i' < \delta$ it is already known that they are polynomials in the elements (3.42); and we let $T = T(p^{\delta_1}, \ldots, p^{\varepsilon_n})$ be an element of the indicated form with $\delta_i + \varepsilon_i = \delta$. If $\delta_1 \geqslant 1$, then, by Lemma 3.8, $T = \Delta^{\delta_1} T(p^{\delta_1 - \delta_1}, \ldots, p^{\varepsilon_n - \delta_1})$. Since $(\delta_i - \delta_1) + (\varepsilon_i - \delta_1) < \delta$ and $\Delta = \Delta_n(p) = T_n(p^2)$, it follows by the induction assumption that this element is a polynomial in the elements (3.42). Now let $\delta_1 = 0$. By the first induction assumption, the element

$$\Psi(T) = T(p^{\delta_2}, \ldots, p^{\delta_n}, p^{\varepsilon_2}, \ldots, p^{\varepsilon_n}),$$

where Ψ is the homomorphism (3.38), can be written in the form

$$\Psi(T) = F(\Psi(T(p)), \psi(T_1(p^2)), \ldots, \Psi(T_{n-1}(p^2))),$$

where

$$F(x_1, \ldots, x_n) = \sum_{\alpha = (\alpha_1, \ldots, \alpha_n)} a_\alpha x_1^{\alpha_1} \cdots x_n^{\alpha_n}.$$

The expansion of the monomial $\Psi(T(p))^{\alpha_1} \cdots \Psi(T_{n-1}(p^2))^{\alpha_n}$ only contains double cosets (H) modulo $\Gamma_0^{n-1}(q)$ for which $r(H) = p^{|\alpha|}$ with $|\alpha| = \alpha_1 + 2\alpha_2 + \cdots + 2\alpha_n$. Thus, after combining similar terms in F, we may assume that only coefficients with $|\alpha| = \delta$ are nonzero. Hence, the element

$$T_1 = T - \sum_{|\alpha| = \delta} a_\alpha (T(p))^{\alpha_1} (T_1(p^2))^{\alpha_2} \cdots (T_{n-1}(p^2))^{\alpha_n}$$

lies in the kernel of Ψ. This means that T_1 is imprimitive. On the other hand, T_1 is a linear combination of elements $T(p^{\beta_1}, \ldots, p^{\gamma_n})$ with $\beta_i + \gamma_i = \delta$. Thus, $T_1 = \Delta \cdot T'$, where T' is a linear combination of elements $T(p^{\beta_1}, \ldots, p^{\gamma_n})$ with $\beta_i + \gamma_i = \delta - 2 < \delta$. By the second induction assumption, T'—and hence also T—is a polynomial in the elements (3.42). This proves part (1).

Part (2) follows from part (1) and Lemma 3.21.

We prove part (3) by induction on n. It was proved above in the case $n = 1$. Suppose that $n > 1$, and part (3) has been proved for all $n' < n$. Suppose that the elements (3.42) are algebraically dependent over $\mathbf{Q}$. Let $F(T(p), \ldots, T_n(p^2)) = 0$ be an algebraic relation between these elements, where the polynomial $F(x_0, x_1, \ldots, x_n)$ has smallest possible degree. Then, applying Ψ, we obtain

$$F(T^{n-1}(p), T_1^{n-1}(p^2), \ldots, T_{n-1}^{n-1}(p^2), 0) = 0.$$

From this and the induction assumption it follows that

$$F(x_0, \ldots, x_n) = x_n F_1(x_0, \ldots, x_n),$$

where F_1 is another polynomial. Since the element $T_n^n(p^2) = \Delta$ is not a zero divisor in $L_p^n(q)$, the equality

$$F(T(p), \ldots, T_n(p^2)) = \Delta F_1(T(p), \ldots, T_n(p^2)) = 0$$

implies that $F_1(T(p), \ldots, T_n(p^2)) = 0$, and this contradicts the choice of F as a polynomial of minimal degree. $\square$

PROBLEM 3.24. State and prove results similar to the results in Problem 2.19 in the case when

$$G = \{ M \in M_{2n}(\mathbf{Q}_p); \, {}^t M J_n M = r(M) J_n, r(M) \neq 0 \}$$

and $\Gamma = G \cap GL_{2n}(\mathbf{Z}_p)$.

3. The spherical map. The procedure described in the previous subsection for expressing an element of a local Hecke ring as a polynomial in the generators is effective, but in general it is not practical. As in the case of the general linear group, we avoid this difficulty by constructing another polynomial realization of the local Hecke rings. Namely, we use rings of polynomials that are invariant under a certain finite group of transformations of the variables.

For later applications it is convenient to carry out all of the constructions for suitable extensions of the local Hecke rings of the symplectic group. The extensions we consider are the Hecke rings of the "triangular" subgroup

$$\Gamma_0 = \Gamma_0^n = \left\{ \begin{pmatrix} A & B \\ 0 & D \end{pmatrix} \in \Gamma^n \right\}$$

of the Siegel modular group Γ^n and the subgroups of the group

$$S_0 = S_0^n = \left\{ \begin{pmatrix} A & B \\ 0 & D \end{pmatrix} \in S^n \right\}$$

(3.43)
$$= \left\{ M = \begin{pmatrix} A & B \\ 0 & D \end{pmatrix} \in M_{2n}(\mathbf{Q}); \, {}^t AD = r(M) E_n, \right.$$

$$\left. r(M) > 0, \, {}^t BD = {}^t DB \right\}.$$

LEMMA 3.25. (1) *Every left coset of* $\Gamma_0^n \setminus S_0^n$ *contains a representative* $M = \begin{pmatrix} A & B \\ 0 & C \end{pmatrix}$, *where D belongs to a fixed left coset of* $\Lambda^n \setminus G^n$, *B belongs to a fixed residue class $B(D)/\mathrm{mod}\, D$ (see (3.24) and (3.25)), and $A = r(M)D^*$.*

(2) *The decomposition of an arbitrary double coset of S_0^n into left cosets of the group* $\Gamma_0 = \Gamma_0^n$ *has the form*

(3.44)
$$\Gamma_0 M \Gamma_0 = \bigcup_{\substack{D_1 \in \Lambda \setminus \Lambda D \Lambda \\ B_1 \in B_M(D_1)/\mathrm{mod}\, D_1}} \Gamma_0 \begin{pmatrix} rD_1^* & B_1 \\ 0 & D_1 \end{pmatrix},$$

where $r = r(M)$, $\Lambda = \Lambda^n$, *and*

$$B_M(D_1) = \left\{ B_1; \begin{pmatrix} rD_1^* & B_1 \\ 0 & D_1 \end{pmatrix} \in \Gamma_0 M \Gamma_0 \right\}.$$

(3) (Γ_0^n, S_0^n) *is a Hecke pair.*

PROOF. Part (1) follows from the definitions. Part (2) follows from Part (1) if we note that for any matrix $\begin{pmatrix} D_2 & * \\ 0 & D_1 \end{pmatrix} \in \Gamma_0 M \Gamma_0$ we have $D_1 \in \Lambda D \Lambda$ and $D_2 = rD_1^*$. To prove Part (3), it is sufficient to verify that every double coset $\Gamma_0 M \Gamma_0$, where $M \in S_0$, consists of finitely many left cosets. Without loss of generality we may assume that M is an integer matrix. Then each set $B_M(D)$ is contained in the set $\{B \in M_n; \,{}^t BD = {}^t DB\}$, which obviously consists of finitely many residue classes modulo D. From this observation and Lemma 2.1 it follows that there are finitely many left cosets in (3.44). $\square$

To construct our extensions of the local Hecke rings for the group S_0^n, we define the subgroups

$$S_{0,p} = S_{0,p}^n = S_0^n \cap GL_{2n}(\mathbf{Z}[p^{-1}]),$$

where p is a prime number. From Lemma 3.25(3) it follows that $(\Gamma_0^n, S_{0,p}^n)$ is a Hecke pair.

LEMMA 3.26. *The Hecke pairs* $(\Gamma_0^n(q), S_p^n(q))$, *where* $(p, q) = 1$, *and* $(\Gamma_0^n, S_{0,p}^n)$ *satisfy the conditions* (1.26). *The following diagram commutes:*

(3.45)

$$
\begin{array}{ccc}
L_p^n & \xrightarrow{\ \varepsilon\ } & \mathbf{L}_{0,p}^n = D_{\mathbf{Q}}(\Gamma_0^n, S_{0,p}^n) \\
{\scriptstyle \varepsilon_{1,q}} \downarrow & & \uparrow {\scriptstyle \varepsilon_q} \\
L_p^n(q) & =\!=\!=\!= & L_p^n(q),
\end{array}
$$

where $\varepsilon = \varepsilon_1$ *and* ε_q *are the imbeddings* (1.27), *and* $\varepsilon_{1,q}$ *is the isomorphism in Lemma* 3.20.

PROOF. The first and third conditions in (1.26) are obvious in the case of our Hecke pairs, and the second condition is a consequence of Lemma 3.4. The commutativity of the diagram follows from the definitions of the three mappings. $\square$

According to this lemma, instead of $L_p^n(q)$ one can study the isomorphic (and independent of q) subring

(3.46) $$\mathbf{L}_p^n = \varepsilon_q(L_q^n(q)) = \varepsilon(L_p^n)$$

of the local Hecke ring $\mathbf{L}_{0,p}^n$ of the group Γ_0^n.

The analogue of Lemma 3.21 that follows allows us to reduce the study of these Hecke rings to that of their integral subrings

(3.47) $$\underline{\mathbf{L}}_{0,p}^n = D_{\mathbf{Q}}(\Gamma_0^n, S_{0,p}^n \cap M_{2n}), \quad \underline{\mathbf{L}}_p^n = \mathbf{L}_p^n \cap \underline{\mathbf{L}}_{0,p}^n.$$

LEMMA 3.27. *The element*

$$(3.48) \qquad\qquad \Delta = \Delta_n(p) = (pE_{2n})_{\Gamma_0^n} \in \underline{\mathbf{L}}_p^n$$

lies in the center of the ring $\mathbf{L}_{0,p}^n$ *and is invertible in the ring* $\mathbf{L}_p^n$; *we have*

$$\Delta^{-1} = (p^{-1}E_{2n})_{\Gamma_0^n}.$$

The ring $\mathbf{L}_p^n$ *(respectively,* $\mathbf{L}_{0,p}^n$*) is generated by the subring* $\underline{\mathbf{L}}_p^n$ *(resp.* $\underline{\mathbf{L}}_{0,p}^n$*) and the element* Δ^{-1}.

PROOF. The lemma is an easy consequence of the definitions, since $\Gamma M \Gamma = \Gamma M = M \Gamma$ for $M = p^{\pm 1} E_{2n}$ and for any subgroup $\Gamma \subset \Gamma^n$. $\qquad \square$

REMARK. The element (3.48) is obviously the image of the element (3.33) under the map ε_q. In general, for simplicity we shall usually use the same notation for elements in $\varepsilon_q(L_p^n(q)) \subset \mathbf{L}_{0,p}^n$ as for their preimages.

The spherical map for the Hecke ring $\mathbf{L}_{0,p}^n$ will be defined in two stages. We first define a map to a suitable extension of the local p-Hecke ring of the general linear group GL_n, and we then use the spherical map of this extension that was defined in §2.3. We start with the left coset space. Let

$$\Gamma_0 M, \quad \text{where } M = \begin{pmatrix} p^\delta D^* & B \\ 0 & D \end{pmatrix} \in S_{0,p},$$

be an arbitrary left coset of the group $S_{0,p}$ modulo Γ_0. By Lemma 3.25, the left coset ΛD of the element $D \in G_p$, along with the exponent δ, is uniquely determined by the original left coset $\Gamma_0 M$. We then set

$$\Phi((\Gamma_0 M)) = x_0^\delta (\Lambda D),$$

where we suppose that all of the powers x_0^i $(i \in \mathbf{Z})$ are linearly independent over the left coset module of G_p modulo Λ. We extend Φ by linearity to a map of the left coset module:

$$\Phi = \Phi_p^n \colon L_{\mathbf{Q}}(\Gamma_0^n, S_{0,p}^n) \to L_{\mathbf{Q}[x_0^{\pm 1}]}(\Lambda^n, G_p^n).$$

PROPOSITION 3.28. *The restriction of* Φ *to the ring* $\mathbf{L}_{0,p}^n$ *is an epimorphism of this ring onto the ring* $H_p^n[x_0^{\pm 1}]$.

PROOF. Let $X \in \mathbf{L}_{0,p}^n$. By definition, X is invariant under right multiplication by any matrix of the form $U(\gamma)$ with $\gamma \in \Lambda$. This implies that $\Phi(X)$ is invariant under right multiplication by any element $\gamma \in \Lambda$, where the multiplication acts only on the left cosets, not on the coefficients. Thus, $\Phi(X) \in H_p^n[x_0^{\pm 1}]$. From the definition of the multiplication in Hecke rings and the definition of Φ it follows that Φ is a ring homomorphism. Finally, if D is an arbitrary matrix in G_p and $\delta \in \mathbf{Z}$, then

$$M = \begin{pmatrix} p^\delta D^* & 0 \\ 0 & D \end{pmatrix} \in S_{0,p}, \text{ and from (3.44) it follows that}$$

$$\Phi((M)_{\Gamma_0}) = \alpha(M) x_0^\delta (D)_\Lambda,$$

where $\alpha(M)$ is a positive integer. This gives us the epimorphism. $\qquad \square$

Now let $\omega = \omega_p^n$ be the $\mathbf{Q}$-linear homomorphism from the ring $H_p^n[x_0^{\pm 1}]$ to the subring $\mathbf{Q}[x_0^{\pm 1}, \dots, x_n^{\pm 1}]$ of the field of rational functions over $\mathbf{Q}$ in the variables

$x_0, x_1, \ldots, x_n$ such that $\omega(x_0) = x_0$ and the restriction of ω to H_p^n coincides with the spherical map ω defined in §2.3. From Theorem 2.20 and the definitions we then have

LEMMA 3.29. *The map* $\omega = \omega_p^n$ *is an isomorphism of the ring* $H_p^n[x_0^{\pm 1}]$ *with the subring* $\mathbf{Q}[x_0^{\pm 1}, \ldots, x_n^{\pm 1}]_s \subset \mathbf{Q}[x_0^{\pm 1}, \ldots, x_n^{\pm 1}]$ *consisting of all symmetric functions in* $x_1, \ldots, x_n$.

Finally, we define the *spherical map* $\Omega = \Omega_p^n$ *from* $\mathbf{L}_{0,p}^n$ *to* $\mathbf{Q}[x_0^{\pm 1}, \ldots, x_n^{\pm 1}]_s$ by setting

$$(3.49) \qquad \Omega(X) = \omega(\Phi(X)) \quad (X \in \mathbf{L}_{0,p}^n).$$

Thus, we obtain a commutative diagram:

$$(3.50) \qquad \begin{array}{ccc} \mathbf{L}_{0,p}^n & \xrightarrow{\ \Omega\ } & \mathbf{Q}[x_0^{\pm 1}, \ldots, x_n^{\pm 1}]_s \\ & {\scriptstyle \Phi}\searrow \quad \nearrow{\scriptstyle \omega} & \\ & H_p^n[x_0^{\pm 1}] & \end{array}$$

Since Φ and ω are $\mathbf{Q}$-linear ring epimorphisms, it follows that Ω is also a $\mathbf{Q}$-linear ring epimorphism.

Let $W = W_n$ be the group of $\mathbf{Q}$-automorphisms of the rational function field $\mathbf{Q}(x_0, x_1, \ldots, x_n)$ that is generated by all permutations of the variables $x_1, \ldots, x_n$ and by the automorphisms $\tau_1, \ldots, \tau_n$, which act according to the rule

$$(3.51) \qquad \tau_i(x_0) = x_0 x_i, \quad \tau_i(x_i) = x_i^{-1}, \quad \tau_i(x_j) = x_j \quad (j \neq 0, i).$$

The reader can easily verify that each of the coefficients $r_a = r_a^n(x_1, \ldots, x_n)$ in the expansion

$$(3.52) \qquad r(x_1, \ldots, x_n; v) = \prod_{i=1}^n (1 - x_i v)(1 - x_i^{-1} v) = \sum_{a=0}^{2n} (-1)^a r_a v^a,$$

as well as the polynomials

$$(3.53) \qquad \rho_a = \rho_a^n(x_0, x_1, \ldots, x_n) = x_0^2 x_1 \ldots x_n r_a^n(x_1, \ldots, x_n)$$

and the polynomial

$$(3.54) \qquad t = t^n(x_0, x_1, \ldots, x_n) = x_0 \prod_{i=1}^n (1 + x_i),$$

are all invariant under the transformations in W_n. The polynomials $t, \rho_0, \ldots, \rho_{n-1}$ play the same role for W_n that the elementary symmetric polynomials play for the symmetric group.

THEOREM 3.30. *Let* $n \in \mathbf{N}$, *and let* p *be a prime number. Then*:

(1) *The restriction of the map* $\Omega = \Omega_p^n$ *to the integral subring* $\underline{\mathbf{L}}_p^n \subset \mathbf{L}_p^n$ *is an isomorphism of this subring with the ring* $\mathbf{Q}[x_0, \ldots, x_n]_W$ *of all* W_n-*invariant polynomials in* $x_0, x_1, \ldots, x_n$ *over* $\mathbf{Q}$.

(2) *Any element in* $\mathbf{Q}[x_0, \ldots, x_n]_W$ *can be written as a polynomial in*

$$(3.55) \qquad t = t^n(x_0, x_1, \ldots, x_n), \quad \rho_a = \rho_a^n(x_0, x_1, \ldots, x_n) \quad (0 \leqslant a \leqslant n - 1),$$

with coefficients in **Q**, *i.e.,*

$$(3.56) \qquad \mathbf{Q}[x_0, x_1, \ldots, x_n]_W = \mathbf{Q}[t, \rho_0, \rho_1, \ldots, \rho_{n-1}].$$

The elements (3.55) *are algebraically independent over* **Q**.

(3) *The restriction of the map* $\Omega = \Omega_p^n$ *to the full subring* $\mathbf{L}_p^n \subset \mathbf{L}_{0,p}^n$ *is an isomorphism of this subring with the ring* $\mathbf{Q}[x_0^{\pm 1}, \ldots, x_n^{\pm 1}]_W$ *of all* W_n-*invariant polynomials in* $x_0^{\pm 1}, x_1^{\pm 1}, \ldots, x_n^{\pm 1}$ *over* **Q**. *The latter ring can be obtained by adjoining the element* $\rho_0^{-1} = (x_0^2 x_1 \cdots x_n)^{-1}$ *to the polynomial ring* (3.56):

$$(3.57) \qquad \mathbf{Q}[x_0^{\pm 1}, \ldots, x_n^{\pm 1}]_W = \mathbf{Q}[x_0, \ldots, x_n]_W[(x_0^2 x_1 \cdots x_n)^{-1}].$$

COROLLARY 3.31. *The restriction of* $\Phi = \Phi_p^n$ *to the subring* $\mathbf{L}_p^n \subset \mathbf{L}_{0,p}^n$ *is a monomorphism.*

The plan of proof of Theorem 3.30 is similar to that for Theorem 2.20. By computing the Ω-images of the generators of $\underline{\mathbf{L}}_p^n$, we obtain generators of the ring $\Omega(\underline{\mathbf{L}}_p^n)$. This enables us to study the algebraic features of this ring and, in particular, to prove that the restriction of Ω to $\underline{\mathbf{L}}_p^n$ is a monomorphism. The ring $\mathbf{L}_p^n$ is investigated using Lemma 3.27. However, in the case of the symplectic group some preliminary work is necessary in order to compute the Ω-images of the generators of $\underline{\mathbf{L}}_p^n$. This is the purpose of Lemmas 3.32–3.34.

LEMMA 3.32. *In the Hecke ring* $\mathbf{L}_p^n \subset \mathbf{L}_{0,p}^n$ *the elements* (3.42) *have the following expansion into left cosets modulo* $\Gamma_0 = \Gamma_0^n$:

$$(3.58) \qquad \mathbf{T}(p) = \mathbf{T}^n(p) = \sum_{a=0}^n \Pi_a,$$

where

$$(3.59) \qquad \Pi_a = \Pi_a^n(p) = \sum_{\substack{D \in \Lambda \backslash \Lambda D_a \Lambda \\ B \in B_0(D)/\mathrm{mod}\, D}} \left(\Gamma_0 \begin{pmatrix} p^2 D^* & B \\ 0 & D \end{pmatrix} \right),$$

$D_a = D_a^n(p)$ *are the matrices* (2.28), $\Lambda = \Lambda^*$,

$$(3.60) \qquad B_0(D) = \{B \in M_n; \ {}^t BD = {}^t DB\},$$

and, as before, the congruence modulo D is understood in the sense of (3.25);

$$(3.61) \qquad \mathbf{T}_i(p^2) = \mathbf{T}_i^n(p^2) = \sum_{\substack{a+b \leqslant n \\ a \geqslant i}} \Pi_{a,b}^{(a-i)} \quad (0 \leqslant i \leqslant n),$$

where

$$(3.62) \qquad \Pi_{a,b}^{(r)} = \sum_{\substack{D \in \Lambda \backslash \Lambda D_{a,b} \Lambda \\ B \in B_0(D)/\mathrm{mod}\, D \\ r_p\left(\begin{pmatrix} p^2 D^* & B \\ 0 & D \end{pmatrix} \right) = n-a+r}} \left(\Gamma_0 \begin{pmatrix} p^2 D^* & B \\ 0 & D \end{pmatrix} \right),$$

$D_{a,b} = D_{a,b}^n(p)$ are the matrices (2.31), and $r_p(M)$ denotes the rank of M over the field of p elements. The sum of the left cosets Π_a $(0 \leqslant a \leqslant n)$ and the sum of the left cosets $\Pi_{a,b}^{(r)}$ $(a + b \leqslant n, r \leqslant a)$ belong to the Hecke ring $\mathbf{L}_{0,p}^n$, and

$$(3.63) \qquad \Pi_a = (M_a)_{\Gamma_0}, \quad \text{where } M_a = \begin{pmatrix} pD_a^{-1} & 0 \\ 0 & D_a \end{pmatrix}.$$

PROOF. Without loss of generality we may consider the elements (3.42) in the case $q = 1$.

From Lemma 3.6 it follows that the double coset $\Gamma M_n \Gamma$, where $\Gamma = \Gamma^n$, coincides with the set $\mathrm{SM}(p) = \mathrm{SM}_n(p, 1)$ (see (3.21)). From the definition of this set we see that it contains the matrix $\begin{pmatrix} A & B \\ 0 & D \end{pmatrix}$ if and only if $A, D \in M_n(\mathbf{Z})$, ${}^t AD = pE_n$, and $B \in B_0(D)$. Lemma 2.2 implies that, if D is a fixed nonsingular integer matrix, then ${}^t A = pD^{-1}$ is an integer matrix if and only if D lies in one of the double cosets $\Lambda D_0 \Lambda$ for $0 \leqslant a \leqslant n$. Thus, by Lemma 3.11 we have the decomposition

$$\Gamma \begin{pmatrix} E_n & 0 \\ 0 & pE_n \end{pmatrix} \Gamma = \bigcup_{a=0}^n \bigcup_{\substack{D \in \Lambda \backslash \Lambda D_a \Lambda \\ B \in B_0(D)/\bmod D}} \Gamma \begin{pmatrix} p^2 D^* & B \\ 0 & D \end{pmatrix}.$$

From this and the definition of the map ε we obtain (3.58).

If we apply Lemma 3.6 to the set $\mathrm{SM}(p^2) = \mathrm{SM}_n(p^2, 1)$, we obtain the decomposition

$$\mathrm{SM}(p^2) = \bigcup_{i=0}^n \mathrm{SM}^{(i)}(p^2),$$

where

$$\mathrm{SM}^{(i)}(p^2) = \Gamma \begin{pmatrix} D_i & 0 \\ 0 & p^2 D_i^{-1} \end{pmatrix} \Gamma.$$

On the other hand, just as in the earlier case of the set $\mathrm{SM}(p)$, we see that Lemmas 2.2 and 3.11 give us the decomposition

$$(3.64) \qquad \mathrm{SM}(p^2) = \bigcup_{\substack{a+b \leqslant n \\ D \in \Lambda \backslash \Lambda D_{a,b} \Lambda \\ B \in B_0(D)/\bmod D}} \Gamma \begin{pmatrix} p^2 D^* & B \\ 0 & D \end{pmatrix}.$$

Since each set $\mathrm{SM}^{(i)}(p^2)$ consists of a single double coset modulo $\Gamma \subset \Lambda^{2n}$, it follows that all of the matrices in such a set have the same rank over the field of p elements; and this rank is obviously $n - i$. Thus, $\mathrm{SM}^{(i)}(p^2)$ is the union of all left cosets ΓM in (3.64) for which $r_p(M) = n - i$. From this and the definitions we obtain (3.61).

We set

$$S_a = \left\{ \begin{pmatrix} p^2 D^* & B \\ 0 & D \end{pmatrix} ; D \in \Lambda D_a \Lambda, B \in B_0(D) \right\}$$

and

$$S_{a,b}^{(r)} = \left\{ M = \begin{pmatrix} p^2 D^* & B \\ 0 & D \end{pmatrix} ; D \in \Lambda D_{a,b} \Lambda, B \in B_0(D), r_p(M) = n - a + r \right\}.$$

Then by Lemma 3.25 we have the expansions

$$(3.65) \qquad\qquad \Pi_a = \sum_{M \in \Gamma_0 \backslash S_a} (\Gamma_0 M)$$

and

$$(3.66) \qquad\qquad \Pi_{a,b}^{(r)} = \sum_{M \in \Gamma_0 \backslash S_{a,b}^{(r)}} (\Gamma_0 M).$$

Since obviously $\Gamma_0 S_a \Gamma_0 = S_a$ and $\Gamma_0 S_{a,b}^{(r)} \Gamma_0 = S_{a,b}^{(r)}$, it follows from the above decompositions that the elements Π_a and $\Pi_{a,b}^{(r)}$ are invariant under any right multiplication by elements of Γ_0; hence, they belong to the ring $\mathbf{L}_{0,p}^n$. Finally, let $M = \begin{pmatrix} pD^* & B \\ 0 & D \end{pmatrix}$ be an arbitrary element of S_a. If we replace M by $\gamma M \gamma_1$ with suitable $\gamma, \gamma_1 \in \Gamma_0$, we may suppose that $D = D_a$. Then B is an integer matrix of the form (B_{ij}) $(i, j = 1, 2)$, where $B_{11} \in S_{n-a}(\mathbf{Z})$, $B_{22} \in S_a(\mathbf{Z})$, and $B_{12} = p \cdot {}^t B_{21}$. This implies that

$$M = T(S)\mathrm{diag}(pD_a^{-1}, D_a)T(S_1) \in \Gamma_0 M_a \Gamma_0,$$

where $S = \begin{pmatrix} B_{11} & {}^t B_{21} \\ B_{21} & 0 \end{pmatrix}$, $S_1 = \begin{pmatrix} 0 & 0 \\ 0 & B_{22} \end{pmatrix}$. Thus, $S_a = \Gamma_0 M_a \Gamma_0$, and (3.63) follows from (3.65). $\qquad\square$

We now describe the sets of the form $B_0(D)/\mathrm{mod}\, D$, and we compute the number of elements they have. It will be more convenient for later applications if we do this in a general form.

LEMMA 3.33. *Suppose that $D \in M_n(\mathbf{Z})$ and $\det D \neq 0$. Then:*

(1) *If $\alpha, \beta \in \Lambda^n$, then $B_0(\alpha D \beta) = \alpha^* B_0(D) \beta$, and one can take the set $\alpha^* \{B_0(D)/\mathrm{mod}\, D\} \beta$ as representatives of $B_0(\alpha D \beta)/\mathrm{mod}\, \alpha D \beta$. In particular, if $b_0(D)$ denotes the number of elements in $B_0(D)/\mathrm{mod}\, D$, then $b_0(\alpha D \beta) = b_0(D)$.*

(2) *Suppose that $D = \mathrm{ed}(D) = \mathrm{diag}(d_1, \ldots, d_n)$ is an elementary divisor matrix (see (2.4)). Then one can take*

$$B_0(D)/\mathrm{mod}\, D = \{B = (b_{ij}); b_{ij} = d_j d_i^{-1} b_{ji} \ (1 \leqslant i < j \leqslant n),$$
$$0 \leqslant b_{ji} < d_i \ (1 \leqslant i \leqslant j \leqslant n)\}.$$

In particular, $b_0(D) = d_1^n d_2^{n-1} \cdots d_n$.

PROOF. The equality ${}^t BD = {}^t DB$ is obviously equivalent to the equality

$$ {}^t(\alpha^* B \beta)(\alpha D \beta) = {}^t(\alpha D \beta)(\alpha^* B \beta);$$

and for $B, B_1 \in B_0(D)$ the congruence $B \equiv B_1 (\mathrm{mod}\, D)$ is equivalent to the congruence $\alpha^* B \beta \equiv \alpha^* B_1 \beta (\mathrm{mod}\, \alpha D \beta)$. This implies the first part of the lemma. The second part follows easily from the definitions. $\qquad\square$

We are now ready to compute the images of elements of the form Π_a and $\Pi_{a,b}^{(r)}$ under the maps $\Phi = \Phi_p^n$ and $\Omega = \Omega_p^n$.

LEMMA 3.34. *Let $n \in \mathbf{N}$, and let p be a prime number. Then*

$$\Phi(\Pi_a^n(p)) = p^{\langle a \rangle} x_0 \pi_a^n(p)$$

and

$$\Omega(\Pi_a^n(p)) = x_0 s_a(x_1, \ldots, x_n),$$

where $\pi_a^n(p) \in H_p^n$ are the elements (2.26), and s_a is the ath elementary symmetric function, $0 \leqslant a \leqslant n$;

$$\Phi(\Pi_{a,b}^{(r)}) = p^{b(a+b+1)} l_p(r,a) x_0^2 \pi_{a,b}^n(p)$$

and

$$\Omega(\Pi_{a,b}^{(r)}) = p^{b(a+b+1)} l_p(r,a) x_0^2 \omega(\pi_{a,b}^n(p)),$$

where $l_p(r,a)$ is the number of $a \times a$ symmetric matrices of rank r over the field of p elements, $\pi_{a,b}^n(p) \in H_p^n$ are the elements (2.31), and ω is the spherical map for the ring H_p^n, $a + b \leqslant n$, $r \leqslant a$. In particular, the following formulas hold for the elements (3.48):

$$\Phi(\Delta(p)) = x_0^2 \pi_n^n(p) \quad and \quad \Omega(\Delta_n(p)) = p^{-\langle n \rangle} x_0^2 x_1 \cdots x_n.$$

PROOF. Using the expansion (3.59), Lemma 3.33, and the definitions, we obtain

$$\Phi(\Pi_a) = \sum_{D \in \Lambda \backslash \Lambda D_a \Lambda} x_0 b_0(D)(\Lambda D) = x_0 b_0(D_a) \sum_{D \in \Lambda \backslash \Lambda D_a \Lambda} (\Lambda D) = x_0 p^{\langle a \rangle} \pi_a,$$

which proves the first formula. The second formula follows from the first one and from Lemma 2.21.

Now suppose that $D \in \Lambda D_{a,b} \Lambda$, where $a + b \leqslant n$. Then $D = \alpha D_{a,b} \beta$ with $\alpha, \beta \in \Lambda$, and by Lemma 3.33 we can take

$$B_0(D)/\mathrm{mod}\, D = \alpha^* \{ B_0(D_{a,b})/\mathrm{mod}\, D_{a,b} \} \beta$$

(3.67)
$$= \alpha^* \left\{ b = \begin{pmatrix} 0 & 0 & 0 \\ 0 & B_{22} & B_{23} \\ 0 & B_{32} & B_{33} \end{pmatrix}; B_{22} \in S_a(\mathbf{Z})/\mathrm{mod}\, p, \right.$$

$$\left. B_{33} \in S_b(\mathbf{Z})/\mathrm{mod}\, p^2, B_{23} = p \cdot {}^t B_{32}, B_{32} \in M_{b,a}(\mathbf{Z})/\mathrm{mod}\, p \right\} \beta.$$

It is not hard to see that for a fixed matrix $B' = \alpha^* B \beta$ in this set we have

$$M = \begin{pmatrix} p^2 D^* & B' \\ 0 & D \end{pmatrix} \in \Gamma_0 \begin{pmatrix} p^2 D_{a,b}^{-1} & B_0 \\ 0 & D_{a,b} \end{pmatrix} \Gamma_0,$$

where

(3.68)
$$B_0 = \begin{pmatrix} 0 & 0 & 0 \\ 0 & B_{22} & 0 \\ 0 & 0 & 0 \end{pmatrix}.$$

This obviously implies that $r_p(M) = b + r_p(B_{22}) + n - a - b$. Thus,

$$\Phi(\Pi_{a,b}^{(r)}) = \sum_{D \in \Lambda \backslash \Lambda D_{a,b} \Lambda} x_0^2 p^{ba + b(b+1)} l_p(r,a)(\Lambda D) = p^{b(a+b+1)} l_p(r,a) x_0^2 \pi_{a,b}^n(p),$$

which proves the third formula. The fourth formula follows from the third one and the definition of the map Ω. Since obviously $\Delta_n(p) = \Pi_{n,0}^{(0)}$ and $\pi_{n,0}^n(p) = \pi_n^n(p)$, the last formula is a consequence of the formulas already proved and Lemma 2.21. $\quad\square$

PROOF OF THEOREM 3.30. We first show that

$$(3.69) \qquad\qquad \Omega(\underline{\mathbf{L}}_p^n) = \mathbf{Q}[t, \rho_0, \rho_1, \ldots, \rho_{n-1}].$$

By Theorem 3.23, the elements $\mathbf{T}(p)$, $\mathbf{T}_1(p^2), \ldots, \mathbf{T}_n(p^2)$ of the Hecke ring $\mathbf{L}_{0,p}^n$ generate the ring $\underline{\mathbf{L}}_p^n$ over $\mathbf{Q}$. Hence, the ring $\Omega(\underline{\mathbf{L}}_p^n)$ is generated by the images $\Omega(\mathbf{T}(p))$, $\Omega(\mathbf{T}_1(p^2)), \ldots, \Omega(\mathbf{T}_n(p^2))$. Using (3.58) and Lemma 3.34, we obtain

$$(3.70) \qquad \Omega(\mathbf{T}(p)) = \sum_{a=0}^{n} \Omega(\Pi_a) = \sum_{a=0}^{n} x_0 s_a(x_1, \ldots, x_n) = x_0 \prod_{i=1}^{n}(1 + x_i) = t.$$

Thus, to prove (3.69) it suffices to verify that the vector spaces

$$V_1 = \left\{ \sum_{i=1}^{n} \alpha_i \Omega(\mathbf{T}_i(p^2)); \alpha_i \in \mathbf{Q} \right\}$$

and

$$V_2 = \left\{ \sum_{j=0}^{n-1} \beta_j \rho_j; \beta_j \in \mathbf{Q} \right\}$$

coincide. From (3.61) and Lemma 3.34 we obtain

$$\Omega(\mathbf{T}_i(p^2)) = \sum_{\substack{a+b \leqslant n \\ a \geqslant i}} p^{b(a+b+1)} l_p(a-i, a) x_0^2 \omega(\pi_{a,b}(p)) = \sum_{a=i}^{n} l_p(a-i, a) x_0^2 \Psi_a,$$

where

$$\Psi_a = \sum_{b=0}^{n-a} p^{b(a+b+1)} \omega(\pi_{a,b}(p)).$$

We set

$$V_3 = \left\{ \sum_{a=1}^{n} \gamma_a x_0^2 \Psi_a; \gamma_a \in \mathbf{Q} \right\}.$$

The above formulas for $\Omega(\mathbf{T}_i(p^2))$ imply that $V_1 \subset V_3$. The same formulas also imply that the coefficient matrix for the expansions of $\Omega(\mathbf{T}_1(p^2)), \ldots, \Omega(\mathbf{T}_n(p^2))$ with respect to $x_0^2 \Psi_1, \ldots, x_0^2 \Psi_n$ is a triangular matrix, has integer entries, and has entries $l_p(0, a) = 1$ $(a = 1, \ldots, n)$ on the main diagonal. Hence, this matrix has an inverse matrix of the same form, and this implies that each $x_0^2 \Psi_a$ $(a = 1, \ldots, n)$ is an integer linear combination of $\Omega(\mathbf{T}_1(p^2)), \ldots, \Omega(\mathbf{T}_n(p^2))$. In particular, $V_3 \subset V_1$. Thus, $V_1 = V_3$. On the other hand, returning to the polynomials ρ_a, by (3.52) we have

$$\sum_{a=0}^{2n} (-1)^a r_a v^a = \left(\sum_{i=0}^{n} (-1)^i s_i(x_1, \ldots, x_n) v^i \right) \left(\sum_{j=0}^{n} (-1)^j s_j(x_1^{-1}, \ldots, x_n^{-1}) v^j \right);$$

if we take into account that $s_j(x_1^{-1}, \ldots, x_n^{-1}) = (x_1 \cdots x_n)^{-1} s_{n-j}(x_1, \ldots, x_n)$, we obtain

$$\rho_a = x_0^2 x_1 \cdots x_n r_a = x_0^2 \sum_{i+j=a} s_i(x_1, \ldots, x_n) s_{n-j}(x_1, \ldots, x_n).$$

Using the spherical map $\omega = \omega_p^n$ and Lemma 2.21, we can rewrite these formulas in the form

$$\rho_a = x_0^2 \omega \left(\sum_{i+j=a} p^{\langle i \rangle + \langle n-j \rangle} \pi_i \pi_{n-j} \right),$$

where $\pi_\alpha = \pi_\alpha^n(p)$. We use the formulas (2.30) to compute the products $\pi_j \pi_{n-j}$, and we substitute the resulting expressions in the last formula for ρ_a. We obtain

$$\rho_a = x_0^2 \omega \left(\sum_{i+j=a} p^{\langle i \rangle + \langle n-j \rangle} \sum_{\substack{\alpha+\beta=i \\ 0 \leqslant \alpha \leqslant j \\ 0 \leqslant \beta \leqslant n-j}} \frac{\varphi_{\alpha+n-j-b}}{\varphi_\alpha \varphi_{n-j-\beta}} \pi_{\alpha+n-j-\beta,\beta} \right).$$

If we set $i = \alpha + \beta$, $j = a - \alpha - \beta$, and note that the conditions on α and β in the summation are then equivalent to the inequalities $\alpha, \beta \geqslant 0$, $2\alpha + \beta \leqslant a$, we obtain

$$\rho_a = x_0^2 \sum_{\substack{2\alpha+\beta \leqslant a \\ \alpha,\beta \geqslant 0}} p^{\langle \alpha+\beta \rangle + \langle \alpha+\beta+n-a \rangle} \frac{\varphi_{2\alpha+n-a}}{\varphi_\alpha \varphi_{\alpha+n-a}} \omega(\pi_{2\alpha+n-a,\beta})$$

$$= x_0^2 \sum_{0 \leqslant \alpha \leqslant a/2} p^{\langle \alpha \rangle + \langle \alpha+n-a \rangle} \frac{\varphi_{2\alpha+n-a}}{\varphi_\alpha \varphi_{\alpha+n-a}} \Psi_{2\alpha+n-a},$$

where, in accordance with our earlier notation,

$$\Psi_{2\alpha+n-a} = \sum_{\beta=0}^{n-(2\alpha+n-a)} p^{\beta(2\alpha+n-a+\beta+1)} \omega(\pi_{2\alpha+n-a,\beta}).$$

Setting $2\alpha + n - a = b$ and replacing a by $n - a$, we obtain

$$\rho_{n-a} = \sum_{\substack{a \leqslant b \leqslant n \\ b \equiv a \,(\mathrm{mod}\, 2)}} p^{(b-a)(b-a+2)/8 + (b+a)(b+a+2)/8} \frac{\varphi_b}{\varphi_{(b-a)/2} \varphi_{(b+a)/2}} x_0^2 \Psi_b.$$

From these formulas it follows, in particular, that the polynomials $\rho_{n-1}, \rho_{n-2}, \ldots, \rho_0$ can be expressed as linear combinations of $x_0^2 \Psi_1, x_0^2 \Psi_2, \ldots, x_0^2 \Psi_n$; and the matrix of the expansion is a triangular matrix with rational entries and nonzero entries on the main diagonal. As before, we conclude from this that the vector spaces spanned by these sets over $\mathbf{Q}$ are the same, i.e., $V_2 = V_3$. Thus, $V_1 = V_2$, and (3.69) is proved.

We now prove (3.56). From the definitions it easily follows that the elements (3.55) are invariant under all transformations in W_n. Thus, the right side of (3.56) is contained in the left side. We prove the reverse inclusion by induction on n, and for fixed n by induction on the degree in x_0 of the W-invariant polynomial. If $n = 1$, then the change of variables $x_0 = z_1$, $x_0 x_1 = z_2$ obviously takes the ring $\mathbf{Q}[t, \rho_0]$ to $\mathbf{Q}[z_1 + z_2, z_1 z_2]$, and takes $\mathbf{Q}[x_0, x_1]_W$ to the ring $\mathbf{Q}[z_1, z_2]_s$ of all polynomials in z_1 and z_2 over $\mathbf{Q}$ that do not change when z_1 and z_2 are permuted (note that if $F(x_0, x_1) = \sum_{i,j \geqslant 0} a_{ij} x_0^i x_1^j$ is W-invariant, i.e., if $F(x_0 x_1, x_1^{-1}) = F(x_0, x_1)$, then $a_{ij} \neq 0$ implies that $i \geqslant j$, and hence $F(z_1, z_2/z_1)$ is a polynomial in z_1, z_2). By the fundamental theorem on symmetric polynomials, the latter ring is $\mathbf{Q}[z_1 + z_2, z_1 z_2]$.

Now suppose that $n > 1$, and (3.56) has been proved for smaller values of n. We use induction on m to prove that any W-invariant polynomial $F(x_0, x_1, \ldots, x_n)$ of degree m in x_0 is a polynomial in $t, \rho_0, \ldots, \rho_{n-1}$. If $m = 0$, then $F = F(x_1, \ldots, x_n)$ is a symmetric polynomial that satisfies, for example, the relation $F(x_1^{-1}, x_2, \ldots, x_n) = F$, and so it clearly must be a constant. Suppose that $m \geqslant 1$, and our claim has been proved for polynomials whose degree in x_0 is less than m. Let

$$F(x_0, x_1, \ldots, x_n) = \sum_{i=0}^{m} x_0^i \varphi_i(x_1, \ldots, x_n)$$

be a W-invariant polynomial of degree m in x_0. Using the definition of the action of the group W_n and the algebraic independence of $x_0, x_1, \ldots, x_n$, we see that each of the polynomials $x_0^i \varphi_i(x_1, \ldots, x_n)$ is also W-invariant. Thus, without loss of generality we may assume that $F = x_0^m \varphi(x_1, \ldots, x_n)$. The W-invariance of F clearly implies that F is also invariant with respect to the group W_{n-1} acting on the variables $x_0, x_1, \ldots, x_{n-1}$. In particular, the polynomial $F(x_0, x_1, \ldots, x_{n-1}, 0)$ is W_{n-1}-invariant. By the first induction assumption, there exists a polynomial

$$P(y_1, \ldots, y_n) = \sum_{i=(i_1, \ldots, i_n)} a_i y_1^{i_1} \cdots y_n^{i_n}$$

with coefficients in $\mathbf{Q}$ such that

$$F(x_0, x_1, \ldots, x_{n-1}, 0) = P(t^{(n-1)}, \rho_0^{(n-1)}, \ldots, \rho_{n-2}^{(n-1)}).$$

If we take into account the special form of F and the algebraic independence of $x_0, x_1, \ldots, x_{n-1}$, we may suppose that in the polynomial P

$$a_i = 0, \quad \text{if } i_1 + 2(i_2 + \cdots + i_r) \neq m$$

(the polynomials t and ρ_a are homogeneous in x_0 of degree 1 and 2, respectively). We now show that when x_n is set equal to zero the polynomials $t^n, \rho_1^n, \ldots, \rho_{n-1}^n$ become $t^{(n-1)}, \rho_0^{(n-1)}, \ldots, \rho_{n-2}^{(n-1)}$, respectively. For t^n this is obvious; as for ρ_a^n, by definition we have

$$\sum_{a=0}^{2n} (-1)^a \rho_a^n v^a = (1 - x_n v)(x_n - v) \sum_{b=0}^{2n-2} (-1)^b \rho_b^{(n-1)} v^b,$$

and so, setting $x_n = 0$, we obtain

$$\sum_{a=0}^{2n} (-1)^a \rho_a^n|_{x_n=0} v^a = \sum_{a=1}^{2n-1} (-1)^a \rho_{a-1}^{(n-1)} v^a,$$

which gives the desired relations. From the previous argument it follows that the polynomial

$$F_1(x_0, x_1, \ldots, x_n) = F(x_0, x_1, \ldots, x_n) - P(t^n, \rho_1^n, \ldots, \rho_{n-1}^n)$$

is identically zero when $x_n = 0$. Hence, it is divisible by x_n. On the other hand, by construction, F_1 is a W-invariant polynomial. In particular, it is symmetric in $x_1, \ldots, x_n$. Thus, F_1 is divisible by the product $x_1 \cdots x_n$:

$$F_1(x_0, x_1, \ldots, x_n) = x_1 \cdots x_n F_2(x_0, x_1, \ldots, x_n).$$

Expanding F_2 in powers of x_0, we obtain

$$F_1(x_0, x_1, \ldots, x_n) = \sum_{s \geq 0} x_0^2 x_1 \cdots x_n f_s(x_1, \ldots, x_n),$$

where the f_s are polynomials. Since F_1 is a W-invariant polynomial, it is invariant under τ_1 (see (3.51)), i.e., it satisfies the relation $F_1(x_0 x_1, x_1^{-1}, x_2, \ldots, x_n) = F_1(x_0, x_1, \ldots, x_n)$. From this and the above sum for F_1 it clearly follows that the polynomials f_s satisfy the relations

$$x_1^{s-2} f_s(x_1^{-1}, x_2, \ldots, x_n) = f_s(x_1, \ldots, x_n).$$

If $s = 0$ or $s = 1$ and f_s is nonzero, then the expression on the left contains negative powers of x_1, and so is not a polynomial. Hence, $f_0(x_1, \ldots, x_n) = f_1(x_1, \ldots, x_n) = 0$, and

$$F_1(x_0, x_1, \ldots, x_n) = x_0^2 x_1 \cdots x_n F'(x_0, x_1, \ldots, x_n),$$

where F' is a polynomial which is obviously W-invariant. In addition, from our assumptions concerning F and P it follows that F' is homogeneous of degree $m - 2$ in x_0. Hence, either $F' = 0$, or else F' is a polynomial in $t, \rho_0, \ldots, \rho_{n-1}$. This proves (3.56).

We now use induction on n to prove that the polynomials (3.55) are algebraically independent over $\mathbf{Q}$. For $n = 1$ they are $x_0(1 + x_1) = z_1 + z_2$ and $x_0^2 x_1 = z_1 z_2$, where $z_1 = x_0$ and $z_2 = x_0 x_1$, and their algebraic independence is obvious. Suppose that $n > 1$, and our claim has been proved for $n' < n$. We use proof by contradiction. Suppose that $G(y, y_0, \ldots, y_{n-1})$ is a polynomial of minimal degree that vanishes when we substitute $y = t^n$, $y_0 = \rho_0^n, \ldots, y_{n-1} = \rho_{n-1}^n$. We expand G in powers of y_0:

$$G = \sum_i g_i(y, y_1, \ldots, y_{n-1}) y_0^i.$$

If $g_0 = 0$, then G is divisible by y_0, and $G y_0^{-1}$ is a polynomial of lower degree that also vanishes under the above substitution. Hence $g_0 \neq 0$. By assumption,

$$\sum_i g_i(t^n, \rho_1^n, \ldots, \rho_{n-1}^n)(\rho_0^n)^i = 0.$$

Since this is an identity in the variables $x_0, x_1, \ldots, x_n$, we can set $x_n = 0$ in it. As we saw before, $t^n, \rho_1^n, \ldots, \rho_{n-1}^n$ then become $t^{(n-1)}, \rho_0^{(n-1)}, \ldots, \rho_{n-2}^{(n-1)}$, and ρ_0^n obviously goes to zero. We thus obtain the identity

$$g_0(t^{(n-1)}, \rho_0^{(n-1)}, \ldots, \rho_{n-2}^{(n-1)}) = 0,$$

which, by the induction assumption, implies that $g_0 = 0$. This contradiction proves that the elements (3.55) are algebraically independent.

At the beginning of the proof of the theorem we saw that $\Omega(\mathbf{T}(p)) = t^n$, and the images $\Omega(\mathbf{T}_i(p^2))$ $(1 \leq i \leq n)$ can be expressed as linear combinations of $\rho_0^n, \ldots, \rho_{n-1}^n$, and conversely. This implies that the elements $\Omega(\mathbf{T}(p)), \Omega(\mathbf{T}_1(p^2)), \ldots, \Omega(\mathbf{T}_n(p^2))$ are algebraically independent over $\mathbf{Q}$; and since $\mathbf{T}(p), \mathbf{T}_1(p^2), \ldots, \mathbf{T}_n(p^2)$ generate the ring $\underline{\mathbf{L}}_p^n$, it follows that the restriction of Ω to $\underline{\mathbf{L}}_p^n$ is a monomorphism. This completes the proof of parts (1) and (2) of the theorem.

The third part follows from parts (1) and (2), Lemma 3.27, and the formula for $\Omega(\Delta_n(p))$ in Lemma 3.34. $\qquad \square$

The theorem just proved enables us to reduce computations in the local Hecke rings of the symplectic group to computations in polynomial rings. To show how this is done, we consider, for example, the problem of summing the formal generating series for elements of the form (3.19), where m runs through the powers of a fixed prime p, $(p, q) = 1$. Thus, we consider the formal power series

$$(3.71) \qquad \sum_{\delta=0}^{\infty} \mathbf{T}^n(p^\delta) v^\delta,$$

where $T^n(p^\delta) \in L_p^n(q)$ are the elements (3.19). From (3.22), Lemma 3.11 and the definitions it follows that

$$(3.72) \qquad \mathbf{T}^n(p^\delta) = \sum_{D,B} \left(\Gamma_0^n \begin{pmatrix} p^\delta D^* & B \\ 0 & D \end{pmatrix} \right),$$

where $D \in \Lambda \setminus \Lambda \mathrm{diag}(p^{\delta_1}, \ldots, p^{\delta_n}) \Lambda$, $0 \leqslant \delta_1 \leqslant \cdots \leqslant \delta_n \leqslant \delta$, and $B \in B_0(D)/\mathrm{mod}\, D$, since $\begin{pmatrix} p^\delta D^* & B \\ 0 & D \end{pmatrix}$ is an integer matrix if and only if B and D are integer matrices and all of the elementary divisors of D divide p^δ. Then from the definition of the map Ω and Lemma 3.33 we obtain the formal identity

$$\sum_{\delta=0}^{\infty} \Omega(\mathbf{T}^n(p^\delta)) v^\delta = \sum_{0 \leqslant \delta_1 \leqslant \cdots \leqslant \delta_n \leqslant \delta} p^{n\delta_1 + (n-1)\delta_2 + \cdots + \delta_n} \omega(t(p^{\delta_1}, \ldots, p^{\delta_n}))(x_0 v)^\delta,$$

where $t(p^{\delta_1}, \ldots, p^{\delta_n}) = (\mathrm{diag}(p^{\delta_1}, \ldots, p^{\delta_n}))_\Lambda \in H_p^n$. The summation of the series on the right in this relation for arbitrary n is based on explicit formulas for the polynomials $\omega(t(p^{\delta_1}, \ldots, p^{\delta_n}))$ and is beyond the scope of this book (see Andrianov [1, 2]). Here we shall limit ourselves to the cases $n = 1$ and $n = 2$. When $n = 1$, from the definitions we obtain

$$(3.73) \qquad \begin{aligned} \sum_{\delta=0}^{\infty} \Omega(\mathbf{T}^1(p^\delta)) v^\delta &= \sum_{\delta_1, \alpha = 0}^{\infty} p^{\delta_1} (x_1 p^{-1})^{\delta_1} (x_0 v)^{\delta_1 + \alpha} \\ &= [(1 - x_0 v)(1 - x_0 x_1 v)]^{-1}. \end{aligned}$$

When $n = 2$, if we set $\delta_2 = \delta_1 + \alpha$ and $\delta = \delta_2 + \beta$ and use Lemmas 2.4 and 2.21, we obtain

$$\sum_{0 \leqslant \delta_1 \leqslant \delta_2 \leqslant \delta} p^{2\delta_1 + \delta_2} \omega(t(p^{\delta_1}, p^{\delta_2}))(x_0 v)^\delta$$

$$= \sum_{\delta_1, \alpha, \beta = 0}^{\infty} p^{2\delta_1 + \delta_2 + \alpha} (p^{-3} x_1 x_2)^{\delta_1} \omega(t(1, p^\alpha))(x_0 v)^{\delta_1 + \alpha + \beta}$$

$$= [(1 - x_0 v)(1 - x_0 x_1 x_2 v)]^{-1} \sum_{\alpha = 0}^{\infty} \omega(t(1, p^\alpha))(p x_0 v)^\alpha,$$

and it remains to compute the last series. This computation easily reduces to our earlier calculation of the generating series for the polynomials $\omega(t_2(p^\delta))$, where $t_n(m)$

are the elements (2.10). First of all, using the definitions and Lemmas 2.4 and 2.21, we have

$$\sum_{\delta=0}^{\infty} \omega(t_2(p^\delta))v_1^\delta = \sum_{\gamma,\alpha=0}^{\infty} \omega(t(p^\gamma, p^{\gamma+\alpha}))v_1^{2\gamma+\alpha}$$

$$= \sum_{\gamma=0}^{\infty} \omega(\pi_2^2(p)^\gamma)v_1^{2\gamma} \sum_{\alpha=0}^{\infty} \omega(t(1, p^\alpha))v_1^\alpha$$

$$= (1 - p^{-3}x_1x_2v_1^2)^{-1} \sum_{\alpha=0}^{\infty} \omega(t(1, p^\alpha))v_1^\alpha.$$

From this formula and (2.38) with $n = 2$ we conclude that

(3.74) $$\sum_{\alpha=0}^{\infty} \omega(t(1, p^\alpha))v_1^\alpha = [(1 - p^{-1}x_1v_1)(1 - p^{-1}x_2v_1)]^{-1}(1 - p^{-3}x_1x_2v_1^2).$$

Finally, we obtain

(3.75)
$$\sum_{\delta=0}^{\infty} \Omega(\mathbf{T}^2(p^\delta))v^\delta = [(1 - x_0v)(1 - x_0x_1v)(1 - x_0x_2v)(1 - x_0x_1x_2v)]^{-1}$$

$$\times (1 - p^{-1}x_0^2x_1x_2v^2).$$

The denominators of the resulting expressions are the special cases when $n = 1$ and $n = 2$ of the polynomial

(3.76)
$$q(x_0, \ldots, x_n; v) = (1 - x_0v) \prod_{r=1}^{n} \prod_{1 \leqslant i_1 < \cdots < i_r \leqslant n} (1 - x_0x_{i_1} \cdots x_{i_r}v)$$

$$= \sum_{i=0}^{m} (-1)^i q_i^n(x_0, \ldots, x_n)v^i \quad (m = 2^n).$$

It is not hard to see that any transformation in $W = W_n$ with respect to the variables $x_0, x_1, \ldots, x_n$ only permutes the factors of q. Hence, all of the coefficients q_i^n are in the ring $\mathbf{Q}[x_0, \ldots, x_n]_W$, and so, by Theorem 3.30, they are the Ω-images of uniquely determined $\mathbf{q}_i^n(p) \in \underline{\mathbf{L}}_p^n$:

(3.77) $$q_i^n(x_0, \ldots, x_n) = \Omega_p^n(\mathbf{q}_i^n(p)), \quad \text{where } \mathbf{q}_i^n(p) \in \underline{\mathbf{L}}_p^n.$$

We set

(3.78) $$\mathbf{Q}(v) = \mathbf{Q}_p^n(v) = \sum_{i=0}^{m} (-1)^i \mathbf{q}_i^n(p)v^i \in \underline{\mathbf{L}}_p^n[v].$$

We note that the obvious relation

$$q(x_0, \ldots, x_n; v) = v^n(x_0^2x_1 \cdots x_n)^{m/2}q(x_0, \ldots, x_n; (x_0^2x_1 \cdots x_nv)^{-1})$$

and the formula for $\Omega(\Delta_n(p))$ in Lemma 3.34 together imply the relation

$$\mathbf{Q}_p^n(v) = v^n(p^{\langle n \rangle}\Delta_n(p))^{m/2}\mathbf{Q}_p^n((p^{\langle n \rangle}\Delta_n(p)v)^{-1}),$$

which, in turn, implies the following relations for the coefficients of $\mathbf{Q}$:

(3.79) $$\mathbf{q}_{m-i}^n(p) = (p^{\langle n \rangle}\Delta_n(p))^{m/2-i}\mathbf{q}_i^n(p) \quad (0 \leqslant i \leqslant m).$$

In particular,

(3.80) $\mathbf{q}_m^n(p) = (p^{\langle n \rangle} \Delta_n(p))^{m/2}.$

Finally, from (3.70) we have

(3.81) $\mathbf{q}_1^n(p) = \mathbf{T}^n(p).$

We now turn to the problem of summing series of the form (3.71).

PROPOSITION 3.35. *The following formal power series identities hold for any prime* p:

$$\sum_{\delta=0}^{\infty} \mathbf{T}^1(p^\delta) v^\delta = \mathbf{Q}_p^1(v)^{-1},$$

$$\sum_{\delta=0}^{\infty} \mathbf{T}^2(p^\delta) v^\delta = \mathbf{Q}_p^2(v)^{-1} \cdot (1 - p^2 \Delta_2(p) v^2),$$

where $\mathbf{T}^n(p^\delta)$ *are elements of the form* (3.19), *regarded as elements in* $\underline{\mathbf{L}}_p^n$, *and* $\mathbf{Q}_p^n(v)$ *are the polynomials* (3.78). *One has the formulas*

$$\mathbf{Q}_p^1(v) = 1 - \mathbf{T}^1(p) v + p \Delta_1(p) v^2,$$
$$\mathbf{Q}_p^2(v) = 1 - \mathbf{T}^2(p) v^2 + \mathbf{q}_2^2(p) v^2 - p^3 \Delta_2(p) \mathbf{T}^2(p) v^3 + p^6 \Delta_2(p) v^4,$$

where

$$\mathbf{q}_2^2(p) = (\Omega_p^n)^{-1}(x_0^2 x_1 x_2 (x_1 + x_2 + x_1^{-1} + x_2^{-1} + 2)).$$

PROOF. From (3.73) and the definitions it follows that the isomorphism Ω maps the constant term of the power series

$$\left(\sum_{\delta=0}^{\infty} \mathbf{T}^1(p^\delta) v^\delta \right) \mathbf{Q}_p^1(v)$$

to one, and takes all of the other coefficients to zero. Hence, the constant term of this series is the unit of the ring $\mathbf{L}_p^1$, and the other coefficients are zero. In a similar way we find that the second identity is a consequence of (3.75). The formulas for the coefficients of $\mathbf{Q}_p^1$ and $\mathbf{Q}_p^2$ follow from (3.81), (3.79), (3.80), and the definitions. $\square$

It is clear that similar identities hold over any ring isomorphic to $\underline{\mathbf{L}}_p^n$, for example, over the ring $\underline{\mathbf{L}}_p^n(q)$, where $p \nmid q$.

Theorem 3.30 enables us to parameterize the set of all nonzero $\mathbf{Q}$-linear homomorphisms from the ring $\mathbf{L}_p^n$ to $\mathbf{C}$.

PROPOSITION 3.36. *Every nonzero* $\mathbf{Q}$-*linear homomorphism* λ *from the ring* $\mathbf{L}_p^n$ *to* $\mathbf{C}$ *has the form*

(3.82) $\mathbf{T} \to \lambda_A(\mathbf{T}) = \Omega_p^n(\mathbf{T})_{(x_0,\dots,x_n)=A}$ $(\mathbf{T} \in \mathbf{L}_p^n),$

where $A = (\alpha_0, \dots, \alpha_n)$ *is a set of nonzero complex numbers that depends on* λ.

PROOF. According to Theorem 3.30, to prove the proposition it suffices to verify that any nonzero $\mathbf{Q}$-linear homomorphism $\mu\colon \mathbf{Q}[x_0^{\pm 1}, \ldots, x_n^{\pm 1}]_W \to \mathbf{C}$ can be obtained by setting $x_0 = \alpha_0, \ldots, x_n = \alpha_n$, where $\alpha_0, \ldots, \alpha_n$ are nonzero complex numbers. Let $r_a = r_a^n$ be the coefficients of the polynomial (3.52), let $\rho_0 = x_0^2 x_1 \cdots x_n$, and let t be the polynomial (3.54). We set $\mu(r_a) = \beta_a$, $\mu(\rho_0) = \delta$, and $\mu(t) = \gamma$. Since $\mu \neq 0$, it follows that μ takes 1 to 1; hence, $\mu(\rho_0)\mu(\rho_0^{-1}) = 1$, and so $\delta = \mu(\rho_0) \neq 0$. Theorem 3.30 implies that every element of the ring $\mathbf{Q}[x_0^{\pm 1}, \ldots, x_n^{\pm 1}]_W$ is a polynomial in $r_1, \ldots, r_{n-1}, \rho_0^{\pm 1}, t$ with coefficients in $\mathbf{Q}$. Thus, it suffices to prove that the system of equations

(3.83)
$$\begin{cases} r_a(x_1, \ldots, x_n) = \beta_a & (1 \leqslant a \leqslant n-1), \\ \rho_0(x_0, \ldots, x_n) = \delta, \\ t(x_0, \ldots, x_n) = \gamma, \end{cases}$$

where $\delta \neq 0$, has a solution in nonzero complex numbers.

Since the polynomial (3.52) obviously satisfies the relation $r(v) = v^{2n} r(v^{-1})$, its coefficients satisfy the relation

(3.84)
$$r_a^n(x_1, \ldots, x_n) = r_{2n-a}^n(x_1, \ldots, x_n) \quad (0 \leqslant a \leqslant 2n),$$

and hence $\beta_a = \mu(r_a) = \mu(r_{2n-a}) = \beta_{2n-a}$ for $a = 0, 1, \ldots, 2n$. From these last relations it follows that the polynomial

$$(\mu r)(v) = \sum_{a=0}^{2n} (-1)^a \beta_a v^a \in \mathbf{C}[v]$$

satisfies the equality $(\mu r)(v) = v^{2n}(\mu r)(v^{-1})$. Since obviously $\beta_0 = \beta_{2n} = 1$, the polynomial μr factors over $\mathbf{C}$ into linear factors of the form

$$(\mu r)(v) = \prod_{i=1}^{2n} (1 - \gamma_i v), \quad \text{where } \gamma_1 \cdots \gamma_{2n} = 1.$$

Applying the above relation, we have

$$\prod_{i=1}^{2n} (1 - \gamma_i v) = v^{2n} \prod_{i=1}^{2n} (1 - \gamma_i v^{-1}) = \prod_{i=1}^{2n} (1 - \gamma_i^{-1} v),$$

from which it follows that the numbers $\gamma_1^{-1}, \ldots, \gamma_{2n}^{-1}$ are the same as the numbers $\gamma_1, \ldots, \gamma_{2n}$ except for their order, i.e., $\gamma_i^{-1} = \gamma_{\sigma(i)}$, where σ is some permutation of the numbers $1, 2, \ldots, 2n$. If $\sigma(i) = i$ for some i, then $\gamma_i^2 = 1$, and $\gamma_i = \pm 1$. We let $i_1, \ldots, i_k$ denote all indices i for which $\sigma(i) = i$ and $\gamma_i = 1$, and we let $j_1, \ldots, j_s$ denote all indices j for which $\sigma(j) = j$ and $\gamma_j = -1$. All of the other indices can be partitioned into pairs $(i, \sigma(i))$ where $\sigma(i) \neq i$. We let $l_1, \ldots, l_t$ denote the first components of these pairs. Then $k + s + 2t = 2n$, and the relation $\gamma_1 \cdots \gamma_{2n} = 1$ implies that $\gamma_{j_1} \cdots \gamma_{j_s} = 1$. Hence, s and k are even numbers. We let $\alpha_1, \ldots, \alpha_n$ denote the numbers γ_i with $i = i_1, \ldots, i_{k/2}, j_1, \ldots, j_{s/2}, l_1, \ldots, l_t$, respectively. We then have

$$(\mu r)(v) = \sum_{a=0}^{2n} (-1)^a \beta_a v^a = \prod_{i=1}^{n} (1 - \alpha_i v)(1 - \alpha_i^{-1} v),$$

and hence

(3.85) $\beta_a = r_a(\alpha_1, \ldots, \alpha_n) \quad (0 \leqslant a \leqslant n).$

In particular, $\alpha_1, \ldots, \alpha_n$ is a nonzero solution of the first $n-1$ equations of the system (3.83). If we substitute these numbers in the last two equations, we obtain the system

$$\begin{cases} x_0^2 \alpha_1 \cdots \alpha_n = \delta, \\ x_0(1+\alpha_1)\cdots(1+\alpha_n) = \gamma \end{cases}$$

in the unknown x_0. It is clear that this system is solvable if (and only if) the following relation holds:

$$\delta^{-1}\gamma^2 = (\alpha_1 \cdots \alpha_n)^{-1}\prod_{i=1}^{n}(1+\alpha_i)^2 = \prod_{i=1}^{n}(1+\alpha_i)(1+\alpha_i^{-1}).$$

Using the definitions, we have the identity

$$\frac{t^n(x_0, \ldots, x_n)^2}{\rho_0^n(x_0, \ldots, x_n)} = (x_1 \cdots x_n)^{-1}\prod_{i=1}^{n}(1+x_i)^2$$

(3.86)

$$= \prod_{i=1}^{n}(1+x_i)(1+x_i^{-1}) = r(-1) = \sum_{a=0}^{2n}r_a^n(x_1, \ldots, x_n),$$

from which, if we apply the homomorphism μ and the equalities (3.85), we obtain

$$\delta^{-1}\gamma^2 = \sum_{a=0}^{2n}\beta_a = \sum_{a=0}^{2n}r_a^n(\alpha_1, \ldots, \alpha_n) = \prod_{i=1}^{n}(1+\alpha_i)(1+\alpha_i^{-1}). \quad \square$$

We call $\alpha_0, \alpha_1, \ldots, \alpha_n$ the *parameters of the homomorphism* $\lambda = \lambda_{(\alpha_0, \ldots, \alpha_n)}$. Clearly, if a set of parameters is obtained from another set by the action of a transformation in W_n, then the corresponding homomorphisms are the same.

PROBLEM 3.37. Prove that the order of the group W_n is equal to $2^n n!$.

PROBLEM 3.38. Prove the following formulas for the middle coefficient $q_2^2(p)$ of $Q_p^2(v)$:

$$q_2^2(p) = p\mathbf{T}(1, p, p^2, p) + p(p^2+1)\Delta_2(p) = (\mathbf{T}^2(p))^2 - \mathbf{T}^2(p^2) - p^2\Delta_2(p),$$

where $\mathbf{T}(1, p, p^2, p) = T_1(p^2)$.

[**Hint:** Using (3.61) and Lemma 3.34, compute the Ω-images of the right side and then apply Theorem 3.30.]

PROBLEM 3.39. Suppose that $q \in \mathbf{N}$, and $\lambda: L^2(q) \to \mathbf{C}$ is a $\mathbf{Q}$-linear homomorphism of the global Hecke ring. Further suppose that $\lambda(T^2(m)) = O(m^\sigma)$ for some $\sigma \in \mathbf{R}$. Show that the zeta-function of the ring $L^2(q)$ with character λ, i.e., the Dirichlet series

$$Z(s, \lambda) = \sum_{m \in \mathbf{N}_{(q)}} \lambda(T^2(m))m^{-s}$$

converges absolutely and uniformly in any region of the form $\{s \in \mathbf{C}; \operatorname{Re} s \geqslant 1+\sigma+\varepsilon\}$, where $\varepsilon > 0$, and in that region it has an Euler product of the form

$$Z(s,\lambda) = \prod_{p \in \mathbf{P}_{(q)}} (1 - \alpha_0(p)^2 \alpha_1(p)\alpha_2(p) p^{-2s-1}) \mathbf{q}(\alpha_0(p), \alpha_1(p), \alpha_2(p); p^{-s})^{-1},$$

where $\alpha_0(p), \alpha_1(p), \alpha_2(p)$ are the parameters of the homomorphism

$$\mathbf{L}_p^2 \xrightarrow{\varepsilon_q^{-1}} L_p^2(q) \xrightarrow{\lambda} \mathbf{C}$$

and q is the polynomial (3.76) for $n = 2$. Prove that the parameters $\alpha_i(p)$ satisfy the inequalities

$$\max(|\alpha_0(p)|, |\alpha_0(p)\alpha_1(p)|, |\alpha_0(p)\alpha_2(p)|, |\alpha_0(p)\alpha_1(p)\alpha_2(p)|) \leqslant p^\sigma.$$

[**Hint:** Use (3.20) and (3.75).]

PROBLEM 3.40. Let $N: \mathbf{L}_p^n \to \mathbf{C}$ be the homomorphism in Problem 1.16. Show that one can take as parameters for N the set $(\alpha_0, \alpha_1, \dots, \alpha_n) = (1, p, \dots, p^n)$.

PROBLEM 3.41. In the notation of Proposition 3.36, prove that $\lambda_A = \lambda_{A'}$, where $A, A' \in (\mathbf{C}^*)^{n+1}$, if and only if $A' = wA$ for some $w \in W_n$.

§4. Hecke rings for the symplectic covering group

In this section we study the Hecke rings $\widehat{L}^n(q)$ of the symplectic covering group $\mathfrak{G}$ that are obtained by lifting the Hecke rings $L^n(q)$. This lifting was described in §1.2 for abstract Hecke rings. In $\widehat{L}^n(q)$ we shall examine the subring $\widehat{E}^n(q)$ generated by the double cosets of elements $\xi = (M, \varphi) \in \mathfrak{G}$ such that $M \in S^n(q)$ and $r(M)$ is the square of a rational number. We pay particular attention to this "even" subring $\widehat{E}^n(q)$ because—and here is the principal difference between the Hecke rings of the symplectic covering group $\mathfrak{G}$ and the Hecke rings of the symplectic group itself—the Hecke operators on spaces of Siegel modular forms of half-integer weight that correspond to all other elements of the ring $\widehat{L}^n(q)$ are the zero operator. Concerning this ring $\widehat{E}^n(q)$ we shall prove that it splits into the tensor product of pairwise commuting local subrings $\widehat{E}_p^n(q)$, where $(p, q) = 1$. These local subrings have one important drawback: they have too many elements, i.e., they have many different elements that represent the same Hecke operator on the spaces of modular forms. Hence, inside $\widehat{E}_p^n(q)$ we look at the minimal commuting subring $\widehat{E}_p^n(q, \varkappa)$ that is analogous to the subring $E_p^n(q) \subset L_p^n(q)$ and has the property that its image under the spherical map Ω coincides with the image of $E_p^n(q)$.

For the duration of this section we suppose that $n, q \in \mathbf{N}$ and q is divisible by 4.

1. Global rings. In §3 we defined and studied the Hecke rings $L^n(q)$ of the symplectic group $S^n = S_{\mathbf{Q}}^n$. We can proceed in the same way in the case of the symplectic covering group $\mathfrak{G}$. However, we reach our goal more quickly if we obtain the Hecke rings of $\mathfrak{G}$ by lifting the Hecke rings of S^n. According to §1.2, to do this we must define two homomorphisms P and δ (see (1.13)). For P we take the epimorphism

(4.1) $$\mathfrak{G} \xrightarrow{P} S_R^n : (M, \varphi) \to M,$$

and for δ we take the monomorphism (3.23) of Chapter 2:

$$(4.2) \qquad \Gamma_0^n(q) \xrightarrow{j^n} \mathfrak{G} \colon M \to \widehat{M} = (M, j_{(2)}^n(M, Z)).$$

The Hecke rings of $\mathfrak{G}$ that we shall be interested in are the rings

$$(4.3) \qquad \widehat{L}(q) = \widehat{L}^n(q) = D_Q(\widehat{\Gamma}_0^n(q), \widehat{S}^n(q)),$$

where

$$(4.4) \qquad \widehat{\Gamma}_0^n(q) = j^n(\Gamma_0^n(q)), \quad \widehat{S}^n(q) = P^{-1}(S^n(q)).$$

A basic role in studying the Hecke rings $\widehat{L}^n(q)$ is played by the following *lifting homomorphism* (compare with the definition (3.26) of Chapter 2):

$$(4.5) \qquad t = t_M \colon (\Gamma_0(q))_M = \Gamma_0^n(q) \cap M^{-1}\Gamma_0^n(q)M \to \mathbf{C}_1,$$

where $M \in S^n(q)$; this map is defined for any $\gamma \in (\Gamma_0^n(q))_M$ by the equality

$$(4.6) \qquad \widehat{M\gamma M^{-1}} = \xi \widehat{\gamma} \xi^{-1}(E_{2n}, t(\gamma)),$$

where for any $\alpha \in \Gamma_0^n(q)$ we let $\widehat{\alpha}$ denote its image $j^n(\alpha)$ in the group $\mathfrak{G}$, and we let ξ denote any P-preimage of M in $\mathfrak{G}$. From Lemma 3.4 of Chapter 2 it follows that $(t_M)^4 \equiv 1$; hence, the kernel of t_M has finite index in the group $(\Gamma_0^n(q))_M$. By assumption, $M \in S^n(q) \subset \widetilde{\Gamma}_0^n(q)$, and so $\operatorname{Ker} t_M$ also has finite index in the group $\Gamma_0^n(q)$. Consequently, by Lemma 1.7 the pair (4.4) is a Hecke pair, and so $\widehat{L}^n(q)$ really is a Hecke ring.

In this section we investigate the algebraic structure of the Hecke rings $\widehat{L}^n(q)$ to roughly the same extent that we studied the structure of the Hecke rings $L^n(q)$. Our investigation will be based on Lemma 1.8, which describes the connection between double cosets of $\widehat{L}^n(q)$ and double cosets of $L^n(q)$, and Proposition 1.9, which enables one to deduce certain properties of $\widehat{L}^n(q)$ from the analogous properties of $L^n(q)$, by comparing their images in the Hecke ring of the triangular subgroup of S^n. In order to use Lemma 1.8, we must know for which matrices M is the homomorphism t_M trivial. This question is answered by the next two lemmas. Before giving those lemmas, we make some preliminary remarks.

Because of Lemma 3.6, without loss of generality we may assume that in any double coset $\widehat{\Gamma}_0^n(q)\xi\widehat{\Gamma}_0^n(q)$, where $\xi = (M, \varepsilon) \in \widehat{S}^n(q)$, the matrix M has been chosen in the canonical form (3.9):

$$(4.7) \qquad K = \begin{pmatrix} P & 0 \\ 0 & Q \end{pmatrix} = \operatorname{diag}(d_1, \dots, d_n, e_1, \dots, e_n),$$

where $d_i, e_j > 0$, $d_i | d_{i+1}$, $d_n | e_n$, $e_{i+1} | e_i$, $d_i e_i = r(K)$.

We compute the value $t_M(N)$ of the homomorphism (4.5) for a matrix $N = \begin{pmatrix} A & B \\ C & D \end{pmatrix}$ in $(\Gamma_0^n(q))_M$ in the case when $M = K$. Using (3.19) of Chapter 2 and the equality $\xi^{-1} = (M^{-1}, \varepsilon^{-1})$, by (4.6) we have

$$(MNM^{-1}, j_{(2)}^n(MNM^{-1}, Z)) = \widehat{MNM^{-1}} = \xi\widehat{N}\xi^{-1}(E_{2n}, t_M(N))$$

$$= (MNM^{-1}, j_{(2)}^n(N, M^{-1}\langle Z \rangle)t_M(N)).$$

From this and (4.37) of Chapter 1 we find that

$$(4.8) \qquad \begin{aligned} t_M(N) &= j_{(2)}^n(MNM^{-1}, Z) j_{(2)}^n(N, M^{-1}\langle Z\rangle)^{-1} \\ &= \chi_{(2)}^n(MNM^{-1}) \chi_{(2)}^n(N)^{-1} I(Z), \end{aligned}$$

where $I(Z) = j(MNM^{-1}, Z)^{1/2} j(N, M^{-1}\langle Z\rangle)^{-1/2}$ and $j(N, Z) = \det(CZ + D)$. Furthermore, if we take into account the definition of the square root $j(N, Z)^{1/2}$ in (4.7) of Chapter 1, we obtain

$$\lim_{Z \to 0} I(Z) = (\det QDQ^{-1})^{-1/2} (\det D)^{-1/2} = 1,$$

where we suppose that $Z \in \mathbf{H}_n$. Since the value $t_M(N)$ does not depend on Z, it follows from (4.8) and the last equality that

$$(4.9) \qquad t_M(N) = \chi_{(2)}^n(MNM^{-1}) \chi_{(2)}^n(N)^{-1},$$

if $M \in S^n(q)$ is a matrix of the form (4.7).

LEMMA 4.1. *Let* $M = \begin{pmatrix} P & 0 \\ 0 & Q \end{pmatrix}$ *be a canonical matrix of the form* (4.7) *in* $S^n(q)$, *where q is divisible by* 4. *If the lifting homomorphism t_M is trivial, then $r(M)$ is the square of a rational number.*

PROOF. Since M can be written in the form $M = m^{-1}M_0$, where $m \in \mathbf{Z}$ and M_0 is an integer matrix in $S^n(q)$, with $r(M) = m^{-2n}r(M_0)$, and since $t_M = t_{M_0}$ by (4.6), it follows that from the beginning we may assume that M is an integer matrix. For $i = 1, \ldots, n$ we define an imbedding $\varphi_i \colon S_\mathbf{R}^1 \to S_\mathbf{R}^n$ by setting

$$(4.10) \qquad \varphi_i \colon \begin{pmatrix} a & b \\ c & d \end{pmatrix} \to$$

Let $M^{(i)} = \begin{pmatrix} d_i & 0 \\ 0 & c_i \end{pmatrix}$, where $d_i e_i = r(M)$, and let

$$(4.11) \qquad \alpha = \begin{pmatrix} a & b \\ c & d \end{pmatrix} \in (\Gamma_0^1(q))_{M^{(i)}}.$$

Then $N = \varphi_i(\alpha) \in (\Gamma_0^n(q))_M$, and we have

$$(4.12) \qquad MNM^{-1} = \varphi_i(M^{(i)}\alpha(M^{(i)})^{-1}) \in \Gamma_0^n(q).$$

From this and from (4.38) and (4.40) of Chapter 1 we have

$$(4.13) \qquad \chi_{(2)}^n(N) = \varepsilon_{\text{sign }d}^{-1} |d|^{1/2-n} \sum_r \exp(2\pi i b r_i^2/d),$$

$$(4.14) \qquad \chi_{(2)}^n(MNM^{-1}) = \varepsilon_{\text{sign }d}^{-1} |d|^{1/2-n} \sum_r \exp(2\pi i b' r_i^2/d),$$

where $r = {}^t(r_1, \ldots, r_i, \ldots, r_n)$ runs through $M_{n,1}(\mathbf{Z}/d\mathbf{Z})$ and $b' = d_i b/e_i$. Since α satisfies the condition (4.11), it follows that b and b' are integers prime to d, where $(d, q) = 1$, and so d is odd. Hence, if we use the formula for the Gauss sum

$$(4.15) \qquad G_d(k) = \sum_{r \in \mathbf{Z}/d\mathbf{Z}} \exp(2\pi i k r^2/d) = \varepsilon_d \left(\frac{k}{d}\right) d^{1/2},$$

where d is a positive odd number and $(k, d) = 1$, which follows from Lemmas 4.13 and 4.14 of Chapter 1, and if we further suppose that $(d, r(M)) = 1$, then from (4.13)–(4.14) and the formula (4.9) for $t_M(N)$ we obtain

$$(4.16) \qquad t_M(N) = \left(\frac{b'}{d}\right)\left(\frac{b}{d}\right) = \left(\frac{d_i e_i}{d}\right) = \left(\frac{r(M)}{d}\right).$$

It is not hard to see that for any $d \in \mathbf{Z}$ prime to $r(M)q$ there exists a matrix $\alpha = \begin{pmatrix} * & * \\ * & d \end{pmatrix}$ satisfying (4.11). Thus, from (4.16) we conclude that

$$(4.17) \qquad \left(\frac{r(M)}{d}\right) = 1 \quad \text{for all } d > 0 \text{ and } (d, r(M)q) = 1,$$

since $t_M(N) = 1$ by assumption. Let $p_1, \ldots, p_s$ be all of the prime numbers that occur to odd powers in the prime factorization of $r(M)$, i.e., $r(M) = r_1^2 p_1 \cdots p_s$. Suppose that this set of primes is nonempty. Since $M \in S^n(q)$, it follows that $(r(M), q) = 1$; and since q is divisible by 4, it follows that $r(M)$ is odd, and so $p_i \neq 2$. We choose $d > 0$ in such a way that the following congruences hold:

$$d \equiv x_1 \pmod{p_1} \quad \text{and} \quad d \equiv 1 \pmod{p_1^{-(2\alpha+1)} r(M) q},$$

where x_1 is a quadratic nonresidue modulo p_1 and $2\alpha + 1$ is the power with which p_1 occurs in $r(M)$. Such a d clearly exists, since the two moduli are relatively prime and $p_1 \neq 2$. Then $(d, r(M)q) = 1$, and from quadratic reciprocity for the Jacobi symbol we have

$$\left(\frac{r(M)}{d}\right) = \left(\frac{p_1 \cdots p_s}{d}\right) = \left(\frac{d}{p_1 \cdots p_s}\right) = \left(\frac{d}{p_1}\right) \cdots \left(\frac{d}{p_s}\right) = -1,$$

which contradicts (4.17). Thus, $r(M) = r_1^2$. $\qquad\square$

The converse of Lemma 4.1 is also true.

LEMMA 4.2. *Let* $M = \begin{pmatrix} P & 0 \\ 0 & Q \end{pmatrix}$ *be a canonical matrix of the form* (4.7) *in the group* $S^n(q)$, *where q is divisible by 4. If $r = r(M)$ is the square of a rational number, then the homomorphism t_M is trivial.*

PROOF. We let

(4.18)
$$N = \begin{pmatrix} A & B \\ C & D \end{pmatrix} \in (\Gamma_0^n(q))_M,$$

and we first consider the case when $(r, \det D) = 1$. Since QDQ^{-1} is an integer matrix by (4.18), it follows that in (4.38) and (4.40) of Chapter 1, which give the value of the multiplier $\chi_{(2)}^n$ for $MNM^{-1} \in \Gamma_0^n(q)$, we can set

(4.19)
$$d = \det D = \det QDQ^{-1},$$

after which these formulas give us

(4.20)
$$\chi_{(2)}^n(MNM^{-1}) = \varepsilon_{\operatorname{sign} d}^{-1} |d|^{1/2-n} \sum_{s \in M_{n,1}(\mathbf{Z}/d\mathbf{Z})} e\{2PBD^{-1}Q^{-1}[s]\}.$$

Since $PQ = rE_n$ and $(r, d) = 1$ by assumption, it follows that the last sum is equal to

(4.21)
$$\sum_s e\{2PBD^{-1}[r^{-1/2}Ps]\} = \sum_{s' \in M_{n,1}(\mathbf{Z}/d\mathbf{Z})} e\{2PBD^{-1}[s']\}.$$

If we set $M = E_{2n}$ in (4.20), we obtain a formula for $\chi_{(2)}^n(N)$. Comparing this formula with (4.20) and (4.21), we conclude that $\chi_{(2)}^n(MNM^{-1}) = \chi_{(2)}^n(N)$. From this and (4.9) it follows that $t_M(N) = 1$ for any $N \in (\Gamma_0^n(q))_M$.

Now suppose that $(r, \det D) = \delta > 1$, p is a prime divisor of δ, and $r = p^{2\beta} r_1$, where $(p, r_1) = 1$. Further suppose that the blocks P and Q of the matrix M have the form

(4.22)
$$P = \operatorname{diag}(P_1, \ldots, P_s), \quad Q = \operatorname{diag}(Q_1, \ldots, Q_s),$$
$$P_i = p^{\alpha_i} P_i', \quad \alpha_1 < \cdots < \alpha_s, \quad \alpha_s = \beta, \quad Q_i = rP_i^{-1},$$

where the P_i' are integer diagonal matrices with $(\det P_i', p) = 1$. Of course, the block P_s might not exist.

The inclusion $MNM^{-1} \in \Gamma_0^n(q)$ implies the congruences

(4.23)
$$A_{ij} \equiv D_{ij} \equiv 0 \pmod p \quad \text{for } i > j,$$
$$B_{ij} \equiv 0 \pmod p \quad \text{for } (i, j) \neq (s, s),$$

where $A = (A_{ij})$, $B = (B_{ij})$, $C = (C_{ij})$, and $D = (D_{ij})$ are divided in blocks that are analogous to (4.22). Using the other inclusion $N \in \Gamma_0^n(q)$ and (2.7) of Chapter 1, we obtain a new series of congruences:

(4.24)
$$\sum_{j \leqslant \min(k,l)} {}^t A_{jk} D_{jl} - \sum_{j \leqslant \min(k,l)} {}^t C_{jk} B_{jl} \equiv E \quad \text{or} \quad 0 \pmod p$$

for $k = l = i$ or $k \neq l$, respectively, where E and 0 are the identity or the zero matrix, as the case may be. If we now consider the congruences (4.24) successively for $(k, l) = (1, 1), (1, 2), (2, 2), \ldots$, then we obtain

(4.25)
$$D_{ij} \equiv 0 \quad \text{for } i \neq j,$$
$$\det D_{ii} \not\equiv 0 \pmod p \quad \text{for } i \neq s \text{ and } \det D_{ss} \equiv 0 \pmod p,$$
$${}^t A_{ss} D_{ss} - {}^t C_{ss} B_{ss} \equiv E \pmod p, \quad {}^t B_{ss} D_{ss} \equiv {}^t D_{ss} B_{ss} \pmod p.$$

We choose matrices $U, V \in SL(\mathbf{Z})$ of the same size as D_{ss} so that the following congruence holds:

$$(4.26) \qquad D'_{ss} = UD_{ss}V \equiv \begin{pmatrix} D_1 & 0 \\ 0 & 0 \end{pmatrix} \pmod{p},$$

where $(\det D_1, p) = 1$. We suppose that all of the (s, s)-blocks are divided into sub-blocks in a way analogous to the matrix on the right in (4.26). Let

$$P_U = \operatorname{diag}(E, U^*, E, U) \in \Gamma_0^n(q),$$

let P_V be defined in the analogous way, and let the matrix

$$N' = P_U N P_V = \begin{pmatrix} A' & B' \\ C' & D' \end{pmatrix}$$

be divided into the same blocks as N. In this notation it follows from (4.26) and (4.25) that

$$D'_{ss} \equiv \begin{pmatrix} D_1 & 0 \\ 0 & 0 \end{pmatrix} \pmod{p}, \quad B'_{ss} \equiv \begin{pmatrix} B_1 & 0 \\ B_3 & B_4 \end{pmatrix} \pmod{p},$$

$$C'_{ss} = \begin{pmatrix} C_1 & C_2 \\ C_3 & C_4 \end{pmatrix}, \quad \text{where} \; - {}^tC_4 B_4 \equiv E \pmod{p};$$

this implies that $(\det B_4, p) = 1$, and, if we set $T_{ss} = \begin{pmatrix} 0 & 0 \\ 0 & E \end{pmatrix}$, then we have

$$(4.27) \qquad \det(T_{ss}B'_{ss} + D'_{ss}) \equiv \det \begin{pmatrix} D_1 & 0 \\ 0 & B_4 \end{pmatrix} \not\equiv 0 \pmod{p}.$$

If we define $T = \begin{pmatrix} E & 0 \\ C_T & E \end{pmatrix} \in \Gamma^n$ by setting $C_T = \operatorname{diag}(0, \ldots, 0, T_{ss})$, then by (4.25) and (4.27) in the matrix

$$(4.28) \qquad N' = TP_U N P_V = \begin{pmatrix} A' & B' \\ C' & D' \end{pmatrix}$$

the block D' is nonsingular modulo p, since

$$(4.29) \qquad \det D' \equiv \left(\prod_{i=1}^{s-1} \det D_{ii} \right) \det(T_{ss}B'_{ss} + D'_{ss}) \not\equiv 0 \pmod{p}.$$

Note that MTM^{-1} is an integer matrix, and the matrices $MP_U M^{-1}$ and $MP_V M^{-1}$ are p-integral. Since, by assumption, p does not divide r_1, it follows from Lemma 3.2(1) that there exists a matrix $U' \in SL(\mathbf{Z})$ of the same size as U such that

$$U' \equiv U \pmod{p^{2\beta}} \quad \text{and} \quad U' \equiv E \pmod{r_1}.$$

An analogous matrix V' exists for V. Next, we define T' to be a matrix similar to T, except that the block T_{ss} is replaced by a symmetric integer matrix T'_{ss} that satisfies the congruences

$$T'_{ss} \equiv T_{ss} \pmod{p^{2\beta}} \quad \text{and} \quad T'_{ss} \equiv 0 \pmod{r_1 q}.$$

Such a matrix exists, since $M \in S^n(q)$, so that $(r, q) = 1$, and hence p does not divide $r_1 q$. From the above definitions it follows that $P_{U'}$, $P_{V'}$, T', and the transformed matrix of (4.28)

$$(4.30) \qquad N'' = T' P_{U'} N P_{V'} = \begin{pmatrix} A'' & B'' \\ C'' & D'' \end{pmatrix}$$

all lie in $(\Gamma_0^n(q))_M$, and the block D'' is still nonsingular modulo p. If we now use (4.30) and the fact that $t = t_M$ is a homomorphism of the group $(\Gamma_0^n(q))_M$, we can write

$$t(N'') = t(T')t(P_{U'})t(N)t(P_{V'}) = t(N),$$

since, by (4.9) and (4.38) of Chapter 1, the homomorphism t takes value 1 on T', $P_{U'}$, and $P_{V'}$. The decomposition (4.30) also implies that $D'' \equiv D' (\bmod\, p)$; hence, by (4.29), we have $(r, \det D'') = p^{-\alpha}\delta$, where α is the power to which p appears in the prime factorization of δ. If we make a similar argument for N successively with the different prime divisors of δ, we see that N can be transformed to a matrix $N_1 = \begin{pmatrix} * & * \\ * & D_1 \end{pmatrix}$ such that $t(N_1) = t(N)$ and $(r, \det D_1) = 1$. Then the first part of the proof of the lemma implies that $t(N) = 1$. $\qquad\square$

Using Lemmas 4.1 and 4.2, we can now prove the following

PROPOSITION 4.3. *Let $M \in S^n(q)$, where q is divisible by 4, and let t_M be the lifting homomorphism (4.5) associated to M. Then t_M is trivial if and only if $r(M)$ is the square of a rational number.*

PROOF. From Lemmas 4.1 and 4.2 it follows that the proposition is true when M is a canonical matrix of the form (4.7). If M is arbitrary, then, by Lemma 3.6, it can be written in the form $M = \varepsilon K \eta$, where $\varepsilon, \eta \in \Gamma_0^n(q)$ and K is a canonical matrix of the form (4.7). Now suppose that in Lemma 1.6 $\Gamma = \Gamma_0^n(q)$ and $\rho_M(N) = (E_{2n}, t_M(N))$ for $N \in (\Gamma_0^n(q))_M$. Then from (1.17) we find that

$$(4.31) \qquad t_M(N) = t_{\varepsilon^{-1}M\eta^{-1}}(\eta N \eta^{-1}) = t_K(\eta N \eta^{-1}).$$

Since $r(M) = r(K)$ and the map $\gamma \to \eta \gamma \eta^{-1}$ is a group isomorphism from $(\Gamma_0^n(q))_M$ to $(\Gamma_0^n(q))_K$, the proposition for M follows from the proposition for the canonical matrix K and from the relation (4.31). $\qquad\square$

We now consider the product of concrete double cosets in the Hecke ring $\widehat{L}^n(q)$. In this ring, as in $L^n(q)$, the multiplication formula for double cosets takes a particularly simple form when one of the double cosets is generated by the P-preimage of a matrix of the form rE_{2n}. Namely, we have

LEMMA 4.4. *Suppose that $\widehat{M} \in \widehat{S}^n(q)$, $\widehat{rE_{2n}}$ is any P-preimage in $\mathfrak{G}$ of the matrix rE_{2n}, where $r \in \mathbf{Z}_q^*$, and $\widehat{\Gamma} = \widehat{\Gamma}_0^n(q)$. Then the following relations hold in the ring $\widehat{L}^n(q)$:*

$$(4.32) \qquad (\widehat{rE_{2n}})_{\widehat{\Gamma}}(\widehat{M})_{\widehat{\Gamma}} = (\widehat{rE_{2n}M})_{\widehat{\Gamma}} = (\widehat{M})_{\widehat{\Gamma}}(\widehat{rE_{2n}})_{\widehat{\Gamma}}.$$

The **proof** follows immediately from the definitions.

We let $S^n(q)^+$ denote the subgroup of $S^n(q)$ consisting of matrices M for which $r(M)$ is the square of a rational number, and we let

$$(4.33) \qquad \widehat{E}^n(q) = D_Q(\widehat{\Gamma}_0^n(q), \widehat{S}^n(q)^+).$$

We call $\widehat{E}^n(q)$ the *even subring of the Hecke ring* $\widehat{L}^n(q)$. As we noted before, only the even subring is important for applications to modular forms. Hence, for the rest of this chapter we shall only be examining $\widehat{E}^n(q)$ and its local subrings.

PROPOSITION 4.5. *Let* $\xi_i = (M_i, t_i)$, *where* $i = 1, 2$, *be elements of the group* $\widehat{S}^n(q)^+$, *and let* $\widehat{\Gamma} = \widehat{\Gamma}_0^n(q)$. *Suppose that the ratios of symplectic divisors* $e_1(M_1)/d_1(M_1)$ *and* $e_1(M_2)/d_1(M_2)$ *are relatively prime. Then the following relations hold in the Hecke ring* $\widehat{E}^n(q)$:

$$(4.34) \qquad (\xi_1)_{\widehat{\Gamma}}(\xi_2)_{\widehat{\Gamma}} = (\xi_1\xi_2)_{\widehat{\Gamma}} = (\xi_2)_{\widehat{\Gamma}}(\xi_1)_{\widehat{\Gamma}}.$$

PROOF. According to Lemma 3.6, without loss of generality we may assume that the M_i are canonical matrices of the form (4.7). Since in this case $\xi_1\xi_2 = \xi_2\xi_1$, the second equality in (4.34) follows from the first equality, which we shall now prove.

From (1.10) it follows that

$$(\xi_1)_{\widehat{\Gamma}}(\xi_2)_{\widehat{\Gamma}} = (\xi_1\xi_2)_{\widehat{\Gamma}} + \sum_j a_j(\eta_j)_{\widehat{\Gamma}},$$

where $(\eta_j)_{\widehat{\Gamma}}$ are double cosets that are distinct from $(\xi_1\xi_2)_{\widehat{\Gamma}}$, and the a_j are nonnegative integers. Using (1.10) again, we see that the last sum here is zero if and only if

$$(4.35) \qquad \mu_{\widehat{\Gamma}}(\xi_1)\mu_{\widehat{\Gamma}}(\xi_2) = \mu_{\widehat{\Gamma}}(\xi_1\xi_2).$$

On the other hand, by Proposition 3.9 we have $(M_1)_\Gamma(M_2)_\Gamma = (M_1M_2)_\Gamma$, and hence

$$(4.36) \qquad \mu_\Gamma(M_1)\mu_\Gamma(M_2) = \mu_\Gamma(M_1M_2).$$

Since $r(M_i)$ and $r(M_1M_2)$ are squares of rational numbers, it follows from Proposition 4.3 and Lemma 1.8 that

$$\mu_{\widehat{\Gamma}}(\xi_i) = \mu_\Gamma(M_i) \quad \text{for } i = 1, 2, \quad \text{and} \quad \mu_{\widehat{\Gamma}}(\xi_1\xi_2) = \mu_\Gamma(M_1M_2),$$

and this together with (4.36) implies (4.35). $\qquad \square$

Just as in the case of Hecke rings for the symplectic group, Proposition 4.5 makes it possible to reduce the study of the even Hecke ring $\widehat{E}^n(q)$ to that of its *local subrings*

$$(4.37) \qquad \widehat{E}_p^n(q) = D_Q(\widehat{\Gamma}_0^n(q), \widehat{S}_p^n(q)^+),$$

where p is a prime not dividing q, and (see (3.26))

$$(4.38) \qquad S_p^n(q)^+ = \{M \in S_p^n(q); r(M) = p^{2\delta}, \delta \in \mathbf{Z}\}.$$

THEOREM 4.6. *The Hecke ring* $\widehat{E}^n(q)$, *where* $n, q \in \mathbf{N}$ *and* q *is divisible by* 4, *is generated by the local Hecke rings* $\widehat{E}_p^n(q)$, *where* p *runs through the primes not dividing* q. *Elements of different local subrings commute with one another.*

PROOF. The theorem follows from the equalities in (4.34) and the proof of Theorem 3.12. $\qquad \square$

PROBLEM 4.7. Suppose that $\xi_i = (M_i, t_i) \in \widehat{S}^n(q)$, $i = 1, 2$, have the property that the ratios of symplectic divisors $e_1(M_1)/d_1(M_1)$ and $e_1(M_2)/d_1(M_2)$ are relatively prime, and let $\Gamma = \Gamma_0^n(q)$. Show that the following multiplication formula holds in the Hecke ring $\widehat{L}^n(q)$:

$$(\xi_1)_{\widehat{\Gamma}}(\xi_2)_{\widehat{\Gamma}} = a \sum_{j=1}^{h} (\xi_1 \xi_2 (E_{2n}, \varepsilon_j))_{\widehat{\Gamma}},$$

where $a \in \mathbf{N}$ and all $\varepsilon_j \in \mathbf{C}_1$; and here one has

$$ah = h(M_1)h(M_2)h(M_1 M_2)^{-1},$$

where $h(M)$ for $M \in S^n(q)$ denotes the index of the kernel of the lifting homomorphism t_M in the group $(\Gamma)_M$.

[**Hint:** Use Proposition 3.9 and the formulas (1.24) and (1.25).]

PROBLEM 4.8. Determine whether or not the map

$$\widehat{L}^n(q) \xrightarrow{P} L^n(q): (\xi)_{\widehat{\Gamma}} \to (P\xi)_{\Gamma},$$

where $\Gamma = \Gamma_0^n(q)$, is a homomorphism of Hecke rings. Answer the same question for the even subring $\widehat{E}^n(q)$.

2. Local rings. Theorem 4.6 enables us to reduce the study of the Hecke ring $\widehat{E}^n(q)$ to that of its local components $\widehat{E}_p^n(q)$. In order to compute in the local rings, one needs to have an explicit description of representatives M_α of the left cosets ΓM_α (here $\Gamma = \Gamma_0^n(q)$) into which the double cosets $\Gamma M \Gamma$ decompose, where M is a matrix in $\mathrm{SM}_n(p^2, q)$ (see (3.21)). More precisely, one must know how to write M_α in the form $M_\alpha = \gamma K \delta$, where $\gamma, \delta \in \Gamma$ and K is a canonical matrix of the form (4.7). The next two lemmas are devoted to this question.

LEMMA 4.9. Let $\mathrm{SM}_n(p^2, q)$ be the subset (3.21) of the group $S^n(q)$, where p is a prime not dividing q, let

$$(4.39) \qquad M_{a,b}(B_0) = \begin{pmatrix} p^2 D_{a,b}^* & B \\ 0 & D_{a,b} \end{pmatrix}, \qquad \text{where } B = \begin{pmatrix} 0 & 0 & 0 \\ 0 & B_0 & 0 \\ 0 & 0 & 0 \end{pmatrix} \begin{matrix} (a), \\ (b) \end{matrix}$$

and let $R_{a,b}^n = \Lambda_{D_{a,b}}^n \setminus \Lambda^n$ with $\Lambda^n = GL_n(\mathbf{Z})$. Then we have the following partition into disjoint left cosets:

$$(4.40) \qquad \mathrm{SM}_n(p^2, q) = \bigcup_{M \in R(p^2)} \Gamma_0^n(q) M,$$

where

$$(4.41) \qquad R(p^2) = \left\{ M_{a,b}(B_0, S, V); B_0 \in S_a(\mathbf{Z}/p\mathbf{Z}), S = \begin{pmatrix} 0 & 0 & 0 \\ 0 & 0 & {}^t S_1 \\ 0 & S_1 & S_2 \end{pmatrix}, \right.$$

$$\left. S_1 \in M_{a,b}(\mathbf{Z}/p\mathbf{Z}), S_2 \in S_b(\mathbf{Z}/p^2\mathbf{Z}), V \in R_{a,b}^n, a + b \leqslant n \right\},$$

and, in the notation (2.2)–(2.3) of Chapter 1,

$$(4.42) \qquad M_{a,b}(B_0, S, V) = M_{a,b}(B_0) T(S) U(V).$$

PROOF. Using an argument similar to the one used to derive (3.64), we can show that

$$\mathrm{SM}_n(p^2, q) = \bigcup_{a,b,B,V} \Gamma_0^n(q) \begin{pmatrix} p^2 D_{a,b}^* & B \\ 0 & D_{a,b} \end{pmatrix} U(V),$$

where $a + b \leqslant n$, $B \in B_0(D_{a,b})/\mathrm{mod}\, D_{a,b}$, $V \in R_{a,b}^n$. The decomposition (4.40) follows from this, along with the definition (3.60) of $B_0(D_{a,b})$. □

Theorems 1.2 and 1.3 of the Appendix tell us that any symmetric matrix $B_0 \in S_a(\mathbf{Z}/p\mathbf{Z})$ of rank $r_p(B_0) = r$, where $p \neq 2$, can be written in the form

(4.43) $B_0 = B_0'['W] = WB_0' {}^tW,$

where $W \in GL_a(\mathbf{Z}/p\mathbf{Z})$, $B_0' = \mathrm{diag}(\lambda_1, \ldots, \lambda_r, 0, \ldots, 0)$, and $\lambda_i \not\equiv 0 (\mathrm{mod}\, p)$. If $\det W = d \neq 1$, then we divide the first column of W by d, we replace λ_1 by $d^2\lambda_1$ in the matrix B_0', and we keep our earlier notation when working with the transformed matrices. After these transformations (4.43) obviously still holds, and $\det W = 1$. Now let $B_0 = {}^tB_0 \in S_a$. Then we may suppose that $B_0' \in M_a$, and, by Lemma 3.2(1), the matrix W lies in $SL_a(\mathbf{Z})$. In this case (4.43) can be written as a congruence

(4.44) $B_0 \equiv B_0'['W](\mathrm{mod}\, p).$

From this and (4.41) we conclude that we can always take $B_0 = B_0'['W]$ in (4.42); consequently, the matrix (4.39) can be represented in the form of a product

(4.45) $M_{a,b}(B_0) = U(W^*)M_{a,b}(B_0')U(W^*)^{-1}.$

We consider the special case when $n = 1$. Then one easily verifies that the following decompositions hold for the matrices $\beta_\lambda = M_{1,0}(\lambda)$ with $(\lambda, p) = 1$ and $\sigma = M_{0,0}(0)$:

$$(4.46)\, \beta_\lambda = \begin{pmatrix} p & \lambda \\ 0 & p \end{pmatrix} = \begin{pmatrix} 1 & 0 \\ pqs & 1 \end{pmatrix} \begin{pmatrix} 1 & 0 \\ 0 & p^2 \end{pmatrix} \begin{pmatrix} p & \lambda \\ -qs & r \end{pmatrix} = \beta_\lambda' \begin{pmatrix} 1 & 0 \\ 0 & p^2 \end{pmatrix} \beta_\lambda'',$$

$$(4.47)\, \sigma = \begin{pmatrix} p^2 & 0 \\ 0 & 1 \end{pmatrix} = \begin{pmatrix} p^2 & -t \\ q & d \end{pmatrix} \begin{pmatrix} 1 & 0 \\ 0 & p^2 \end{pmatrix} \begin{pmatrix} p^2 d & t \\ -q & 1 \end{pmatrix} = \sigma' \begin{pmatrix} 1 & 0 \\ 0 & p^2 \end{pmatrix} \sigma'',$$

where all of the matrices have integer entries, $pr + q\lambda s = 1$ with $r > 0$, and $p^2 d + qt = 1$ with $d > 0$.

Using the imbeddings (4.10), we introduce some additional matrices:

(4.48) $P_{a,b}'(B_0') = \prod_{k=1}^{r} \varphi_{n-a-b+k}(\beta_{\lambda_k}'),$

and $P_{a,b}''(B_0')$ is defined similarly, with β_{λ_k}' replaced by β_{λ_k}'' (see (4.46));

(4.49) $M_{a,b}^r = U(D)$ for $D = \mathrm{diag}(E_{n-a-b}, p^2 E_r, pE_{a-r}, p^2 E_b),$

(4.50) $P_{a,b}(\alpha) = \prod_{k=1}^{n-a-b} \varphi_k(a)$ for $\alpha \in SL_2(\mathbf{R}),$

(4.51) $P_{a,b}^r = U(R)$ for $R = \mathrm{diag}(E_{n-a-b}, E_r, R_1),$

where for $a - r \leqslant b$ or $a - r > b$ the matrix R_1 is respectively equal to

$$\begin{pmatrix} 0 & 0 & E_{a-r} \\ 0 & E_{b-a+r} & 0 \\ -E_{a-r} & 0 & 0 \end{pmatrix} \quad \text{or} \quad \begin{pmatrix} 0 & 0 & E_b \\ 0 & E_{a-r-b} & 0 \\ -E_b & 0 & 0 \end{pmatrix},$$

finally, for $i = 0, 1, \ldots, n$ we define the following canonical matrices of the form (4.7):

$$(4.52) \qquad\qquad K_i = \operatorname{diag}(E_{n-i}, pE_i, p^2 E_{n-i}, pE_i).$$

LEMMA 4.10. *Let B_0, S, V be the matrices in (4.41), let $r = r_p(B_0)$, and let p be any odd prime. Then the matrices (4.42) can be represented in the form*

$$(4.53) \qquad M_{a,b}(B_0, S, V) = U(W^*) M_{a,b}(B_0) U(W^*)^{-1} T(S) U(V),$$

where B_0' and W satisfy the congruence (4.44) and, in addition,

$$(4.54) \qquad\qquad M_{a,b}(B_0') = P_{a,b}'(B_0') M_{a,b}^r P_{a,b}''(B_0'),$$

$$(4.55) \qquad\qquad M_{a,b}^r = P_{a,b}(\sigma') P_{a,b}^r K_{a-r}(P_{a,b}^r)^{-1} P_{a,b}(\sigma''),$$

where all of the matrices on the right in (4.53)–(4.54) except for $M_{a,b}(B_0')$, $M_{a,b}^r$, and K_{a-r} lie in the group $\Gamma_0^n(q)$; here q is any natural number not divisible by p.

PROOF. (4.53) is a consequence of (4.42) and (4.45), and (4.54) is merely the "direct sum" of the equations (4.46). Similarly, (4.55) can be obtained from (4.47), (4.50), and (4.51). The other parts of the lemma follow directly from the definitions. $\square$

We have thus found representatives of all of the left cosets of $\mathrm{SM}_n(p^2, q)$ modulo $\Gamma_0^n(q)$, and we have expressed each representative as a product $\gamma K \delta$ of matrices $\gamma, \delta \in \Gamma_0^n(q)$ and a canonical matrix K of the form (4.52). However, to compute in the Hecke ring $\widehat{E}_p^n(q)$ we still need to know the second component of the product $\widehat{\gamma} \widetilde{K} \widehat{\delta}$, where $\widehat{\gamma} = j^n(\gamma)$, $\widehat{\delta} = j^n(\delta)$, and $\widetilde{K}$ is an arbitrary P-preimage of K in $\mathfrak{G}$. To do this we prove a result that in certain cases makes it possible to reduce the calculation of the multiplier $\chi_{(2)}^n$ of degree n to that of the multiplier $\chi_{(2)}^1$ of degree 1. But we must first give some more definitions.

Let $q \in \mathbf{N}$ be divisible by 4, and let p be a prime not dividing q. For every matrix $M = \begin{pmatrix} * & * \\ 0 & D \end{pmatrix}$ in the group $S_{0,p}^n$ (see §3.3) we fix a P-preimage $\widetilde{M}$ in the symplectic covering group $\mathfrak{G}$, by setting

$$(4.56) \qquad\qquad \widetilde{M} = (M, |\det D|^{1/2}).$$

If M lies in the subgroup

$$(4.57) \qquad (S_{0,p}^n)^+ = \{M \in S_{0,p}^n; r(M) = p^{2\delta}, \delta \in \mathbf{Z}\} \subset S_{0,p}^n$$

or even in the less restrictive subgroup $S_p^n(q)^+$ of $S_p^n(q)$ (see (4.38)) and $M = \gamma K \delta$, where $\gamma, \delta \in \Gamma_0^n(q)$ and K is a canonical matrix of the form (4.7), then we define a second P-preimage of M in $\mathfrak{G}$ as follows:

$$(4.58) \qquad \widehat{M} = \widehat{\gamma} \widetilde{K} \widehat{\delta} \quad \text{with } \widehat{\gamma} = j^n(\gamma) \text{ and } \widehat{\delta} = j^n(\delta).$$

We show that the element $\widehat{M} \in \mathfrak{G}$ does not depend on the above choice of representation of M. In fact, suppose that $M = \gamma_i K \delta_i$ ($i = 1, 2$) are two such representations. Then, by Lemma 3.6, $K_1 = K_2$, and so it is enough to show that one has $\widehat{\gamma}_1 \widetilde{K} \widehat{\delta}_1 = \widehat{\gamma}_2 \widetilde{K} \widehat{\delta}_2$, or, equivalently,

$$(4.59) \qquad \widehat{\gamma} \widetilde{K} \widehat{\delta} = \widetilde{K}, \quad \text{where } \gamma = \gamma_2^{-1} \gamma_1 \text{ and } \delta = \delta_1 \delta_2^{-1} \in \Gamma_0^n(q).$$

By assumption, the factor $r(M)$ is an even power of the prime p. Hence, Proposition 4.3 implies that

$$(\widehat{K \alpha K^{-1}}) = \widetilde{K} \widehat{\alpha} \widetilde{K}^{-1} \quad \text{for } \alpha \in (\Gamma_0^n(q))_K.$$

Hence for $\delta = K^{-1} \gamma^{-1} K \in (\Gamma_0^n(q))_K$ we have

$$\widetilde{K} \widehat{\delta} \widetilde{K}^{-1} = (\widehat{K \alpha K^{-1}}) = (\widehat{\gamma^{-1}}) = \widehat{\gamma}^{-1},$$

and this proves (4.59).

The above argument shows that, in particular, if $M = K \in (S_{0,p}^n)^+$, then

$$(4.60) \qquad\qquad\qquad \widehat{K} = \widetilde{K}.$$

Finally, for an arbitrary element $\xi = (M, \varphi(Z)) \in \mathfrak{G}$ we set

$$(4.61) \qquad\qquad t(\xi) = \varphi(Z) \quad \text{and} \quad s(\xi) = t(\xi) |t(\xi)|^{-1}.$$

LEMMA 4.11. *For $i = 1, \ldots, n$ suppose that the matrices $R_i, S_i \in (S_{0,p}^1)^+$ and $\gamma_i, \delta_i \in \Gamma_0^1(q)$ are connected by the relations $R_i = \gamma_i S_i \delta_i$, and the elements $d(R_i)$, $d(S_i)$, $d(\gamma_i)$, and $d(\delta_i)$ in the lower-right corner of these matrices are all positive. Furthermore, let*

$$R = \prod_{i=1}^{n} \varphi_i(R_i) \in (S_{0,p}^n)^+,$$

where φ_i is the imbedding (4.10), and let $S \in (S_{0,p}^n)^+$ and $\gamma, \delta \in \Gamma_0^n(q)$ be defined analogously. Then $\widehat{R}$ and $\widehat{S}$ satisfy the relations

$$(4.62) \qquad\qquad t(\widehat{R}) = \chi_{(2)}^n(\gamma) \chi_{(2)}^n(\delta) s(\widehat{S}) t(\widehat{R}),$$

$$(4.63) \qquad\qquad s(\widehat{R}) = \chi_{(2)}^n(\gamma) \chi_{(2)}^n(\delta) s(\widehat{S}),$$

where $\chi_{(2)}^n$ is the multiplier (4.38) of Chapter 1.

PROOF. From the definition (4.61) we see that (4.63) is a consequence of (4.62); we now prove the latter relation. By (4.58) we can write

$$(4.64) \qquad\qquad\qquad \widehat{R} = \widehat{\gamma} \widehat{S} \widehat{\delta}.$$

In fact, let $S = \alpha K \beta$, where $\alpha, \beta \in \Gamma_0^n(q)$ and K is a canonical matrix. Then the equality $R = \gamma S \delta$ along with (4.58) implies that

$$\widehat{R} = (\widehat{\gamma \alpha}) \widetilde{K} (\widehat{\beta \delta}) = \widehat{\gamma} (\widehat{\alpha} \widetilde{K} \widehat{\beta}) \widehat{\delta} = \widehat{\gamma} \widehat{S} \widehat{\delta}.$$

Hence, from (4.2) and (3.19) of Chapter 2 we obtain

$$(4.65) \qquad t(\widehat{R}) = \chi_{(2)}^n(\gamma) \chi_{(2)}^n(\delta) t(\widehat{S}) j(\gamma, S\langle Z \rangle)^{1/2} j(\delta, Z)^{1/2},$$

where $j(\cdot, \cdot)$ is the automorphy factor (4.3) of Chapter 1, and the branches of the square roots are determined by (4.7) of Chapter 1. We find the limit as $Z \to 0$ of the right side of (4.65). To do this, we define two holomorphic functions of $z_1, \ldots, z_n \in \mathbf{H}_1$ by setting

$$\Psi(\gamma; z_1, \ldots, z_n) = \prod_{i=1}^{n} s(j(\gamma_i, z_i)),$$

(4.66)

$$\Psi(\delta; z_1, \ldots, z_n) = \prod_{i=1}^{n} s(j(\delta_i, z_i)),$$

where for any $\alpha = \begin{pmatrix} * & * \\ c & d \end{pmatrix} \in \Gamma_0^1(q)$ with $d > 0$ we let $s(j(\alpha, z))$ denote the function $s_+(j(\alpha, z))$ or $s_-(j(\alpha, z))$ depending on whether $c \geqslant 0$ or $c < 0$. Here $s_\pm(w)$ are the holomorphic functions on $((\pm 1)\mathbf{H}_1) \cup \mathbf{R}$ defined by the conditions: $s_\pm(w)^2 = w$ and $s_\pm(w) > 0$ for $w \in \mathbf{R}$ and $w > 0$. Note that the restrictions of the functions $j(\gamma, Z)^{1/2}$ and $j(\delta, Z)^{1/2}$ to the main diagonal $\mathbf{H}_1 \times \cdots \times \mathbf{H}_1 \subset \mathbf{H}_n$ coincide with the corresponding functions in (4.66), since the assumptions in the lemma imply that they coincide for $Z = \mathrm{diag}(z_1, \ldots, z_n)$ sufficiently close to zero. According to the definition (4.61), the function $t(\widehat{R})$ does not depend on Z. Hence, if we set $Z = \mathrm{diag}(z_1, \ldots, z_n)$ and use the functions (4.66), we can rewrite (4.65) in the form

(4.67) $\qquad t(\widehat{R}) = \chi_{(2)}^n(\gamma) \chi_{(2)}^n(\delta) \Psi(\delta, z_1, \ldots, z_n) \Psi(\gamma, S_1 \delta_1 \langle z_1 \rangle, \ldots, S_n \delta_n \langle z_n \rangle)$

and pass to the limit as $z_i \to 0$ ($z_i \in \mathbf{H}_1$, $i = 1, \ldots, n$) on the right of this equality. According to (4.66) and Lemma 4.2 of Chapter 1, this problem reduces to computing the limits of expressions of the form

(4.68) $\qquad s(j(\gamma_i, S_i \delta_i \langle z_i \rangle)) s(j(\delta_i, z_i)) = s(j(R_i, z_i) j(S_i \delta_i, z_i)^{-1}) s(j(\delta_i, z_i)).$

If we again use the definition (4.66), we find that the desired limit of (4.68) is equal to

$$(d(R_i)(d(S_i)d(\delta_i))^{-1})d(\delta_i)^{1/2} = d(R_i)^{1/2} d(S_i)^{-1/2},$$

where all of the square roots are positive. From this, (4.67), and (4.68) we obtain (4.62). $\square$

Lemmas 4.10 and 4.11 make it possible for us to find the P-preimages in $\mathfrak{G}$ of the matrices (4.42). To do this, we must introduce a certain special function $\varkappa$ that is defined on the set of symmetric integer matrices and is closely related to the multiplier $\chi_{(2)}^n$. We now give the definition of $\varkappa$.

As we noted earlier, for any matrix $A \in S_a(\mathbf{Z})$ of rank $r_p(A) = r$, where p is an odd prime, there exists a matrix $V \in M_a(\mathbf{Z})$ that is nonsingular modulo p and satisfies the congruence

(4.69) $\qquad A \equiv \begin{pmatrix} A' & 0 \\ 0 & 0 \end{pmatrix} [V] \pmod{p},$

where $A' \in S_r(\mathbf{Z})$ is a matrix that is nonsingular modulo p. If A satisfies (4.69), then we set

(4.70) $\qquad \varkappa(A) = \begin{cases} \varepsilon_p^{-r} \left(\dfrac{(-1)^r \det A'}{p} \right), & \text{if } r > 0, \\ 1, & \text{if } r = 0, \end{cases}$

where ε_p is the function (4.39) of Chapter 1. It is easy to see that the value $\varkappa(A)$ does not depend on the choice of matrices V and A' with the indicated properties.

PROPOSITION 4.12. *Let $M_{a,b}(B_0, S, V)$ be the matrices (4.42), where $a + b \leqslant n$ and B_0, S, and V run through the set of matrices in (4.41). Then we have the following formula for the P-preimages of these matrices in $\mathfrak{G}$ as defined in (4.58):*

$$(4.71) \qquad \widehat{M}_{a,b}(B_0, S, V) = (M_{a,b}(B_0, S, V); \varkappa(B_0) p^{(a+2b)/2}).$$

PROOF. From the formulas (4.37)–(4.38) of Chapter 1 it follows that $j^n_{(2)}(\gamma, Z) = 1$ for matrices γ of the form $U(V)$ or $T(S)$ in the group $\Gamma^n_0(q)$, where q is divisible by 4, as usual. Hence, by (4.2), we have $\widehat{\gamma} = (\gamma, 1)$ for such matrices γ. If we now use (4.53) and (4.64), we obtain

$$\widehat{M}_{a,b}(B_0, S, V) = \widehat{U}(W^*) \widehat{M}_{a,b}(B'_0) \widehat{U}(W^*)^{-1} \widehat{T}(S) \widehat{U}(V),$$

and from this and (4.61) it follows that

$$(4.72) \qquad s(\widehat{M}_{a,b}(B_0, S, V)) = s(\widehat{M}_{a,b}(B'_0)).$$

Arguing in an analogous way, we also have (see (4.55))

$$s(P^r_{a,b} \widehat{K_{a-r}(P^r_{a,b})^{-1}}) = s(\widehat{K}_{a-r}) = 1,$$

and hence, applying Lemma 4.11 to the equalities (4.54) and (4.55), we obtain

$$(4.73) \qquad s(\widehat{M}_{a,b}(B'_0)) = \chi^n_{(2)}(P'_{a,b}(B'_0)) \chi^n_{(2)}(P''_{a,b}(B'_0)) \chi^n_{(2)}(P_{a,b}(\sigma')) \chi^n_{(2)}(P_{a,b}(\sigma'')).$$

All of the matrices on the right in this equality have the form

$$(4.74) \qquad \gamma = \prod_{i=1}^{n} \varphi_i(\gamma_i) \quad \text{with } \gamma_i = \begin{pmatrix} a_i & b_i \\ c_i & d_i \end{pmatrix} \in \Gamma^1_0(q).$$

From the definitions (4.48) and (4.50) and from the equalities (4.46)–(4.47) it follows that the entries d_i in all of these matrices are positive. By (4.38) of Chapter 1, we have the following formula for the matrices (4.74):

$$(4.75) \qquad \chi^n_{(2)}(\gamma) = \prod_{i=1}^{n} \chi^1_{(2)}(\gamma_i).$$

We use this formula to compute the value of the multiplier $\chi^n_{(2)}$ at each of the matrices in (4.73). By Proposition 4.15 and the relation (4.37) of Chapter 1, we have

$$(4.76) \qquad \chi^1_{(2)}(\gamma) = \varepsilon_d^{-1} \left(\frac{c}{d} \right) \quad \text{for } \gamma = \begin{pmatrix} a & b \\ c & d \end{pmatrix} \in \Gamma^1_0(q).$$

Using (4.76), we show that the following equalities hold for the matrices in (4.46) and (4.47):

$$(4.77) \qquad \chi^1_{(2)}(\beta'_\lambda) = \chi^1_{(2)}(\sigma') = \chi^1_{(2)}(\sigma'') = 1, \quad \chi^1_{(2)}(\beta''_\lambda) = \varepsilon_p^{-1} \left(\frac{-\lambda}{d} \right).$$

For β'_λ and σ'' the equalities are obvious. In the case of σ' the equality follows from the congruence $p^2 d \equiv 1 \pmod{q}$, where q is divisible by 4, which implies that $d \equiv 1 \pmod 4$, and from the usual properties of the Jacobi symbol. Next, in the case of β''_λ the relation (4.76) leads to the equality

$$\chi^1_{(2)}(\beta''_\lambda) = \varepsilon_p^{-1} \left(\frac{-qs}{r} \right) = \varepsilon_p^{-1} \left(\frac{-qs}{p} \right),$$

since by (4.46) we have $pr \equiv 1 (\mathrm{mod}\, qs)$, and hence $p \equiv r (\mathrm{mod}\, 4)$. If we recall that $(-qs)(-\lambda) \equiv 1 (\mathrm{mod}\, p)$, we obtain the desired equality in (4.77).

Now (4.73), (4.75), and (4.77) imply that

$$(4.78) \qquad s(\widehat{M}_{a,b}(B_0')) = \prod_{i=1}^{r} \varepsilon_p^{-1}\left(\frac{-\lambda_i}{p}\right) = \varepsilon_p^{-r}\left(\frac{(-1)^r \det B_0'}{p}\right).$$

Finally, by (4.56) and (4.39) we have

$$|t(\widehat{M}_{a,b}(B_0, S, V))| = (\det D_{a,b})^{1/2} = p^{(a+2b)/2},$$

and we conclude from this and from (4.72) and (4.78) that (4.71) holds. $\qquad\square$

At the beginning of the first subsection we mentioned that there is an essential analogy between the Hecke rings of the symplectic group and the symplectic covering group. This analogy enabled us, in particular, to prove that the even Hecke ring $\widehat{E}^n(q)$, like $E^n(q)$, is generated by pairwise commuting local subrings $\widehat{E}_p^n(q)$. If we now try to follow the same analogy to define the spherical map Ω in the way that we did in §3.3, it will turn out that Ω no longer gives an isomorphic imbedding of $\widehat{E}_p^n(q)$ in a polynomial ring; thus, we lose one of the most powerful instruments for studying Hecke rings. In order to deal with this situation, we have to choose a path that is the exact reverse of what we did in §3: we first find generators of the even Hecke ring $E_p^n(q)$ and determine the corresponding elements in $\widehat{E}_p^n(q)$, and we then investigate the subring of $\widehat{E}_p^n(q)$ that is generated by these elements.

By definition, the ring $E_p^n(q)$ is generated by the double cosets $(M)_\Gamma$ of elements $M \in S_p^n(q)^+$ relative to the group $\Gamma = \Gamma_0^n(q)$. Thus, by Theorem 3.23,

$$(4.79) \qquad E_p^n(q) = \mathbf{Q}[T(p)^2, T_1(p^2), \ldots, T_{n-1}(p^2), T_n(p^2)^{\pm 1}],$$

where the expression on the right is the ring of polynomials in the elements in brackets with coefficients in $\mathbf{Q}$. The formula (1.10) shows that the only double cosets that appear in the product $T(p)^2 = T(p) \cdot T(p)$ are those of matrices $M \in S_p^n(q)^+$ with $r(M) = p^2$. But since, by Lemma 3.6, such matrices M can all be taken to be canonical matrices of the form (4.52), it follows that

$$(4.80) \qquad T(p)^2 = \sum_{i=0}^{n} a_i T_i(p^2),$$

where the a_i are nonnegative integers and $a_0 > 0$, since

$$K_0 \in \Gamma \begin{pmatrix} E_n & 0 \\ 0 & pE_n \end{pmatrix} \Gamma \begin{pmatrix} E_n & 0 \\ 0 & pE_n \end{pmatrix} \Gamma.$$

If we now compare (4.79) and (4.80), we find another system of generators for the ring $E_p^n(q)$, one that is more convenient for our purposes:

$$(4.81) \qquad E_p^n(q) = \mathbf{Q}[T_0(p^2), \ldots, T_{n-1}(p^2), T_n(p^2)^{\pm 1}].$$

We now proceed to the next step, and determine the elements in $\widehat{E}_p^n(q)$ that correspond to the elements in (4.81). There are, of course, many ways to do this. To be definite, we set

$$(4.82) \qquad \widehat{T}_i(p^2) = (\widehat{K}_i)_{\widehat{\Gamma}} \quad (i = 0, 1, \ldots, n),$$

where K_i are the matrices (4.52) and the elements $\widehat{K}_i \in \mathfrak{G}$ are determined by (4.58). All of the other P-preimages of the matrices K_i in $\mathfrak{G}$ are of the form $\widehat{K}_i\widehat{E}$, where $\widehat{E} = (E_{2n}, \varepsilon)$ with $\varepsilon \in \mathbf{C}_1$, and

$$(\widehat{K}_i\widehat{E})_{\widehat{\Gamma}} = (\widehat{K}_i)_{\widehat{\Gamma}}(\widehat{E})_{\widehat{\Gamma}}.$$

Moreover, the double cosets of the form $(\widehat{E})_{\widehat{\Gamma}}$ are contained in the center of the ring $\widehat{E}_p^n(q)$, and so the degree of choice in the elements in (4.82) in no way affects the algebraic properties of the subring they generate. From the point of view of applications of the Hecke rings to modular forms, the choices made in (4.82) are also of no importance, since the Hecke operators for $(\widehat{E})_{\widehat{\Gamma}}$ are operators of multiplication by a power of ε. We thus come to the conclusion that the natural analogue of the even Hecke ring $E_p^n(q)$ is not the entire ring $\widehat{E}_p^n(q)$, but rather the subring

$$(4.83) \qquad \widehat{E}_p^n(q, \varkappa) = \mathbf{Q}[\widehat{T}_0(p^2), \ldots, \widehat{T}_{n-1}(p^2), \widehat{T}_n(p^2)^{\pm 1}].$$

We show that this subring is commutative. Recall that in the case of the ring $L^n(q)$ the proof of commutativity was based on: (1) the existence of the anti-automorphism $*$ of the ring $L^n(q)$, and (2) the invariance of elements of $L^n(q)$ relative to $*$. Thus, we begin by defining an anti-automorphism $*$ for the ring $\widehat{L}_p^n(q)$.

For any $\xi = (M, \varphi) \in \mathfrak{G}$ we set

$$(4.84) \qquad \xi_0 = (r(\xi)E_{2n}, r(\xi)^{n/2}) \quad \text{and} \quad r(\xi) = r(M).$$

Then the map $\xi \to \xi_0$ is a homomorphism from the group $\mathfrak{G}$ to its center, and $\widehat{\Gamma}_0^n(q)$ is contained in the kernel of this homomorphism. This implies that the map

$$(4.85) \qquad \mathfrak{G} \xrightarrow{*} \mathfrak{G}: \xi \to \xi^* = \xi_0 \cdot \xi^{-1}$$

has the properties

$$(4.86) \qquad (\xi^*)^* = \xi \quad \text{and} \quad (\xi_1\xi_2)^* = \xi_2^*\xi_1^*,$$

i.e., it is an anti-automorphism of order two of the group $\mathfrak{G}$; here we have

$$(4.87) \qquad \widehat{\Gamma}^* = \widehat{\Gamma}, \quad \text{where } \Gamma = \Gamma_0^n(q).$$

Using Proposition 1.11 and (4.85), we now define an anti-automorphism I of the ring $\widehat{L}_p^n(q)$ by setting

$$(4.88) \qquad (\xi)_{\widehat{\Gamma}} \xrightarrow{*} (\xi^*)_{\widehat{\Gamma}}$$

and extending this map by $\mathbf{Q}$-linearity to all of $\widehat{L}_p^n(q)$.

THEOREM 4.13. *The ring $\widehat{E}_p^n(q, \varkappa)$, where q is divisible by 4 and p is a prime not dividing q, is commutative.*

PROOF. Because of (4.83), to prove the theorem it suffices to show that the generators (4.82) commute with one another in pairs. This, in turn, will follow if we show that both the generators (4.82) themselves and also their pairwise products are invariant relative to the anti-automorphism (4.88), i.e.,

$$(4.89) \qquad \widehat{T}_i(p^2)^* = \widehat{T}_i(p^2) \quad \text{and} \quad \widehat{X}^* = \widehat{X},$$

where $\widehat{X} = \widehat{T}_i(p^2)\widehat{T}_j(p^2)$ and $i, j = 0, 1, \ldots, n$. The first equality in (4.89) is a consequence of a more general fact.

LEMMA 4.14. *Let* $\widehat{M} \in \mathfrak{G}$, *where* $M \in S_p^n(q)^+$, *be defined by* (4.58), *and let* $\widehat{T}(M) = (\widehat{M})_{\widehat{\Gamma}}$, *where* $\Gamma = \Gamma_0^n(q)$. *Then* $\widehat{T}(M)^* = \widehat{T}(M)$.

PROOF. By Lemma 3.6 we can write $M = \gamma K \delta$, where $\gamma, \delta \in \Gamma$ and K is a canonical matrix of the form (4.7); hence $\widehat{M} = \widehat{\gamma}\widehat{K}\widehat{\delta}$, and so $\widehat{T}(M) = \widehat{T}(K)$. Thus, we may suppose that $M = K$.

Let $n = 1$. Since $(p, q) = 1$, it follows that for any $m \in \mathbf{N}$ there exist integers t and $d > 0$ such that $p^m d + qt = 1$. If m is even and

$$K_m = \begin{pmatrix} 1 & 0 \\ 0 & p^m \end{pmatrix}, \quad \sigma_m' = \begin{pmatrix} p^m & -t \\ q & d \end{pmatrix}, \quad \sigma_m'' = \begin{pmatrix} p^m d & t \\ -q & 1 \end{pmatrix},$$

then obviously $p^m K_m^{-1} = \sigma_m' K_m \sigma_m''$. Hence, from Lemma 4.11 and (4.76) we obtain

$$(\widehat{p^m K_m^{-1}}) = \widehat{\sigma}_m' \widehat{K} \widehat{\sigma}_m'' = (p^m K_m^{-1}, 1).$$

On the other hand, (4.85) implies that

$$(\widehat{K}_m)^* = (K_m, p^{m/2})^* = (p^m K_m^{-1}, 1),$$

and so $(\widehat{K}_m)^* = \widehat{\sigma}_m' \widehat{K}_m \widehat{\sigma}_m''$. From this and (4.88) we have

$$\widehat{T}(K_m)^* = (\widehat{K}_m^*)_{\widehat{\Gamma}} = (\widehat{\sigma}_m' \widehat{K}_m \widehat{\sigma}_m'')_{\widehat{\Gamma}} = (\widehat{K}_m)_{\widehat{\Gamma}},$$

i.e., $\widehat{T}(K_m)^* = \widehat{T}(K_m)$. Since the elements $(p^\delta E_2, p^{\delta/2})$ belong to the center of $\mathfrak{G}$ and are *-invariant, it follows from this and the last equality that the lemma holds in the case $n = 1$. We pass to arbitrary n using Lemma 4.11 and (4.75). Since we already went through this sort of argument when proving Proposition 4.12, we shall omit it here. ☐

Following the analogy with the Hecke rings $L_p^n(q)$, we might expect that the second equality in (4.89) would also follow from Lemma 4.14, since $\widehat{X}$ is equal to a sum of double cosets of elements $\xi = (M, \varphi) \in \widehat{S}_p^n(q)^+$. However, the whole point is that ξ and $\widehat{M}$ are not necessarily the same, and in that case $(\xi)_{\widehat{\Gamma}}^* \neq (\xi)_{\widehat{\Gamma}}$. This also seems to be what explains the complication in the proof that the rings $\widehat{E}_p^n(q, \varkappa)$ are commutative.

PROOF THAT $\widehat{X}^* = \widehat{X}$. Using Lemma 4.9, we can rewrite (3.61)–(3.62) in the form

$$(4.90) \qquad T_i(p^2) = \sum_{\substack{a,b,B_0,S,V \\ a+b \leqslant n, r_p(B_0) = a-i}} (\Gamma_0^n(q) M_{a,b}(B_0, S, V)),$$

where the matrices B_0, S, and V run through the sets in (4.41). From (4.58) and (4.90) it follows that the elements $\widehat{M}_{a,b}(B_0, S, V)$ lie in $\widehat{\Gamma}\widehat{K}_i\widehat{\Gamma}$. On the other hand, from Lemma 1.8 and Proposition 4.3 we find that the map $P \colon \widehat{M} \to M$ gives a one-to-one correspondence between the double cosets $\widehat{\Gamma}\widehat{K}_i\widehat{\Gamma}$ and $\Gamma K_i \Gamma$. This implies that

$$(4.91) \qquad \widehat{T}_i(p^2) = \sum_{\substack{a,b,B_0,S,V \\ a+b \leqslant n, r_p(B_0) = a-i}} (\widehat{\Gamma}_0^n(q) \widehat{M}_{a,b}(B_0, S, V)),$$

where, by (4.42), we have

$$(4.92) \qquad \widehat{M}_{a,b}(B_0, S, V) = \widehat{M}_{a,b}(B_0) \widehat{T}(S) \widehat{U}(V)$$

and $\widehat{T}(S)\widehat{U}(V) = \widehat{U}(V)\widehat{T}(S_1)$, where $S_1 = S[V]$. But since the matrix $T = K_j^{-1}T(S_1)K_j$ lies in Γ_{K_j}, it follows that $\widehat{T}(S)\widehat{U}(V)\widehat{K}_j = \widehat{U}(V)\widehat{K}_j\widehat{T}$ and $\widehat{T} \in \widehat{\Gamma} = \widehat{\Gamma}_0^n(q)$. Let $a, b, B_0, S,$ and V be the same as in (4.90). Then by Lemma 1.5 and (4.91) we have

$$
\begin{aligned}
(4.93) \quad \widehat{X} &= \sum_{a,b,B_0,S,V} \mu(\widehat{K}_j)\mu(\widehat{M}_{a,b}(B_0, S, V)\widehat{K}_j)^{-1}(\widehat{M}_{a,b}(B_0, S, V)\widehat{K}_j)_{\widehat{\Gamma}} \\
&= \sum_{a,b,B_0,S,V} \mu(\widehat{K}_j)\mu(\widehat{M}_{a,b}(B_0)\widehat{U}(V)\widehat{K}_j)^{-1}(\widehat{M}_{a,b}(B_0)\widehat{U}(V)\widehat{K}_j)_{\widehat{\Gamma}} \\
&= \sum_{a,b,B_0,V} \mu_{a,b}(B_0, V)(\xi_{a,b}(B_0, V))_{\widehat{\Gamma}},
\end{aligned}
$$

where

$$
\begin{aligned}
(4.94) \quad \mu_{a,b}(B_0, V) &= p^{b(a+b+1)}\mu(\widehat{K}_j)\mu(\xi_{a,b}(B_0, V))^{-1}, \\
\xi_{a,b}(B_0, V) &= \widehat{M}_{a,b}(B_0)\widehat{U}(V)\widehat{K}_j = (Y_{a,b}(B_0, V); t_{a,b}(B_0, V)). \qquad \Box
\end{aligned}
$$

REMARK 4.15. If the elements $\xi = \xi_{a,b}(B_0, V)$ on the right in (4.93) are replaced by elements of the form $\widehat{\gamma}_1\xi\widehat{\gamma}_2$, where $\gamma_i \in \Gamma_0^n$, then it is not hard to verify that this does not change either $t_{a,b}(B_0, V)$ or $\widehat{X}$.

Using this remark, we prove the following property of the double cosets in (4.93).

LEMMA 4.16. *Let $\xi_{a,b}(B_0, V)$ be the elements* (4.94), *let $\Gamma = \Gamma_0^n(q)$, and let $*$ be the anti-automorphism* (4.88). *Then*

$$
(\xi_{a,b}(B_0, V))_{\widehat{\Gamma}}^* = (\xi_{a,b}(-B_0, V))_{\widehat{\Gamma}}.
$$

PROOF. First of all, using (4.39) and (4.52), we find that $Y_{a,b}(B_0, V) = \begin{pmatrix} p^4 D^* & N \\ 0 & D \end{pmatrix}$, where

$$
D = \mathrm{diag}(E_{n-a-b}, pE_a, p^2E_b)V\,\mathrm{diag}(p^2E_{n-j}, pE_j)
$$

and

$$
N = \mathrm{diag}(0_{n-a-b}, B_0, 0_b)V\,\mathrm{diag}(p^2E_{n-j}, pE_j).
$$

We now choose V_1 and $V_2 \in \Lambda^n$ in such a way that in the matrix

$$
Y_{a,b}^{(1)}(B_0, V) = U(V_1)Y_{a,b}(B_0, V)U(V_2) = \begin{pmatrix} p^4 D_1^* & N_1 \\ 0 & D_1 \end{pmatrix}
$$

the block D_1 is a diagonal matrix. Since p^4D^* is an integer matrix, it follows from Lemma 2.2 and the expression for D that we can take

$$
D_1 = \mathrm{diag}(pE_{v_1}, p^2E_{v_2}, p^3E_{v_3}, p^4E_{v_4}) \quad \text{and} \quad v_1 + \cdots + v_4 = n.
$$

Since $N_1 = V_1^* N V_2$ and ${}^tD_1N_1 = {}^tN_1D_1$, we have

$$
N_1 = (pA_{ij})_{4\times4} \quad \text{and} \quad {}^tA_{ij} = p^{j-i}A_{ji} \in M_{v_j,v_i}.
$$

We next define matrices S_1 and S_2 by setting

$$-p^{-1}S_1 = \begin{pmatrix} A_{11} & {}^tA_{21} & {}^tA_{31} & {}^tA_{41} \\ A_{21} & & & \\ A_{31} & & 0 & \\ A_{41} & & & \end{pmatrix}.$$

$$-S_2 = \begin{pmatrix} 0 & & & 0 \\ & 0 & {}^tA_{32} & p\,{}^tA_{42} \\ 0 & A_{32} & A_{33} & p\,{}^tA_{43} \\ & pA_{42} & pA_{43} & pA_{44} \end{pmatrix}.$$

Using this notation, we find that in the matrix

$$Y^{(2)}_{a,b}(B_0, V) = T(S_1)\,Y^{(1)}_{a,b}(B_0, V)\,T(S_2) = \begin{pmatrix} p^4 D^*_1 & N_2 \\ 0 & D_1 \end{pmatrix}$$

the block N_2 is $\mathrm{diag}(0, pA_{22}, 0, 0)$. We choose a matrix $V'_3 \in SL_{v_2}(\mathbf{Z})$ that satisfies the congruence

$$A_{22}[V'_3] \equiv \mathrm{diag}(\lambda_1, \dots, \lambda_\rho, 0, \dots, 0)(\mathrm{mod}\,p)$$

with $\lambda_1, \dots, \lambda_\rho$ not divisible by p (compare with (4.43) and (4.44)). If we now set $V_3 = \mathrm{diag}(E_{v_1}, V'_3, E_{v_3}, E_{v_4})$, then $Y^{(2)}_{a,b}(B_0, V)$ can be transformed to

$$Y^{(3)}_{a,b}(B_0, V) = U(V_3)^{-1}\,Y^{(2)}_{a,b}(B_0, V)\,U(V_3) = \begin{pmatrix} p^4 D^*_1 & N_3 \\ 0 & D_1 \end{pmatrix},$$

where $N_3 \doteq \mathrm{diag}(0, pA'_{22}, 0, 0)$, and we may assume here that

$$A'_{22} = \mathrm{diag}(\lambda_1, \dots, \lambda_\rho, 0, \dots, 0),$$

since this can always be achieved by multiplying $Y^{(3)}_{a,b}(B_0, V)$ by a suitable matrix of the form $T(S) \in \Gamma$, which is permissible by Remark 4.15. Thus, we may assume that in (4.93)

$$\xi_{a,b}(B_0, V) = (Y^{(3)}_{a,b}(B_0, V); t_{a,b}(B_0, V)).$$

Using the notation (4.10) and (4.46), we define the following matrices in Γ:

$$P'(A'_{22}) = \prod_{i=v_1+1}^{v_1+\rho} \varphi_i(\beta'_{\lambda_i}), \qquad P''(A'_{22}) = \prod_{i=v_1+1}^{v_1+\rho} \varphi_i(\beta''_{\lambda_i}).$$

Then the matrix $Y^{(3)}_{a,b}(B_0, V)$ can be written in the form $Y^{(3)}_{a,b}(B_0, V) = P'(A'_{22}) \times Y^{(4)}_{a,b}(B_0, V)P''(A'_{22})$, where

(4.95) $$Y^{(4)}_{a,b}(B_0, V) = U(p^4, D_2) = \mathrm{diag}(p^4 D^*_2, D_2),$$

in which $D_2 = \mathrm{diag}(pE_{v_1}, p^3 E_\rho, p^2 E_{v_2-\rho}, p^3 E_{v_3}, p^4 E_{v_4})$.

We note—and this is essential for the rest of the proof—that the matrix $Y^{(4)}_{a,b}(B_0, V)$ does not change if B_0 is replaced by $-B_0$. This property is not satisfied by any of the matrices $Y_{a,b}(B_0, V)$ or $Y^{(k)}_{a,b}(B_0, V)$ for $k = 1, 2, 3$. In all of those matrices the block N or N_k would change to $-N$ or $-N_k$ if B_0 were replaced by $-B_0$.

According to Lemma 4.11, we obtain the following relations from the last equality for $Y_{a,b}^{(3)}(B_0, V)$:

$$\widehat{Y}_{a,b}^{(3)}(B_0, V) = (Y_{a,b}^{(3)}(B_0, V); t(\widehat{Y}_{a,b}^{(3)}(B_0, V))) = \widehat{P}'(A_{22}')\, \widehat{Y}_{a,b}^{(4)}(B_0, V)\widehat{P}''(A_{22}'),$$

$$s(\widehat{Y}_{a,b}^{(3)}(B_0, V)) = \chi_{(2)}^n(P'(A_{22}'))\chi_{(2)}^n(P''(A_{22}'))s(\widehat{Y}_{a,b}^{(4)}(B_0, V)).$$

It is not hard to see that, if we multiply the matrix $Y_{a,b}^{(4)}(B_0, V)$ by suitable matrices of the form $U(W) \in \Gamma$ and $\varphi_i(\sigma)$, where σ is either σ' or σ'' (see (4.47)), we can reduce it to the canonical form (4.7). Hence, using Lemma 4.11, the relation $s(\widehat{U}(W)) = 1$, and (4.77), we conclude that $s(\widehat{Y}_{a,b}^{(4)}(B_0, V)) = 1$. From this, (4.75), and (4.77) we finally obtain

$$(4.96) \qquad\qquad s(\widehat{Y}_{a,b}^{(3)}(B_0, V)) = \varkappa(A_{22}).$$

Since, by the last equality for $\xi_{a,b}(B_0, V)$, we can rewrite this element as the product

$$\widehat{Y}_{a,b}^{(3)}(B_0, V)(E_{2n}; t_{a,b}(B_0, V)t(\widehat{Y}_{a,b}^{(3)}(B_0, V))^{-1})$$

and since the elements $\widehat{Y}_{a,b}^{(k)}(B_0, V)$ for $k = 3, 4$ lie in the same $\widehat{\Gamma}$-double coset, it follows that in (4.93) we can take

$$(4.97) \qquad\qquad \xi_{a,b}(B_0, V) = \widehat{Y}_{a,b}^{(4)}(B_0, V)(E_{2n}; s_{a,b}(B_0, V)),$$

in which $s_{a,b}(B_0, V) = t_{a,b}(B_0, V)t(\widehat{Y}_{a,b}^{(3)}(B_0, V))^{-1} \in \mathbf{C}_1$. In addition, by Proposition 4.12, (4.42), and (4.94) we have $t_{a,b}(B_0, V) = \varkappa(B_0)p^\delta$, where $\delta \in \mathbf{Z}$; hence, by (4.96), we obtain

$$s_{a,b}(B_0, V) = \varkappa(B_0)\varkappa(A_{22}^{-1}).$$

When B_0 is replaced by $-B_0$, the corresponding matrix A_{22} is also transformed to $-A_{22}$, since, by definition, A_{22} is a linear function of B_0 (for fixed V). From this and the obvious relation for the function in (4.70)

$$(4.98) \qquad\qquad \varkappa(-A) = \overline{\varkappa}(A) = \varkappa(A)^{-1},$$

where $A = {}^t A$ is an arbitrary integer matrix, it follows that $s_{a,b}(-B_0, V) = s_{a,b}(B_0, V)^{-1}$. From Lemma 4.14, (4.88), and (4.97) we now have

$$(\xi_{a,b}(B_0, V))_{\widehat{\Gamma}}^* = (Y_{a,b}^{(4)}(B_0, V)^*(E_{2n}; s_{a,b}(B_0, V)^{-1}))_{\widehat{\Gamma}}$$

$$= (Y_{a,b}^{(4)}(B_0, V)(E_{2n}; s_{a,b}(-B_0, V)))_{\widehat{\Gamma}} = (\xi_{a,b}(-B_0, V))_{\widehat{\Gamma}},$$

since $Y_{a,b}^{(4)}(B_0, V) = Y_{a,b}^{(4)}(-B_0, V)$. $\qquad\qquad\square$

We now prove the second equality in (4.89). The relation (4.97) shows that the value $\mu(\xi_{a,b}(B_0, V))$ does not change if B_0 is replaced by $-B_0$. But because the map $B_0 \to -B_0$ is obviously an automorphism of the space of matrices in $S_a(\mathbf{Z}/p\mathbf{Z})$ with fixed r_p-rank, by Lemma 4.16 and (4.93) this implies that $\widehat{X}^* = \widehat{X}$. $\qquad\qquad\square$

PROBLEM 4.17. Suppose that $\xi = (M, \varphi) \in \widehat{S}_p^n(q)^+$, $\widehat{M}$ is the element defined by (4.58), and $\Gamma = \Gamma_0^n(q)$. Show that $\xi = \widehat{M}(E_{2n}, \varepsilon)$, where $\varepsilon \in \mathbf{C}_1$, and that $(\xi)_{\widehat{\Gamma}}^* \neq (\xi)_{\widehat{\Gamma}}$ if $\varepsilon^2 \neq 1$.

[**Hint:** Use (4.88) and (3.19) of Chapter 2.]

PROBLEM 4.18. Prove that $(\widehat{K})_{\widehat{\Gamma}}^* \neq (\widehat{K})_{\widehat{\Gamma}}$, where

$$K = \begin{pmatrix} 1 & 0 \\ 0 & p \end{pmatrix}, \quad p \equiv 3 \pmod 4, \quad \text{and} \quad \Gamma = \Gamma_0^1(q).$$

[**Hint:** Use (4.65) and (4.76).]

3. The spherical map. In this section we define an imbedding of the commutative ring $\widehat{E}_0^n(q, \varkappa)$ in a polynomial ring. We do this by first imbedding it in the triangular Hecke ring $\mathbf{L}_{0,p}^n$ (more precisely, in its complexification) and then using the spherical map Ω.

We begin by defining the Hecke ring (see §3.3)

$$(4.99) \qquad \widehat{\mathbf{L}}_{0,p}^n = D_{\mathbf{Q}}(\widehat{\Gamma}_0^n \widehat{S}_{0,p}^n).$$

For any matrix $M \in S_{0,p}^n$ the homomorphism $t_M \colon (\Gamma_0^n)_M \to \mathbf{C}_1$ defined by (4.6) is trivial. Hence, by Lemma 1.7, $(\widehat{\Gamma}_0^n, \widehat{S}_{0,p}^n)$ is a Hecke pair, and $\widehat{\mathbf{L}}_{0,p}^n$ is actually a Hecke ring.

If we now let $\widehat{L}_p^n(q)$ denote the local subring of the ring $\widehat{L}^n(q)$ in (4.3), then, using Lemma 3.4, we easily see that the Hecke pairs that determine the rings $\widehat{L}_p^n(q)$ and $\widehat{\mathbf{L}}_{0,p}^n$ satisfy the conditions (1.26). Hence, Proposition 1.9 enables us to define an imbedding

$$(4.100) \qquad \widehat{L}_p^n(q) \xrightarrow{\widehat{\varepsilon}_q} \widehat{\mathbf{L}}_{0,p}^n.$$

Furthermore, for any odd integer k we define the map

$$(4.101) \qquad \widehat{\mathbf{L}}_{0,p}^n \xrightarrow{P_k} \overline{\mathbf{L}}_{0,p}^n = \mathbf{L}_{0,p}^n \otimes_{\mathbf{Q}} \mathbf{C}$$

by mapping double cosets

$$(\widehat{\Gamma}_0^n \xi \widehat{\Gamma}_0^n) \xrightarrow{P_k} s(\xi)^{-k}(\Gamma_0^n P(\xi) \Gamma_0^n),$$

where $s(\xi)$ is the function (4.61), and then extending P_k by **Q**-linearity to the entire ring $\widehat{\mathbf{L}}_{0,p}^n$. Since $t_m \equiv 1$ for all $M \in S_{0,p}^n$, it follows from (1.24)–(1.25) and (1.10) that P_0 is a homomorphism. According to (3.19) of Chapter 2 and (4.2), the map $s \colon \widehat{S}_{0,p}^n \to \mathbf{C}_1$ is a homomorphism, whose kernel contains the group $\widehat{\Gamma}_0^n$. Consequently, for any k the map P_k is also a homomorphism. We note that, although k can be any integer in the definition (4.101), only the case of odd k is important for applications. Hence, in what follows we shall always suppose that k is odd.

Suppose that $\widehat{X} = (\xi)_{\widehat{\Gamma}}$, where $\Gamma = \Gamma_0^n(q)$, belongs to the ring $\widehat{L}_p^n(q)$, and $r(\xi)$ is not an even power of the prime p. Then from Proposition 4.3 it follows that $t_M \neq 1$, where $M = P(\xi)$, and so the partition $\Gamma_M = \bigcup_j (\operatorname{Ker} t_M) \beta_j$ contains more than one coset. If we are also given a second partition $\Gamma = \bigcup_i \Gamma_M \alpha_i$, then

$$\widehat{X} = \bigcup_i \bigcup_j \widehat{\Gamma} \xi \widehat{\beta}_j \widehat{\alpha}_i$$

is also a partition into disjoint cosets. Since we have $\xi\widehat{\beta}_j = \widehat{\beta}'_j(E_{2n}, t_M(\beta_j)^{-1})\xi$ by (4.6), where $\beta'_j \in \Gamma$, it follows that the last decomposition can be rewritten in the form

$$\widehat{X} = \bigcup_i \bigcup_j \widehat{\Gamma}(E_{2n}, t_M(\beta_j)^{-1})\xi\widehat{\alpha}_i,$$

where, by Lemma 3.4, we may suppose that $\xi\widehat{\alpha}_i \in \widehat{S}^n_{0,p}$. We now let

(4.102) $$\qquad\qquad\qquad\qquad \widehat{\varepsilon}_{q,k} = \widehat{\varepsilon}_q \cdot P_k$$

denote the composition of the maps (4.100) and (4.101). We obtain

$$\widehat{\varepsilon}_{q,k}(\widehat{X}) = \left(\sum_j t_M(\beta_j)^k \right) \sum_i s(\xi\widehat{\alpha}_i)^{-k}(\Gamma^n_0 P(\xi\widehat{\alpha}_i)).$$

Since the set $\{t_M(\beta_j)\}$ is a nontrivial subgroup of the group of fourth roots of unity (because $t^4_M \equiv 1$ by Lemma 3.4 of Chapter 2), it now follows that, since k is odd, $\varepsilon_{q,k}(X) = 0$; thus,

(4.103) $$\qquad\qquad\qquad\qquad \widehat{\mathbf{L}}^n_p(q) = \widehat{\mathbf{E}}^n_p(q),$$

where $\widehat{\mathbf{L}}^n_p(q) = \widehat{\varepsilon}_{q,k}(\widehat{L}^n_p(q))$ and $\widehat{\mathbf{E}}^n_p(q) = \widehat{\varepsilon}_{q,k}(\widehat{E}^n_p(q))$. In Chapter 4 we shall show that the homomorphism $\widehat{\varepsilon}_{q,k}$ commutes with the representation of the Hecke rings on the corresponding spaces of modular forms. Hence, (4.103) shows that in the theory of Siegel modular forms of half-integer weight $k/2$, where k is the same as in (4.101), it is only the even Hecke rings $\widehat{E}^n_p(q)$ (or the rings $\widehat{E}^n_p(q, \varkappa)$, which do not differ from them in any essential way) that are of importance.

LEMMA 4.19. *In the ring*

(4.104) $$\qquad\qquad\qquad \widehat{\mathbf{E}}^n_p(q, \varkappa) = \widehat{\varepsilon}_{q,k}(E^n_p(q, \varkappa)) \subset \overline{\mathbf{L}}^n_{0,p}$$

the images of the elements (4.82) have the following left coset decompositions:

(4.105) $$\qquad\qquad \widehat{\mathbf{T}}_i(p^2) = \widehat{\varepsilon}_{q,k}(\widehat{T}_i(p^2)) = \sum_{\substack{a+b\leqslant n \\ a\geqslant i}} \Pi^{(a-i)}_{a,b}(k)$$

for $i = 0, 1, \ldots, n$, where, in the notation (4.42), one has

(4.106) $$\qquad \Pi^{(r)}_{a,b}(k) = \sum_{\substack{B_0, S, V \\ r_p(B_0)=r}} \varkappa(B_0)^{-k}(\Gamma^n_0 M_{a,b}(B_0)T(S)U(V)),$$

in which $\varkappa$ is the function (5.70) and the matrices B_0, S, and V run through the sets in (4.41).

PROOF. The lemma follows from (4.91) and Proposition 4.12. $\square$

In §3.3 we defined the maps Φ, ω, and Ω. We shall use the same letters to denote the extensions by linearity to the complexifications of the corresponding rings in (3.50).

LEMMA 4.20. *Let* $\pi_{a,b}^n(p)$ *be the elements* (2.31) *of the ring* H_p^n, *and let* Φ *be the homomorphism in the diagram* (3.50). *Then the following relation holds in the ring* $H_p^n[x_0^{\pm 1}]$:

$$(4.107) \qquad \Phi(\widehat{E}_p^n(q, \varkappa)) = \mathbf{Q}[x_0^2 \Lambda_0, \dots, x_0^2 \Lambda_{n-1}, (x_0^2 \Lambda_n)^{\pm 1}],$$

where Λ_a *denotes the element of* H_p^n *of the form*

$$(4.108) \qquad \Lambda_a = \sum_{0 \leqslant b \leqslant n-a} p^{b(a+b+1)} \pi_{a,b}^n(p).$$

PROOF. (4.106) shows that

$$(4.109) \quad \Phi(\Pi_{a,b}^{(r)}(k)) = x_0^2 \sum_{\substack{B_0, S, V \\ r_p(B_0)=r}} \varkappa(B_0)^{-k} (\Lambda^n D_{a,b} V) = p^{b(a+b+1)} l_p^k(r, a) x_0^2 \pi_{a,b}^n(p),$$

where for $0 \leqslant r \leqslant a$ *we have used the abbreviated notation*

$$(4.110) \qquad l_p^k(r, a) = \sum_{B \in S_a(\mathbf{F}_p), r_p(B)=r} \varkappa(B)^{-k}.$$

Now from (4.105) and (4.109) we obtain the following formula for the Φ-images of the elements $\widehat{T}_i(p^2)$:

$$(4.111) \qquad \Phi(\widehat{T}_i(p^2)) = x_0^2 \sum_{i \leqslant a \leqslant n} l_p^k(a - i, a) \Lambda_a.$$

From the definition (4.110) it follows that $l_p^k(0, a) = 1$. In addition, if we replace B by $-B$ in the sum (4.110) and use (4.98), we see that the sums (4.110) have real values. But by (4.70) they lie in the field $\mathbf{Q}(\sqrt{-1})$; hence, the $l_p^k(r, a)$ are rational numbers. From this and from (4.111) and (4.83) we obtain (4.107). □

It turns out that the image under Φ of the ring $\mathbf{E}_p^n(q)$ coincides with the ring on the right side of (4.107). The proof of the next basic result of this section relies upon this fact.

THEOREM 4.21. *Let* $n, q \in \mathbf{N}$, *where* q *is divisible by* 4, *and let* p *be a prime not dividing* q. *Then, in the notation of Theorem* 3.30:
(1) *The restriction of the map* Ω *to the subring*

$$(4.112) \qquad \underline{\widehat{E}}_p^n(q, \varkappa) = \widehat{\varepsilon}_{q,k}(\underline{\widehat{E}}_p^n(q, \varkappa)) \subset \overline{\mathbf{L}}_{0,p}^n,$$

where k *is an arbitrary odd integer and*

$$(4.113) \qquad \underline{\widehat{E}}_p^n(q, \varkappa) = \mathbf{Q}[\widehat{T}_0(p^2), \dots, \widehat{T}_n(p^2)]$$

is the integral subring of the ring (4.83), *gives an isomorphism of this ring with the ring* $\mathbf{Q}[x_0, \dots, x_n]_{W^2}$ *of polynomials that are invariant under the group of automorphisms* $W^2 = W_n^2$, *which is obtained by adjoining to* W_n *the automorphism* τ_0: $\tau_0(x_0) = -x_0$, $\tau_0(x_i) = x_i$ *for* $i = 1, \dots, n$.
(2) *The ring* $\mathbf{Q}[x_0, \dots, x_n]_{W^2}$ *is generated over* $\mathbf{Q}$ *by the polynomials*

$$(4.114) \quad t^2 = (t^n(x_0, x_1, \dots, x_n))^2, \quad \rho_a = \rho_a^n(x_0, x_1, \dots, x_n) \quad (0 \leqslant a \leqslant n - 1).$$

The polynomials (4.114) are algebraically independent over **Q**.

(3) *The restriction of* Ω *to the full subring* $\widehat{E}_p^n(q, \varkappa)$ *gives an isomorphism of this ring with the polynomial ring*

$$(4.115) \qquad \mathbf{Q}[x_0^{\pm 1}, \ldots, x_n^{\pm 1}]_{W^2} = (\mathbf{Q}[x_0, \ldots, x_n]_{W^2})[\rho_0^{-1}],$$

and hence the restriction of Ω *to the subring* $\widehat{E}_p^n(q, \varkappa)$ *of the ring* $\overline{\mathbf{L}}_{0,p}^n$ *is a monomorphism.*

(4) *The elements* $\widehat{T}_0(p^2), \ldots, \widehat{T}_n(p^2)$ *of the ring* $\widehat{E}_p^n(q, \varkappa)$ *are algebraically independent over* **Q**, *and the map* $\widehat{\varepsilon}_{q,k}$ *gives an isomorphism of this ring with the ring* $\widehat{\mathbf{E}}_p^n(q, \varkappa)$.

PROOF. The plan of proof is as follows: we compute the Φ-image of $\mathbf{E}_p^n(q)$, compare it with the analogous image of $\widehat{\mathbf{E}}_p^n(q, \varkappa)$, and then apply Theorem 3.30.

According to Theorem 3.23, (4.79), and (4.81), we can write

$$(4.116) \qquad \underline{E}_p^n(q) = \mathbf{Q}[T, T_1(p^2), \ldots, T_n(p^2)],$$

where $T = T(p)^2$ or $T_0(p^2)$, and

$$(4.117) \qquad E_p^n(q) = \underline{E}_p^n(q)[T_n(p^2)^{-1}].$$

Let $\mathbf{E}_p^n(q)$ be the ε_q-image of the ring (4.117). From (3.61) and Lemma 3.34 it follows that Φ takes the generators $\mathbf{T}_i(p^2)$ of this ring to the elements

$$(4.118) \qquad \Phi(\mathbf{T}_i(p^2)) = x_0^2 \sum_{i \leqslant a \leqslant n} l_p(a-i, a)\Lambda_a,$$

where Λ_a is given by (4.108). Since the $l_p(r, a)$ are rational numbers and $l_p(0, a) = 1$, it follows from this and from (4.116) that $\Phi(\underline{\mathbf{E}}_p^n(q)) = \mathbf{Q}[x_0^2 \Lambda_0, \ldots, x_0^2 \Lambda_n]$, which, together with (4.107), implies that

$$(4.119) \qquad \Omega(\underline{\widehat{\mathbf{E}}}_p^n(q, \varkappa)) = \Omega(\underline{\mathbf{E}}_p^n(q)).$$

We now apply Theorem 3.30. Since $\Omega(\mathbf{T}(p)) = t$ (by (3.70)), it follows from (4.116) that

$$(4.120) \qquad \Omega(\underline{\mathbf{E}}_p^n(q)) = \mathbf{Q}[t^2, \rho_0, \ldots, \rho_{n-1}].$$

If we take into account the definitions (3.52)–(3.54) of the polynomials t and ρ_a, we see that the right side of the last equality coincides with the polynomial ring $\mathbf{Q}[x_0, \ldots, x_n]_{W^2}$. From this, (4.119), (4.120), and the commutativity of the ring $\widehat{\mathbf{E}}_p^n(q, \varkappa)$ (which follows from Theorem 4.13), we obtain the first and second parts of the theorem.

The third part follows from (4.117), the analogous equality for the ring $\widehat{E}_p^n(q, \varkappa)$, and the fact that

$$(4.121) \qquad \Omega(\widehat{\mathbf{T}}_n(p^2)) = \Omega(\mathbf{T}_n(p^2)) = p^{-\langle n \rangle} \rho_0.$$

Here the first equality follows from (4.111) and (4.118), and the second equality is a consequence of Lemma 3.34.

Finally, the fourth part follows from the second and third parts and from the commutativity of the ring $\widehat{E}_p^n(q, \varkappa)$. $\square$

Just as in the case of $\mathbf{L}_p^n$, this theorem enables us to parameterize the set of all **Q**-linear homomorphisms from the ring $\widehat{\mathbf{E}}_p^n(q, \varkappa)$ to the field **C**.

PROPOSITION 4.22. *Every nonzero* **Q**-*linear homomorphism* λ *from the ring* $\widehat{\mathbf{E}}_p^n(q, \varkappa)$ *to* **C** *has the form*

(4.122) $$\mathbf{T} \to \lambda_A(\mathbf{T}) = \Omega_p^n(\mathbf{T})_{(x_0,\dots,x_n)=A},$$

where $\mathbf{T} \in \widehat{\mathbf{E}}_p^n(q, \varkappa)$ *and* $A = (\alpha_0, \dots, \alpha_n)$ *is a set of nonzero complex numbers that depends on* λ. *This set is called the parameters of the homomorphism* λ. *If one set of parameters is obtained from another by the action of a transformation in* W_n^2, *then the two sets of parameters correspond to the same homomorphism.*

PROOF. The proposition is an immediate consequence of Theorem 4.21 and the proof of Proposition 3.36. □

PROBLEM 4.23. Let $n = 1$ and $\Delta = \widehat{\mathbf{T}}_1(p^2)$. Prove that Ω takes the polynomials over $\widehat{\mathbf{E}}_p^1(q, \varkappa)$

$$\widehat{\mathbf{R}}_p^1(v) = 1 - (p\Delta)^{-T}\widehat{\mathbf{T}}_0(p^2) + v^2, \quad \widehat{\mathbf{Q}}_p^1(v) = 1 - \widehat{\mathbf{T}}_0(p^2)v + (p\Delta)^2 v^2$$

respectively to the polynomials $r(x_1; v)$ and $q(x_0^2, x_1^2; v)$ in $\mathbf{Q}[x_0^{\pm 1}, x_1^{\pm 1}]_{W^2}$ (see (3.52) and (3.76)).

[Hint: Use (4.111).]

§5. Hecke rings for the triangular subgroup of the symplectic group

When studying elements of a Hecke ring of the symplectic group, it is sometimes convenient to decompose them into suitable components which, however, do not themselves belong to this Hecke ring. The place where all of these components lie is a suitable Hecke ring of the triangular subgroup Γ_0^n of the modular group Γ^n.

1. Global rings. According to Lemma 3.25(3), we can define the *global Hecke ring* for Γ_0^n

(5.1) $$\mathbf{L}_0^n = D_Q(\Gamma_0^n, S_0^n)$$

and for any $q \in \mathbf{N}$ we can define its q-subring

(5.2) $$\mathbf{L}_0^n(q) = D_Q(\Gamma_0^n, S_0^n(q)),$$

where $S_0^n(q) = S_0^n \cap GL_{2n}(\mathbf{Z}_q)$. It is clear that the local rings $\mathbf{L}_{0,p}^n$ that were introduced in §3.3 are contained in $\mathbf{L}_0^n(q)$ if $(p, q) = 1$.

By analogy with the local case, it follows from Lemma 3.4 that the Hecke pairs $(\Gamma_0^n(q), S^n(q))$ and $(\Gamma_0^n, S_0^n(q))$ and the Hecke pairs obtained from them in the case $4|q$ by lifting by means of the homomorphisms j^n and P (see (4.3) and (4.99)) satisfy the conditions (1.26). Thus, one can determine imbeddings (1.27) of the corresponding Hecke rings:

(5.3) $$L^n(q) \xrightarrow{\varepsilon_q} \mathbf{L}_0^n(q), \quad \widehat{L}^n(q) \xrightarrow{\widehat{\varepsilon}_q} \widehat{L}_0^n(q),$$

which enables us in place of $L^n(q)$ and (by Theorems 4.6 and 4.21)

(5.4) $$\widehat{E}^n(q, \varkappa) = \bigotimes_{p \nmid q} \mathbf{Q} \widehat{E}_p^n(q, \varkappa) \subset \widehat{L}^n(q)$$

to consider the isomorphic subrings

$$(5.5) \qquad \mathbf{L}^n(q) = \varepsilon_q(L^n(q)), \quad \widehat{\mathbf{E}}^n(q, \varkappa) = \widehat{\varepsilon}_{q,k}(\widehat{E}^n(q, \varkappa)),$$

inside the global ring $\overline{\mathbf{L}}_0^n(q) = \mathbf{L}_0^n(q) \otimes_{\mathbf{Q}} \mathbf{C}$, where $\widehat{\varepsilon}_{q,k} = \widehat{\varepsilon} \cdot P_k$.

We shall examine certain multiplicative properties of the rings (5.2). Unlike the Hecke rings of the symplectic group and the symplectic covering group, the Hecke rings of the triangular subgroup are noncommutative and contain zero divisors.

PROBLEM 5.1. Let $n = 1$, let p be an odd prime, and let X be an element of $\mathbf{L}_{0,p}^1$ of the form

$$X = \sum_{i=1}^{p-1} (-1)^i \left(\begin{pmatrix} p & i \\ 0 & p \end{pmatrix} \right)_\Gamma \quad (\Gamma = \Gamma_0^1).$$

Show that the following relations hold in the ring $\mathbf{L}_{0,p}^1$:

$$\left(\begin{pmatrix} p & 0 \\ 0 & p \end{pmatrix} \right)_\Gamma \cdot X = 0, \quad X \cdot \left(\begin{pmatrix} p & 0 \\ 0 & p \end{pmatrix} \right)_\Gamma \neq 0.$$

[**Hint:** Show that the double cosets

$$\left(\begin{pmatrix} p & i \\ 0 & p \end{pmatrix} \right)_\Gamma \quad \text{and} \quad \left(\begin{pmatrix} p^2 & i \\ 0 & p \end{pmatrix} \right)_\Gamma \quad \text{for } i = 1, \dots, p-1$$

are pairwise distinct, and each of them consists of a single left coset modulo Γ.]

However, several important properties of the rings $L^n(q)$ and $\widehat{E}^n(q, \varkappa)$ do carry over to the rings $\mathbf{L}_0^n(q)$. Moreover, in practice we shall have need only of certain subrings and submodules of the rings $\mathbf{L}_0^n(q)$ and $\mathbf{L}_{0,p}^n$.

LEMMA 5.2. *For any $M \in S_0^n(q)$ and $r \in \mathbf{Z}_q^*$, where $n, q \in \mathbf{N}$, the following relations hold in the ring $\mathbf{L}_0^n(q)$:*

$$\Delta_n(r)(M)_{\Gamma_0} = (M)_{\Gamma_0}\Delta_n(r) = (rM)_{\Gamma_0},$$

where

$$(5.6) \qquad \Delta_n(r) = (rE_{2n})_{\Gamma_0} = (\Gamma_0 r E_{2n})$$

and $\Gamma_0 = \Gamma_0^n$. In particular, $\Delta_n(r_1)\Delta_n(r_2) = \Delta_n(r_1 r_2)$ for $r_1, r_2 \in \mathbf{Z}_q^*$.

PROOF. The decomposition (5.6) is obvious. The lemma follows from this decomposition and the definition of multiplication in Hecke rings. □

The lemma implies that elements of the form $\Delta_n(r)$ lie in the center of $\mathbf{L}_0^n(q)$ and are invertible in this ring. As in the case of the analogous lemmas for the Hecke rings considered earlier, in practical calculations this lemma enables us to reduce the case of arbitrary double cosets to the case of double cosets of integer matrices.

The map j in §1.4 allows us to define an important anti-automorphism $*$ of the ring $\overline{\mathbf{L}}_0^n(q)$.

PROPOSITION 5.3. *Let $\Gamma_0 = \Gamma_0^n$, and let*

$$X = \sum_i a_i (M_i)_{\Gamma_0}, \quad \text{where } a_i \in \mathbf{C}, M_i \in S_0^n(q),$$

*be an arbitrary element of $\overline{\mathbf{L}}_0^n(q)$. Then the **C**-antilinear map*

$$(5.7) \qquad X \to X^* = \sum_i \overline{a}_i (r(M_i) M_i^{-1})_{\Gamma_0},$$

where $\overline{a}_i$ is the complex conjugate of a_i, is an anti-automorphism of order 2 of the ring $\overline{\mathbf{L}}_0^n(q)$, i.e.,

$$(5.8) \qquad (XY)^* = Y^* X^*, \quad (X^*)^* = X \quad (X, Y \in \overline{\mathbf{L}}_0^n(q)).$$

Every element $\mathbf{T}$ in the subring (5.5) of $\overline{\mathbf{L}}_0^n(q)$ is invariant relative to the anti-automorphism (5.7):

$$(5.9) \qquad X^* = X \quad \text{for } X \in \mathbf{L}^n(q) \text{ or } \widehat{\mathbf{E}}^n(q, \varkappa).$$

PROOF. We consider the diagram

$$(5.10) \qquad \begin{array}{ccccc}
\widehat{E}^n(q, \varkappa) & \xrightarrow{\widehat{\varepsilon}_q} & \widehat{\mathbf{L}}_0^n(q) & \xrightarrow{P_k} & \overline{\mathbf{L}}_0^n(q) \\
\downarrow * & & \downarrow * & & \downarrow * \\
\widehat{E}^n(q, \varkappa) & \xrightarrow{\widehat{\varepsilon}_q} & \widehat{\mathbf{L}}_0^n(q) & \xrightarrow{P_k} & \overline{\mathbf{L}}_0^n(q)
\end{array}$$

in which only the first two vertical arrows have not been defined. We define the first of these maps in such a way that it acts like (4.88) on the local components $\widehat{E}_p^n(q, \varkappa)$. By Theorems 4.6 and 4.13 and the formula (4.89), this map $*$ is an anti-automorphism, and $\widehat{X}^* = \widehat{X}$ for any $\widehat{X} \in \widehat{E}^n(q, \varkappa)$. The second map is defined on double cosets by setting $(\xi)_{\widehat{\Gamma}_0} \xrightarrow{*} (\xi^*)_{\widehat{\Gamma}_0}$, where $\Gamma_0 = \Gamma_0^n$, and by extending the map by **Q**-linearity onto all of $\widehat{\mathbf{L}}_0^n(q)$. The fact that this is an anti-automorphism will be shown by examining the third $*$-map, which is equal to the composition $i \cdot j = j \cdot i$ of the **C**-antilinear automorphism i of $\overline{\mathbf{L}}_0^n(q)$ and the **C**-linear anti-automorphism j of $\overline{\mathbf{L}}_0^n(q)$ (both of order 2), which, in turn, are defined on double cosets by the conditions

$$(M)_{\Gamma_0} \xrightarrow{i} (r(M^{-1})M)_{\Gamma_0}, \quad (M)_{\Gamma_0} \xrightarrow{j} (M^{-1})_{\Gamma_0} \quad (M \in S_0^n(q)).$$

Now the first part of the proposition follows from Proposition 1.11. From Lemma 1.13 and the definition of all of the maps in (5.10) we conclude that the diagram commutes. This implies (5.9) for $X \in \widehat{\mathbf{E}}^n(q, \varkappa)$. For $X \in \mathbf{L}^n(q)$ the equality (5.9) is obtained from (3.14). $\qquad \square$

We now turn our attention to subrings of $\mathbf{L}_0^n(q)$. It turns out that, in addition to the Hecke rings of the symplectic group, this ring also contains commutative subrings that can be obtained as the centralizers of certain sets of elements and that are naturally isomorphic to Hecke rings of the general linear group of order n (more precisely, to certain extensions of them). This circumstance makes it possible to reduce various questions in the theory of Hecke rings and Hecke operators for the symplectic group to analogous questions for the general linear group.

We consider the following elements of the Hecke ring $\mathbf{L}_0^n(q)$:

$$(5.11) \qquad \Pi_-^n(m) = \left(\begin{pmatrix} mE_n & 0 \\ 0 & E_n \end{pmatrix} \right)_{\Gamma_0}, \quad \Pi_+^n(m) = \left(\begin{pmatrix} E_n & 0 \\ 0 & mE_n \end{pmatrix} \right)_{\Gamma_0},$$

where $m \in \mathbf{N}$ and $(m, q) = 1$. We have the following

LEMMA 5.4. *The decomposition of the elements* $\Pi_{\pm}(m) = \Pi_{\pm}^n(m)$ *into left cosets modulo* $\Gamma_0 = \Gamma_0^n$ *has the form*

$$(5.12) \quad \Pi_-(m) = \left(\Gamma_0 \begin{pmatrix} mE & 0 \\ 0 & E \end{pmatrix} \right), \quad \Pi_+(m) = \sum_{B \in S_m/\text{mod } m} \left(\Gamma_0 \begin{pmatrix} E & B \\ 0 & mE \end{pmatrix} \right),$$

where $E = E_n$. *We have the relations:*

$$(5.13) \qquad \Pi_-(m_1)\Pi_-(m_2) = \Pi_-(m_1 m_2), \quad \Pi_+(m_1)\Pi_+(m_2) = \Pi_+(m_1 m_2),$$

$$(5.14) \qquad\qquad\qquad \Pi_-(m)\Pi_+(m) = m^{\langle n \rangle} \Delta_n(m),$$

where Δ_n *is the element* (5.6);

$$(5.15) \qquad \Pi_-(m_1)\Pi_+(m_2) = \Pi_+(m_2)\Pi_-(m_1), \quad \text{if } (m_1, m_2) = 1,$$

$$(5.16) \qquad\qquad \Pi_-(m)^* = \Pi_+(m), \quad \Pi_+(m)^* = \Pi_-(m).$$

PROOF. The decompositions in (5.12) follow from (3.44). The relations (5.13)–(5.16) are obtained directly from the definitions and (5.12). $\qquad \square$

We now consider the subsets of $\mathbf{L}_0^n(q)$ consisting of all elements that commute with all elements of the form $\Pi_-(m)$ and all elements of the form $\Pi_+(m)$, respectively:

$$(5.17) \qquad C_-^n = \{ X \in \mathbf{L}_0^n(q); \Pi_-(m)X = X\Pi_-(m), (m, q) = 1 \},$$

$$(5.18) \qquad C_+^n = \{ X \in \mathbf{L}_0^n(q); \Pi_+(m)X = X\Pi_+(m), (m, q) = 1 \}.$$

These are clearly subrings of $\mathbf{L}_0^n(q)$. From (5.16) it follows that the anti-automorphism $*$ takes each of these subrings into the other one:

$$(5.19) \qquad\qquad C_-^n(q)^* = C_+^m(q), \quad C_+^n(q)^* = C_-^n(q).$$

PROPOSITION 5.5. *The ring* $C_-^n(q)$ *(resp.* $C_+^n(q)$*) is spanned by the double cosets modulo* $\Gamma_0 = \Gamma_0^n$ *of elements of the form*

$$(5.20) \qquad\qquad M = U(r, D) \in S_0^n(q), \quad \text{where } d_n(D)^2 \mid r$$

(resp., of the form

$$(5.21) \qquad\qquad M = U(r, D) \in S_0^n(q), \quad \text{where } r \mid d_1(D)^2),$$

where $U(r, D) = \begin{pmatrix} rD^* & 0 \\ 0 & D \end{pmatrix}$ *and* $d_i(D)$ *denotes the* i*th elementary divisor of the matrix* D.

We first describe the decomposition of the Γ_0-double cosets of elements of the form (5.20) and (5.21) into left cosets modulo Γ_0.

LEMMA 5.6. *If M is an element of the form* (5.20), *then*

$$(5.22) \qquad (M)_{\Gamma_0} = \sum_{D_i \in \Lambda \backslash \Lambda D \Lambda} (\Gamma_0 U(r, D_i)),$$

and if M is of the form (5.21), *then*

$$(5.23) \qquad (M)_{\Gamma_0} = \sum_{\substack{D_i \in \Lambda \backslash \Lambda D \Lambda \\ S \in S_n / S_n^i}} \left(\Gamma_0 \begin{pmatrix} rD_i^* & rD_i^* S \\ 0 & D_i \end{pmatrix} \right),$$

where $\Lambda = \Lambda^n$, and in the last condition under the summation the set $S_n^i = r^{-1}\, {}^t D_i S_n D_i$ is contained in the group $S_n = S_n(\mathbf{Z})$ and is regarded as a subgroup there.

PROOF OF THE LEMMA. It is easy to see that

$$\Gamma_0 U(r, D)\Gamma_0 = \left\{ \begin{pmatrix} rD_i^* & B_{ij} \\ 0 & D_i \end{pmatrix} ; D_i \in \Lambda D \Lambda, B_{ij} \in rD_i^* S_n + S_n D_i \right\}$$

and that matrices in this set lie in the same Γ_0-left coset if and only if the corresponding D_i lie in the same Λ-left coset and the corresponding B_{ij} are congruent on the right modulo D_i. If $d_n(D)^2 | r$, then $rS[D_i^{-1}]$ is an integer matrix for any $D_i \in \Lambda D \Lambda$ and $S \in S_n$, since $d_n(D)D_i^{-1}$ is an integer matrix. Consequently, in this case all of the matrices in the set $rD_i^* S_n + S_n D_i$ are divisible on the right by D_i, and in choosing Γ_0-left coset representatives we may suppose that all $B_{ij} = 0$. This gives (5.22). In the general case, when choosing left coset representatives we may always take $B_{ij} = rD_i^* S_j$, where the $S_j \in S_n(\mathbf{Z})$ are such that the matrices B_{ij} are pairwise noncongruent modulo D_i. The congruence $B_{ij'} \equiv B_{ij} \pmod{D_i}$ means that $B_{ij} - B_{ij'} = S'D_i$ with $S' \in S_n(\mathbf{Z})$, i.e., $S_j - S_{j'} = r^{-1}S'[D_i] \in r^{-1}\,{}^t D_i S_n(\mathbf{Z})D_i$. If $r | d_1(D)^2$, then this last set is contained in $S_n(\mathbf{Z})$. $\square$

PROOF OF THE PROPOSITION. We first note that from the uniqueness of elementary divisors it follows that the following relations hold for any nonsingular rational $n \times n$-matrix D:

$$(5.24) \qquad d_i(D^{-1}) = d_{n+1-i}(D)^{-1} \quad \text{for } i = 1, \dots, n.$$

These relations imply, in particular, that $d_1(rD^{-1}) = rd_n(D)^{-1}$, and so we find that the matrix M has the form (5.20) if and only if the matrix $rM^{-1} = U(r, rD^{-1})$ has the form (5.21). Thus, the anti-automorphism $*$ takes linear combinations of double cosets of elements of the form (5.20) to similar linear combinations of double cosets of elements of the form (5.21), and conversely. Taking (5.19) into account, we see that it suffices to prove the proposition for one of the two rings $C_{\pm}^n(q)$, say, for $C_-^n(q)$.

If M has the form (5.20), then from (5.11) and (5.22) we obtain

$$\Pi_-(m)(M)_{\Gamma_0} = \sum_{D_i \in \Lambda \backslash \Lambda D \Lambda} (\Gamma_0 U(mr, D_i)) = (M)_{\Gamma_0}\Pi_-(m).$$

Conversely, suppose that

$$X = \sum_{\alpha} a_\alpha \left(\Gamma_0 \begin{pmatrix} r_\alpha D_\alpha^* & B_\alpha \\ 0 & D_\alpha \end{pmatrix} \right) \in C_-^n(q),$$

where the left cosets are pairwise distinct and all a_α are nonzero. We choose an integer m prime to q for which all of the matrices $mB_\alpha D_\alpha^{-1}$ are integer matrices. Then (see (5.11))

$$\Pi_-(m)X = \sum_\alpha a_\alpha(\Gamma_0 U(mr_\alpha, D_\alpha)).$$

On the other hand,

$$\Pi_-(m)X = X\Pi_-(m) = \sum_\alpha a_\alpha \left(\Gamma_0 \begin{pmatrix} mr_\alpha D_\alpha^* & B_\alpha \\ 0 & D_\alpha \end{pmatrix}\right).$$

Comparing these sums, we conclude that for each α the matrix B_α is divisible on the right by D_α. Hence, we may assume that all $B_\alpha = 0$. Since X lies in the Hecke ring of the group Γ_0, it is invariant under right multiplication by matrices in Γ_0 of the form $T(S)$ with $S \in S_n(\mathbf{Z})$. This implies that for any such S and for any α the matrix $r_\alpha D_\alpha^* S$ is also divisible on the right by D_α, i.e.,

$$(5.25) \qquad\qquad r_\alpha D_\alpha^* S_n(\mathbf{Z}) D_\alpha^{-1} \in S_n(\mathbf{Z}).$$

Let $\mathrm{diag}(d_1, \dots, d_n) = \mathrm{ed}(D_\alpha)$ be the elementary divisor matrix of D_α. Since $D_\alpha = \gamma(\mathrm{ed}D_\alpha)\delta$, where $\gamma, \delta \in \Lambda$, it follows that D_α can be replaced by $\mathrm{ed}(D_\alpha)$ in (5.25). Then this condition obviously means that all of the ratios $r_\alpha/d_i d_j$ are integers, and this is equivalent to the condition $d_n(D_\alpha)^2 | r_\alpha$. Finally, because X is invariant under right multiplication by matrices in Γ_0 of the form $U(\gamma)$ with $\gamma \in \Lambda$, it follows that the expansion of X can be rewritten in the form

$$X = \sum_\beta a'_\beta \left\{ \sum_{D_\alpha \in \Lambda \backslash \Lambda D'_\beta \Lambda} (\Gamma_0 U(r'_\beta, D_\alpha)) \right\},$$

where D'_β runs through D_α lying in pairwise distinct Λ-double cosets, and any of the Λ-left cosets in $\Lambda D'_\beta \Lambda$ is equal to one of the ΛD_α with $D_\alpha \in \Lambda D'_\beta \Lambda$. Then the relation $d_n(D_\alpha)^2 | r_\alpha$ and (5.22) imply that the expression in braces is the decomposition of some double coset $(M_\beta)_{\Gamma_0}$, where M_β has the form (5.20). $\qquad \square$

An important tool for studying the global ring $\mathbf{L}_0^n$ and its subrings is the global analogue of the map Φ that was defined in §3.3. We shall associate variables y_p to the prime numbers p, and for different p we suppose that they commute with one another. We let $\mathfrak{Q} = \mathbf{Q}[\dots, y_p^{\pm 1}, \dots]$ be the ring of polynomials over $\mathbf{Q}$ in the variables $y_p^{\pm 1}$ ($p = 2, 3, 5, \dots$). We define the $\mathbf{Q}$-linear map $\Phi = \Phi^n$ from the module $L_{\mathbf{Q}}(\Gamma_0^n, S_0^n)$ to the module $L_{\mathfrak{Q}}(\Lambda^n, G^n)$ by setting

$$(5.26) \qquad\qquad \Phi\left(\Gamma_0 \begin{pmatrix} rD^* & B_\alpha \\ 0 & D \end{pmatrix}\right) = y_{p_1}^{\alpha_1} \cdots y_{p_r}^{\alpha_r} (\Lambda^n D),$$

if $r = p_1^{\alpha_1} \cdots p_r^{\alpha_r}$. It is clear that this map does not depend on the choice of left coset representatives.

PROPOSITION 5.7. *The restriction of the map Φ^n to the Hecke ring $\mathbf{L}_0^n \subset L_{\mathbf{Q}}(\Gamma_0^n, S_0^n)$ gives an epimorphism of this ring onto the Hecke ring $D_{\mathfrak{Q}}(\Lambda^n, G^n)$ of the Hecke pair (Λ^n, G^n) over $\mathfrak{Q}$:*

$$(5.27) \qquad\qquad \Phi = \Phi^n \colon \mathbf{L}_0^n \to D_{\mathfrak{Q}}(\Lambda^n, G^n) = H^n[\dots, y_p^{\pm 1}, \dots].$$

PROOF. The invariance of elements $X \in \mathbf{L}_0^n$ under right multiplication by matrices $U(\gamma)$ $(\gamma \in \Lambda^n)$ implies the invariance of their images $\Phi(X)$ under right multiplication by matrices $\gamma \in \Lambda^n$. Hence, $\Phi(\mathbf{L}_0^n) \subset D_\Omega(\Lambda^n, G^n)$. From the definition of multiplication in Hecke rings it then follows that (5.27) is a homomorphism. Finally, from (5.11) we have

$$\Phi(\Pi_-(p_1^{\alpha_1} \cdots p_r^{\alpha_r})) = y_{p_1}^{\alpha_1} \cdots y_{p_r}^{\alpha_r},$$

and from (3.44) we have $\Phi((U(D)_{\Gamma_0})) = \alpha(D)(D)_\Lambda$, where $D \in G^n$ and $\alpha(D) \neq 0$. These equalities imply that we have an epimorphism.   □

THEOREM 5.8. *The restrictions of the map Φ^n to the subrings $C_-^n(q)$ and $C_+^n(q)$ of $\mathbf{L}_0^n$, where $n, q \in \mathbf{N}$, are monomorphisms. In particular, $C_-^n(q)$ and $C_+^n(q)$ are commutative rings with no zero divisors.*

PROOF. The proof is similar for $C_-^n(q)$ and $C_+^n(q)$; to be definite, we consider the case of $C_+^n(q)$. By Proposition 5.5, every nonzero $X \in C_+^n(q)$ can be written in the form

$$X = \sum_i a_i (U(r_i, D_i))_{\Gamma_0},$$

where all of the $U(r_i, D_i)$ have the form (5.21), the double cosets are pairwise distinct, and all a_i are nonzero. Then from (5.23) it follows that

$$\Phi(X) = \sum_i a_i \alpha_i \left(\prod_i y_p^{v_p(r_i)} \right) (D_i)_\Lambda,$$

where the α_i are positive integers, p runs through the prime numbers, and $v_p(r)$ denotes the exact power of p that occurs in the prime factorization of r. The terms corresponding to different i have either different r_i, or else different $(D_i)_\Lambda$, and hence do not cancel one another. Consequently, $\Phi(X) \neq 0$.   □

According to this theorem, the rings $C_\pm^n(q)$ may be regarded as extensions of the global Hecke ring of the general linear group. Since the ring $\mathbf{L}_0^n$ contains the global Hecke rings of the symplectic group, this makes it possible for us to examine the connections between Hecke rings of the symplectic group and the general linear group.

2. Local rings. In earlier sections we have already studied the local Hecke ring $\overline{\mathbf{L}}_{0,p}^n$ for each prime p, and also its local subrings $\mathbf{L}_p^n$ and $\widehat{\mathbf{E}}_p^n(q, \varkappa)$ (see (3.45) and (4.112)). It is clear that

(5.28) $\mathbf{L}_{0,p}^n \subset \mathbf{L}_0^n(q),$ if $(p, q) = 1$.

We now introduce local analogues of the rings $C_\pm^n(q)$. We set

(5.29) $C_{-p}^n = C_-^n(q) \cap \mathbf{L}_{0,p}^n,$ $C_{+p}^n = C_+^n(q) \cap \mathbf{L}_{0,p}^n.$

From Proposition 5.5 it follows that these rings do not depend on q with $(p, q) = 1$. The global Theorem 5.8 implies the following local variant.

THEOREM 5.9. *The rings C_{-p}^n and C_{+p}^n, where $n \in \mathbf{N}$ and p is a prime number, are commutative rings with no zero divisors. Moreover, the restrictions to either C_{-p}^n or C_{+p}^n of the maps Φ_p^n and $\Omega_p^n = \omega_p^n \cdot \Phi_p^n$ that were defined in §3.3 are monomorphisms.*

PROOF. We note that we can obtain the map Φ_p^n on $\mathbf{L}_{0,p}^n$ if we take the restriction to $\mathbf{L}_{0,p}^n$ of the map Φ^n and then set $y_p = x_0$. Thus, it follows from Theorem 5.8 that the restriction of Φ_p^n to either C_{-p}^n or C_{+p}^n is a monomorphism. From this and Lemma 3.29 we see that the restrictions of Ω_p^n are also monomorphisms.   □

In the next section we make a more detailed study of the properties of the local rings for fixed p, in connection with the problem of factoring polynomials over $\mathbf{L}_p^n$ or $\widehat{\mathbf{E}}_p^n(q, \varkappa)$. For now we limit ourselves to a discussion of some of the connections between the local rings corresponding to different primes.

THEOREM 5.10. *The ring $C_-^n(q)$ (resp. $C_+^n(q)$), where $n, q \in \mathbf{N}$, is generated by the subrings C_{-p}^n (resp. C_{+p}^n), where p runs through all prime numbers not dividing q.*

PROOF. From (5.19) it follows that it suffices to treat, say, the case of $C_-^n(q)$. Proposition 5.5 implies that for this it is enough to verify that, given an arbitrary M of the form (5.20), the double coset $(M)_{\Gamma_0}$ is a finite product of double cosets $(M_p)_{\Gamma_0} \in C_{-p}^n$, where p runs through a set of distinct primes not dividing q. Let $M = U(r, D)$. By Lemma 2.2, if we replace M by another representative of the same Γ_0-double coset, we may assume that D is equal to its elementary divisor matrix $\mathrm{ed}(D) = \mathrm{diag}(d_1, \ldots, d_n)$. For each p we set

$$D_p = \mathrm{diag}(p^{v_p(d_1)}, \ldots, p^{v_p(d_n)}), \quad r_p = p^{v_p(r)}, \quad \text{and} \quad M_p = U(r_p, D_p),$$

where $v_p(a)$ is the exponent of p in the prime factorization of the rational number a. Clearly, M_p is not equal to the identity matrix for only finitely many p, and none of these p divide q. Each matrix M_p lies in $S_{0,p}^n$. Since $d_n^2 | r$, it follows that $d_n(D_p)^2$ divides r_p for each p; hence, $(M_p)_{\Gamma_0} \in C_{-p}^n$. Because $\prod_p r_p = r$ and, by Proposition 2.5, $\prod_p (D_p)_\Lambda = \left(\prod_p D_p\right)_\Lambda = (D)_\Lambda$, we conclude from Lemma 5.6 and the definitions that the double coset $(M)_{\Gamma_0}$ is equal to the product of the double cosets $(M_p)_{\Gamma_0}$.   □

Our next task is to prove that elements of the rings $\mathbf{L}_p^n$ and $\widehat{\mathbf{E}}_p^n(q, \varkappa)$ commute with elements of the rings $C_{\pm p_1}^n$ if $(p, p_1) = 1$. To do this, we examine the double cosets $\mathbf{T}_i(p^2)$ and $\widehat{\mathbf{T}}_i(p^2)$ in more detail. According to (3.61) and (4.105), these double cosets can be represented as sums of the Γ_0-double cosets $\Pi_{a,b}^{(r)}$ and $\Pi_{a,b}^{(r)}(k)$, where $\Gamma_0 = \Gamma_0^n$. Just as in the case of (4.106), from (4.90) we obtain the expansion

$$(5.30) \qquad \Pi_{a,b}^{(r)} = \sum_{B_0, S, V; r_p(B_0) = r} (\Gamma_0^n M_{a,b}(B_0) T(S) U(V)),$$

and hence $\Pi_{a,b}^{(r)}$ is a sum of double cosets

$$(5.31) \qquad (M_{a,b}(B_0))_{\Gamma_0}, \quad \text{where } B_0 \in S_a(\mathbf{Z})/\bmod p, \quad r_p(B_0) = r.$$

Suppose that the matrix $\varepsilon = (\varepsilon_{ij}) \in \Lambda_{(D)}$ is divided into blocks in the same way as $D = D_{a,b}$ (see (2.31)). Then $\varepsilon = D^{-1} \eta^{-1} D$ (i.e., $\eta D \varepsilon = D$) with $\eta = (\eta_{ij}) \in \Lambda$; this implies the congruences

$$(5.32) \qquad \begin{array}{c} \varepsilon_{12} \equiv 0 (\bmod p), \quad \varepsilon_{23} \equiv 0 (\bmod p), \quad \varepsilon_{13} \equiv 0 (\bmod p^2), \\ \det \varepsilon_{ii} \not\equiv 0 (\bmod p) \quad \text{for } i = 1, 2, 3. \end{array}$$

Using these congruences, (4.39), and the relation $\eta^* = D^{-1} \cdot {}^t \varepsilon D$, we obtain

$$(5.33) \qquad (U(\eta)M_{a,b}(B_0)U(\varepsilon))_{\Gamma_0} = (M_{a,b}(B_0[\varepsilon_{22}]))_{\Gamma_0}.$$

From Lemma 3.2 it follows that for $a < n$ and for any $U \in GL_a(\mathbf{F}_p)$ there exists $\varepsilon \in \Lambda_{(D)}$ such that $\varepsilon_{22} \equiv U(\mathrm{mod}\ p)$. If $a = n$, then $\varepsilon_{22} \equiv \varepsilon$, so that $\det \varepsilon_{22} = \det \varepsilon = \pm 1$, and the matrix U must satisfy the condition $\det U \equiv \pm 1(\mathrm{mod}\ p)$. From this and (5.33) it follows that the matrix B_0 in the double coset (5.31) can be any matrix of the set

$$(5.34) \qquad \{B_0\}_p^n = \begin{cases} \{B_0[U]/\mathrm{mod}\ p;\ U \in GL_a(\mathbf{F}_p)\} \\ \{B_0[U]/\mathrm{mod}\ p;\ U \in GL_a(\mathbf{F}_p),\ \det U \equiv \pm 1(\mathrm{mod}\ p)\} \end{cases}$$

for $a < n$ and for $a = n$, respectively, where $\mathbf{F}_p = \mathbf{Z}/p\mathbf{Z}$.

One can also show that, if the double cosets (5.31) for B_0 and B_0' are the same, then $\{B_0\}_p^n = \{B_0'\}_p^n$. Hence, taking into account (5.33) and (5.30), we obtain

$$(5.35) \qquad (M_{a,b}(B_0))_{\Gamma_0} = \sum_{B_0' \in \{B_0\}_p^n, S, V} (\Gamma_0 M_{a,b}(B_0')T(S)U(V)),$$

where S and V run through the sets of matrices in (4.41). From this we easily see that the equality

$$(5.36) \qquad (M_{a,b}(B_0))_{\Gamma_0} = (M_{a,b}(B_0'))_{\Gamma_0}$$

holds if and only if $a = a_1$, $b = b_1$, and $\{B_0\}_p^n = \{B_0'\}_p^n$.

LEMMA 5.11. *Let* $0 \leqslant r \leqslant a$, $a + b \leqslant n$. *Then* $\Pi_{a,b}^{(r)}$ *and* $\Pi_{a,b}^{(r)}(k)$ *have the following decompositions into* Γ_0-*double cosets, where* $\Gamma_0 = \Gamma_0^n$:

$$(5.37) \qquad \Pi_{a,b}^{(r)} = \sum_{\{B_0\}_p^n, r_p(B_0)=r} (M_{a,b}(B_0))_{\Gamma_0},$$

$$(5.38) \qquad \Pi_{a,b}^{(r)}(k) = \sum_{\{B_0\}_p^n, r_p(B_0)=r} \varkappa(B_0)^{-k}(M_{a,b}(B_0))_{\Gamma_0},$$

where the summation is taken over the set (5.34) *in* $S_a(\mathbf{Z})/\mathrm{mod}\ p$. *The action of the anti-automorphism* $*$ *on these elements is given by the formulas*

$$(5.39) \qquad (\Pi_{a,b}^{(r)})^* = \Pi_{a,n-a-b}^{(r)}, \qquad (\Pi_{a,b}^{(r)}(k))^* = \Pi_{a,n-a-b}^{(r)}(k).$$

PROOF. Since all of the left cosets in (5.30) and (4.106) and all of the double cosets on the right in (5.37) and (5.38) are pairwise distinct, and since these double cosets occur in $\Pi_{a,b}^{(r)}$ and $\Pi_{a,b}^{(r)}(k)$, it follows that (5.37) and (5.38) are consequences of (5.35).

By the definition of the anti-automorphism $*$ (see (5.7)) we have

$$(M_{a,b}(B_0))_{\Gamma_0}^* = \left(\begin{pmatrix} D_{a,b} & -B \\ 0 & p^2 D_{a,b}^{-1} \end{pmatrix} \right)_{\Gamma_0},$$

where B is the matrix in (4.39). We let $I = I_n$ denote the $n \times n$-matrix with 1's on the anti-diagonal and 0's everywhere else. Then the map $A \to IAI$ of any $n \times n$-matrix A reverses the order of its rows and columns; from this it easily follows that

$$U(I) \begin{pmatrix} D_{a,b} & -B \\ 0 & p^2 D_{a,b}^{-1} \end{pmatrix} U(I) = M_{a,n-a-b}(-I_a B_0 I_a).$$

Since $U(I) \in \Gamma_0$, we thus obtain the relation

$$(M_{a,b}(B_0))^*_{\Gamma_0} = (M_{a,n-a-b}(-I_a B_0 I_a))_{\Gamma_0}.$$

The equalities in (5.39) follow from these relations and from (5.37)–(5.38), since the map $B_0 \to -I_a B_0 I_a$ merely permutes the classes $\{B_0\}^n_p$ with $r_p(B_0) = r$, and since, by (4.70) and (4.98), we have $\overline{\varkappa}(B_0) = \varkappa(-B_0) = \varkappa(-I_a B_0 I_a)$.   □

PROPOSITION 5.12. *Let p, p_1 be distinct primes, and let $n \in \mathbf{N}$. Then:*
(1) *every element in $\mathbf{L}^n_p$ or $\widehat{\mathbf{E}}^n_p(q, \varkappa)$ commutes with every element in $C^n_{-p_1}$ or $C^n_{+p_1}$;*
(2) *every element in C^n_{-p} commutes with every element in $C^n_{+p_1}$.*

PROOF. By Proposition 5.3 and (5.19), it suffices to prove part (1) for, say, $C^n_{-p_1}$. According to Proposition 5.5, to do this it is enough to verify that any double coset $(M)_{\Gamma_0}$, where $M = U(r, A^0) \in S^n_{0,p_1}$ and $d_n(A_0)^2|r$, commutes with all of the generators $\mathbf{T}(p)$ and $\mathbf{T}_i(p^2)$ $(1 \leqslant i \leqslant n)$ of the ring $\mathbf{L}^n_p$ and with all of the generators $\widehat{\mathbf{T}}_i(p^2)$ $(1 \leqslant i \leqslant n)$ of the ring $\widehat{\mathbf{E}}^n_p(q, \varkappa)$. From (3.58), (3.61), (4.105), and Lemma 5.11 we see that this, in turn, will follow if we show that $(M)_{\Gamma_0}$ commutes with all elements of $\mathbf{L}^n_{0,p}$ of the form Π_a $(0 \leqslant a \leqslant n)$ or $(M_{a,b}(B_0))_{\Gamma_0}$ $(a + b \leqslant n)$. From (3.59) and (5.22) we obtain

$$(5.40) \qquad (M)_{\Gamma_0}\Pi_a = \sum_A \sum_{D,B} \left(\Gamma_0 \begin{pmatrix} rp(AD)^* & rA^*B \\ 0 & AD \end{pmatrix} \right),$$

where $A \in \Lambda \setminus \Lambda A_0 \Lambda$, $D \in \Lambda \setminus \Lambda D_a \Lambda$, and $B \in B_0(D)/\mathrm{mod}\, D$. We need some preliminaries in order to transform this expression. Given a subring K of the field $\mathbf{Q}$ and a matrix $D \in GL_n(\mathbf{Q})$, we set

$$B_K(D) = \{B \in M_n(K);\ {}^t BD = {}^t DB\}.$$

Just as in the first part of Lemma 3.33, it is not hard to verify that

$$B_K(\alpha D\beta) = \alpha^* B_K(D)\beta \quad \text{if } \alpha, \beta \in GL_n(K),$$

and that in this case we can take the set $\alpha^*\{B_K(D)/S_n(K)D\}\beta$ as a set of representatives of the residue classes $B_K(\alpha D\beta)/S_n(K)\alpha D\beta$. Now let $K = \mathbf{Z}[p_1^{-1}]$, and let D be an integer matrix all of whose elementary divisors are prime to p_1. Then each residue class $B_K(D)/S_n(K)D$ contains an integer matrix. Namely, if we write an arbitrary matrix B in $B_K(D)$ in the form $q^{-1}B_0$ with B_0 an integer matrix and with $q = p_1^\alpha$, and if we choose $S_0 \in S_n(\mathbf{Z})$ so that $B_0 + S_0 D \equiv 0 \pmod{q}$ (D is invertible modulo q), then we obtain $B + q^{-1}S_0 D \in B_0(D)$. This implies that in our case we can take

$$B_K(D)/S_n(K)D = B_0(D)/\mathrm{mod}\, D.$$

Returning to (5.40), if we use the above considerations and Proposition 2.5, we can write this expression in the form

$$(M)_{\Gamma_0}\Pi_a = \sum_{C,B} \left(\Gamma_0 \begin{pmatrix} rpC^* & B \\ 0 & C \end{pmatrix} \right),$$

where $C \in \Lambda \setminus \Lambda A_0 D_\alpha \Lambda$, $B \in r\{B_K(C)/S_n(K)C\}$, and $K = \mathbf{Z}[p_1^{-1}]$. Since r is invertible in K, the factor r can be omitted in the last condition. Furthermore, using the commutativity of the Hecke ring of the group Λ and Proposition 2.5, we see that

$$(A_0 D_a)_\Lambda = (A_0)_\Lambda (D_a)_\Lambda = (D_a)_\Lambda (A_0)_\Lambda = (D_a A_0)_\Lambda.$$

Hence, $\Lambda A_0 D_\alpha \Lambda = \Lambda D_\alpha A_0 \Lambda$. Thus, in the last sum we may suppose that $C \in \Lambda \setminus \Lambda D_a A_0 \Lambda$; and, again using (5.22) and the above properties of the sets B_K, we can rewrite this sum in the form

$$\sum_{D,A,B} \left(\Gamma_0 \begin{pmatrix} {}^r p(DA)^* & BA \\ 0 & DA \end{pmatrix} \right) = \Pi_a(M)_{\Gamma_0}$$

with the same A, B, and D as in (5.40).

We now prove that $(M)_{\Gamma_0}$ commutes with the double cosets $(M_{a,b}(B_0))_{\Gamma_0}$. According to (5.35), we have

$$(5.41) \qquad (M_{a,b}(B_0))_{\Gamma_0} = \sum_{D \in \Lambda \setminus \Lambda D_{a,b} \Lambda, B} \left(\Gamma_0 \begin{pmatrix} p^2 D^* & B \\ 0 & D \end{pmatrix} \right),$$

where B runs through the matrices in the set

$$(5.42) \qquad B_{\mathbf{Z}}(D) = \eta^* B_{\mathbf{Z}}(D_{a,b}) \varepsilon,$$

where $D = \eta D_{a,b} \varepsilon$, $\eta, \varepsilon \in \Lambda$, and

$$(5.43) \qquad B_{\mathbf{Z}}(D_{a,b}) = \left\{ B = \begin{pmatrix} 0 & 0 & 0 \\ 0 & B_{22} & p\,{}^t B_{32} \\ 0 & B_{32} & B_{33} \end{pmatrix} ; B_{22} \in \{B_0\}_p^n, \right.$$

$$\left. B_{32} \in M_{b,a}(\mathbf{Z})/\mathrm{mod}\, p, B_{33} \in S_b(\mathbf{Z})/\mathrm{mod}\, p^2 \right\}.$$

The definition (5.42) is correct, i.e., it does not depend on how $D \in \Lambda D_{a,b} \Lambda$ is represented in the form $\eta D_{a,b} \varepsilon$. To see this, it suffices to verify that, if $\eta D_{a,b} \varepsilon = D_{a,b}$ with $\eta, \varepsilon \in \Lambda$, then

$$(5.44) \qquad \eta^* B_{\mathbf{Z}}(D_{a,b}) \varepsilon = B_{\mathbf{Z}}(D_{a,b}).$$

Using (5.32), the relation $\eta^* = D_{a,b}^{-1} \cdot {}^t\varepsilon D_{a,b}$, and (5.43), we find the following congruences for the blocks $B_{ij}(\varepsilon)$ in the matrix $B(\varepsilon) = \eta^* B \varepsilon$:

$$(5.45) \qquad \begin{aligned} B_{22}(\varepsilon) &\equiv B_{22}[\varepsilon_{22}](\mathrm{mod}\, p), \\ B_{32}(\varepsilon) &\equiv B_1 + B_{32}^0(\varepsilon)(\mathrm{mod}\, p), \quad B_{33}(\varepsilon) \equiv B_2 + B_{33}^0(\varepsilon)(\mathrm{mod}\, p^2), \end{aligned}$$

where B_1 and $B_2 = {}^t B_2$ are integer matrices whose explicit form is not important now, and

$$B_{32}^0(\varepsilon) \equiv {}^t\varepsilon_{33} B_{32} \varepsilon_{22} + {}^t\varepsilon_{33} B_{33} \varepsilon_{32}(\mathrm{mod}\, p),$$

$$B_{33}^0(\varepsilon) \equiv {}^t\varepsilon_{33} B_{32} \varepsilon_{23} + {}^t\varepsilon_{23} \,{}^t B_{32} \varepsilon_{33} + B_{33}[\varepsilon_{33}](\mathrm{mod}\, p^2).$$

If $B_{32}^0 \equiv 0(\mathrm{mod}\, p)$ and $B_{33}^0 \equiv 0(\mathrm{mod}\, p^2)$, then (5.32) implies that $B_{32} \equiv 0(\mathrm{mod}\, p)$ and $B_{33} \equiv 0(\mathrm{mod}\, p^2)$. From this, (5.45), and (5.43) we conclude that $B \rightarrow B(\varepsilon)$ is a one-to-one map of the set $B_{\mathbf{Z}}(D_{a,b})$, and this proves (5.44).

Given the ring $K = \mathbf{Z}[p_1^{-1}]$ and a matrix $D = \eta D_{a,b}\varepsilon$, where $\eta, \varepsilon \in GL_n(K)$, in the double coset $GL_n(K)D_{a,b}GL_n(K)$ we define the following set:

$$(5.46) \qquad\qquad B_K(D) = \eta^* B_K(D_{a,b})\varepsilon,$$

where the definition of the set $B_K(D_{a,b})$ is similar to (5.43), except that B_{32} and B_{33} have entries in K/pK and K/p^2K, respectively, and the matrices B_{22} belong to the class $\{B_0\}_K^n$ determined by the equations (5.34) with $\mathbf{F}_p = \mathbf{Z}/p\mathbf{Z}$ replaced by K/pK. Since $(p_1, p) = 1$ by assumption, the proof that the definition (5.46) is correct is exactly the same as in the case when the ring is $\mathbf{Z}$. We obtain the following relation directly from the definition (5.46):

$$(5.47) \qquad\qquad B_K(\alpha D\beta) = \alpha^* B_K(D)\beta, \quad \text{if } \alpha, \beta \in GL_n(K).$$

But since the residue ring $K/p^m K$, $m \in \mathbf{N}$, is isomorphic to $\mathbf{Z}/p^m\mathbf{Z}$, it follows that every class in the set $B_K(D)$ contains an integer matrix; in this sense we may regard $B_K(D)$ as equal to $B_{\mathbf{Z}}(D)$ for any $D \in \Lambda D_{a,b}\Lambda$.

Next, if we use (5.41) and the properties of the sets $B_{\mathbf{Z}}(D)$ and $B_K(D)$, then we can prove that the double cosets $(M)_{\Gamma_0}$ and $(M_{a,b}(B_0))_{\Gamma_0}$ commute, in exactly the same way as we did for elements of Π_a.

To prove part (2) of the proposition, according to Proposition 5.5 and Lemma 5.2 it is sufficient to verify that the double cosets $(M_0)_{\Gamma_0}$ and $(N_0)_{\Gamma_0}$ commute, where $M_0 = U(r, A_0)$ is an integer matrix in $S_{0,p}^n$ satisfying (5.20), and $N_0 = U(t, D_0)$ is an integer matrix in S_{0,p_1}^n satisfying (5.21). By Lemma 5.6, we can take matrices of the form $M = U(r, A)$ and $N = U(t, D) \cdot T(S)$, where $A \in \Lambda \setminus \Lambda A_0\Lambda$, $D \in \Lambda \setminus \Lambda D_0\Lambda$, and $S \in S_n(\mathbf{Z})/t^{-1} \cdot {}^tDS_n(\mathbf{Z})D$, as representatives of the left cosets contained in these double cosets. Since $p \neq p_1$, the matrix S can obviously be chosen in the corresponding residue class in such a way that $S \equiv 0 \pmod{r}$. Then

$$MN = \begin{pmatrix} rt(AD)^* & rt(AD)^* S \\ 0 & AD \end{pmatrix} \in \Gamma_0 U(rt, AD)\Gamma_0 \subset \Gamma_0 N_0 M_0 \Gamma_0,$$

since $\Lambda AD\Lambda = \Lambda A_0\Lambda \cdot \Lambda D_0\Lambda = \Lambda D_0\Lambda \cdot \Lambda A_0\Lambda = \Lambda D_0 A_0\Lambda$, by Proposition 2.5 and Theorem 2.3. Similarly, we have

$$NM = \begin{pmatrix} rt(DA)^* & rt(DA)^* t^{-1} S[A] \\ 0 & DA \end{pmatrix} \in \Gamma_0 U(rt, AD)\Gamma_0 \subset \Gamma_0 N_0 M_0 \Gamma_0.$$

This implies that $(M_0)_{\Gamma_0}(N_0)_{\Gamma_0} = \alpha(N_0 M_0)_{\Gamma_0}$ and $(N_0)_{\Gamma_0}(M_0)_{\Gamma_0} = \beta(N_0 M_0)_{\Gamma_0}$ for certain constants α and β. A count of the left cosets on the left and right sides of these equalities shows that $\alpha = \beta$. $\qquad\qquad\square$

PROBLEM 5.13. Prove that $\hat{C}_{-p}^n$ (resp. C_{+p}^n) is the centralizer of $\Pi_-(p)$ (resp. $\Pi_+(p)$) in $\mathbf{L}_{0,p}^n$.

3. Expansion of $\mathbf{T}^n(m)$ for $n = 1, 2$. At the beginning of this section we mentioned that by passing from Hecke rings of Γ^n to Hecke rings of the subgroup Γ_0^n one can often decompose elements of the former rings into more elementary components. In §6 we shall consider these questions in more detail for the case of local Hecke rings. Here we shall remain in the global situation, and for $n = 1, 2$ we shall obtain expansions of the images $\mathbf{T}^n(m) \in \mathbf{L}_0^n(q)$ of the elements (3.19) under the map (5.3).

PROPOSITION 5.14. *Let $m, q \in \mathbf{N}$ with $(m, q) = 1$. Then*

$$(5.48) \qquad \mathbf{T}^1(m) = \sum_{d_i \in \mathbf{N}, d_1 d_2 = m} \Pi_+^1(d_1) \Pi_-^1(d_2),$$

$$(5.49) \qquad \mathbf{T}^2(m) = \sum_{d_i \in \mathbf{N}, d_1 d_2 d_3 = m} \Pi_+^2(d_1) \Pi(d_2) \Pi_-^2(d_3),$$

where $\Pi_\pm^n(d)$ are the elements (5.11), *and*

$$(5.50) \qquad \Pi(d) = \Pi_1^2(d) = (U(d, D(d)))_{\Gamma_0^2} \quad with\ D(d) = \begin{pmatrix} 1 & 0 \\ 0 & d \end{pmatrix}.$$

PROOF. From (3.22), Lemma 3.11, and the definitions we obtain:

$$\mathbf{T}^1(m) = \sum_{d_1 d_2 = m} \sum_{b \bmod d_1} \left(\Gamma_0 \begin{pmatrix} d_2 & b \\ 0 & d_1 \end{pmatrix} \right),$$

$$\mathbf{T}^2(m) = \sum_{d_1 d_2 d_3 = m} \sum_{D, B'} \left(\Gamma_0 \begin{pmatrix} d_3 d_2 D^* & B' \\ 0 & d_1 D \end{pmatrix} \right),$$

where $D \in \Lambda \setminus \Lambda D(d_2)\Lambda$, $B' \in B_0(D)/\bmod d_1 D$. On the other hand, by (5.12) we have

$$\sum_{d_1 d_2 = m} \Pi_+^1(d_1) \Pi_-^1(d_2) = \sum_{d_1 d_2 = m} \sum_{b \bmod d_1} \left(\Gamma_0 \begin{pmatrix} 1 & b \\ 0 & d_1 \end{pmatrix} \begin{pmatrix} d_2 & 0 \\ 0 & 1 \end{pmatrix} \right),$$

which, combined with the above expansion, proves (5.48). Next, we use (5.12) again, and we note that the following relation is easily verified using Lemma 3.33:

$$(5.51) \qquad \Pi(d) = \sum_{D, B} \left(\Gamma_0 \begin{pmatrix} dD^* & B \\ 0 & D \end{pmatrix} \right),$$

where $D \in \Lambda \setminus \Lambda D(d)\Lambda$ and $B \in B_0(D)/\bmod D$. As a result we find that the sum on the right in (5.49) is equal to

$$\sum_{d_1 d_2 d_3 = m} \sum_{D, B, S} \left(\Gamma_0 \begin{pmatrix} d_3 d_2 D^* & B + SD \\ 0 & d_1 D \end{pmatrix} \right),$$

where $D \in \Lambda \setminus \Lambda D(d_2)\Lambda$, $B \in B_0(D)/\bmod D$, $S \in S_2(\mathbf{Z})/\bmod d$. Comparing this expression with the above expansion for $\mathbf{T}^2(m)$, we see that (5.49) will be proved if we show that the matrix $B + SD$ runs through the set $B_0(D)/\bmod d_1 D$ if B runs through the set $B_0(D)/\bmod D$ and S runs through the set $S_2(\mathbf{Z})/\bmod d_1$. The details of this verification, which is easily carried out using Lemma 3.33, will be left to the reader. $\square$

PROBLEM 5.15. Prove the following formal identities:

$$\sum_{m \in \mathbf{N}_{(q)}} \mathbf{T}^1(m) m^{-s} = \left(\sum_{d \in \mathbf{N}_{(q)}} \Pi_+^1(d) d^{-s} \right) \left(\sum_{d \in \mathbf{N}_{(q)}} \Pi_-^1(d) d^{-s} \right)$$

$$= \prod_{p \in \mathbf{P}_{(q)}} (1 - \Pi_+^1(p) p^{-s})^{-1} \prod_{p \in \mathbf{P}_{(q)}} (1 - \Pi_-^1(p) p^{-s})^{-1}$$

$$= \prod_{p \in \mathbf{P}_{(q)}} \{(1 - \Pi_+^1(p) p^{-s})(1 - \Pi_-^1(p) p^{-s})\}^{-1}.$$

PROBLEM 5.16. (1) Show that

$$\Pi^2(d)\Pi^2(d_1) = \Pi^2(dd_1), \quad \text{if } (d, d_1) = 1.$$

(2) Show that the following formal identity holds for p a prime:

$$\sum_{\delta=0}^{\infty} \Pi^2(p^\delta)v^\delta = (1 - \Pi_+^2(p)v)\mathbf{Q}_p^2(v)^{-1}(1 - \Pi_-^2(p)v)(1 - p^2\Delta_2(p)v^2),$$

where $\mathbf{Q}_p^2(v)$ is the polynomial in Proposition 3.35.

[**Hint:** Use Proposition 3.35.]

§6. Hecke polynomials for the symplectic group

In this section we develop a technique that enables us to reduce several questions in the theory of Hecke rings and Hecke operators for the symplectic group to the analogous questions for the general linear group. Because of the multiplicativity of these theories, it suffices to limit ourselves to the case of local rings for a fixed prime p. The setting in which the action will be played out is the local Hecke ring $\mathbf{L}_{0,p}^n$ of the triangular subgroup Γ_0^n of the modular group. The rings $\mathbf{L}_p^n$ or the rings $\mathbf{L}_p^n(q)$ and $\widehat{\mathbf{E}}_p^n(q, \varkappa)$ will play the role of the local rings of Γ^n or of $\Gamma_0^n(q)$ and $\widehat{\Gamma}_0^n(q)$, respectively. The local Hecke ring of the group Λ^n will appear in one of the two dual extensions C_{-p}^n or C_{+p}^n.

In this section we suppose that $n \in \mathbf{N}$, p is a fixed prime number, and $(p, q) = 1$. When we consider Hecke rings for the symplectic covering group, we further suppose that q is divisible by 4. The indices n and p will often be omitted.

1. Negative powers of Frobenius elements. Because of their number theoretic associations, the elements

$$\Pi_- = \Pi_-^n(p) \quad \text{and} \quad \Pi_+ = \Pi_+^n(p)$$

are called the *Frobenius elements of the ring* $\mathbf{L}_0 = \mathbf{L}_{0,p}^n$. In §5 we defined the integral domains $C_\pm = C_{\pm p}^n$, which can also be characterized by the conditions

$$(6.1) \qquad C_- = \{X \in \mathbf{L}_0; \Pi_- X = X\Pi_-\}, \quad C_+ = \{X \in \mathbf{L}_0; \Pi_+ X = X\Pi_+\}.$$

In fact, the left sides are contained in the right sides by definition, and the reverse inclusions follow from the local variant of Proposition 5.5, the proof of which we leave to the reader. From (5.19) it follows that the rings C_- and C_+ are dual to one another relative to the anti-automorphism $*$:

$$(6.2) \qquad\qquad\qquad C_-^* = C_+, \quad C_+^* = C_-.$$

We now show that any element of $\mathbf{L}_0$ can be projected onto either C_- or C_+. Let

$$(6.3) \qquad\qquad X = \sum_i a_i \left(\Gamma_0 \begin{pmatrix} A_i & B_i \\ 0 & D_i \end{pmatrix} \right)$$

be a left coset decomposition of a nonzero element X of $\mathbf{L}_0$ without cancellation (i.e., the left cosets are pairwise distinct and all of the coefficients are nonzero). Then all of the elementary divisors of the matrices D_i and all of the denominators of entries in the matrices B_i are powers of p. Hence, there exist nonnegative $\delta \in \mathbf{Z}$ such that $p^\delta B_i D_i^{-1}$ are integer matrices in all of the terms of the decomposition. We call the smallest such

δ the *left exponent* of X, denoted $\delta_-(X)$. The number $\delta_+(X) = \delta_-(X^*)$ will be called the *right exponent* of X.

PROPOSITION 6.1. *Let $X \in \mathbf{L}_0$, and let d be an integer that is $\geqslant \delta_-(X)$ (resp. $\geqslant \delta_+(X)$). Then $\Pi_-^d X \in C_-$ (resp. $X\Pi_+^d \in C_+$).*

PROOF. From the definitions it follows that, under the conditions of the proposition, $\delta_-(\Pi_-^d X) = 0$ (resp. $\delta_+(X\Pi_+^d) = 0$). Hence, the proposition follows from the next lemma:

LEMMA 6.2. *One has:*

$$C_- = \{X \in \mathbf{L}_0; \delta_-(X) = 0\},$$
$$C_+ = \{X \in \mathbf{L}_0; \delta_+(X) = 0\}.$$

PROOF. By (6.2) and the definition of the left and right exponents it suffices to verify the first equality. Proposition 5.5 and the decomposition (5.22) imply that $\delta_-(X) = 0$ for any $X \in C_-$. Conversely, let X be an element of $\mathbf{L}_0$ written as in (6.3) with no cancellation. Suppose that $\delta_-(X) = 0$. Then for all i the matrix $V_i = B_i D_i^{-1}$ is a symmetric integer matrix. Hence,

$$X = \sum_i a_i \left(\Gamma_0 \begin{pmatrix} E & -V_i \\ 0 & E \end{pmatrix} \begin{pmatrix} A_i & B_i \\ 0 & D_i \end{pmatrix} \right) = \sum_i a_i \left(\Gamma_0 \begin{pmatrix} A_i & 0 \\ 0 & D_i \end{pmatrix} \right)$$

$$= \sum_i a_i \left(\Gamma_0 \begin{pmatrix} A_i & 0 \\ 0 & D_i \end{pmatrix} \begin{pmatrix} E & S \\ 0 & E \end{pmatrix} \right) = \sum_i a_i \left(\Gamma_0 \begin{pmatrix} A_i & A_i S \\ 0 & D_i \end{pmatrix} \right),$$

where S is an arbitrary matrix in $S_n(\mathbf{Z})$. If we again use the fact that $\delta_-(X) = 0$, we conclude that $A_i S D_i^{-1} = r_i D_i^* S D_i^{-1}$ are all integer matrices. Thus, X is a linear combination of double cosets of elements of the form $M_i = U(r_i, D_i)$ that satisfy (5.25), and hence also the condition $d_n(D_i)^2 | r_i$. Then $X \in C_-$ by Proposition 5.5. □

The next lemma gives an easy and practical method for finding exponents d that satisfy Proposition 6.1 for elements in the subrings $\mathbf{L} = \mathbf{L}_p^n$ and $\widehat{\mathbf{E}} = \widehat{\mathbf{E}}_p^n(q, \varkappa)$ of the ring $\overline{\mathbf{L}}_0$ in the case when the left coset decomposition is not known, but the image under $\Omega = \Omega_p^n$ is known.

LEMMA 6.3. *For any $X \in \overline{\mathbf{L}}_0$ and $t \in \mathbf{Z}$ one has*

$$\delta_-(\Delta' X) = \delta_-(X), \quad \delta_+(\Delta' X) = \delta_+(X),$$

where $\Delta = \Delta_n(p)$ is the element (3.48). If $X \in \mathbf{L} = \mathbf{L}_p^n$ or if $X \in \underline{\widehat{\mathbf{E}}} = \underline{\widehat{\mathbf{E}}}_p^n(q, \varkappa)$ (the integral subrings of $\mathbf{L}$ and $\widehat{\mathbf{E}}$), then

(6.4) $$\delta_-(X) = \delta_+(X) \leqslant \deg_{x_0} \Omega_p^n(X),$$

where the expression on the right is the degree in x_0 of the polynomial $\Omega(X)$.

PROOF. The first part follows immediately from the definitions. If $X \in \mathbf{L}$ or $X \in \widehat{\mathbf{E}}$, then $X^* = X$ by Proposition 5.3, and hence $\delta_+(X) = \delta_-(X^*) = \delta_-(X)$. By the definition of the map Ω and Lemma 4.19, we have

$$(6.5) \qquad \delta_-(X_{(\alpha)}) \leqslant \sum_{i=0}^{n} 2\alpha_i = \deg_{x_0} \Omega(X_{(\alpha)}),$$

where $X_{(\alpha)} = \prod_{i=0}^{n} \widehat{\mathbf{T}}_i(p^2)^{\alpha_i}$ with $\alpha_i \in \mathbf{Z}$ and $\alpha_i \geqslant 0$. According to (4.112) and (4.113), each $X \in \widehat{\mathbf{E}}$ can be written in the form $X = \sum_{(\alpha)} a_{(\alpha)} X_{(\alpha)}$, where all of the $a_{(\alpha)}$ are nonzero and the (α) are pairwise distinct. Thus, by Theorem 4.21(1), the polynomials $\Omega(X_{(\alpha)})$ are linearly independent over $\mathbf{Q}$. From this and (6.5) we obtain (6.4) for $\delta_-(X)$. The proof of the same inequality for $X \in \underline{\mathbf{L}}$ is similar; one uses Theorems 3.23 and 3.30 and Lemma 3.32. $\square$

Elements of the form $\Pi_-^d X \in \overline{C}_-$ and $X \Pi_+^d \in \overline{C}_+$ (where $\overline{C}_\pm = C_\pm \otimes_{\mathbf{Q}} \mathbf{C}$) will be called, respectively, the *left* and *right projections of* $X \in \overline{\mathbf{L}}_0$. Since the elements in $\overline{C}_\pm$ have simple left coset decompositions (see Lemma 5.6), computation in these subrings is much simpler than in the full ring $\overline{\mathbf{L}}_0$. Hence, it is natural to try to reduce actions on various elements of $\overline{\mathbf{L}}_0$ to actions on their projections. In this connection the problem arises of recovering a given $X \in \overline{\mathbf{L}}_0$ from its left or right projection. In general, this cannot be done, since Π_- and Π_+ are left and right divisors of zero, respectively, in $\overline{\mathbf{L}}_0$. However, it turns out that in several important cases—for example, when $X \in \mathbf{L}$ or $X \in \widehat{\mathbf{E}}$—$X$ can be uniquely recovered if one knows either of its projections. The possibility of recovering elements of $\mathbf{L}$ and $\widehat{\mathbf{E}}$ is based on the remarkable fact that Π_- and Π_+ are in some sense algebraic over the rings $\mathbf{L}$ and $\widehat{\mathbf{E}}$. More precisely, we have the following

PROPOSITION 6.4. *The following relations hold in the ring* $\overline{\mathbf{L}}_0 = \overline{\mathbf{L}}_{0,p}^n$:

$$(6.6) \qquad \sum_{i=0}^{m} (-1)^i \Pi_-^i \mathbf{q}_{m-i} = 0, \qquad \sum_{i=0}^{m} (-1)^i \mathbf{q}_{m-i} \Pi_+^i = 0,$$

$$(6.7) \qquad \sum_{i=0}^{m} (-1)^i \Pi_-^{2i} \widehat{\mathbf{q}}_{m-i} = 0, \qquad \sum_{i=0}^{m} (-1)^i \widehat{\mathbf{q}}_{m-i} \Pi_+^{2i} = 0,$$

where $m = 2^n$, $\mathbf{q}_j = \mathbf{q}_j^n(p)$ *are the elements* (3.77) *of the ring* $\mathbf{L} = \mathbf{L}_p^n$, *and* $\widehat{\mathbf{q}}_j = \widehat{\mathbf{q}}_j^n(p)$ *are elements of* $\widehat{\mathbf{E}} = \widehat{\mathbf{E}}_p^n(q, \varkappa)$ *such that*

$$(6.8) \qquad \Omega_p^n(\widehat{\mathbf{q}}_j^n(p)) = q_j^n(x_0^2, \ldots, x_n^2),$$

where $q_j^n(x_0, \ldots, x_n)$ *are the coefficients of the polynomial* (3.76).

PROOF. From (5.9) and (5.16) it follows that the anti-automorphism $*$ transforms the first equalities in (6.6) and (6.7) to the second ones, and conversely. Hence, it suffices to prove the first equalities. Let Y be the left side of (6.6), and let $\widehat{Y}$ be the left side of (6.7). Using (3.79) and the analogous relations

$$\widehat{\mathbf{q}}_{m-i} = (p^{\langle n \rangle} \Delta)^{m-2i} \widehat{\mathbf{q}}_i \qquad (0 \leqslant i \leqslant m)$$

for the elements $\widehat{\mathbf{q}}_j \in \widehat{\mathbf{E}}$ whose existence and uniqueness are guaranteed by Theorem 4.21, we can rewrite Y and $\widehat{Y}$ in the form

$$Y = \sum_{i=0}^{m} (-1)^i (p^{\langle n \rangle} \Delta)^{m/2-i} \Pi_-^i \mathbf{q}_i, \quad \widehat{Y} = \sum_{i=0}^{m} (-1)^i (p^{\langle n \rangle} \Delta)^{m-2i} \Pi_-^{2i} \widehat{\mathbf{q}}_i.$$

By definition, $\Omega(\mathbf{q}_i)$ and $\Omega(\widehat{\mathbf{q}}_i)$ are polynomials in $x_0, x_1, \ldots, x_n$, and these polynomials have degree i and $2i$, respectively, in the variable x_0. Thus, from Theorems 3.30(1) and 4.21(1) and Lemma 6.3 we conclude that $\mathbf{q}_i \in \mathbf{L}$, $\delta_-(\mathbf{q}_i) \leqslant i$, and $\widehat{\mathbf{q}}_i \in \widehat{\mathbf{E}}$, $\delta_-(\widehat{\mathbf{q}}_i) \leqslant 2i$. Then, by Proposition 6.1, each of the products $\Pi_-^i \mathbf{q}_i$ and $\Pi_-^{2i} \widehat{\mathbf{q}}_i$ is contained in $\overline{C}_-$. Since obviously $\Delta \in C_-$, this means that Y and $\widehat{Y}$ also lie in $\overline{C}_-$. On the other hand, by Lemma 3.34 we have $\Omega(\Pi_-) = \Omega(\Pi_0^n(p)) = x_0$, so that, if we use (3.76) and the definition of $\mathbf{q}_i$ and $\widehat{\mathbf{q}}_i$, we obtain

$$\Omega(Y) = \sum_{i=0}^{m} (-1)^i x_0^i q_i(x_0, \ldots, x_n) = x_0^m q(x_0, \ldots, x_n; x_0^{-1}) = 0 \ .$$

and similarly $\Omega(\widehat{Y}) = 0$. Hence, by Theorem 5.9, we have $Y = \widehat{Y} = 0$. $\qquad\square$

If we multiply (6.6) and (6.7) by $\Pi_\pm^d$ and $(\Pi_\pm^2)^d$ with $d \in \mathbf{N}$, we obtain the relations

(6.9)
$$\sum_{i=0}^{m} (-1)^i \Pi_-^{d+i} \mathbf{q}_{m-i} = 0, \quad \sum_{i=0}^{m} (-1)^i \mathbf{q}_{m-i} \Pi_+^{d+i} = 0,$$

$$\sum_{i=0}^{m} (-1)^i (\Pi_-^2)^{d+i} \widehat{\mathbf{q}}_{m-i} = 0, \quad \sum_{i=0}^{m} (-1)^i \widehat{\mathbf{q}}_{m-i} (\Pi_+^2)^{d+i} = 0,$$

which may be regarded as recursive relations for the sequences of nonnegative powers of $\Pi_\pm$ and $\Pi_\pm^2$. Since $\mathbf{q}_0 = \widehat{\mathbf{q}}_0 = 1$, the relations (6.9) give high powers of these two elements as linear combinations of smaller powers with (right or left) coefficients in the rings $\mathbf{L}$ and $\widehat{\mathbf{E}}$. On the other hand, by (3.80), (6.8), and Lemma 3.27, the coefficients $\mathbf{q}_m$ and $\widehat{\mathbf{q}}_m$ are invertible in $\overline{\mathbf{L}}_0$. Hence, the relations (6.9) can also be used to determine small powers of the Frobenius elements in terms of higher powers. Namely, if $\Pi_\pm^\delta$ and $(\Pi_\pm^2)^\delta$ for $\delta > d$ have already been determined, then we set

(6.10)
$$\Pi_-^d = \mathbf{q}_m^{-1} \left(\sum_{i=1}^{m} (-1)^i \Pi_-^{i+d} \mathbf{q}_{m-i} \right),$$

$$\Pi_+^d = \mathbf{q}_m^{-1} \left(\sum_{i=1}^{m} (-1)^i \mathbf{q}_{m-i} \Pi_+^{i+d} \right),$$

(6.11)
$$(\Pi_-^2)^d = \widehat{\mathbf{q}}_m^{-1} \left(\sum_{i=1}^{m} (-1)^i (\Pi_-^2)^{i+d} \widehat{\mathbf{q}}_{m-i} \right),$$

$$(\Pi_+^2)^d = \widehat{\mathbf{q}}_m^{-1} \left(\sum_{i=1}^{m} (-1)^i \widehat{\mathbf{q}}_{m-i} (\Pi_+^2)^{i+d} \right).$$

The elements $\Pi_\pm^d$ and $(\Pi_\pm^2)^d$ that are obtained in this way for $d < 0$ are not the negative powers of $\Pi_\pm$ and $\Pi_\pm^2$, since these elements are not invertible in $\mathbf{L}_0$. They are not even the powers of $\Pi_\pm^{-1}$ and $(\Pi_\pm^2)^{-1}$, since, for example, $\Pi_-^{-2} \neq (\Pi_-^{-1})^2$ and $\Pi_+^{-2} \neq (\Pi_+^{-1})^2$. Nevertheless, for brevity we shall sometimes speak of *negative powers of the Frobenius elements*. Note that, if we use (5.9), (5.16), and induction on d, then from (6.10) and (6.11) we find that the negative powers of the Frobenius elements, together with the positive powers, are dual with respect to the anti-automorphism $*$:

$$(6.12) \qquad \begin{array}{l} (\Pi_-^d)^* = \Pi_+^d, \quad (\Pi_+^d)^* = \Pi_-^d \quad \text{for all } d \in \mathbf{Z}, \\[4pt] ((\Pi_-^2)^d)^* = (\Pi_+^2)^d, \quad ((\Pi_+^2)^d)^* = (\Pi_-^2)^d \quad \text{for all } d \in \mathbf{Z}. \end{array}$$

We consider the following subspaces of $\overline{\mathbf{L}}_0$:

$$(6.13) \qquad \begin{array}{l} O_- = O_{-p}^n = \overline{C}_{-p}^n \cdot \mathbf{L}_p^n = \left\{ \sum_\alpha X_\alpha T_\alpha; X_\alpha \in \overline{C}_{-p}^n, T_\alpha \in \mathbf{L}_p^n \right\}, \\[10pt] O_+ = O_{+p}^n = \mathbf{L}_p^n \cdot \overline{C}_{+p}^n = \left\{ \sum_\alpha T_\alpha Y_\alpha; T_\alpha \in \mathbf{L}_p^n, Y_\alpha \in \overline{C}_{+p}^n \right\}, \end{array}$$

and the similarly defined subspaces

$$(6.14) \qquad \widehat{O}_- = \widehat{O}_{-p}^n = \overline{C}_{-p}^n \cdot \widehat{\mathbf{E}}_p^n(q,\varkappa), \quad \widehat{O}_+ = \widehat{O}_{+p}^n = \widehat{\mathbf{E}}_p^n(q,\varkappa) \cdot \overline{C}_{+p}^n.$$

According to (5.9) and (6.2), these spaces are dual with respect to $*$:

$$(6.15) \qquad O_-^* = O_+, \quad O_+^* = O_- \qquad \text{and} \qquad \widehat{O}_-^* = \widehat{O}_+, \quad \widehat{O}_+^* = \widehat{O}_-.$$

From the definition of $\Pi_\pm^d$ and $(\Pi_\pm^2)^d$ one can show by induction on d that

$$(6.16) \qquad \Pi_\pm^d \in O_\pm \quad \text{and} \quad (\Pi_\pm^2)^d \in \widehat{O}_\pm$$

for all $d \in \mathbf{Z}$.

THEOREM 6.5. *For* $\delta \in \mathbf{N}$ *let the elements* $\Pi_\pm^{-\delta} \in O_{\pm p}^n$ *and* $(\Pi_\pm^2)^{-\delta} \in \widehat{O}_{\pm p}^n$ *be defined by the recursive relations* (6.10) *and* (6.11), *respectively. Then:*

(1) *Every element in* O^n *(respectively, every element in* $\widehat{O}_{\pm p}^n$) *satisfies the relations*

$$(6.17) \qquad X = \Pi_-^\delta X \Pi_-^{-\delta} \quad \text{for all } X \in O_{-p}^n \text{ and } \delta \geqslant \delta_-(X),$$

$$(6.18) \qquad X = \Pi_+^{-\delta} X \Pi_+^\delta \quad \text{for all } X \in O_{+p}^n \text{ and } \delta \geqslant \delta_+(X)$$

(*respectively,*

$$(6.19) \qquad \widehat{X} = (\Pi_-^2)^\delta \widehat{X} (\Pi_-^2)^{-\delta} \quad \text{for all } \widehat{X} \in \widehat{O}_{-p}^n \text{ and } 2\delta \geqslant \delta_-(\widehat{X}),$$

$$(6.20) \qquad \widehat{X} = (\Pi_+^2)^{-\delta} \widehat{X} (\Pi_+^2)^\delta \quad \text{for all } \widehat{X} \in \widehat{O}_{+p}^n \text{ and } 2\delta \geqslant \delta_+(\widehat{X})).$$

Conversely, every X *or* $\widehat{X} \in \overline{\mathbf{L}}_0$ *that satisfies any of the relations* (6.17), (6.18) *or* (6.19), (6.20), *is contained in the corresponding space* O_{-p}^n, O_{+p}^n *or* $\widehat{O}_{-p}^n$, $\widehat{O}_{+p}^n$.

(2) *The restrictions of the maps* Φ_p^n *and* Ω_p^n *(see §§3.3 and 4.3) to the spaces* $O_{\pm p}^n$ *and* $\widehat{O}_{\pm p}^n$ *are all monomorphisms.*

We first prove a lemma.

LEMMA 6.6. *For any* $\mathbf{T} \in \mathbf{L} = \mathbf{L}_p^n$ *(respectively, for any* $\widehat{\mathbf{T}} \in \widehat{\mathbf{E}} = \widehat{\mathbf{E}}_p^n(\mathbf{q}, \varkappa)$*) one has the relations*

$$(6.21) \qquad \Pi_-^\delta \mathbf{T} \Pi_-^d = \Pi_-^{\delta+d} \mathbf{T} \quad \text{for all } \delta \geqslant \delta_-(\mathbf{T}) \text{ and } d \in \mathbf{Z},$$

$$(6.22) \qquad \Pi_+^\delta \mathbf{T} \Pi_+^d = \Pi_+^{d+\delta} \mathbf{T} \quad \text{for all } \delta \geqslant \delta_+(\mathbf{T}) \text{ and } d \in \mathbf{Z}$$

(respectively,

$$(6.23) \qquad (\Pi_-^2)^\delta \widehat{T} (\Pi_-^2)^d = (\Pi_-^2)^{\delta+d} \widehat{\mathbf{T}} \quad \text{for all } 2\delta \geqslant \delta_-(\widehat{\mathbf{T}}) \text{ and } d \in \mathbf{Z},$$

$$(6.24) \qquad (\Pi_+^2)^\delta \widehat{T} (\Pi_+^2)^d = (\Pi_+^2)^{d+\delta} \widehat{\mathbf{T}} \quad \text{for all } 2\delta \geqslant \delta_+(\widehat{\mathbf{T}}) \text{ and } d \in \mathbf{Z}).$$

PROOF. The anti-automorphism $*$ takes (6.21) to (6.22) and (6.23) to (6.24). Hence, it suffices to prove, say, the former relations. We use descending induction on d. If $d \geqslant 0$, then Π_-^d and $(\Pi_-^2)^d \in \overline{C}_-$, and the relations (6.21) and (6.23) follow from Proposition 6.1 and the commutativity of the ring C_-. Suppose that (6.21) and (6.23) have been proved for all $\delta \geqslant \delta_-(\mathbf{T})$ and $2\delta \geqslant \delta_-(\widehat{\mathbf{T}})$ and all $d' > d$. Then, using (6.10)–(6.11) and the commutativity of $\mathbf{L}$ and $\widehat{\mathbf{E}}$, we obtain

$$\Pi_-^\delta \mathbf{T} \Pi_-^\delta = -\mathbf{q}_m^{-1} \left(\sum_{i=1}^m (-1)^i \Pi_-^\delta \mathbf{T} \Pi_-^{i+d} \mathbf{q}_{m-i} \right)$$

$$= \mathbf{q}_m^{-1} \left(\sum_{i=1}^m (-1)^i \Pi_-^{\delta+i+d} \mathbf{q}_{m-i} \mathbf{T} \right) = \Pi_-^{\delta+d} \mathbf{T}$$

and in exactly the same way $(\Pi_-^2)^\delta \widehat{\mathbf{T}} (\Pi_-^2)^d = (\Pi_-^2)^{\delta+d} \widehat{\mathbf{T}}$.   □

PROOF OF THE THEOREM. By the duality relations (6.15) and (6.12), it suffices to prove the first part of the theorem for, say, O_- and $\widehat{O}_-$. Let $X \in O_-$, and let $\widehat{X} \in \widehat{O}_-$. Then, by definition,

$$X = \sum_\alpha Y_\alpha \mathbf{T} \quad \text{and} \quad \widehat{T} = \sum_\alpha \widehat{Y}_\alpha \widehat{\mathbf{T}}_\alpha,$$

where $Y_\alpha \in C_-$, $\mathbf{T}_\alpha \in \mathbf{L}$ and $\widehat{Y}_\alpha \in \overline{C}_-$, $\widehat{\mathbf{T}}_\alpha \in \widehat{\mathbf{E}}$. If δ' is no smaller than any of the exponents $\delta_-(\mathbf{T}_\alpha)$, or $2\delta'$ is no smaller than any of the exponents $\delta_-(\widehat{\mathbf{T}}_\alpha)$, then, using (6.21) or (6.23), respectively, we obtain

$$\Pi_-^{\delta'} X \Pi_-^{-\delta'} = \sum_\alpha Y_\alpha \Pi_-^{\delta'} \mathbf{T}_\alpha \Pi_-^{-\delta'} = \sum_\alpha Y_\alpha \mathbf{T}_\alpha = X,$$

$$(\Pi_-^2)^{\delta'} \widehat{X} (\Pi_-^2)^{-\delta'} = \sum_\alpha \widehat{Y}_\alpha (\Pi_-^2)^{\delta'} \widehat{\mathbf{T}}_\alpha (\Pi_-^2)^{-\delta'} = \sum_\alpha \widehat{Y}_\alpha \widehat{\mathbf{T}}_\alpha = \widehat{X}.$$

Now suppose that $\delta \geqslant \delta_-(X)$. Then by what was already proved and by Proposition 6.1, we have

$$X = \Pi_-^{\delta+\delta'} X \Pi_-^{-(\delta+\delta')} = \Pi_-^\delta X \Pi_-^{\delta'} \Pi_-^{-(\delta+\delta')} = \Pi_-^\delta X \Pi_-^{-\delta},$$

where we used (6.21) with $\mathbf{T} = 1$ in the last step. Similarly, for $2\delta \geqslant \delta_-(\widehat{X})$ we use (6.23) to obtain

$$\widehat{X} = (\Pi_-^2)^{\delta+\delta'} \widehat{X} (\Pi_-^2)^{-(\delta+\delta')} = (\Pi_-^2)^\delta \widehat{X} (\Pi_-^2)^{-\delta}.$$

Conversely, if we have elements X and $\widehat{X} \in \overline{\mathbf{L}}_0$ written in the form $X = \Pi_-^\delta X \Pi_-^{-\delta}$ for some $\delta \geqslant \delta_-(X)$ and $\widehat{X} = (\Pi_-^2)^\delta \widehat{X}(\Pi_-^2)^{-\delta}$ for some $2\delta \geqslant \delta_-(\widehat{X})$, then from Proposition 6.1 and the inclusions (6.16) it follows that $X \in O_-$ and $\widehat{X} \in \widehat{O}_-$. The first part of the theorem is proved.

We shall prove the second part, for example, for $\Omega = \Omega_p^n$. First note that we have the formulas

$$
\text{(6.25)} \qquad
\begin{aligned}
\Omega(\Pi_-^d) &= x_0^d, \quad \Omega(\Pi_+^d) = (x_0 x_1 \cdots x_n)^d \quad (d \in \mathbf{Z}), \\
\Omega((\Pi_-^2)^d) &= x_0^{2d}, \quad \Omega((\Pi_+^2)^d) = (x_0 x_1 \cdots x_n)^{2d} \quad (d \in \mathbf{Z}),
\end{aligned}
$$

where $\Pi_\pm^d$ and $(\Pi_\pm^2)^d$ for $d < 0$ are defined by the recursive relations (6.10) and (6.11), respectively. Namely, these formulas were proved for nonnegative d in Lemma 3.34, while for $d = -\delta < 0$ we have by Lemma 6.6:

$$
\Pi_-^\delta \Pi_-^{-\delta} = \Pi_+^{-\delta} \Pi_+^\delta = 1, \quad (\Pi_-^2)^\delta (\Pi_-^2)^{-\delta} = (\Pi_+^2)^{-\delta} (\Pi_+^2)^\delta = 1,
$$

so that $\Omega(\Pi_-^d) = \Omega(\Pi_-^\delta)^{-1} = x_0^d$, and similarly for Π_+^d and $(\Pi_\pm^2)^d$. Now suppose that $X \in O_-$ and $\Omega(X) = 0$. We take $\delta \geqslant \delta_-(X)$. Then by (6.17) we have

$$
0 = \Omega(X) = \Omega(\Pi_-^\delta X \Pi_-^{-\delta}) = \Omega(\Pi_-^\delta X) x_0^{-\delta},
$$

so that $\Omega(\Pi_-^\delta X) = 0$. By Theorem 5.9, the last equality implies that $\Pi_-^\delta X = 0$, and then also $X = (\Pi_-^\delta X)\Pi_-^{-\delta} = 0$. The cases of O_+ and $\widehat{O}_\pm$ are similar. $\qquad \square$

PROBLEM 6.7. (1) Show that the map

$$
X \to (p^{\langle n \rangle} \Delta)^{-d} \Pi_-^d X \Pi_+^d,
$$

where $d \geqslant \min(\delta_-(X), \delta_+(X))$ and $\Delta = \Delta_n(p)$, does not depend on the choice of d, commutes with the anti-automorphism $*$, and maps the entire space $\mathbf{L}_0 = \mathbf{L}_{0,p}^n$ onto the subspace

$$
C_- \cdot C_+ = \left\{ \sum_\alpha X_\alpha Y_\alpha; X_\alpha \in C_-, Y_\alpha \in C_+ \right\}.
$$

Show that this subspace is the set of all elements in $\mathbf{L}_0$ that are invariant relative to the above map. Then deduce that the restrictions to $C_- \cdot C_+$ of Φ and Ω are monomorphisms.

(2) Show that

$$
\mathbf{T}(p), \Pi_- \mathbf{T}_i(p^2), \mathbf{T}_i(p^2)\Pi_+ \in C_- \cdot C_+,
$$

where $\mathbf{T}(p)$, $\mathbf{T}_i(p^2)$ are the images in $\mathbf{L}_0$ of the elements (3.42), $0 \leqslant i \leqslant n$. Then deduce that, if $\mathbf{T} \in \mathbf{L}$ and the image $\Omega(\mathbf{T})$ is a polynomial in $x_0, x_1, \ldots, x_n$ having degree δ in x_0, then $\Pi_-^a \mathbf{T} \Pi_+^b \in C_- \cdot C_+$ for any $a, b \geqslant 0, a + b \geqslant \delta - 1$.

[**Hint:** Use the first assertion and (3.58), (3.61). For the second assertion use the fact that $\mathbf{T}$ is a polynomial in $\mathbf{T}(p)$, $\mathbf{T}_i(p^2)$.]

(3) Show that $\Pi_\pm^{-1} \in C_- \cdot C_+$, and then deduce the relations $\Pi_-^{-1} = (p^{\langle n \rangle} \Delta)^{-1} \Pi_+$, $\Pi_+^{-1} = (p^{\langle n \rangle} \Delta)^{-1} \Pi_-$.

[**Hint:** Use the first two parts of the problem and the definition of negative powers.]

2. Factorization of Hecke polynomials. By "Hecke polynomials" we mean polynomials over the local Hecke rings of the symplectic group or the symplectic covering group. These polynomials arise naturally when summing various generating series (formal local zeta-functions) over these rings, and appear as denominators of the resulting fractions (see, for example, Proposition 3.35). Although usually irreducible over the original local ring, these polynomials often factor when one extends that local ring to a local Hecke ring of the triangular subgroup of the symplectic group. These factorizations enable one to express both the coefficients of the polynomials themselves, and also the coefficients of the corresponding generating series, in terms of a simpler type of element. This turns out to be essential both for computations in the local rings and for the study of their representations on modular forms. Here we examine the simplest scheme for factoring Hecke polynomials, and give some important examples.

THEOREM 6.8. *Let*

$$P(v) = \sum_{i=0}^{N} \mathbf{p}_i v^i$$

be a polynomial with coefficients in the subring $\mathbf{L} = \mathbf{L}_p^n$ *(respectively the subring* $\widehat{\mathbf{E}} = \widehat{\mathbf{E}}_p^n(q, \varkappa)$*) of the ring* $\overline{\mathbf{L}}_0 = \overline{\mathbf{L}}_{0,p}^n$. *Suppose that the polynomial*

$$\Omega(P)(v) = \sum_{i=0}^{N} \Omega(\mathbf{p}_i) v^i,$$

where $\Omega = \Omega_p^n$ *is the spherical map* (3.49), *factors into a product of two polynomials with coefficients in* $\mathbf{C}[x_0^{\pm 1}, \ldots, x_n^{\pm 1}]$:

$$\Omega(P)(v) = F(v)G(v),$$

where

$$F(v) = \sum_{i=0}^{N_1} f_i v^i, \quad G(v) = \sum_{j=0}^{N_2} g_j v^j,$$

$$f_j, g_j \in \mathbf{C}[x_0^{\pm 1}, \ldots, x_n^{\pm 1}].$$

(1) *If all of the coefficients of the first polynomial belong to the image* $\Omega(\overline{C}_-)$ *of the ring* $\overline{C}_- = \overline{C}_{-p}^n$:

$$f_i = \Omega(f_i'), \quad f_i' \in \overline{C}_-, \quad \text{and} \quad f_0 = 1,$$

then all of the coefficients of the second polynomial belong to the image $\Omega(O_-)$ *of* $O_- = O_{-p}^n$ *(respectively the image* $\Omega(\widehat{O}_-)$ *of* $\widehat{O}_- = \widehat{O}_{-p}^n$*):*

$$g_j = \Omega(g_j'), \quad g_j' \in O_- \quad \text{(respectively } g_j' \in \widehat{O}_-),$$

and over $\overline{\mathbf{L}}_0$ *one has the factorization*

(6.26) $$P(v) = \left(\sum_{i=0}^{N_1} f_i' v^i \right) \left(\sum_{j=0}^{N_2} g_j' v^j \right).$$

(2) *If, on the other hand, all of the coefficients of the second polynomial belong to the image* $\Omega(C_+)$ *of* $C_+ = C_{+p}^n$:

$$g_j = \Omega(g_j'), \quad g_j' \in C_+, \quad \text{and} \quad g_0 = 1,$$

then all of the coefficients of the first polynomial belong to the image $\Omega(O_+)$ *of* $O_+ = O_{+p}^n$ *(respectively the image* $\Omega(\widehat{O}_+)$ *of* $\widehat{O}_+ = \widehat{O}_{+p}^n$*):*

$$f_j = \Omega(f_j'), \quad f_j' \in O_+ \quad (\text{respectively } f_j' \in \widehat{O}_+),$$

and one again has the factorization (6.26).

PROOF. The two cases are dual to one another with respect to the anti-automorphism $*$, and so the proofs are analogous. We shall treat, say, the first case. Since the restriction of Ω to C_- is a monomorphism (by Theorem 5.9) and $f_0 = 1$, it follows that $f_0' = 1$. Then the polynomial $\sum_i f_i' v^i$ is invertible in the ring of formal power series over the commutative ring $\overline{C}_-$, i.e., there exist $a_t' \in \overline{C}_-$ such that

$$\left(\sum_{t=0}^{\infty} a_t' v^t \right) \left(\sum_{i=0}^{N_1} f_i' v^i \right) = 1.$$

We consider the formal power series over $\overline{\mathbf{L}}_0$

$$\left(\sum_{t=0}^{\infty} a_t' v^t \right) P(v) = \sum_{j=0}^{\infty} \left(\sum_{t+i=j} a_t' \mathbf{p}_i \right) v^j = \sum_{j=0}^{\infty} g_j' v^j.$$

By construction, the coefficients g_j' of this series lie in the space $O_- = \overline{C}_- \cdot \mathbf{L}$ (resp. in $\widehat{O}_- = \overline{C}_- \cdot \widehat{\mathbf{E}}$). On the other hand, if we replace these coefficients by their images under Ω, we obtain the series

$$\left(\sum_{t=0}^{\infty} \Omega(a_t') v^t \right) \left(\sum_{i=0}^{N_1} f_i v^i \right) \left(\sum_{j=0}^{N_2} g_j v^j \right) = \sum_{j=0}^{N_2} g_j v^j,$$

and hence $\Omega(g_j') = g_j$ for $j = 0, 1, \ldots, N_2$, and $\Omega(g_j') = 0$ for $j > N_2$. By Theorem 6.5, the last equality implies that $g_j' = 0$ for $j > N_2$. Thus, we obtain the factorization (6.26) for $P(v)$, since the elements $g_j' \in O_-$ (resp. $g_j' \in \widehat{O}_-$) are uniquely determined by their images under Ω. $\square$

REMARK. In practice, when looking for a preimage in O_- of some element $g \in \Omega(O_-)$, one can take an arbitrary image g_1 of this element in $\mathbf{L}_0$ and then set

$$g' = \Pi_-^d g_1 \Pi_-^{-d}, \quad \text{where } d \geqslant \delta_-(g_1).$$

Then by (6.16) and Proposition 6.1 one has $g' \in O_-$; and by (6.14) one has $\Omega(g') = x_0^d \Omega(g_1) x_0^{-d} = \Omega(g_1)$. One can proceed similarly to find preimages in O_+ and $\widehat{O}_\pm$. It is usually not hard to find preimages in $\mathbf{L}_0$ or $\overline{\mathbf{L}}_0$. Thus, the problem of finding preimages in $O_\pm$ and $\widehat{O}_\pm$ reduces to the computation of suitable negative powers of the Frobenius elements.

We now consider some examples of applications of Theorem 6.8. In §3.3 we defined the polynomial $r(x_1, \ldots, x_n; v)$, all of whose coefficients are invariant relative to any transformation in the group W_n^2. By Theorems 3.30 and 4.21, there exist uniquely determined polynomials

$$(6.27) \qquad \mathbf{R}_p^n(v) = \sum_{a=0}^{2n} (-1)^a \mathbf{r}_a^n(p) v^a, \quad \widehat{\mathbf{R}}_p^n(v) = \sum_{a=0}^{2n} (-1)^a \widehat{\mathbf{r}}_a^n(p) v^a,$$

with $\mathbf{r}_a^n(p) \in \mathbf{L}_p^n$ and $\widehat{\mathbf{r}}_a^n(p) \in \widehat{\mathbf{E}}_p^n(q, \varkappa)$ such that

(6.28)

$$\Omega(\mathbf{R}_p^n)(v) = \sum_{a=0}^{2n} (-1)^a \Omega(\mathbf{r}_a^n(p)) v^a = r(x_1, \ldots, x_n; v),$$

$$\Omega(\widehat{\mathbf{R}}_p^n)(v) = \sum_{a=0}^{2n} (-1)^a \Omega(\widehat{\mathbf{r}}_a^n(p)) v^a = r(x_1, \ldots, x_n; v).$$

PROPOSITION 6.9. *The polynomials* $\mathbf{R}_p^n(v)$ *and* $\widehat{\mathbf{R}}_p^n(v)$ *factor as follows over the ring* $\overline{\mathbf{L}}_{0,p}^n$:

(6.29)

$$\mathbf{R}_p^n(v) = \left(\sum_{i=0}^{n} (-1)^i (p^{\langle n \rangle} \Delta)^{-1} \Pi_- \Pi_{n-i} v^i \right) \left(\sum_{i=0}^{n} (-1)^i \Pi_- \Pi_i \Pi_-^{-2} v^i \right),$$

(6.30)

$$\mathbf{R}_p^n(v) = \left(\sum_{i=0}^{n} (-1)^i \Pi_+^{-2} \Pi_{n-i} \Pi_+ v^i \right) \left(\sum_{i=0}^{n} (-1)^i (p^{\langle n \rangle} \Delta)^{-1} \Pi_i \Pi_+ v^i \right)$$

and

(6.31)

$$\widehat{\mathbf{R}}_p^n(v) = \left(\sum_{i=0}^{n} (-1)^i (\mathbf{p}^{\langle n \rangle} \Delta)^{-1} \Pi_- \Pi_{n-i} v^i \right) \left(\sum_{i=0}^{n} (-1)^i \Pi_- \Pi_i (\Pi_-^2)^{-1} v^i \right),$$

(6.32)

$$\mathbf{R}_p^n(v) = \left(\sum_{i=0}^{n} (-1)^i (\Pi_+^2)^{-1} \Pi_{n-i} \Pi_+ v^i \right) \left(\sum_{i=0}^{n} (-1)^i (p^{\langle n \rangle} \Delta)^{-1} \Pi_i \Pi_+ v^i \right),$$

where $\Delta = \Delta_n(p)$ *is the element* (3.48); $\Pi_- = \Pi_-^n(p) = \Pi_0^n(p)$, $\Pi_+ = \Pi_+^n(p) = \Pi_n^n(p)$, $\Pi_a = \Pi_a^n(p)$ *are the elements* (3.59); *and* $\Pi_\pm^{-2}$ *and* $(\Pi_\pm^2)^{-1}$ *are determined from the recursive relations* (6.10) *and* (6.11), *respectively.*

PROOF. We first show that

(6.33) $$\Pi_a^n(p)^* = \Pi_{n-a}^n(p) \quad \text{for } a = 0, 1, \ldots, n.$$

In fact, if D_a is the matrix (2.28), then obviously $\Lambda^n p D_a^{-1} \Lambda^n = \Lambda^n D_{n-a} \Lambda^n$, and hence from the definition of the anti-automorphism $*$ and the relations (3.63) we obtain

$$\Pi_a^n(p)^* = (pM_a^{-1})_{\Gamma_0} = \left(\begin{pmatrix} pD_{n-a}^{-1} & 0 \\ 0 & D_{n-a} \end{pmatrix} \right)_{\Gamma_0} = \Pi_{n-a}^n(p).$$

From these equalities and (6.12) it follows that the anti-automorphism $*$ takes the factorization (6.29) to (6.30), takes (6.31) to (6.32), and conversely. Hence, it suffices to prove, say, (6.29) and (6.31). By definition, we have

$$\Omega(\mathbf{R}_p^n)(v) = \prod_{i=1}^{n} (1 - x_i^{-1} v) \prod_{i=1}^{n} (1 - x_i v)$$

$$= \left(\sum_{i=0}^{n} (-1)^i s_i(x_1^{-1}, \ldots, x_n^{-1}) v^i \right) \left(\sum_{i=0}^{n} (-1)^i s_i(x_1, \ldots, x_n) v^i \right).$$

From Lemma 3.34 it follows that

$$\Omega((p^{\langle n \rangle}\Delta)^{-1}\Pi_-\Pi_{n-i}) = s_i(x_1^{-1}, \ldots, x_n^{-1}).$$

It follows from the definitions that $\delta_-(\Pi_i) = 1$. From Proposition 6.1 we then obtain the inclusions

(6.34) $\Pi_-\Pi_i, (p^{\langle n \rangle}\Delta)^{-1}\Pi_-\Pi_{n-i} \in C_-.$

Next, from Lemma 3.34 and (6.25) we find that

(6.35) $\Omega(\Pi_-\Pi_i\Pi_-^{-2}) = \Omega(\Pi_-\Pi_i(\Pi_-^2)^{-1}) = s_i(x_1, \ldots, x_n),$

and the inclusions (6.16) and (6.34) imply that

(6.36) $\Pi_-\Pi_i\Pi_-^{-2} \in O_-, \quad \Pi_-\Pi_i(\Pi_-^2)^{-1} \in \widehat{O}_-.$

The factorizations (6.29) and (6.31) follow from these relations and inclusions and from Theorem 6.8.   □

We now turn to the polynomial $\mathbf{Q}(v) = \mathbf{Q}_p^n(v)$ defined by the conditions (3.76)–(3.78). Since $\Omega(\Pi_-) = x_0$ and $\Omega(\Pi_+) = x_0x_1 \cdots x_n$, the next proposition is an immediate consequence of Theorem 6.8.

PROPOSITION 6.10. *One has the following factorizations over the ring* $\mathbf{L}_{0,p}^n$:

$$\mathbf{Q}_p^n(v) = (1 - \Pi_-v)Q_-(v) = Q_+(v)(1 - \Pi_+v),$$

where Q_- *and* Q_+ *are polynomials of degree* $2^n - 1$ *with coefficients in* O_{-p}^n *and* O_{+p}^n, *respectively.*

In order to use the factorizations of the Hecke polynomials, one must be able to compute the coefficients of the factors in the form of linear combinations of Γ_0-left or double cosets. The rest of this section is devoted to these calculations for the polynomials $\mathbf{Q}_p^n$ with $n = 1, 2$ and the polynomials $\mathbf{R}_p^n$ and $\widehat{\mathbf{R}}_p^n$ with $n \in \mathbf{N}$.

PROBLEM 6.11. Let $F(v)$ be a polynomial of degree N with coefficients in C_{-p}^n and $F(0) = 1$. Show that there exists a polynomial $G(v)$ of degree $\leq N(2^n - 1)$ with coefficients in O_{-p}^n and $G(0) = 1$, such that all of the coefficients of the polynomial $F(v)G(v)$ lie in the ring $\mathbf{L}_p^n$. From this deduce that every $X \in C_{-p}^n$ satisfies an equation of the form $\sum_{i=0}^{N} X^i\mathbf{T}_i$, where $\mathbf{T}_i \in \mathbf{L}_p^n$, $\mathbf{T}_N = 1$, and $N \leq 2^n$. State and prove similar results for the ring C_{+p}^n.

[Hint: In the polynomial $f(v) = \Omega(F)(v)$ that is obtained from F by replacing its coefficients by their images under Ω, all of the coefficients are symmetric in the variables $x_1, \ldots, x_n$. Consequently, there exists a polynomial $g(v)$ of degree $\leq N(2^n - 1)$ over the ring $\mathbf{Q}[x_0^{\pm 1}, \ldots, x_n^{\pm 1}]$ such that all of the coefficients in the product fg are invariant with respect to W_n. Hence, $fg = \Omega(\mathbf{P})(v)$, where $\mathbf{P}$ is a polynomial of degree $\leq N \cdot 2^n$ over $\mathbf{L}_p^n$. Apply Theorem 6.8 to $\mathbf{P}$. To prove the second assertion, apply the first part to the polynomial $(1 - Xv)$.]

3. Symmetric factorization of the polynomials $\mathbf{Q}_p^n(v)$ for $n = 1, 2$. We obtain factorizations of $\mathbf{Q}_p^n(v), n = 1, 2$, that are invariant with respect to the anti-automorphism $*$, and we compute the coefficients of the polynomial factors.

PROPOSITION 6.12. *Over the ring* $\mathbf{L}^1_{0,p}$ *one has the factorization*

$$\mathbf{Q}^1_p(v) = (1 - \Pi_- v)(q - \Pi_+ v),$$

where $\Pi_- = \Pi^1_-(p) = \Pi^1_0(p)$ *and* $\Pi_+ = \Pi^1_+(p) = \Pi^1_1(p)$.

PROOF. According to (3.58) and (5.14), for $n = 1$ we have

$$(1 - \Pi_- v)(q - \Pi_+ v) = (1 - \mathbf{T}^1(p)v + p\Delta_1(p)v^2).$$

The last polynomial is equal to $\mathbf{Q}^1_p(v)$, by Proposition 3.35. $\qquad\square$

PROPOSITION 6.13. *Over the ring* $\mathbf{L}^2_{0,p}$ *one has the factorization*

$$\mathbf{Q}^2_p(v) = (1 - \Pi_- v)(1 - \Pi_1 v + p(\Pi^{(1)}_{2,0} + \Pi^{(0)}_{2,0})v^2)(1 - \Pi_+ v),$$

where $\Pi_- = \Pi^2_-(p) = \Pi^2_0(p)$, $\Pi_+ = \Pi^2_+(p) = \Pi^2_2(p)$, *and* $\Pi_1 = \Pi^2_1(p)$ *are the elements* (3.59), *and* $\Pi^{(r)}_{2,0}$ *are the elements* (3.62) *for* $n = 2$.

PROOF. We shall need the following formulas for the products of the elements of $\mathbf{L}^2_{0,p}$ listed in the proposition:

(6.37) $$\Pi_- \Pi_1 = p\Pi^0_{1,0}, \quad \Pi_1 \Pi_+ = p\Pi^0_{1,1},$$

(6.38) $$\Pi_- \Pi_1 \Pi_+ = p^3 \Delta \Pi_1,$$

where $\Delta = \Delta_2(p) = \Pi^{(0)}_{2,0}$, and

(6.39) $$\Pi_- \Pi^{(1)}_{2,0} = (p^2 - 1)\Delta\Pi_-, \quad \Pi^{(1)}_{2,0}\Pi_+ = (p^2 - 1)\Delta\Pi_+.$$

To prove these relations, we note that in each case the expression on the right is an integer multiple of a certain double coset modulo $\Gamma_0 = \Gamma^2_0$. Hence, it suffices to verify that any left coset in the product on the left is contained in the double coset on the right, and then verify that the coefficient is correct, by comparing the number of left cosets on both sides or by applying Ω to both sides. For example, in the first relation in (6.39) we have $(p^2 - 1)\left(\begin{pmatrix} p^2 E & 0 \\ 0 & pE \end{pmatrix} \right)_{\Gamma_0}$ on the right. On the left, by definition, we have

$$\Pi_- \Pi^{(1)}_{2,0} = \sum_{\substack{B \in B_0(pE_2)/\mathrm{mod}\, p \\ r_p(B)=1}} \left(\Gamma_0 \begin{pmatrix} p^2 E & pB \\ 0 & pE \end{pmatrix} \right).$$

Since $pB \equiv 0 (\mathrm{mod}\, pE_2)$, it follows that any left coset in the last expansion is contained in $\Gamma_0 \begin{pmatrix} p^2 E & 0 \\ 0 & pE \end{pmatrix} \Gamma_0$. Since $\Pi_- \Pi^{(1)}_{2,0}$ is an element of the Hecke ring, by the same token we have

$$\Pi_- \Pi^{(1)}_{2,0} = \alpha \left(\begin{pmatrix} p^2 E & 0 \\ 0 & pE \end{pmatrix} \right)_{\Gamma_0} = \alpha\Delta\Pi_-.$$

If we apply the map $\Omega = \Omega^2_p$ to both sides of this relation and use Lemma 3.34, we find that

$$x_0 l_p(1, 2)x^2_0 p^{-3} x_1 x_2 = \alpha p^{-3} x^2_0 x_1 x_2 x_0,$$

so that $\alpha = l_p(1, 2)$ is the number of symmetric 2×2-matrices of rank 1 over the field of p elements. It is not hard to list all such matrices: $\begin{pmatrix} a & 0 \\ 0 & 0 \end{pmatrix}$ and $a \begin{pmatrix} b^2 & b \\ b & 1 \end{pmatrix}$, where

$a \in \mathbf{F}_p^*$ and $b \in \mathbf{F}_p$. Hence, $\alpha = p^2 - 1$, and the desired equality is proved. The details of the verification of the other relations will be left to the reader as an easy exercise in preparation for the much more difficult computations in the next subsection.

If we multiply the polynomials on the right in the claimed factorization, we obtain the polynomial

$$1 - (\Pi_- + \Pi_1 + \Pi_+)v + (\Pi_-\Pi_1 + \Pi_-\Pi_+ + \Pi_1\Pi_+ + pv)v^2$$
$$- (p\Pi_-v + pv\Pi_+ + \Pi_-\Pi_1\Pi_+)v^3 + p\Pi_-v\Pi_+v^4,$$

where $v = \Pi_{2,0}^{(1)} + \Pi_{2,0}^{(0)}$. Using the definitions of the elements in the above expression, the relations (6.37)–(6.39), and (5.14), we can rewrite this polynomial in the form

$$1 - (\Pi_0 + \Pi_1 + \Pi_2)v + (p(\Pi_{1,0}^{(0)} + \Pi_{1,1}^{(0)} + \Pi_{2,0}^{(1)}) + (p^3 + p)\Pi_{2,0}^{(0)})v^2$$
$$- p^3\Delta(\Pi_0 + \Pi_1 + \Pi_2)v^3 + p^6\Delta^2 v^4$$
$$= 1 - \mathbf{T}(p)v + (p\mathbf{T}_1(p^2) + (p^3 + p)\Delta)v^2 - p^3\Delta\mathbf{T}(p)v^3 + p^6\Delta^2 v^4,$$

where in the last step we used the expressions (3.58) and (3.61) for $\mathbf{T}(p) = \mathbf{T}^2(p)$ and $\mathbf{T}_1(p^2) = \mathbf{T}_1^2(p^2)$, respectively. According to the formula for $\mathbf{Q}_p^2(v)$ in Proposition 3.35, to complete the proof it suffices to verify that

(6.40) $\mathbf{q}_2^2(p) = p\mathbf{T}_1^2(p^2) + (p^3 + p)\Delta_2(p),$

where $\mathbf{q}_2^2(p)$ is the element of $\mathbf{L}_p^2$ determined by the condition

$$\Omega(\mathbf{q}_2^2(p)) = x_0^2 x_1 x_2 (x_1 + x_2 + x_1^{-1} + x_2^{-1} + 2).$$

Since the map Ω is a monomorphism on $\mathbf{L}_p^2$, to do this it is enough to verify that the right side of (6.40) has the same Ω-image as the left side. We compute the Ω-image of the right side by replacing $\mathbf{T}_1(p^2)$ by its expression in (3.61), and using Lemma 3.34 and then Lemma 2.21 to calculate the polynomials $\omega(\pi_{1,0}^2(p)) = \omega(\pi_1^2(p))$, $\omega(\pi_{1,1}^2(p)) = \omega(\pi_2^2(p)\pi_1^2(p))$, and $\omega(\pi_{2,0}^2(p)) = \omega(\pi_2^2(p))$. The reader can easily see that the result is the polynomial that gives the left side.   □

PROBLEM 6.14. For any $n \in \mathbf{N}$ and any prime p, prove the following factorization over the ring $\mathbf{L}_{0,p}^n$:

$$\mathbf{Q}_p^n(v) = (1 - \Pi_-v)Q'(v)(1 - \Pi_+v),$$

where $\Pi_- = \Pi_-^n(p)$, $\Pi_+ = \Pi_+^n(p)$, and Q' is a polynomial of degree $2^n - 2$.

[Hint: Using Problem 6.7(2) and (3.79), show that all of the coefficients in the formal power series $(1 - \Pi_-v)^{-1}\mathbf{Q}_p^n(v)(1 - \Pi_+v)^{-1}$ lie in $C_{-p}^n \cdot C_{+p}^n$, and then use the fact that Ω_p^n is a monomorphism on this space.]

4. Coefficients in the factorization of Rankin polynomials. Here we compute the Γ_0^n-left and double coset expansions of the coefficients in the factorizations (6.29)–(6.32) of the Rankin polynomials $\mathbf{R}_p^n(v)$ and $\widehat{\mathbf{R}}_p^n(v)$. To do this, we must find certain products of elements of the form $\Pi_a = \Pi_a^n(p)$, $\Pi_{a,b}^{(r)}$, and $\Pi_{a,b}^{(r)}(k)$ in the ring $\overline{\mathbf{L}}_{0,p}^n$ (see Lemmas 3.32 and 4.19).

PROPOSITION 6.15. *The following relations hold in* $\overline{\mathbf{L}}_0 = \overline{\mathbf{L}}_{0,p}^n$:

$$(6.41) \qquad \Pi_b \Pi_+ = p^{\langle n-b \rangle} \Pi_{n-b,b}^{(0)} = p^{\langle n-b \rangle} \Pi_{n-b,b}^{(0)}(k),$$

where $0 \leqslant b \leqslant n$; *and for* $0 \leqslant r \leqslant a \leqslant n$, $0 \leqslant b \leqslant n$, *and* $r + b \leqslant a$,

$$(6.42) \qquad \Pi_{a,0}^{(r)} \Pi_{n-b,b}^{(0)} = \sum_{\substack{\max(0,a+b-n) \leqslant t \leqslant b \\ 0 \leqslant s \leqslant r}} c(a,b,r,t,s) \Delta \Pi_{a+b-2t,t}^{(r-s)},$$

(6.43)

$$\Pi_{a,0}^{(r)}(k) \Pi_{n-b,b}^{(0)}(k) = \sum_{\substack{\max(0,a+b-n) \leqslant t \leqslant b \\ 0 \leqslant s \leqslant r}} c(k;a,b,r,t,s) \Delta \Pi_{a+b-2t,t}^{(r-s)}(k),$$

in which

$$(6.44) \qquad \begin{aligned} c(k;a,b,r,t,s) &= p^{t(r+t-a-b-s-1)+b(n+1)} \\ &\times l_p^k (O_{a+s-r-t}; s, a+s-r) \frac{\varphi_{a+b+s-r-2t}(p)}{\varphi_{a+s-r-t}(p) \varphi_{b-t}(p)}, \end{aligned}$$

where φ_s *is the function* (2.29),

$$(6.45) \qquad l_p^k(O_m; c, d) = \sum_{M \in S_d(\mathbf{F}_p), r_p(M)=c}' \varkappa(M)^{-k},$$

k *is a fixed odd integer, and the prime over the summation means that it is taken over the set of matrices with zero* $m \times m$-*block in the upper-left corner; the coefficients* $c(a,b,r,t,s)$ *are obtained from the coefficients* (6.44) *by setting* $k = 0$.

PROOF. From (3.59), (6.33), and the definitions it follows that the left and right exponents of the Π_b are at most 1. Then, by Proposition 6.1, the left side of (6.41) lies in the ring $C_+ = C_{+p}^n$. Since, by (5.37), $\Pi_{n-b,b}^{(0)} = (M_{n-b,b}(0))_{\Gamma_0}$, it follows from Proposition 5.5 that this element lies in C_+. Thus,

$$(6.46) \qquad \Pi_b \Pi_+, \Pi_{n-b,b}^{(0)} = \Pi_{n-b,b}^{(0)}(k) \in C_+ \quad (0 \leqslant b \leqslant n).$$

From these inclusions and Theorem 5.9 we see that to prove (6.41) it suffices to verify that both sides have the same image under Φ or Ω. But this follows immediately from Lemma 3.34. (6.41) is proved.

Things are not so simple in the case of (6.42) and (6.43), where the reader should expect some rather tedious computations. First of all, we note that it is enough to compute products of the form

$$(6.47) \qquad (M_{a,0}(B_0))_{\Gamma_0} \cdot \Pi_{n-b,b}^{(0)},$$

where $(M_{a,0}(B_0))_{\Gamma_0}$ with $B_0 \in S_a(\mathbf{Z})$ and $r_p(B_0) = r$ is one of the double cosets in the expansion (5.37) of $\Pi_{a,0}^{(r)}$. From (5.37) it follows that $\Pi_{n-b,b}^{(0)} = (M_{n-b,b}(0))_{\Gamma_0}$. Thus, using the second formula in Lemma 1.5 and the expansion (5.35) of the double coset $(M_{a,0}(B_0))_{\Gamma_0}$, we see that the computation of the product (6.47) requires that we find the double cosets to which products of the form

$$(6.48) \qquad M_{a,0}(B_h) U(\xi) M_{n-b,b}(0) \quad \text{for } B_h \in \{B_0\}_p^n, \xi \in \Lambda_{(D_a)} \setminus \Lambda$$

$(D_a = D_{a,0})$ belong (and with what multiplicity). To do this we need a special set of representatives of the left cosets $\Lambda_{(D_a)} \setminus \Lambda$.

We introduce some notation. We set

$$(6.49) \qquad I_{n-a,n} = \{\mathbf{i} = (i_1, \ldots, i_{n-a}) \in \mathbf{N}^{n-a}; 1 \leqslant i_1 < \cdots < i_{n-a} \leqslant n\}.$$

For $\mathbf{i} \in I_{n-a,n}$ we let $\hat{\mathbf{i}}$ denote the set $(j_\beta) \in I_{a,n}$ that is the complement of $(i_1, \ldots, i_{n-a})$ in the set $(1, 2, \ldots, n)$. To every permutation σ of the numbers $1, 2, \ldots, n$ we associate the $n \times n$-matrix

$$M(\sigma) = (\delta_{\sigma^{-1}(i),j}),$$

where $\delta_{\alpha\alpha} = 1, \delta_{\alpha\beta} = 0$ for $\alpha \neq \beta$, is the Kronecker symbol. It is easy to see that

$$M(\sigma\tau) = M(\sigma)M(\tau) \quad \text{and} \quad M(\sigma^{-1}) = M(\sigma)^{-1} = {}^t M(\sigma).$$

Next, to every $\mathbf{i} \in I_{n-a,n}$ we associate the permutation $\sigma(\mathbf{i})$ and the matrix $M(\mathbf{i})$ by setting

$$(6.50) \qquad \sigma(\mathbf{i}) = \begin{pmatrix} i_1, \ldots, i_{n-a} & j_1, \ldots, j_a \\ 1, \ldots, n-a & n-a+1, \ldots, n \end{pmatrix} \quad \text{and} \quad M(\mathbf{i}) = M(\sigma(\mathbf{i})),$$

where $(j_\beta) = \hat{\mathbf{i}}$. Note that right multiplication of any n-column matrix by a matrix of the form $M(\sigma)$ is equivalent to performing the permutation σ of its columns. In particular,

$$(6.51) \qquad (\mathbf{t}_1, \ldots, \mathbf{t}_n)M(\mathbf{i})^{-1} = (\mathbf{t}_{i_1}, \ldots, \mathbf{t}_{i_{n-a}}, \mathbf{t}_{j_1}, \ldots, \mathbf{t}_{j_a}),$$

where the columns $\mathbf{t}_\alpha$ are all of the same size. Finally, for $\mathbf{i} \in I_{n-a,n}$ with $\hat{\mathbf{i}} \in I_{a,n}$ we define the sets

$$
\begin{aligned}
V(\mathbf{i}) = \{ & V = (v_{\alpha\beta}) \in M_{n-a,n}; 0 \leqslant v_{\alpha\beta} < p, \text{ if } i_\alpha < j_\beta, \\
(6.52) \qquad & \qquad\qquad\qquad v_{\alpha\beta} = 0, \text{ if } i_\alpha > j_\beta \}, \\
W(\mathbf{i}) = \Big\{ & \varepsilon = \begin{pmatrix} E_{n-a} & V \\ 0 & E_a \end{pmatrix} M(\mathbf{i}); V \in V(\mathbf{i}) \Big\},
\end{aligned}
$$

and

$$W_a = W_a^n(p) = \bigcup_{\mathbf{i} \in I_{n-a,n}} W(\mathbf{i}).$$

We now show that W_a is a complete set of left coset representatives of $\Lambda = \Lambda^n$ modulo the subgroup $\Lambda_{(D_a)} = \Lambda \cap D_a^{-1} \Lambda D_a$:

$$(6.53) \qquad W_a = \Lambda_{(D_a)} \setminus \Lambda.$$

To do this, we first note that the number of elements in $W(\mathbf{i})$ is

$$(6.54) \qquad |W(\mathbf{i})| = |V(\mathbf{i})| = \mathbf{p}^{j_1 + \cdots + j_a - \langle a \rangle},$$

where $(j_\beta) = \hat{\mathbf{i}}$, since for fixed β there are exactly $j_\beta - \beta$ indices i_α satisfying the inequality $i_\alpha < j_\beta$. From this and (2.33) we conclude that the number of elements in the set W_a is

$$p^{-\langle a \rangle} \sum_{1 \leqslant j_1 < \cdots < j_a \leqslant n} p^{j_1 + \cdots + j_a} = \frac{\varphi_n}{\varphi_a \varphi_{n-a}}, \qquad \text{where } \varphi_s = \varphi_s(p).$$

On the other hand, according to Lemma 1.2 and (2.28), the index $\mu_\Lambda(D_a)$ of $\Lambda_{(D_a)}$ in Λ is equal to the same number. Hence, in order to verify (6.53) it suffices to show that all of the matrices in W_a are in pairwise distinct $\Lambda_{(D_a)}$-left cosets. From the definition it follows that

$$\Lambda_{(D_a)} = \left\{ \lambda = \begin{pmatrix} \lambda_1 & \lambda_2 \\ \lambda_3 & \lambda_4 \end{pmatrix} \in \Lambda; \lambda_2 \equiv 0 (\mathrm{mod}\, p) \right\},$$

where λ_2 is an $(n-a) \times a$-block. We hence find that the $(n-a) \times (n-a)$-block λ_1 of any matrix $\lambda \in \Lambda_{(D_a)}$ is nonsingular modulo p, and if two matrices $\varepsilon, \varepsilon' \in \Lambda$ belong to the same $\Lambda_{(D_a)}$-left coset, then the matrices $\varepsilon_{(n-a)}$ and $\varepsilon'_{(n-a)}$ consisting of the first $n-a$ rows satisfy the relation

(6.55) $\lambda_1 \varepsilon_{(n-a)} \equiv \varepsilon'_{(n-a)} (\mathrm{mod}\, p)$,   where $\lambda_1 \in GL_{n-a}(\mathbf{Z}/p\mathbf{Z})$.

If $\varepsilon \in W(\mathbf{i})$, then from (6.51) and the definition of $V(\mathbf{i})$ it follows that the set $(i_1, \ldots, i_{n-a})$ is the first (in the sense of lexicographical order) set of indices of $n-a$ columns of $\varepsilon_{(n-a)}$ that are linearly independent modulo p. Hence, if (6.55) holds for two matrices $\varepsilon, \varepsilon' \in W_a$, then they both belong to the same subset $W(\mathbf{i}) \subset W_a$. It then follows from (6.55) that $\lambda_1 \equiv E_{n-a}(\mathrm{mod}\, p)$, from which we conclude that the corresponding matrices $V, V' \in V(\mathbf{i})$ are congruent modulo p, and hence coincide, i.e., $\varepsilon = \varepsilon'$. This proves (6.53).

We return to the products (6.48), using the set (6.51) as our set of representatives of $\Lambda_{(D_a)} \setminus \Lambda$. Let $\xi = \varepsilon \in W_a$. If we multiply out the matrices in (6.48) and take into account that $D_{a,0} = D_a$ and $D_{n-b,b} = pD_b$, then we obtain the matrix

(6.56) $p \begin{pmatrix} p^2(D_a \varepsilon D_b)^* & B\varepsilon D_b \\ 0 & D_a \varepsilon D_b \end{pmatrix}$,   where $B = \begin{pmatrix} 0 & 0 \\ 0 & B_h \end{pmatrix}$.

Suppose that $\varepsilon \in W(\mathbf{i})$, where $\mathbf{i} \in I_{n-a,n}$, i.e., $\varepsilon = \begin{pmatrix} E_{n-a} & V \\ 0 & E_a \end{pmatrix} M$ with $M = M(\mathbf{i})$ and $V \in V(\mathbf{i})$. If we set $d(k)$ equal to 0 or 1 according to the formula $D_b = \mathrm{diag}(p^{d(1)}, \ldots, p^{d(n)})$, then, using (6.51) and the analogous relation for the rows that comes from the transpose of (6.51), we conclude that $MD_b M^{-1} = C = \begin{pmatrix} C_1 & 0 \\ 0 & C_2 \end{pmatrix}$, where $C_1 = \mathrm{diag}(p^{d(i_1)}, \ldots, p^{d(i_{n-a})})$, $C_2 = \mathrm{diag}(p^{d(j_1)}, \ldots, p^{d(j_a)})$, and $(j_\beta) = \hat{\mathbf{i}}$. These formulas imply that

$$D_a \varepsilon D_b = D_a \begin{pmatrix} E_{n-a} & V \\ 0 & E_a \end{pmatrix} MD_b M^{-1} M = D_a C \eta,$$

where

$$\eta = C^{-1} \begin{pmatrix} E_{n-a} & V \\ 0 & E_a \end{pmatrix} CM = \begin{pmatrix} E_{n-a} & C_1^{-1} VC_2 \\ 0 & E_a \end{pmatrix} M.$$

The matrix $C_1^{-1} VC_2 = (p^{-d(i_\alpha)+d(j_\beta)} v_{\alpha\beta})$ is an integer matrix, since $d(j_\beta) = 0$ and $d(i_\alpha) = 1$ in the case $d(j_\beta) < d(i_\alpha)$; hence, $i_\alpha > n - b \geqslant j_\beta$, and from the definition of $V(\mathbf{i})$ it follows that $v_{\alpha\beta} = 0$. Thus, $\eta \in \Lambda$, and the matrix (6.56) belongs to the same Γ_0-right coset as the matrix

(6.57) $p \begin{pmatrix} p^2(D_a C)^* & B\varepsilon D_b \eta^{-1} \\ 0 & D_a C \end{pmatrix} = p \begin{pmatrix} p^2(D_a C)^{-1} & BC \\ 0 & D_a C \end{pmatrix}$

$(\varepsilon D_b \eta^{-1} = C)$. We let $t = t(\mathbf{i})$ denote the number of j_β in the set $(j_\beta) = \hat{\mathbf{i}}$ which satisfy the inequality $j_\beta > n - b$. From the definition of the matrices C_1 and C_2 it follows that they have the form

$$
C_1 = \begin{pmatrix} E_{n-a-b+t} & 0 \\ 0 & pE_{b-t} \end{pmatrix}, \quad C_2 = \begin{pmatrix} E_{a-t} & 0 \\ 0 & pE_t \end{pmatrix},
$$

and hence $D_a C = D_{a+b-2t,t}$ and $BC = \begin{pmatrix} 0 & 0 \\ 0 & B_h C_2 \end{pmatrix}$. For any matrix A, we shall
let $A^{(s)}$ denote the $s \times s$-block in the upper-left corner of A. From the form of our
matrices and the expansion (5.35) it follows that the matrix (6.57) is contained in the
Γ_0-double coset of the matrix

$$
pM_{a+b-2t,t}(\widehat{B}_h^{(a-t)}) \quad \text{for } \widehat{B}_h^{(a-t)} = \begin{pmatrix} 0 & 0 \\ 0 & B_h^{(a-t)} \end{pmatrix}
$$

($\widehat{B}_h^{(a-t)}$ is obviously an $(a + b - 2t) \times (a + b - 2t)$-matrix). Thus, every product of
the form (6.56) lies in the double coset of some matrix of the form $pM_{a+b-2t,t}(K)$,
where $K \in S_{a+b-2t}(\mathbf{Z})$, and it falls in a particular double coset of this form if and only
if $\varepsilon \in W(\mathbf{i})$, where $\mathbf{i} \in I_{n-a,n}$ and $t(\mathbf{i}) = t$, and the matrix $\widehat{B}_h^{(a-t)}/\bmod p$ is contained
in $\{K\}_p^n$ (see (5.34)). Using the formula (6.54) for the number of elements in the
set $W(\mathbf{i})$, we find that the number of products (6.56) contained in the double coset
$\Gamma_0 p M_{a+b-2t,t}(K) \Gamma_0$ is equal to $\sigma(a,b,t)v(\{B_0\},\{K\}, a - t)$, where

$$
\sigma(a,b,t) = \sum_{\substack{1 \leqslant j_1 < \cdots < j_{a-t} \leqslant n-b \\ n-b < j_{a-t+1} < \cdots < j_a \leqslant n}} p^{j_1 + \cdots + j_a - \langle a \rangle}
$$

and $v(\{B_0\}, \{K\}, s)$ denotes the number of matrices $B_h \in \{B_0\}_p^n$ for which the matrix
$\widehat{B}_h^{(s)}/\bmod p$ (of the same size as K) lies in the class $\{K\}_p^n$. Since t must obviously be
$\geqslant 0$ and $\geqslant a + b - n$, by Lemma 1.5 we obtain the formula

$$
(6.58) \qquad (M_{a,0}(B_0))_{\Gamma_0} \cdot \Pi_{n-b,b}^{(0)} = \sum_{t,\{K\}_p^n} c(a,b,t; B_0, K) \Delta(M_{a+b-2t,t}(K))_{\Gamma_0},
$$

where $\max(0, a+b-n) \leqslant t \leqslant \min(a,b)$, $\{K\}_p^n \in S_{a+b-2t}(\mathbf{Z}/p\mathbf{Z})$, and the coefficients
$c(a,b,t; B_0, K)$ have the form

$$
\sigma(a,b,t)v(\{B_0\},\{K\}, a - t)\mu(M_{n-b,b}(0))\mu(M_{a+b-2t,t}(K))^{-1},
$$

where $\mu(M)$ is the number of Γ_0-left cosets in the double coset $\Gamma_0 M \Gamma_0$. The terms in
the last expression can be computed using the formulas we already know.

First, by (2.33), $\sigma(a,b,t)$ is equal to

$$
\sum_{\substack{1 \leqslant j_1 < \cdots < j_{a-t} \leqslant n-b \\ 1 \leqslant \beta_1 < \cdots < \beta_t \leqslant b}} p^{j_1 + \cdots + j_{a-t} + (\beta_1 + (n-b)) + \cdots + (\beta_t + (n-b)) - \langle a \rangle}
$$

$$
= p^{t(n+t-a-b)} \frac{\varphi_{n-b}\varphi_b}{\varphi_{a-t}\varphi_{n+t-a-b}\varphi_t\varphi_{b-t}},
$$

where $\varphi_s = \varphi_s(p)$. Next, from (5.35) and (2.28) we have

$$
\mu(M_{n-b,b}(0)) = p^{b(n-b)+b(b+1)}\mu_\Lambda(D_{n-b,b}) = p^{b(n+1)}\mu_\Lambda(D_b) = p^{b(n+1)} \frac{\varphi_n}{\varphi_b\varphi_{n-b}}.
$$

Finally, from (5.35) and (2.32) we obtain

$$\mu(M_{a+b-2t,t}(K)) = |\{K\}_p^n| p^{t(n+1)} \frac{\varphi_n}{\varphi_{n+t-a-b}\varphi_{a+b-2t}\varphi_t}.$$

If we substitute these expressions in the formula for $c(a,b,t;B_0,K)$ and take the sum of the expansions (6.58), multiplied by $\varkappa(B_0)^{-k}$, over all distinct classes $\{B_0\}_p^n$ in $S_a(\mathbf{Z}/p\mathbf{Z})$ for which $r_p(B_0) = r$, then from (5.38) we obtain the formula

$$\Pi_{a,0}^{(r)}(k)\Pi_{n-b,b}^{(0)} = \sum_{t,\{K\}_p^n} c(k;a,b,r,t;K)\Delta(M_{a+b-2t,t}(K))_{\Gamma_0},$$

where $\max(0, a+b-n) \leqslant t \leqslant b$, $\{K\}_p^n \in S_{a+b-2t}(\mathbf{Z}/p\mathbf{Z})$, and

$$
\begin{aligned}
c(k;a,b,r,t;K) &= p^{t(t-a-b-1)+b(n+1)} \frac{\varphi_{a+b-2t}}{\varphi_{a-t}\varphi_{b-t}} \\
&\quad \times \left(|\{K\}_p^n|^{-1} \sum_{\substack{\{B_0\}_p^n \in S_a(\mathbf{Z}/p\mathbf{Z}) \\ r_p(B_0)=r}} \varkappa(B_0)^{-k} v(\{B_0\}, \{K\}, a-t) \right)
\end{aligned}
$$

(6.59)

(by assumption $r + b \leqslant a$, so that $\min(a,b) = b$).

We now turn our attention to the expression in large parentheses on the right side of (6.59). For brevity we shall denote it $S(k, \{K\})$. By the definition of $v(\{B_0\}, \{K\}, s)$ we obtain

(6.60) $$S(k, \{K\}) = |\{K\}_p^n|^{-1} \sum_V l_p(k; V, r, a),$$

where

(6.61) $$l_p(k; V, r, a) = \sum_{\substack{T \in S_a(\mathbf{Z}/p\mathbf{Z}), r_p(T)=r \\ T \equiv \begin{pmatrix} V & * \\ * & * \end{pmatrix} (\bmod p)}} \varkappa(T)^{-k}.$$

We now show that the sum (6.61) depends only on the rank of V over the field $\mathbf{Z}/p\mathbf{Z}$. More precisely, suppose that V is an $(a-t) \times (a-t)$-matrix of rank $r_p(V) = r - s$, where $t, s \geqslant 0$. Then we have the formula (see (6.45))

(6.62) $$l_p(k; V, r, a) = p^{(r-s)t} \varkappa(V)^{-k} l_p(k; O_{a+s-r-t}; s, a+s-r).$$

To see this, first note that, by (4.70), the sum (6.61) depends only on the class of the matrix V over $\mathbf{Z}/p\mathbf{Z}$. By Theorem 1.3 of Appendix 1, we may suppose that $V = \begin{pmatrix} V_1 & 0 \\ 0 & 0 \end{pmatrix}$, where V_1 is an $(r-s) \times (r-s)$-matrix that is nonsingular over $\mathbf{Z}/p\mathbf{Z}$. In this case, any matrix T satisfying the conditions in (6.61) is congruent modulo p to a matrix of the form

$$\begin{pmatrix} V_1 & 0 & T_{13} \\ 0 & 0 & T_{23} \\ {}^tT_{13} & {}^tT_{23} & T_{33} \end{pmatrix} = \begin{pmatrix} V_1 & 0 \\ 0 & T_1 \end{pmatrix} \left[\begin{pmatrix} E_{r-s} & 0 & V_1^*T_{12} \\ 0 & E_{a+s-r-t} & 0 \\ 0 & 0 & E_t \end{pmatrix} \right],$$

where $T_1 = \begin{pmatrix} 0 & T_{23} \\ {}^tT_{23} & T_{33} \end{pmatrix}$, whose rank over $\mathbf{Z}/p\mathbf{Z}$ is equal to

$$r_p(V_1) + r_p(T_1) = r - s + r_p(T_1).$$

Conversely, any matrix of the above form, with $T_1 \in S_{r-s}(\mathbf{Z}/p\mathbf{Z})$ and $r_p(T_1) = s$, satisfies the conditions in (6.61). Furthermore, since (4.70) implies that $\varkappa(T) = \varkappa\left(\begin{pmatrix} V_1 & 0 \\ 0 & T_1 \end{pmatrix}\right) = \varkappa(V_1)\varkappa(T_1)$ (because for any symmetric integer matrices A_1 and A_2 we have

$$(6.63) \qquad \varkappa\left(\begin{pmatrix} A_1 & 0 \\ 0 & A_2 \end{pmatrix}\right) = \varkappa(A_1)\varkappa(A_2)),$$

it follows that (6.62) is a consequence of the definition (6.45) and the above considerations.

We return to the computation of the sum (6.60). In order for this not to be the empty sum, the matrix $K \in S_{a+b-2t}(\mathbf{Z}/p\mathbf{Z})$ must clearly satisfy the inequality $r_p(K) \leqslant r$. We set $r_p(K) = r - s$. Then any matrix V on the right in (6.60) must satisfy the relation $r_p(V) = r_p(K) = r - s$. Hence, if we apply (6.62), we can rewrite the sum (6.60) in the form

$$(6.64) \qquad S(k, \{K\}) = p^{(r-s)} l_p(k; O_{a+s-r-t}; s, a + s - r)|\{K\}_p^n|^{-1} S_1(k, \{K\}),$$

where

$$(6.65) \qquad S_1(k, \{K\}) = \sum_V \varkappa(V)^{-k}$$

in which $V \in S_{a-t}(\mathbf{Z}/p\mathbf{Z})$, $\begin{pmatrix} 0 & 0 \\ 0 & V \end{pmatrix} / \mathrm{mod}\ p \in \{K\}_p^n$. By Theorem 1.3 of Appendix 1, we may suppose that $K = \begin{pmatrix} 0 & 0 \\ 0 & K_1 \end{pmatrix}$, where $K_1 \in S_{r-s}(\mathbf{Z})$ and $r_p(K_1) = r - s$. We let G denote the group $GL_{a+b-2t}(\mathbf{Z}/p\mathbf{Z})$ if $a + b - 2t < n$; if $a + b - 2t = n$, then we let G denote the subgroup of this group that consists of the matrices of determinant $\pm 1/\mathrm{mod}\ p$. Then, by (5.34), the class $\{K\} = \{K\}_p^n$ consists of matrices of the form $K[U]$ ($U \in G$). We divide the matrix $U \in G$ into blocks U_{ij} ($1 \leqslant i, j \leqslant 3$) with diagonal blocks U_{11}, U_{22}, U_{33} of size $(b - t) \times (b - t)$, $(a + s - t - r) \times (a + s - t - r)$, and $(r - s) \times (r - s)$, respectively. Since U is nonsingular modulo p, by a direct calculation we easily see that, if $K = \begin{pmatrix} 0 & 0 \\ 0 & K_1 \end{pmatrix}$ with K_1 an $(r - s) \times (r - s)$-matrix that is nonsingular modulo p, then

$$(6.66) \qquad K[U] \equiv \begin{pmatrix} 0 & 0 \\ 0 & V \end{pmatrix} \pmod{p}$$

if and only if U belongs to the subset

$$G_1 = \{U = (U_{ij}) \in G; U_{31} \equiv 0 (\mathrm{mod}\ p)\}.$$

From this condition it follows that the rows of the matrix $(U_{32} U_{33})$ are linearly independent over $\mathbf{Z}/p\mathbf{Z}$, and hence this matrix can be filled out to a matrix $U_1 \in GL_{a-t}(\mathbf{Z}/p\mathbf{Z})$. Then (6.66) implies the congruence

$$\begin{pmatrix} 0 & 0 \\ 0 & K_1 \end{pmatrix} [U_1] \equiv V (\mathrm{mod}\ p),$$

and consequently $\varkappa(V) = \varkappa(K_1) = \varkappa(K)$. From this and (6.65) we obtain

$$S_1(k, \{K\}) = \varkappa(K)^{-k} |G_1| \cdot |G_0|^{-1},$$

where

$$G_0 = \left\{ U \in G; \begin{pmatrix} 0 & 0 \\ 0 & K_1 \end{pmatrix} [U] \equiv \begin{pmatrix} 0 & 0 \\ 0 & K_1 \end{pmatrix} \pmod{p} \right\}$$

is the stabilizer of the matrix K in the group G. On the other hand, $|\{K\}| = |G| \cdot |G_0|^{-1}$. If we substitute this expression into (6.64), we find that $S(k, \{K\})$ is equal to

$$(6.67) \qquad p^{(r-s)t} \varkappa(K)^{-k} l_p(k; O_{a+s-r-t}; s, a+s-r) |G|^{-1} |G_1|.$$

To compute $|G|^{-1} |G_1|$ we need the following

LEMMA 6.16. *Let $1 \leqslant c \leqslant d$, and let p be a prime number. Then the number of matrices $V \in M_{d,c}(\mathbf{Z}/p\mathbf{Z})$ satisfying the condition $r_p(V) = c$ is equal to*

$$p^{\langle c-1 \rangle} (p^d - 1) \cdots (p^{d-c+1} - 1) = p^{\langle c-1 \rangle} \frac{\varphi_d(p)}{\varphi_{d-c}(p)}.$$

PROOF. We use induction on c. The formula is obvious for $c = 1$. Suppose that it has already been proved for some c, $1 \leqslant c < d$. Any matrix $V' \in M_{d,c+1}(\mathbf{Z}/p\mathbf{Z})$ with $r_p(V') = c + 1$ can be obtained by taking a suitable $V \in M_{d,c}(\mathbf{Z}/p\mathbf{Z})$ with $r_p(V) = c$ and adding a column which is not a linear combination modulo p of the columns of V. The number of such columns that are distinct modulo p is obviously equal to $p^d - p^c = p^c(p^{d-c} - 1)$. Hence, when passing from c to $c + 1 \leqslant d$, the number of matrices with the desired properties gets multiplied by $p^c(p^{d-c} - 1)$. From this and the induction assumption we obtain the lemma for $c + 1$. $\qquad\square$

We now complete the computation of the sum $S(k, \{K\})$. An argument similar to the one used in the proof of Lemma 6.16 shows that, given any matrix $M \in M_{a+b-2t, b-t}(\mathbf{Z}/p\mathbf{Z})$, where $1 \leqslant b - t < a + b - 2t$ and $r_p(M) = b - t$, there exist matrices $M' \in M_{a+b-2t, a-t}(\mathbf{Z}/p\mathbf{Z})$ for which $(M, M') \in G$, and the number of M' with this property does not depend on M. From this and Lemma 6.16 it follows that this number is equal to $|G| p^{\langle b-t-1 \rangle} \varphi_{a-t} \varphi_{a+b-2t}^{-1}$. On the other hand, the number of matrices M of the above form for which the last $r - s$ rows consist of zeros is equal (by the same lemma) to

$$p^{\langle b-t-1 \rangle} \varphi_{a+b+s-2t-r} \varphi_{a+s-t-r}^{-1}.$$

If we multiply these expressions, we obtain a formula for the number of elements in G_1:

$$|G_1| = |G| \varphi_{a-t} \varphi_{a+b+s-2t-r} \varphi_{a+b-2t}^{-1} \varphi_{a+s-t-r}^{-1}.$$

This formula also holds in the cases $b - t = 0$ and $b - t = a + b - 2t$ that were excluded, since in the first case both sides of the equality are equal to $|G|$, and the second case reduces to the first case because $t \leqslant b \leqslant a$. If we substitute the expression for $|G_1|$ into (6.67) and substitute the resulting expression for $S(k, \{K\})$ into (6.59), after obvious simplifications we obtain the formula

$$c(k; a, b, r, t; K) = p^{t(r+t-a-b-s-1)+b(n+1)} \varkappa(K)^{-k}$$
$$\times l_p(k; O_{a+s-r-t}; s, a+s-r) \frac{\varphi_{a+b+s-r-2t}}{\varphi_{a+s-r-t} \varphi_{b-t}},$$

where s is determined by the condition $r_p(K) = r - s$. If we substitute these expressions into the formula for the product $\Pi_{a,0}^{(r)}(k) \cdot \Pi_{n-b,b}^{(0)}$, by Lemma 5.11 we obtain the formula (6.43). Setting $k = 0$ in this formula, we obtain (6.42).

We are now ready to compute the coefficients in the expansions (6.29) and (6.30) of the polynomial $\mathbf{R}_p^n(v)$.

PROPOSITION 6.17. *In the ring* $\mathbf{L}_{0,p}^n$ *one has*

$$(6.68) \qquad \Pi_- \Pi_i \Pi_-^{-2} = p^{-\langle i \rangle - i(n-i)} \Delta^{-1} \sum_{j=0}^{i} \alpha_{ij} \sum_{a=n-i+j}^{n} \Pi_{a,n-a}^{(i-j+a-n)},$$

$$(6.69) \qquad \Pi_+^{-2} \Pi_{n-i} \Pi_+ = p^{-\langle i \rangle - i(n-i)} \Delta^{-1} \sum_{j=0}^{i} \alpha_{ij} \sum_{a=n-i+j}^{n} \Pi_{a,0}^{(i-j+a-n)},$$

where

$$(6.70) \qquad \alpha_{ij} = \alpha_{ij}^n(p) = \frac{\varphi_{n-i+j}(p)}{\varphi_{n-i}(p)} \sum_{t=0}^{j} \frac{(-p)^t}{\varphi_t(p)\varphi_{j-t}(p)},$$

φ_s *is the function* (2.29), $\Pi_{a,b}^{(r)}$ *is the element* (3.62), *and the rest of the notation is the same as in Proposition 6.9.*

PROOF. The formulas (6.12), (6.33), and (5.39) show that the anti-automorphism $*$ takes (6.68) to (6.69) and conversely; hence, it suffices to prove one of the two. We shall prove (6.69). We first verify that both sides of (6.69) lie in the subspace $O_+ = O_{+p}^n = \mathbf{L}_p^n \cdot \overline{C}_{+p}^n \subset \overline{\mathbf{L}}_{0,p}^n$. We then show that both sides have the same image under the map Ω_p^n. By Theorem 6.5(2), this will imply that they are equal.

From (6.16) and (6.46) we have

$$(6.71) \qquad \Pi_+^{-2} \Pi_{n-i} \Pi_+ \in O_+ \quad (0 \leqslant i \leqslant n).$$

In order to examine the right side of (6.69), we introduce the sums

$$(6.72) \qquad S_{i,c} = \sum_{a=n-i}^{n-c} \Pi_{a,c}^{(i+a-n)}$$

for $0 \leqslant c \leqslant i \leqslant n$. Since the right side of (6.69) is a linear combination of $\Delta^{-1} S_{i-j,0}$ for $j = 0, 1, \ldots, i$, it follows that it lies in O_+ provided that $S_{i,0} \in O_+$ for all $i = 0, 1, \ldots, n$. To prove the latter claim, we use induction on $d = i - c$ to show that

$$(6.73) \qquad S_{i,c} = S_{c+d,c} \in O_+ \quad (0 \leqslant c \leqslant c + d \leqslant n).$$

Since $S_{c,c} = \Pi_{n-c,c}^{(0)}$, it follows by (4.46) that $S_{c,c} \in C_+ \subset O_+$; this proves (6.73) in the case $d = 0$. Now suppose that (6.73) holds for all c, d satisfying the conditions $0 \leqslant d < h$ and $0 \leqslant c \leqslant n - d$, where $0 < h \leqslant n$. From (3.61) and (6.72) it follows that we have

$$\mathbf{T}_{n-h}(p^2) = S_{h,0} + \sum_{c+d=h, c \geqslant 1} S_{c+d,c}.$$

By the induction assumption, the sum $S_{c+d,c}$, where $c + d = h$ and $c \geqslant 1$, i.e., $d < h$, lies in C_+. By the definition of O_+, the element $\mathbf{T}_{n-h}(p^2) \in \mathbf{L}_p^n$ is also contained in that space. We thus have $S_{h,0} \in O_+$. We now use induction on c to prove that

$$(6.74) \qquad S_{c+h,c} \in O_+ \quad \text{for } 0 \leqslant c \leqslant n - h.$$

The case $c = 0$ has already been treated. Suppose that (6.74) holds for c satisfying the inequality $0 \leqslant c < b \leqslant n - h$, where $0 < b \leqslant n - h$. Consider the product

$$S_{h,0} \cdot \Pi^{(0)}_{n-b,b} = \sum_{a=n-h}^{n} \Pi^{(h+a-n)}_{a,0} \Pi^{(0)}_{n-b,b}.$$

Since $b \leqslant n - h$, it follows that $(h + a - n) + b \leqslant a$, and we can apply (6.42) to compute the products in the last sum (this is the only place where we need (6.42)!). Note that the coefficients $c(a, b, r, t, s)$ in (6.42) do not depend on the individual values of a and r, but rather only on the difference $a - r$; so we can set $c(a, b, r, t, s) = \gamma(a - r, b, t, s)$. With this notation we have

$$S_{h,0}\Pi^{(0)}_{n-b,b} = \sum_{a=n-h}^{n} \sum_{\substack{\max(0,a+b-n) \leqslant t \leqslant b \\ 0 \leqslant s \leqslant h+a-n}} \gamma(n - h, b, t, s)\Delta\Pi^{(h+a-n-s)}_{a+b-2t,t}$$

$$= \sum_{t=0}^{b} \sum_{s=0}^{h} \gamma(n - h, b, t, s)\Delta S_{h+2t-b-s,t}.$$

From the inclusion $S_{h,0} \in O_+$ that was proved above and from (6.46) it follows that the left side of the last relation is contained in O_+. By our induction assumptions, $S_{h+2t-b-s,t} \in O_+$ if $h + t - b - s < h$ or if $h + t - b - s = h$ and $t < b$, i.e., for all possible combinations of t and s except $t = b$, $s = 0$. Consequently, the term with $t = b$ and $s = 0$ is also contained in O_+:

$$\gamma(n - h, b, b, 0)\Delta S_{h+b,b} = p^{bh}\Delta S_{h+b,b} \in O_+,$$

and hence $S_{h+b,b} \in O_+$ (see Lemma 3.27). We have thus proved (6.74), and hence also (6.73). Both sides of (6.69) are then contained in O_+, and it remains for us to prove that they have the same image under Ω.

From (6.25) and Lemma 3.34 we have

$$(6.75) \qquad \begin{aligned} \Omega(\Pi^{-2}_+\Pi_{n-i}\Pi_+) &= \Omega(\Pi^{-2}_+)\Omega(\Pi_{n-i})\Omega(\Pi_+) \\ &= (x_1 \cdots x_n)^{-1} s_{n-i}(x_1, \ldots, x_n). \end{aligned}$$

On the other hand, if we again use Lemmas 3.34 and 2.21, we find that the Ω-image of the right side of (6.69) is

$$p^{-\langle i \rangle - i(n-i) + \langle n \rangle}(x_0^2 x_1 \cdots x_n)^{-1}$$

$$\times \sum_{j=0}^{i} \alpha_{ij} \sum_{a=n-i+j}^{n} l_p(i - j + a - n, a) x_0^2 p^{-\langle a \rangle} s_a(x_1, \ldots, x_n)$$

$$= (x_1 \cdots x_n)^{-1} p^{\langle n-i \rangle} \sum_{a=n-i}^{n} p^{-\langle a \rangle} s_a(x_1, \ldots, x_n)$$

$$\times \sum_{a=n-i}^{a+i-n} \alpha_{ij} l_p(i - j + a - n, a).$$

Our goal is to prove that the last expression is the same as (6.75). For this it is clearly sufficient to verify the relations

$$\sum_{j=0}^{a+i-n} \alpha_{ij} l_p(i-j+a-n,a) = \begin{cases} 1, & \text{if } a=n-i, \\ 0, & \text{if } n-i < a \leqslant n, \end{cases}$$

which, if we set $n - a = k$, can be rewritten in the form

$$(6.76) \qquad \sum_{j=0}^{i-k} \alpha_{ij} l_p(i-j-k,n-k) = \begin{cases} 1, & \text{if } k=i, \\ 0, & \text{if } 0 \leqslant k < i. \end{cases}$$

In order to prove these relations, we must analyze the function $l_p(r,a)$ and find a way to compute it. The next two lemmas are devoted to this.

LEMMA 6.18. *For $0 \leqslant r \leqslant a$ and for p a prime, set*

$$L_p(r,a) = \{A \in S_a(\mathbf{F}_p); r_p(A) = r\}, \quad \text{where } \mathbf{F}_p = \mathbf{Z}/p\mathbf{Z},$$

and for $\mathbf{i} \in I_{r,a}$ (see (6.49)) let $L_p(r,a;\mathbf{i})$ denote the subset of matrices in $L_p(r,a)$ for which $\mathbf{i}$ is the first (in lexicographical order) set of indices of r rows (and columns) which are linearly independent modulo p. Then one can take

$$(6.77) \qquad L_p(r,a;\mathbf{i}) = \left\{ \begin{pmatrix} A & AV \\ {}^t VA & A[V] \end{pmatrix} [M(\mathbf{i})]; A \in L_p(r,r), V \in V(\mathbf{i}) \right\},$$

where $M(\mathbf{i})$ is the matrix (6.50) and $V(\mathbf{i})$ is the set defined in (6.52). In particular, the number of elements in the set (6.77) is given by the formula

$$(6.78) \qquad |L_p(r,a,\mathbf{i})| = p^{j_1+\cdots+j_{a-r}-\langle a-r\rangle} l_p(r,r),$$

where $(j_1,\ldots,j_{a-r}) = \hat{\mathbf{i}} \in I_{a-r,a}$ is the set of indices complementary to i in $(1,2,\ldots,n)$. The number of elements in the entire set $L_p(r,a)$ is given by the formula

$$(6.79) \qquad l_p(r,a) = l_p(r,r)\frac{\varphi_a(p)}{\varphi_r(p)\varphi_{a-r}(p)}.$$

PROOF. Let $A' = (a_{st}) \in L_p(r,a;\mathbf{i})$. Then, using (6.51) and the analogous relation for the rows, we obtain

$$A'[M(\mathbf{i})^{-1}] = A'' = \begin{pmatrix} A & B \\ {}^t B & C \end{pmatrix},$$

where $A = (a_{i_\alpha,i_\beta})$, $B = (a_{i_\alpha,j_\beta})$, $C = (a_{j_\alpha,j_\beta})$, and $(j_\beta) = \hat{\mathbf{i}}$. We now show that $r_p(A) = r$. In fact, $r_p(A'') = r_p(A') = r$, and, by construction, the first r columns of A'' are linearly independent modulo p. Hence, all of the columns of A'' are linear combinations modulo p of its first r columns. In particular, the same is true for the matrix (A,B), which also has rank r over $\mathbf{F}_p$. Thus, the first r columns of this last matrix cannot be linearly dependent modulo p. Further note that for each $\beta = 1,\ldots,a-r$ the βth column of B is a linear combination modulo p of the first s columns of A, where s is the largest integer such that $i_s < j_\beta$, because if the columns of (A,B) with index $i_1, \ldots, i_s, j_\beta$—and hence the columns of A' with the same indices—were linearly independent modulo p, then the index i_{s+1} could be replaced by $j_\beta < i_{s+1}$. Thus,

$$^t(a_{i_1,j_\beta}, \ldots, a_{i_r,j_\beta}) \equiv \sum_{\alpha=1}^{s} v_{\alpha\beta} \, ^t(a_{i_1,i_\alpha}, \ldots, a_{i_r,i_\alpha})$$

$$\equiv \sum_{\alpha=1}^{r} v_{\alpha\beta} \, ^t(a_{i_1,i_\alpha}, \ldots, a_{i_r,i_\alpha}) \pmod{p},$$

where $0 \leqslant v_{\alpha\beta} < p$ for $1 \leqslant \alpha \leqslant r$, $1 \leqslant \beta \leqslant a - r$, and $v_{\alpha\beta} = 0$ if $i_\alpha > j_\beta$; hence, $B \equiv AV \pmod{p}$, and $V = (v_{\alpha\beta}) \in V(\mathbf{i})$. Furthermore, since

$$r = r_p\left(\begin{pmatrix} A & AV \\ {}^tVA & C \end{pmatrix}\right) = r_p\left(\begin{pmatrix} A & 0 \\ 0 & C - A[V] \end{pmatrix}\begin{bmatrix}\begin{pmatrix} E_r & V \\ 0 & E_{a-r} \end{pmatrix}\end{bmatrix}\right)$$

and $r_p(A) = r$, it follows that $r_p(C - A[V]) = 0$, and hence $C \equiv A[V] \pmod{p}$. If we carry out the same argument in the opposite order, we readily see that any matrix of the form in (6.77) belongs modulo p to the set $L_p(r, a; \mathbf{i})$; and that two matrices of this type are incongruent modulo p if they correspond to distinct V or to matrices A which are incongruent modulo p. This proves (6.77). (7.78) follows from (6.77) and from the formula (6.54) for the number of elements in $V(\mathbf{i})$; (6.79) follows from (6.78) and (2.33).   □

The next lemma gives us explicit formulas for $l_p(r, r)$.

LEMMA 6.19. *The following formal power series identities over* $\mathbf{Q}$ *hold for any prime* p:

$$\sum_{r=0}^{\infty} \frac{l_p(r, r)}{\varphi_r(p)} v^r = \left(\sum_{a=0}^{\infty} \frac{p^{\langle a \rangle}}{\varphi_a(p)} v^a\right)\left(\sum_{b=0}^{\infty} \frac{(-1)^b p^{\langle b-1 \rangle}}{\varphi_b(p)} v^b\right),$$

$$\left(\sum_{r=0}^{\infty} \frac{l_p(r, r)}{\varphi_r(p)} v^r\right)^{-1} = \left(\sum_{a=0}^{\infty} \frac{(-p)^a}{\varphi_a(p)} v^a\right)\left(\sum_{b=0}^{\infty} \frac{1}{\varphi_b(p)} v^b\right).$$

PROOF OF THE LEMMA. Since the set $S_a(\mathbf{F}_p)$ contains $p^{\langle a \rangle}$ different matrices, it follows from (6.79) that

$$p^{\langle a \rangle} = \sum_{r=0}^{a} l_p(r, a) = \varphi_a \sum_{r=0}^{a} \frac{l_p(r, r)}{\varphi_r} \cdot \frac{1}{\varphi_{a-r}} \quad (\varphi_s = \varphi_s(p)).$$

All of these relations for $a = 0, 1, \ldots$ together are equivalent to the following single formal power series identity:

$$\left(\sum_{r=0}^{\infty} \frac{l_p(r, r)}{\varphi_r} v^r\right)\left(\sum_{c=0}^{\infty} \frac{1}{\varphi_c} v^c\right) = \sum_{a=0}^{\infty} \frac{p^{\langle a \rangle}}{\varphi_a} v^a.$$

Thus, to prove the lemma it suffices to verify the identities

(6.80)
$$\left(\sum_{c=0}^{\infty} \frac{1}{\varphi_c} v^c \right) \left(\sum_{b=0}^{\infty} \frac{(-1)^b p^{\langle b-1 \rangle}}{\varphi_b} v^b \right) = 1,$$

(6.81)
$$\left(\sum_{c=0}^{\infty} \frac{(-p)^c}{\varphi_c} v^c \right) \left(\sum_{b=0}^{\infty} \frac{p^{\langle b \rangle}}{\varphi_b} v^b \right) = 1.$$

If we set $x = p$ and $t = -p^{-1}$ in (2.34), then for $n = 0, 1, \dots$ we obtain

(6.82)
$$\varphi_n \sum_{b,c \geqslant 0, b+c=n} \frac{(-1)^b p^{\langle b-1 \rangle}}{\varphi_b} \cdot \frac{1}{\varphi_c} = \begin{cases} 1, & \text{if } n = 0, \\ 0, & \text{if } n > 0, \end{cases}$$

which together are equivalent to the power series identity (6.80). (6.81) follows from (6.80) if we replace v by $-pv$.   □

We now complete the proof of Proposition 6.17. It suffices to verify (6.76). From (6.70) for the numbers α_{ij} and from the second identity in Lemma 6.19 we obtain the congruence

$$\sum_{j=0}^{i} \alpha_{ij} \frac{\varphi_{n-i}}{\varphi_{n-i+j}} v^j = \sum_{j=0}^{i} \left(\sum_{t=0}^{j} \frac{(-p)^t}{\varphi_t \varphi_{j-t}} \right) v^j \equiv \left(\sum_{a=0}^{\infty} \frac{(-p)^a}{\varphi_a} v^a \right) \left(\sum_{b=0}^{\infty} \frac{1}{\varphi_b} v^b \right)$$

$$= \left(\sum_{r=0}^{\infty} \frac{l_p(r,r)}{\varphi_r} v^r \right)^{-1} \pmod{v^{i+1}},$$

which implies the relations

$$\sum_{j,r \geqslant 0, j+r=i-k} \alpha_{ij} \frac{\varphi_{n-i} l_p(r,r)}{\varphi_{n-i+j} \varphi_r} = \begin{cases} 1, & \text{if } i - k = 0, \\ 0, & \text{if } 0 < i - k \leqslant i, \end{cases}$$

where $0 \leqslant k \leqslant i$. Since the right side is nonzero only when $i = k$, the factor φ_{n-i} can be replaced by φ_{n-k}. If we set $r = i - j - k$, then by (6.79) we obtain

$$\frac{\varphi_{n-k} l_p(r,r)}{\varphi_{n-i+j} \varphi_r} = l_p(i-j-k, i-j-k) \frac{\varphi_{n-k}}{\varphi_{i-j-k} \varphi_{n-k-(i-j-k)}} = l_p(i-j-k, n-k).$$

We now compute the coefficients in the expansions (6.31) and (6.32) of the polynomial $\widehat{\mathbf{R}}_p^n(v)$. To do this we first introduce the new functions

(6.83)
$$\varphi_r^+(v) = \prod_{\substack{1 \leqslant i \leqslant r \\ i \equiv 0 (\bmod 2)}} (v^i - 1), \quad \varphi_r^-(v) = \prod_{\substack{1 \leqslant i \leqslant r \\ i \equiv 1 (\bmod 2)}} (v^i - 1).$$

PROPOSITION 6.20. *In the ring* $\overline{\mathbf{L}}_{0,p}^n$ *one has*:

(6.84)

$$\Pi_-\Pi_i(\Pi_-^2)^{-1} = p^{-\langle i\rangle - i(n-i)}\Delta^{-1}\sum_{j=0}^{i}\widehat{\alpha}_{ij}\sum_{a=n-i+j}^{n}\Pi_{a,n-a}^{(i-j+a-n)}(k),$$

(6.85)

$$(\Pi_+^2)^{-1}\Pi_{n-i}\Pi_+ = p^{-\langle i\rangle - i(n-i)}\Delta^{-1}\sum_{j=0}^{i}\widehat{\alpha}_{ij}\sum_{a=n-i+j}^{n}\Pi_{a,0}^{(i-j+a-n)}(k),$$

where

(6.86) $\qquad \widehat{\alpha}_{ij} = \widehat{\alpha}_{ij}^n(p) = \begin{cases} \dfrac{(-p)^{j/2}\varphi_{n-i+j}(p)}{\varphi_j^+(p)\varphi_{n-i}(p)}, & \text{if } j \equiv 0\,(\mathrm{mod}\,2), \\ 0, & \text{if } j \equiv 1\,(\mathrm{mod}\,2), \end{cases}$

$\Pi_{a,b}^{(r)}(k)$ *is given by* (4.106), *and the rest of the notation is the same as in Proposition* 6.9.

PROOF. In view of (6.12), (6.33), and (5.39), the anti-automorphism $*$ takes (6.84) to (6.85) and conversely. Hence, it suffices to prove, say, (6.85). Just as in the proof of Proposition 6.17, using (4.105) and (6.43) we see that

$$S_{i,c}(k) = \sum_{a=n-i}^{n-c}\Pi_{a,c}^{(i+a-n)}(k) \in \widehat{O}_{+p}^n$$

for $0 \leqslant c \leqslant i \leqslant n$, and the right side of (6.85) can be expressed as a linear combination of the elements $\Delta^{-1}S_{i-j,0}(k)$, $j = 0, 1, \ldots, i$. Since (6.16) and (6.46) imply that the left side of (6.85) is also contained in $\widehat{O}_{+p}^n$, it follows from Theorem 6.5(2) that to prove (6.85) we need only show that both sides have the same image under $\Omega = \Omega_p^n$.

Using (4.109) and Lemma 2.21, we have

$$\Omega(\Pi_{a,0}^{(i-j+a-n)}(k)) = l_p^k(i-j+a-n, a)x_0^2 p^{-\langle a\rangle}s_a(x_1, \ldots, x_n).$$

From this and Lemma 2.21 we find that the Ω-image of the right side of (6.85) is

$$(x_1 \cdots x_n)^{-1}p^{\langle n-i\rangle}\sum_{a=n-i}^{n}p^{-\langle a\rangle}s_a(x_1, \ldots, x_n)\sum_{j=0}^{a+i-n}\widehat{\alpha}_{ij}l_p^k(i-j+a-n, a).$$

Since, by (6.25) and Lemma 3.34,

$$\Omega((\Pi_+^2)^{-1}\Pi_{n-i}\Pi_+) = (x_1 \cdots x_n)^{-1}s_{n-i}(x_1, \ldots, x_n),$$

we see that to prove (6.85) it remains to verify the relations

(6.87) $\qquad \displaystyle\sum_{j=0}^{a+i-n}\widehat{\alpha}_{ij}l_p^k(i-j+a-n, a) = \begin{cases} 1, & \text{if } a = n-i, \\ 0, & \text{if } n-i < a \leqslant n. \end{cases}$

LEMMA 6.21. *If p is an odd prime and $0 \leqslant r \leqslant a$, then*

$$(6.88) \qquad l_p^k(r,a) = l_p^k(r,r) \frac{\varphi_a(p)}{\varphi_r(p)\varphi_{n-r}(p)},$$

and the sum $l_p^k(r,r)$ is equal to

$$(6.89) \qquad \prod_{j=1}^{r/2} p^{2j-1}(p^{2j-1}-1) \quad or \quad 0$$

for even and odd r, respectively.

PROOF. Any matrix M in the set (6.77) satisfies the congruence

$$M \equiv \begin{pmatrix} A & 0 \\ 0 & 0 \end{pmatrix} \left[\begin{pmatrix} E_r & V \\ 0 & E_{a-r} \end{pmatrix} M(\mathbf{i}) \right] \pmod{p}.$$

Hence, $\varkappa(M) = \varkappa(A)$, and (6.88) is a consequence of (6.77)–(6.79). To prove (6.89), we write the sum $l_p^k(r,a)$ (see (4.110)) in the form

$$(6.90) \qquad \sum_{d=0}^{\infty} \sum_{\substack{Z \in S_{a-1}(\mathbf{F}_p) \\ r_p(Z)=d}} \sum_{\substack{A \in S_a(\mathbf{F}_p), r_p(A)=r \\ A^{(a-1)} \equiv Z \pmod{p}}} \varkappa(A)^{-k},$$

where $A^{(s)}$ is the $s \times s$-block in the upper-left corner of A. By Theorem 1.3 of Appendix 1 and the formula

$$(6.91) \qquad \varkappa(A[U]) = \varkappa(A) \quad \text{for } A \in S_a(\mathbf{F}_p) \text{ and } U \in GL_a(\mathbf{F}_p),$$

which follows from (4.70), we may suppose that $Z = \begin{pmatrix} Z_1 & 0 \\ 0 & 0 \end{pmatrix}$ in the inner sum in (6.90), where Z_1 is a $d \times d$-matrix that is nonsingular modulo p. Thus, the matrix A in (6.90) has the form

$$(6.92) \qquad \begin{pmatrix} Z_1 & 0 & X_1 \\ 0 & 0 & X_2 \\ {}^tX_1 & {}^tX_2 & Y \end{pmatrix} = \begin{pmatrix} Z_1 & 0 & X_1 \\ 0 & 0 & X_2 \\ 0 & {}^tX_2 & Y_1 \end{pmatrix} \left[\begin{pmatrix} E_d & 0 & Z_1^{-1}X_1 \\ 0 & E_{a-d-1} & 0 \\ 0 & 0 & E_1 \end{pmatrix} \right],$$

where $Y_1 = Y - Z_1^{-1}[X_1]$, and X_1, X_2, Y are columns of length d, $a-d-1$, 1, respectively.

If $r_p(X_2) = 1$, then ${}^tX_2 U = (0,\dots,0,1)$ for some $U \in GL_{(a-d-1)}(\mathbf{F}_p)$, and hence

$$X \left[\begin{pmatrix} U & 0 \\ 0 & 1 \end{pmatrix} \right] = \begin{pmatrix} 0 & 0 \\ 0 & Y_2 \end{pmatrix} \quad \text{with } X = \begin{pmatrix} 0 & X_2 \\ {}^tX_2 & Y_1 \end{pmatrix}, Y_2 = \begin{pmatrix} 0 & 1 \\ 1 & Y_1 \end{pmatrix}.$$

Since $\varkappa(Y_2) = \varepsilon_p^{-2}(\frac{-1}{p}) = 1$ by (4.70), it follows from the last relation and (6.91) that

$$(6.93) \qquad \varkappa(X) = \varkappa(Y_2) = 1 \quad \text{and} \quad r_p = 2.$$

On the other hand, if $r_p(X_2) = 0$, then we obviously have

$$(6.94) \qquad \varkappa(X) = \varkappa(Y_1) \quad \text{and} \quad r_p(X) = r_p(Y_1).$$

If we now use (6.92) along with (6.30) and (6.91), we find that $\varkappa(A) = \varkappa(Z_1)\varkappa(X)$ $= \varkappa(Z)\varkappa(X)$. This implies that the inner sum in (6.90) is equal to

$$(6.95) \qquad p^d \varkappa(Z)^- \sum_{\substack{X \in S_{a-d}(\mathbf{F}_p) \\ r_p(X)=r-d}} \varkappa(X)^{-k} = p^d \varkappa(Z)^{-k} \sigma(r - d).$$

The relations (6.93) and (6.94) show that the following equalities hold for $\sigma(p)$:

$$\sigma(0) = 1, \quad \sigma(1) = 0, \quad \sigma(2) = p(p^{a-d-1} - 1),$$
$$\sigma(\rho) = 0 \quad \text{for } \rho > 2.$$

From this, (6.90) and (6.95) we find that $l_p^k(r, a)$ is equal to

$$(6.96) \qquad p^{r-1}(p^{a-r-1} - 1)l_p^k(r - 2, a - 1) + p^r l_p^k(r, a - 1).$$

Thus, if we set $a = r$ and use (6.88), we obtain

$$l_p^k(r, r) = p^{r-1}(p^{r-1} - 1)l_p^k(r - 2, r - 2).$$

From this relation, using induction, we obtain (6.89), since $l_p^k(0, 0) = 1$ and $l_p^k(1, 1) = 0$ (recall that k is odd by assumption). $\qquad \square$

We return to the proof of Proposition 6.20. We first make the substitution $a = n - b$ in (6.87) and use (6.88). This transforms the system to the form

$$\sum_{j=0}^{i-b} \widehat{\alpha}_{ij} \frac{l_p^k(i - j - b, i - j - b)\varphi_{n-b}}{\varphi_{i-j-b}\varphi_{n-i+j}} = \begin{cases} 1 & \text{for } b = i, \\ 0 & \text{for } 0 \leqslant b < i. \end{cases}$$

After making another substitution $\varphi_{n-b} \to \varphi_{n-i}$ ($\varphi_s = \varphi_s(p)$) and $i - b - j \to r$, we obtain the new system of equalities

$$\sum_{j,r \geqslant 0, j+r=i-b} \frac{\varphi_{n-i}}{\varphi_{n-i+j}} \widehat{\alpha}_{ij} \cdot \frac{l_p^k(r, r)}{\varphi_r} = \begin{cases} 1 & \text{for } i - b = 0, \\ 0 & \text{for } 0 < i - b \leqslant i, \end{cases}$$

which is obviously equivalent to the congruence

$$(6.97) \qquad \left(\sum_{j=0}^{i} \frac{\varphi_{n-i}}{\varphi_{n-i+j}} \widehat{\alpha}_{ij} v^j \right) \left(\sum_{r=0}^{\infty} \frac{l_p^k(r, r)}{\varphi_r} v^r \right) \equiv 1 \pmod{v^{i+1}}.$$

Hence, the proof of (6.85) reduces to the inversion of the infinite formal series in this congruence. With this in mind, we prove the following identity.

LEMMA 6.22. *In the notation* (4.110) *and* (6.83),

$$\sum_{b=0}^{a} \frac{\varphi_{2a}^+(p)}{\varphi_{2a-2b}(p)} \cdot \frac{l_p^k(2b, 2b)}{\varphi_{2b}(p)} v^{2b} = \prod_{j=1}^{a} (1 + p^{2j-1} v^2).$$

PROOF. From (6.89) and the definition of the polynomials (2.29) and (6.83) it follows that

$$(6.98) \qquad \frac{l_p^k(2b, 2b)}{\varphi_{2b}(p)} = \frac{p^{b^2}}{\varphi_{2b}^+(p)},$$

and from this the lemma is obtained by induction on a. $\qquad\square$

If we now set $v = \sqrt{-1} \cdot p^{-1/2}$ in Lemma 6.22, we obtain the system of equalities

$$\sum_{b=0}^{a} \frac{(-p)^{a-b}}{\varphi_{2a-2b}^+(p)} \cdot \frac{l_p^k(2b, 2b)}{\varphi_{2b}(p)} = \begin{cases} 1 & \text{for } a = 0, \\ 0 & \text{for } a > 0, \end{cases}$$

which, together with (6.89), implies the formal power series identity

$$\left(\sum_{r=0}^{\infty} \frac{l_p^k(r, r)}{\varphi_r(p)} v^p \right) \left(\sum_{c=0}^{\infty} \frac{(-p)^c}{\varphi_{2c}^+(p)} v^{2c} \right) = 1.$$

If we compare this identity with (6.97), by (6.86) we obtain the congruence

$$\sum_{j=0}^{i} \frac{\varphi_{n-i}(p)}{\varphi_{n-i+j}(p)} \widehat{\alpha}_{ij} v^j \equiv \sum_{c=0}^{\infty} \frac{(-p)^c}{\varphi_{2c}^+(p)} v^{2c} \pmod{v^{i+1}}. \qquad\square$$

PROBLEM 6.23. Let $\Pi_{\pm} = \Pi_{\pm}^n(p)$ be the Frobenius elements of the Hecke ring $\mathbf{L}_{0,p}^n$, and for $d \geqslant 1$ let $\Pi_{\pm}^{-d}$ be defined by the recursive relations (6.10). Prove that the negative squares of the Frobenius elements are given by the formulas

$$\Pi_{-}^{-2} = (p^{\langle n \rangle} \Delta)^{-2} \sum_{j=0}^{n} \alpha_{nj}^n(p) \sum_{a=j}^{n} \Pi_{a,n-a}^{(a-j)},$$

$$\Pi_{+}^{-2} = (p^{\langle n \rangle} \Delta)^{-2} \sum_{j=0}^{n} \alpha_{nj}^n(p) \sum_{a=j}^{n} \Pi_{a,0}^{(a-j)},$$

where $\alpha_{ij}^n(p)$ are the coefficients (6.70). For $d > 1$ show that $\Pi_{-}^{-d} \neq (\Pi_{-}^{-1})^d$ and $\Pi_{+}^{-d} \neq (\Pi_{+}^{-1})^d$. For p an odd prime obtain analogous formulas for the elements $(\Pi_{\pm}^2)^{-1}$ that are defined by the recursive relations (6.11).

[**Hint:** Use Propositions 6.17 and 6.20.]

5. Symmetric factorization of Rankin polynomials. Here we complete our study of the factorization of $\mathbf{R}_p^n(v)$ and $\widehat{\mathbf{R}}_p^n(v)$ by obtaining factorizations that are invariant relative to the anti-automorphism $*$. These factorizations will play a fundamental role in the applications. To derive them we use the factorizations in Proposition 6.9, the multiplication formulas in Propositions 6.17 and 6.20, and certain additional considerations.

As in the rest of this section, we suppose that $n \in \mathbf{N}$, p is a fixed prime number, and $(p, q) = 1$.

THEOREM 6.24. *Let* $\mathbf{R}_p^n(v) \in \mathbf{L}_p^n[v]$ *and* $\widehat{\mathbf{R}}_p^n(v) \in \widehat{\mathbf{E}}_p^n(q, \varkappa)[v]$ *be the polynomials defined in* (6.27). *These polynomials have the following factorizations over the Hecke ring* $\overline{\mathbf{L}}_{0,p}^n$:

$$(6.99) \qquad \mathbf{R}_p^n(v) = X_-(v) \left(\sum_{i=0}^n (-1)^i b_i v^i \right) X_+(v),$$

$$(6.100) \qquad \widehat{\mathbf{R}}_p^n(v) = X_-(v) \left(\sum_{i=0}^n (-1)^i \widehat{b}_i v^i \right) X_+(v),$$

where, in the notation (3.62) *and* (4.106), *for* $i = 0, 1, \ldots, n$

$$(6.101) \qquad X_-(v) = X_{-p}^n(v) = \sum_{i=0}^n (-1)^i (p^{\langle n \rangle} \Delta)^{-1} \Pi_- \Pi_{n-i} v^i,$$

$$(6.102) \qquad X_+(v) = X_{+p}^n(v) = \sum_{i=0}^n (-1)^i (p^{\langle n \rangle} \Delta)^{-1} \Pi_i \Pi_+ v^i,$$

$$(6.103) \qquad b_i = b_i^n(p) = p^{-\langle i \rangle - i(n-i)} \Delta^{-1} \sum_{j=0}^i \alpha_{ij} \Pi_{n,0}^{(i-j)}$$

with coefficients $\alpha_{ij} = \alpha_{ij}^n(p)$ *defined by* (6.70), *and*

$$(6.104) \qquad \widehat{b}_i = \widehat{b}_i^n(p) = p^{-\langle i \rangle - i(n-i)} \Delta^{-1} \sum_{j=0}^i \widehat{\alpha}_{ij} \Pi_{n,0}^{(i-j)}(k)$$

with coefficients $\widehat{\alpha}_{ij} = \widehat{\alpha}_{ij}^n(p)$ *defined by* (6.86).

We first prove a lemma.

LEMMA 6.25. *The following formal power series identities hold over the ring* $\mathbf{L}_{0,p}^n$:

$$(6.105) \qquad \begin{aligned} X_-(v)^{-1} &= \sum_{d=0}^\infty p^{-dn} t^-(p^d) v^d, \\ X_+(v)^{-1} &= \sum_{d=0}^\infty p^{-dn} t^+(p^d) v^d, \end{aligned}$$

where $X_-(v)$ *and* $X_+(v)$ *are the polynomials* (6.101) *and* (6.102), *and*

$$(6.106) \qquad t^-(p^d) = t_n^-(p^d) = \sum_{\substack{D \in \Lambda^n \backslash M_n / \Lambda^n \\ \det D = \pm p^d}} (U(D^*))_{\Gamma_0^n},$$

$$(6.107) \qquad t^+(p^d) = t_n^+(p^d) = \sum_{\substack{D \in \Lambda^n \backslash M_n / \Lambda^n \\ \det D = \pm p^d}} (U(D))_{\Gamma_0^n}.$$

PROOF. The two identities in (6.105) have analogous proofs; moreover, the anti-automorphism $*$ (applied coefficient by coefficient) takes one into the other (see (6.33)). Hence, it suffices to prove, say, the second identity in (6.105). From (6.46) and Proposition 5.5 it follows that all of the coefficients on both sides of this identity are contained in the ring $C_+ = C_{+p}^n$. Thus, by Theorem 6.5(2), the identity will be proved if we verify that

$$(6.108) \qquad \left(\sum_{i=0}^n (-1)^i \Phi(x_i) v^i \right)^{-1} = \sum_{d=0}^\infty p^{-dn} \Phi(t^+(p^d)) v^d,$$

where the x_i are the coefficients of the polynomial (6.102) and $\Phi = \Phi_p^n$. By Lemma 3.34, we have $\Phi(x_i) = p^{\langle i \rangle} \pi_i^n(p)$. To compute the coefficients on the right in (6.108) we use the expansions (5.23) of the double cosets; using the definitions, we see that this gives the formulas

$$\Phi(t^+(p^d)) = \sum_{\substack{D \in \Lambda^n \backslash M_n / \Lambda^n \\ \det D = \pm p^d}} |S_n / {}^t D S_n D| (\Lambda^n D).$$

We note that if $D \in M_n$ and $\det D = \pm m$, then

$$(6.109) \qquad\qquad |S_n / {}^t D S_n D| = m^{n+1}.$$

In fact, the index on the left obviously depends only on the double coset $\Lambda^n D \Lambda^n$, and so, by Lemma 2.2, the matrix D can be replaced by its elementary divisor matrix $\mathrm{ed}(D) = \mathrm{diag}(d_1, \ldots, d_n)$, for which the index is

$$\prod_{1 \leqslant i \leqslant j \leqslant n} d_i d_j = (d_1 \cdots d_n)^{n+1} = m^{n+1}.$$

Thus,

$$\Phi(t^+(p^d)) = p^{d(n+1)} t(p^d),$$

where $t(p^d)$ is the element (2.10) of the ring H^n. These formulas show that the identity (6.108) that we want to prove is nothing other than the identity in Proposition 2.22 with v replaced by pv. $\square$

PROOF OF THE THEOREM. Using the notation of Proposition 6.9, we define the following polynomials:

$$Y_-(v) = \sum_{i=0}^n (-1)^i \Pi_+^{-2} \Pi_{n-i} \Pi_+ v^i, \qquad Y_+(v) = \sum_{i=0}^n (-1)^i \Pi_- \Pi_i \Pi_-^{-2} v^i,$$

$$\widehat{Y}_-(v) = \sum_{i=0}^n (-1)^i (\Pi_+^2)^{-1} \Pi_{n-i} \Pi_+ v^i, \qquad \widehat{Y}_+(v) = \sum_{i=0}^n (-1)^i \Pi_- \Pi_i (\Pi_-^2)^{-1} v^i.$$

Then, by Proposition 6.9, we have the factorizations

$$(6.110) \qquad \begin{aligned} \mathbf{R}_p^n(v) &= X_-(v) Y_+(v) = Y_-(v) X_+(v), \\ \widehat{\mathbf{R}}_p^n(v) &= X_-(v) \widehat{Y}_+(v) = \widehat{Y}_-(v) X_+(v), \end{aligned}$$

and hence

$$X_-(v)^{-1} Y_-(v) = Y_+(v) X_+(v)^{-1} \quad \text{and} \quad X_-(v)^{-1} \widehat{Y}_-(v) = \widehat{Y}_+(v) X_+(v)^{-1}.$$

If we let $B(v)$ and $\widehat{B}(v)$ denote these formal power series, and let $(-1)^i b_i$ and $(-1)^i \widehat{b}_i$ denote their coefficients, we obtain the identities

(6.111)
$$B(v) = \sum_{i=0}^{\infty} (-1)^i b_i v^i = X_-(v)^{-1} Y_-(v) = Y_+(v) X_+(v)^{-1},$$

$$\widehat{B}(v) = \sum_{i=0}^{\infty} (-1)^i \widehat{b}_i v^i = X_-(v)^{-1} \widehat{Y}_-(v) = \widehat{Y}_+(v) X_+(v)^{-1},$$

from which, using (6.110), we have:

$$\mathbf{R}_p^n(v) = X_-(v) B(v) X_+(v) \quad \text{and} \quad \widehat{\mathbf{R}}_p^n(v) = X_-(v) \widehat{B}(v) X_+(v).$$

Thus, to prove the theorem it suffices to verify that $B(v)$ and $\widehat{B}(v)$ are actually polynomials of degree n, and that their coefficients are given by (6.103) and (6.104). To do this we need some preliminary observations.

The map

(6.112)
$$S_0^n \ni M = \begin{pmatrix} A & B \\ 0 & D \end{pmatrix} \to s(M) = \det D (\det A)^{-1},$$

where S_0^n is the triangular subgroup (3.43) of $S^n = S_{\mathbf{Q}}^n$, is obviously a homomorphism from S_0^n to the multiplicative group of positive rational numbers. For p a prime, the p-order of $s(M)$ will be referred to as the *p-signature* of M and of the corresponding double coset $(M)_{\Gamma_0}$, and we shall denote it by σ_p:

(6.113)
$$\sigma_p(M) = \sigma_p((M)_{\Gamma_0}) = \nu_p(s(M)) = \nu_p(\det D) - \nu_p(\det A).$$

A linear combination of double cosets $(M_i)_{\Gamma_0}$ is said to be σ_p-*homogeneous* if all of the double cosets that occur with nonzero coefficients have the same p-signature; in that case this p-signature is called the *p-signature* of the linear combination of double cosets, again denoted σ_p. Clearly, if two linear combinations X and Y are σ_p-homogeneous, then so is their product $X \cdot Y$, and we have

(6.114)
$$\sigma_p(X \cdot Y) = \sigma_p(X) + \sigma_p(Y), \quad \text{if } X \cdot Y \neq 0.$$

Returning to the proof of the theorem, we consider the subspaces L_0^-, L_0^+, L_0^0, I^-, and I^+ of the space $\overline{L}_0 = \overline{L}_{0,p}^n$ that consist of all (finite) linear combinations of σ_p-homogeneous elements whose p-signature is, respectively, nonpositive, nonnegative, zero, negative, and positive. The spaces L_0^-, L_0^+, and L_0^0 are clearly subrings of $\overline{L}_0$, and I^- and I^+ are two-sided ideals of the rings L_0^- and L_0^+, respectively. From the definitions, it follows that the following elements of $\overline{L}_0$ are σ_p-homogeneous with p-signature as given below:

(6.115)
$$\sigma_p(\Pi_a) = 2a - n, \quad \sigma_p(\Pi_{a,b}^{(r)}) = \sigma_p(\Pi_{a,b}^{(r)}(k)) = 2(a + 2b - n),$$

$$\sigma_p(\Delta) = 0, \quad \sigma_p(t^-(p^d)) = -2d, \quad \sigma_p(t^+(p^d)) = 2d.$$

In particular, all of the coefficients of the series $X_-(v)^{-1}$ lie in the ring L_0^-, and all of the coefficients of $X_+(v)^{-1}$ lie in L_0^+. On the other hand, the above formulas and Propositions 6.17 and 6.20 show that all of the coefficients of the polynomials $Y_+(v)$ and $\widehat{Y}_+(v)$ lie in L_0^+, and all of the coefficients of the polynomials $Y_-(v)$ and $\widehat{Y}_-(v)$

lie in L_0^-. From these observations and (6.111) it follows that all of the coefficients in the series $B(v)$ and $\widehat{B}(v)$ are contained both in L_0^- and in L_0^+, i.e.,

$$(6.116) \qquad b_i, \widehat{b}_i \in L_0^- \cap L_0^+ = L_0^0 \quad \text{for } i = 0, 1, \ldots.$$

We now examine the coefficients b_i modulo I^-. By (6.106) and (6.115), all of the coefficients of $X_-(v)^{-1}$ except for the constant term lie in the ideal I^-, and the constant term is 1; hence, if we pass to congruence modulo I^- coefficient by coefficient in the equation $B(v) = X_-(v)^{-1} Y_-(v)$, we find that $B(v) \equiv Y_-(v) \pmod{I^-}$, i.e.,

$$(6.117) \qquad b_i \equiv \Pi_+^{-2} \Pi_{n-i} \Pi^+ \pmod{I^-} \quad \text{for } 0 \leqslant i \leqslant n,$$

$$(6.118) \qquad b_i \equiv 0 \pmod{I^-} \quad \text{for } i > n.$$

Since $L_0^0 \cap I^- = \{0\}$, it follows from (6.116) and (6.118) that $b_i = 0$ for $i > n$. Hence, $B(v)$ is a polynomial of degree n. Furthermore, it follows from (6.115) that $\Delta^{-1} \Pi_{a,0}^{(r)} \in I^-$ if $a < n$. Thus, from (6.117) and (6.69) we obtain the following congruences for $0 \leqslant i \leqslant n$:

$$b_i \equiv p^{-\langle i \rangle - i(n-i)} \Delta^{-1} \sum_{j=0}^{i} \alpha_{ij} \Pi_{n,0}^{(i-j)} \pmod{I^-}.$$

If we take into account that in each of these congruences both sides are contained in L_0^0 and if we recall that $L_0^0 \cap I^- = \{0\}$, we see that the two sides are actually equal. We then carry out an analogous argument with the coefficients $\widehat{b}_i$, using (6.85); we find that for $0 \leqslant i \leqslant n$ the coefficients $\widehat{b}_i$ satisfy the inequalities (6.104), and $\widehat{b}_i = 0$ for $i > n$. $\qquad \square$

CHAPTER 4

Hecke Operators

Modular forms arose as a result of abstracting the analytic and group properties of the generating series for the number of integral representations of positive definite integral quadratic forms by one another. Thus, the basic object of arithmetic interest in the theory and application of modular forms was and continues to be the Fourier coefficients regarded as a number-theoretic function. As we saw in §§1.1 and 1.5 of Chapter 3, the Hecke rings of the symplectic group act as rings of linear operators on spaces of modular forms. The Hecke operators, which act on modular forms and hence on their Fourier coefficients, make it possible to carry the various relations between elements of the Hecke rings over to these number-theoretic functions, thereby revealing multiplicative properties of the Fourier coefficients. These properties are reflected in the Euler products of the Dirichlet series (zeta-functions) that are constructed from the Fourier coefficients of eigenfunctions of the Hecke operators.

§1. Hecke operators for congruence subgroups of the modular group

In this section we define the Hecke operators for a broad class of congruence subgroups, and we examine some of their properties.

1. Hecke operators. Just as in §3.1 of Chapter 3, when we speak of Hecke rings for congruence subgroups $K \subset \Gamma^n$, we have to impose certain restrictions on K. In addition, one must restrict the class of characters of the group K in order to define the representations of these rings on modular forms. Thus, let K be a congruence subgroup of Γ^n, and let χ be a *congruence character* of K, i.e., a character whose kernel contains a principal congruence subgroup $\Gamma^n(q) \subset K$. As in §3.1 of Chapter 3, we suppose that K satisfies the q-symmetry condition ((3.4) of Chapter 3). As for χ, we suppose that it can be extended to a homomorphism X from the group $S(K)$ to $\mathbf{C}^*$:

$$(1.1) \qquad\qquad X: S(K) \to \mathbf{C}^*, \quad X|_K = \chi.$$

For brevity, we shall refer to a pair (K, χ) satisfying these conditions as a *q-regular pair* (*of degree n*).

We consider the space $\mathfrak{M}_k(K, \chi)$ of modular forms of weight k and character χ for the group K, where k is an integer and (K, χ) is a q-regular pair. To every element of the Hecke ring $L(K) = D_{\mathbf{Q}}(K, S(K))$ we shall associate a linear operator on this space. According to the scheme in §1.5 of Chapter 3, to do this we first need a suitable automorphy factor of the group $S(K)$. For $M = \begin{pmatrix} A & B \\ C & D \end{pmatrix} \in S(K)$ and $Z \in \mathbf{H}_n$ we set

$$(1.2) \qquad\qquad \varphi_{k,X}(M, Z) = X(M) \det(CZ + D)^k.$$

By Lemmas 4.2 and 4.1(2) of Chapter 1, the function $\varphi_{k,X}$ is an automorphy factor of $S(K)$ on $\mathbf{H}_n$ with values in $\mathbf{C}^*$. Then, by Lemma 4.1(3) of Chapter 1, we can define an action of the group $S(K)$ on functions $F: \mathbf{H}_n \to \mathbf{C}$:

$$S(K) \ni M: F \to F|_{\varphi_{k,X}} M = F|_{k,X} M$$

$$(1.3) \qquad\qquad = \varphi_{k,X}(M, Z)^{-1} F(M\langle Z \rangle) = X(M)^{-1} F|_k M,$$

where $|_k M$ is the operator (3.14) of Chapter 2, which satisfies the relations

$$(1.4) \qquad F|_{k,X} M_1 |_{k,X} M_2 = F|_{k,X} M_1 M_2 \quad (M_i \in S(K)).$$

Since $X(M) = \chi(M)$ if $M \in K$, the condition (2.4) of Chapter 2 in the definition of modular forms of weight k and character χ for K can be rewritten in the above notation in the form

$$(1.5) \qquad F|_{k,X} M = F \quad \text{for all } M \in K.$$

Thus, nothing prevents us from defining the action of the ring $L(K)$ on $\mathfrak{M}_k(K, \chi)$ in the same way as we did for general Hecke rings in §§1.1 and 1.5 of Chapter 3. The only difference is that in the general case the automorphic forms were defined using only functional equations of the type (1.5), while functions in $\mathfrak{M}_k(K, \chi)$ must also satisfy certain analytic conditions. Thus, if $F \in \mathfrak{M}_k(K, \chi)$ and $T = \sum_i \alpha_i(KM_i) \in L(K)$, we set

$$(1.6) \qquad F|T = F|_{k,X} T = \sum_i \alpha_i F|_{k,X} M_i.$$

From (1.4) and (1.5) it immediately follows that the function $F|T$ does not depend on the choice of representatives M_i in the left cosets KM_i; from the definition of the Hecke rings and from (1.4) we then find that

$$(F|T)|_{k,X} M = \sum_i \alpha_i F|_{k,X} M_i |_{k,X} M$$

$$= \sum_i \alpha_i F|_{k,X} M_i M = F|T, \quad \text{if } M \in K,$$

i.e., $F|T$ satisfies all of the functional equations (1.5) for modular forms in $\mathfrak{M}_k(K, \chi)$ whenever F does. From (1.3), (1.6), and Proposition 3.8 of Chapter 2, we see that the operator $|T$ also preserves the analytic properties of modular forms.

We now suppose that $K \subset \Gamma_0^n(4)$, and consider the space $\mathfrak{M}_{k/2}(K, \chi)$ of modular forms of half-integer weight $k/2$. According to (3.19) of Chapter 2, in this case we can take the function

$$(1.7) \qquad\qquad \varphi_{k/2,X}(\widehat{M}, Z) = X(M) \varphi(Z)^k,$$

where $\widehat{M} = (M, \varphi) \in \widehat{S}(K)$, as the automorphy factor of the group $\widehat{S}(K) = P^{-1}(S(K))$, where P is the homomorphism (4.1) of Chapter 3, on $\mathbf{H}_n$ with values in $\mathbf{C}^*$. Using (3.21) and (3.22) of Chapter 2, we see that if for any function F on $\mathbf{H}_n$ we set

$$(1.8) \qquad F|_{k/2,X} \widehat{M} = \varphi_{k/2,X}(\widehat{M}, Z)^{-1} F(M\langle Z \rangle) = X(M)^{-1} F|_{k/2} \widehat{M},$$

then for any $\widehat{M}_1$ and $\widehat{M}_2 \in \widehat{S}(K)$ we have

$$(1.9) \qquad\qquad F|_{k/2,X} \widehat{M}_1 |_{k/2,X} \widehat{M}_2 = F|_{k/2,X} \widehat{M}_1, \widehat{M}_2$$

and the condition (2.5) of Chapter 2 in the definition of a modular form in $\mathfrak{M}_{k/2}(K,\chi)$ can be written in the form

$$(1.10) \qquad F_{k/2,X}\widehat{M} = F \quad \text{for all } \widehat{M} \in \widehat{K} = j^n(K),$$

where j^n is the monomorphism (4.2) of Chapter 3. As for the ring $L(K)$, by Lemma 3.4 of Chapter 2 and Lemma 1.7 of Chapter 3 it can be lifted to the ring $\widehat{L}(K) = D_Q(\widehat{K}, \widehat{S}(K))$. By analogy to (1.6), we find that the formula

$$(1.11) \qquad F|\widehat{T} = F|_{k/2,X}\widehat{T} = \sum_i \alpha_i F|_{k/2,X}\widehat{M}_i$$

for $\widehat{T} = \sum_i \alpha_i(\widehat{K}\widehat{M}_i) \in \widehat{L}(K)$ gives a representation of $\widehat{L}(K)$ in the space of modular forms $\mathfrak{M}_{k/2}(K,\chi)$.

Thus, from the above observations, the definition of modular forms, and Proposition 1.14 of Chapter 3 we have the following

PROPOSITION 1.1. *Let (K,χ) be a q-regular pair of degree $n \geq 1$, let $w = k$ or $k/2$ be an integer or half-integer, and suppose that $K \subset \Gamma_0^n(4)$ if $w = k/2$. Then any operator $|_{w,X}\tau$, where $\tau \in L(K)$ or $\tau \in \widehat{L}(K)$ depending on whether $w = k$ or $w = k/2$, respectively, takes the space $\mathfrak{M}_w(K,\chi)$ to itself. The map $\tau \to |_{w,X}\tau$ is a linear homomorphism from the corresponding Hecke ring to the ring of endomorphisms of $\mathfrak{M}_w(K,\chi)$. In particular,*

$$(1.12) \qquad F|\tau_1|\tau_2 = F|\tau_1\tau_2 \quad \text{for } F \in \mathfrak{M}_w(K,\chi).$$

The operators $|_{w,X}\tau$ on the space $\mathfrak{M}_w(K,\chi)$ are called *Hecke operators.*

Our definition (1.6) and (1.11) of the Hecke operators is somewhat arbitrary, since the extension X of the character χ to $S(K)$ can be chosen in different ways. Although this element of choice has little effect on the Hecke operators, for convenience in later computations we would like to remove it. We first describe all possible extensions of χ to $S(K)$.

LEMMA 1.2. *Let (K,χ) be a q-regular pair of degree $n \geq 1$, and let ρ be an arbitrary homomorphism from $S(\Gamma^n(q))$ to $\mathbf{C}^*$ that is trivial on $\Gamma^n(q)$. Then there exists a unique homomorphism $X = X_{\rho,\chi}$ from $S(K)$ to $\mathbf{C}^*$ whose restriction to K coincides with χ and whose restriction to $S(\Gamma^n(q))$ coincides with ρ.*

PROOF. By assumption, there exists an extension X_0 of the homomorphism χ: $K \to \mathbf{C}^*$ to $S(K)$. This implies that the character χ satisfies the following condition:

$$(1.13) \quad \text{if } M\gamma = \gamma'M', \text{ where } M, M' \in S(\Gamma^n(q)) \text{ and } \gamma, \gamma' \in K, \text{ then } \chi(\gamma) = \chi(\gamma').$$

In fact, the equality $M\gamma = \gamma'M'$ implies that $X_0(M)\chi(\gamma) = \chi(\gamma')X_0(M')$, and Theorem 3.3(3) of Chapter 3 with $K_1 = \Gamma^n(q)$ implies that $X_0(M) = X_0(M')$, since X_0 as well as χ is trivial on $\Gamma^n(q)$; hence, $\chi(\gamma) = \chi(\gamma')$.

According to (3.5) of Chapter 3, any matrix $M \in S(K)$ can be written in the form

$$(1.14) \qquad M = \gamma N, \quad \text{where } \gamma \in K \text{ and } N \in S(\Gamma^n(Q)).$$

We then set

$$(1.15) \qquad X(M) = X_{\rho,\chi}(M) = \chi(\gamma)\rho(N).$$

If $M = \gamma N = \gamma_1 N_1$ are two decompositions (1.14), then $\gamma_1^{-1}\gamma = N_1 N^{-1} \in K \cap S(\Gamma^n(q)) = \Gamma^n(q)$, and hence $\chi(\gamma_1) = \chi(\gamma)$ and $\rho(N_1) = \rho(N)$. Thus, $X(M)$ does not depend on how M is written in the form (1.14). The map $X\colon S(K) \to \mathbf{C}^*$ clearly coincides with χ on K, and with ρ on $S(\Gamma^n(q))$. We check that X is a homomorphism. If $M = \gamma N$ and $M_1 = \gamma_1 N_1$ are two matrices in $S(K)$ written in the form (1.14), then, by (3.4) of Chapter 3, we can write $N\gamma_1 = \gamma_1' N'$, where $\gamma_1' \in K$ and $N' \in S(\Gamma^n(q))$. By (1.13) we have $\chi(\gamma_1') = \chi(\gamma_1)$. From Theorem 3.3(3) of Chapter 3 with $K_1 = \Gamma^n(q)$ it follows that $\rho(N') = \rho(N)$. From these relations and (1.15) we obtain

$$X(MM_1) = X(\gamma N \gamma_1 N_1) = X(\gamma\gamma_1' N' N_1)$$
$$= \chi(\gamma\gamma_1')\rho(N' N_1) = \chi(\gamma)\chi(\gamma_1)\rho(N)\rho(N_1) = X(M)X(M_1).$$

This proves the existence of the homomorphism $X_{\rho,\chi}$. Its uniqueness is obvious. $\square$

According to the lemma, we can fix an extension X of χ to $S(K)$ by arbitrarily giving the restriction ρ of X to $S(\Gamma^n(q))$. We set

$$(1.16) \qquad\qquad \rho_w(M) = r(M)^{-wn + \langle n \rangle}, \quad \text{if } M \in S(\Gamma^n(q)),$$

and we use $\rho = \rho_w$ in (1.15) to fix once and for all the *normalized extension*

$$(1.17) \qquad\qquad\qquad X_{w,\chi} = X_{\rho_w,\chi}$$

to $S(K)$ *of the character* χ *of* K, where (K, χ) is a q-regular pair. The corresponding automorphy factors (1.2) and (1.7) will be denoted $\varphi_{w,\chi}$, and the operators (1.3) and (1.8) will be denoted $|_{w,\chi}$. Finally, the Hecke operators (1.6) and (1.11) on $\mathfrak{M}_w(K, \chi)$ with $X = X_{w,\chi}$ will be denoted $|_{w,\chi}\tau$, or simply $|\tau$ if w and χ are clear from the context:

$$(1.18) \qquad\qquad\qquad |\tau = |_{w,\chi}\tau = |_{w,X}\tau,$$

and we call these the *normalized Hecke operators*.

We now show that the normalized Hecke operators are compatible with the natural imbeddings of spaces of modular forms and with the natural isomorphisms of Hecke rings described in Theorem 3.3(5) of Chapter 3.

PROPOSITION 1.3. *Let* (K, χ) *and* (K_1, χ_1) *be two* q-*regular pairs of degree* n. *Suppose that* $K_1 \subset K$, *and the restriction of* χ *to* K_1 *coincides with* χ_1. *Then the following equality holds for any modular form* $F \in \mathfrak{M}_k(K, \chi) \subset \mathfrak{M}_k(K_1, \chi_1)$, *where* k *is an integer, and for any* $T \in L(K)$:

$$F|_{k,\chi}T = F|_{k,\chi_1}\varepsilon(T),$$

where $\varepsilon\colon L(K) \to L(K_1)$ *is the isomorphism of Hecke rings in Theorem* 3.3(5) *of Chapter* 3. *In particular, the subspace* $\mathfrak{M}_k(K, \chi)$ *of* $\mathfrak{M}_k(K_1, \chi_1)$ *is invariant under all of the Hecke operators in* $L(K_1)$.

PROOF. Let $T = \sum_i a_i(KM_i)$. By part (1) of Theorem 3.3 in Chapter 3, we may suppose that all of the M_i lie in the group $S(K_1)$. Then parts (4) and (5) of the same theorem imply that $\varepsilon(T) = \sum_i a_i(K_1 M_i)$ (recall that ε coincides with the map (1.27) of Chapter 3). Note that the restriction to $S(K_1)$ of the normalized homomorphism $X_{k,\chi}$ coincides with X_{k,χ_1}, since these two maps agree on the subgroups K_1 and $\Gamma^n(q)$ that together generate $S(K_1)$; this implies that $|_{k,\chi}M = |_{k,\chi_1}M$ for $M \in S(K_1)$. Thus, for $F \in \mathfrak{M}_k(K, \chi)$ we obtain

$$F|_{k,\chi}T = \sum_i a_i F|_{k,\chi}M_i = \sum_i a_i F|_{k,\chi_1}M_i = F|_{k,\chi_1}\varepsilon(T). \qquad \square$$

Under the conditions of Proposition 1.3, suppose that q is divisible by 4, $K_1 \subset K \subset \Gamma_0^n(4)$, and the Hecke rings

(1.19) $\widehat{L}(K) = D_Q(\widehat{K}, \widehat{S}(K))$ and $\widehat{L}(K_1) = D_Q(\widehat{K}_1, \widehat{S}(K_1))$

are defined as in (4.3) of Chapter 3. By Theorem 3.3(1)-(2) of Chapter 3, the Hecke pairs for the rings (1.19) satisfy the conditions (1.26) of Chapter 3; hence, there exists a monomorphism of the form (1.27) of Chapter 3:

(1.20) $\widehat{\varepsilon} : \widehat{L}(K) \to \widehat{L}(K_1)$,

which is also compatible with the action of the corresponding Hecke operators on the spaces of modular forms.   □

PROPOSITION 1.4. *Suppose that (K, χ) and (K_1, χ_1) are q-regular pairs, and q is divisible by 4. Further suppose that $K_1 \subset K \subset \Gamma_0^n(4)$, and the restriction of χ to K_1 coincides with χ_1. Then the following equality holds for any modular form $F \in \mathfrak{M}_{k/2}(K, \chi) \subset \mathfrak{M}_{k/2}(K_1, \chi_1)$, where k is odd:*

$$F|_{k/2,\chi}\widehat{T} = F|_{k/2,\chi_1}\widehat{\varepsilon}(\widehat{T})   \text{ for } T \in \widehat{L}(K).$$

Moreover, the map $\widehat{\varepsilon}$ gives an isomorphism between the even subrings $\widehat{E}(K)$ and $\widehat{E}(K_1)$ of the Hecke rings (1.19), and the subspace $\mathfrak{M}_{k/2}(K, \chi)$ of $\mathfrak{M}_{k/2}(K_1, \chi_1)$ is invariant under the Hecke operators of $\widehat{E}(K_1)$.

PROOF. From Lemma 1.8, Proposition 4.3, and Theorem 3.3(4) of Chapter 3 it follows that the restriction of $\widehat{\varepsilon}$ to the even subrings truly is an isomorphism. The other parts of the proposition are proved in the same way as Proposition 1.3.   □

2. Invariant subspaces and eigenfunctions. In the spaces of modular forms we now define some standard subspaces that are invariant under the Hecke operators. In particular, we prove that the spaces of cusp-forms are spanned by eigenfunctions of all of the Hecke operators. We begin with the cusp-forms.

PROPOSITION 1.5. *Let (K, χ) be a q-regular pair of degree n, and let $w = k$ or $k/2$, where k is an integer. In the case $w = k/2$ we suppose that k is odd, q is divisible by 4, and $K \subset \Gamma_0^n(4)$. Then the subspace $\mathfrak{N}_w(K, \chi)$ of cusp-forms in $\mathfrak{M}_w(K, \chi)$ is invariant relative to all of the Hecke operators $|_{w,\chi}\tau$ for $\tau \in L(K)$ or $\tau \in \widehat{L}(K)$, depending on whether $w = k$ or $w = k/2$, respectively.*

PROOF. The proposition follows from Proposition 1.1, the formulas (1.3) and (1.8), and Theorem 3.13(3) of Chapter 2.   □

We now define a multiplicative family of operators on $\mathfrak{M}_w(K, \chi)$ that commute with the Hecke operators. Let $d \in \mathbf{Z}$, $(d, q) = 1$. We let $E(d) = E_n(d)$ denote one of the matrices that satisfies the following conditions (such matrices exist by Lemma 3.2(1) of Chapter 3):

(1.21) $E(d) \in \Gamma_0^n(q)$ and $E(d) \equiv \begin{pmatrix} d^{-1}E_n & 0 \\ 0 & dE_n \end{pmatrix} \pmod{q}$.

For fixed d modulo q, it is clear that all such matrices belong to the same left (or right, or double) coset of $\Gamma_0^n(q)$ modulo $\Gamma^n(q)$. This implies that the operator

$$|_w\tau(d): \mathfrak{M}_w(\Gamma^n(q), 1) \ni F \to F|_w\tau(d),$$

where $\tau(d) = E(d)$ or $\tau(d) = \widehat{E}(d) = j^n(E(d))$, depending on whether $w = k$ or $w = k/2$, respectively, does not depend on the particular choice of $E(d)$ satisfying (1.21), but rather depends only on d modulo q. In addition, we have

$$(1.22) \qquad F|_w\tau(d_1)|_w\tau(d_2) = F|_w\tau(d_1)\tau(d_2) = F|_w\tau(d_1 d_2),$$

since $E(d_1)E(d_2) \equiv E(d_1 d_2)(\mathrm{mod}\ q)$.

LEMMA 1.6. *Suppose that* (K, χ) *is a q-regular pair of degree n,* $w = k$ *or* $k/2$, *and in the latter case q is divisible by 4 and* $K \subset \Gamma_0^n(4)$. *Then for every d,* $(d, q) = 1$:

(1) *the map* $\gamma_1 \to \gamma = E(d)^{-1}\gamma_1 E(d)$ *gives an automorphism of the group K that does not affect the character* χ, *i.e.,* $\chi(\gamma) = \chi(\gamma_1)$;

(2) *the subspaces* $\mathfrak{M}_w(K, \chi)$ *and* $\mathfrak{N}_w(K, \chi)$ *in* $\mathfrak{M}_w(\Gamma^n(q), 1)$ *are invariant relative to the operator* $|_w\tau(d)$;

(3) *the operator* $|_w\tau(d)$ *on* $\mathfrak{M}_w(K, \chi)$ *commutes with all of the Hecke operators* $|_{w,\chi}\tau$ *for* $\tau \in L(K)$ *or* $\tau \in \widehat{E}(K)$, *depending on whether* $w = k$ *or* $w = k/2$, *respectively.*

PROOF. Since obviously $M = d \cdot E(d) \in S(\Gamma^n(q))$ and $\Gamma^n(q)M\Gamma^n(q) = M\Gamma^n(q)$, it follows from Theorem 3.3(4) of Chapter 3 that $KMK = KM = MK$. Thus, for any $\gamma_1 \in K$ there exists $\gamma \in K$ such that $\gamma_1 M = M\gamma$, and conversely. From (1.13) it then follows that $\chi(\gamma) = \chi(\gamma_1)$. Since $\gamma = M^{-1}\gamma_1 M = E(d)^{-1}\Gamma_1 E(d)$, the first part of the lemma is proved.

Since the operator $|_w\tau(d)$ obviously takes cusp-forms to cusp-forms, it suffices to prove the second part for the space $\mathfrak{M}_w(K, \chi)$. In the case $w = k$ this means that if a function $F \in \mathfrak{M}_k(\Gamma^n(q), 1)$ satisfies the condition $F|_k\gamma = \chi(\gamma)F$ for $\gamma \in K$, then the function $F|_k E(d)$ also satisfies this condition. But this is a consequence of part (1) of the lemma and the relations (3.15) of Chapter 2, since if $\gamma \in K$, then $\gamma = E(d)^{-1}\gamma_1 E(d)$ with $\gamma_1 \in K$, and

$$F|_k E(d)|_k \gamma = F|_k E(d)|_k E(d)^{-1}\gamma_1 E(d) = F|_k \gamma_1|_k E(d)$$
$$= \chi(\gamma_1)F|_k E(d) = \chi(\gamma)F|_k E(d).$$

Because the map $j^n: \Gamma_0^n(4) \to \widehat{\Gamma}_0^n(4)$ is an isomorphism, it follows by (3.22) of Chapter 2 that the above argument carries over to the case $w = k/2$.

From part (1) and from Lemma 1.2 it easily follows that the map $\gamma_1 \to \gamma = E(d)^{-1}\gamma_1 E(d)$ gives an automorphism of the group $S(K)$ that does not affect the homomorphism $X = X_{k,\chi}$, i.e., $X(\gamma) = X(\gamma_1)$. If $M \in S(\Gamma^n(q))$, then it follows from Theorem 3.3(3) of Chapter 3 that the matrices M and $E(d)^{-1}ME(d)$ belong to the same $\Gamma^n(q)$-double coset, and hence to the same K-double coset. From this and (3.4) of Chapter 3 we conclude that the above automorphism of $S(K)$ takes every double coset KMK, $M \in S(K)$, to itself. Thus, if $M_1, \ldots, M_\mu$ are any set of representatives of the left cosets $K \setminus KMK$, then so are $E(d)^{-1}M_1 E(d), \ldots, E(d)^{-1}M_\mu E(d)$. Using the definitions (see (1.6), (1.3), (1.18), and (3.15) of Chapter 2), we find the following relation for $F \in \mathfrak{M}_k(K, \chi)$:

$$F|_k E(d)|_{k,\chi}(M)_K = \sum_{i=1}^{\mu} X_{k,\chi}(E(d)^{-1}M_i E(d))^{-1} F|_k M_i E(d)$$

$$= \left(\sum_{i=1}^{\mu} X_{k,\chi}(M_i)^{-1} F|_k M_i \right)\bigg|_k E(d) = F|_{k,\chi}(M)_K|_k E(d),$$

which proves part (3) for $\tau = (M)_K$. The general case follows from this case and from Lemma 1.5 of Chapter 3.

If $w = k/2$, then part (1) implies that $E(d)^{-1}\widehat{K}E(d) = \widehat{K}$, since in this case $E(d)$ and K lie in $\Gamma_0^n(4)$. Suppose that $M \in S(\Gamma^n(q))$ and $r(M)$ is the square of a rational number. Then, as noted before, $E(d)^{-1}ME(d) = \gamma_1 M\gamma_2$ for some $\gamma_1, \gamma_2 \in \Gamma^n(q)$, and we can write $M\delta_2 M^{-1} = \delta_1$, where $\delta_1 = E(d)\gamma_1$ and $\delta_2 = E(d)\gamma_2^{-1} \in \Gamma_0^n(4)$. From this, (4.6), and Proposition 4.3 of Chapter 3 it follows that

$$\widehat{\delta}_1 = \widehat{M\delta_2 M}^{-1} = \widehat{M}\widehat{\delta}_2\widehat{M}^{-1}$$

for any P-preimage $\widehat{M}$ of M; hence,

$$\widehat{K}E(d)^{-1}\widehat{M}E(d)\widehat{K} = \widehat{K}\widehat{\gamma}_1\widehat{M}\widehat{\gamma}_2\widehat{K} = \widehat{K}\widehat{M}\widehat{K}.$$

Thus, if $\widehat{M}_1, \ldots, \widehat{M}_\mu$ is a set of representatives of the left cosets $\widehat{K} \setminus \widehat{K}\widehat{M}\widehat{K}$, then so is $\widehat{E}(d)^{-1}\widehat{M}_1\widehat{E}(d), \ldots, \widehat{E}(d)^{-1}\widehat{M}_\mu\widehat{E}(d)$. Part (3) for $w = k/2$ now follows from (1.11), (1.18), and (3.22) of Chapter 2. □

Based on this lemma, one can define the standard decompositions of our spaces of modular forms. Suppose that $V = \mathfrak{M}_w(K, \chi)$ or $\mathfrak{N}_w(K, \chi)$. The map $d \to |_w\tau(d)$ gives a representation of the abelian group $(\mathbf{Z}/q\mathbf{Z})^*$ on V; hence, V is a direct sum of irreducible invariant subspaces, each of dimension 1. If $F|_w\tau(d) = \psi(d)F$, then ψ is a character of the group $(\mathbf{Z}/q\mathbf{Z})^*$. From this and Lemma 1.6 we obtain

PROPOSITION 1.7. *Suppose that (K, χ) is a q-regular pair of degree n, $w = k$ or $k/2$ is an integer or half-integer, and in the latter case $K \subset \Gamma_0^n(4)$. Then one has the direct sum decompositions*

(1.23)
$$\mathfrak{M}_w(K, \chi) = \bigoplus_\psi \mathfrak{M}_w(K, \chi, \psi),$$
$$\mathfrak{N}_w(K, \chi) = \bigoplus_\psi \mathfrak{N}_w(K, \chi, \psi),$$

where ψ runs through all of the characters of the group $(\mathbf{Z}/q\mathbf{Z})^*$, and for each ψ we set

(1.24)
$$\mathfrak{M}_w(K, \chi, \psi) = \{F \in \mathfrak{M}_w(K, \chi); F|_w\tau(d) = \psi(d)F, (d, q) = 1\},$$
$$\mathfrak{N}_w(K, \chi, \psi) = \mathfrak{M}_w(K, \chi, \psi) \cap \mathfrak{N}_w(K, \chi).$$

Each of the subspaces $\mathfrak{M}_w(K, \chi, \psi)$ and $\mathfrak{N}_w(K, \chi, \psi)$ is invariant under all of the Hecke operators $|_{w,\chi}\tau$ for $\tau \in L(K)$ or $\tau \in \widehat{E}(K)$ in the case $w = k$ or $w = k/2$, respectively.

We now consider the action of the Hecke operators on subspaces of the form (1.24). In $\widehat{E}(K)$ we look at the subring $\widehat{E}(K, \varkappa)$ that is analogous to the ring (5.4) in Chapter 3. Namely, we set

(1.25)
$$\widehat{E}(K, \varkappa) = \widehat{\varepsilon}_1^{-1}(\widehat{\varepsilon}_2(\widehat{E}^n(q, \varkappa))),$$

where $\widehat{\varepsilon}_1$ and $\widehat{\varepsilon}_2$ are monomorphisms of the form (1.20) for the pairs of groups $\Gamma^n(q) \subset K$ and $\Gamma^n(q) \subset \Gamma_0^n(q)$, respectively. Here by $\widehat{\varepsilon}_1^{-1}$ we mean the inverse of the restriction of $\widehat{\varepsilon}_1$ to the even Hecke ring.

PROPOSITION 1.8. *Suppose that the pair* (K, χ) *satisfies the conditions of Proposition 1.7, and* ψ *is a character modulo* q. *Then the following formula holds for any modular forms* $F, G \in \mathfrak{M}_w(K, \chi, \psi)$ *of which at least one is a cusp-form:*

$$(1.26) \qquad (F|_{w,\chi}\tau, G) = \psi(r(M)) \quad (F, G|_{w,\chi}\tau),$$

where $\tau = (M)_K \in L(K)$ *or* $\tau = (\widehat{M})_{\widehat{K}} \in \widehat{E}(K, \varkappa)$ *for* $w = k$ *or* $w = k/2$, *respectively, and* $(\cdot, \cdot)$ *is the scalar product in* (5.7) *of Chapter 2.*

PROOF. By Lemma 1.2 of Chapter 3, the number of K-left cosets in KMK is equal to the index $[K : K_{(M)}]$. The number of K-right cosets in KMK is obviously equal to the number of left cosets in $KM^{-1}K$, i.e., $[K : K_{(M^{-1})}]$. Since these indices are equal, by Lemma 5.4 of Chapter 2, it follows that the number of K-left cosets in KMK is equal to the number of K-right cosets there. On the other hand, every left coset clearly has nonempty intersection with each right coset. Hence, there exists a set of representatives $M_1, \ldots, M_\mu$ of the left cosets $K \setminus KMK$ that is also a set of right coset representatives. Using this set of representatives and the properties of the scalar product in Theorem 5.3 of Chapter 2, we obtain

$$(1.27) \qquad \begin{aligned} (F|_{k,\chi}(M)_K, G) &= \sum_{i=1}^{\mu} X(M_i)^{-1}(F|_k M_i, G|_k M_i^{-1}|_k M_i) \\ &= \left(F, \sum_{i=1}^{\mu} r^{-nk}\overline{X}(M_i)^{-1} G|_k M_i^{-1} \right), \end{aligned}$$

where $X = X_{k,\chi}$ and $r = r(M) = r(M_i)$. Since obviously $G|_k \lambda M' = \lambda^{-nk} G|_k M'$, it follows that the last sum in (1.27) is equal to

$$(1.28) \quad \sum_{i=1}^{\mu} \overline{X}(M_i)^{-1} G|_k E(r)^{-1} E(r) r M_i^{-1} = \overline{\psi}(r) \sum_{i=1}^{\mu} \overline{X}(M_i)^{-1} G|_k E(r) r M_i^{-1}.$$

Because $M_1, \ldots, M_\mu$ is a set of representatives of the right cosets KMK/K, it follows from this and Lemma 1.6(1) that the set $E(r)rM_1^{-1}, \ldots, E(r)rM_\mu^{-1}$ is a set of representatives of the K-left cosets in the double coset $KE(r)rM^{-1}K$. Furthermore, since (1.13) implies that any homomorphism X of the form (1.17) satisfies the relation

$$\overline{X}(M) = X(E(r)rM^{-1}) \quad \text{for } M \in S(K) \quad \text{with } r(M) = r,$$

it follows from the above considerations that the expression (1.28) can be rewritten in the form

$$\overline{\psi}(r)G|_{k,\chi}(M')_K, \quad \text{where } M' = E(r)rM^{-1}.$$

The two matrices M and M' obviously have the same symplectic divisors, and, by (3.5) of Chapter 3, we may suppose that they lie in $S(\Gamma^n(q))$; hence, by Lemma 3.6 of Chapter 3, $\Gamma^n M \Gamma^n = \Gamma^n M' \Gamma^n$. If we now apply Theorem 3.3(3) of Chapter 3, we conclude that these matrices belong to the same $\Gamma^n(q)$-double coset, and, in particular,

$$(1.29) \qquad (E(r)rM^{-1})_K = (M)_K \quad (M \in S(K), r = r(M)).$$

Thus, $\overline{\psi}(r)G|_{k,\chi}(M')_K = \overline{\psi}(r)G|_{k,\chi}(M)_K$, and, if we substitute this expression in the sum in (1.27), we obtain (1.26) for $w = k$.

By Lemma 1.8 and Proposition 4.3 of Chapter 3, the number of $\widehat{K}$-left and $\widehat{K}$-right cosets in $\widehat{K}\widehat{N}\widehat{K}$, where $\widehat{N} = \widehat{M}^{\pm 1}$, is the same as the number of K-left and K-right cosets in KNK. Hence, there exist elements $\widehat{M}_1, \ldots, \widehat{M}_\mu$ in $\widehat{K}\widehat{M}\widehat{K}$ that are a set of representatives simultaneously for the $\widehat{K}$-left and the $\widehat{K}$-right cosets in this double coset. Setting $\widehat{rE} = (rE_{2n}, r^{n/2})$ and using Lemma 1.6(1), we find that

$$\widehat{K}\widehat{M}'K = \sum_{i=1}^{\mu} \widehat{K}\widehat{E}(r)\widehat{rE}\widehat{M}_i^{-1}, \quad \text{where } \widehat{M}' = \widehat{E}(r)\widehat{rE}\widehat{M}^{-1}.$$

Just as in the case of (1.27) and (1.28), we now obtain the relation

$$(F|_{k/2,\chi}(\widehat{M})_{\widehat{K}}, G) = \psi(r(M))(F, G|_{k/2,\chi}(\widehat{M}')_{\widehat{K}}),$$

so that in order to prove (1.26) for $w = k/2$ we must show that

(1.30) $\qquad (\widehat{E}(r)\widehat{rE}\widehat{M}^{-1})_{\widehat{K}} = (\widehat{M})_{\widehat{K}}, \quad \text{where } M \in S(\Gamma^n(q))^+.$

Since $E(r) \in \Gamma = \Gamma_0^n(4)$ and $\widehat{rE} \cdot \widehat{M}^{-1} = \widehat{M}^*$ (see (4.85) of Chapter 3), from (4.34) and Lemma 4.14 of Chapter 3 it follows that $\widehat{\Gamma}\widehat{M}'\widehat{\Gamma} = \widehat{\Gamma}\widehat{M}\widehat{\Gamma}$, or, equivalently, $\widehat{M}' = \widehat{\gamma}\widehat{M}\widehat{\delta}$ with $\gamma, \delta \in \Gamma$. We shall show that this implies the equality of double cosets

(1.31) $\qquad \widehat{\Gamma}_1\widehat{M}'\widehat{\Gamma}_1 = \widehat{\Gamma}_1\widehat{M}\widehat{\Gamma}_1, \quad \text{where } \Gamma_1 = \Gamma^n(q),$

which, in turn, implies (1.30). We choose an integer q_1 prime to q such that $q_1 M^{\pm 1}$ are integer matrices. By Lemma 3.2(2) of Chapter 3, the matrix γ can be represented in the form $\gamma = \gamma_1\gamma_2$ with $\gamma_1 \in \Gamma_1$ and $\gamma_2 \in \Gamma^n(q_1^2)$. Moreover, γ_2 and $\delta_1 = M^{-1}\gamma_2 M \in \Gamma$, since q is divisible by 4 and $M \in S(\Gamma_1)$. By assumption, $r(M)$ is the square of a rational number; hence, by (4.6) and Proposition 4.3 of Chapter 3, we have $\widehat{\delta}_1 = \widehat{M}^{-1}\widehat{\gamma}_2\widehat{M}$, and so

$$\widehat{M}' = \widehat{\gamma}_1\widehat{M}\widehat{\delta}_2, \quad \text{where } \gamma_1 \in \Gamma_1 \text{ and } \delta_2 = \delta_1\delta \in \Gamma.$$

Since $\delta_2 = M^{-1}\gamma_1^{-1}M' \in S(\Gamma_1) \cap \Gamma$, we have proved (1.31). $\qquad \square$

We are now ready to prove the following

THEOREM 1.9. *Suppose that (K, χ) is a q-regular pair of degree n, where $n, q \in \mathbf{N}$, $w = k$ or $k/2$ is an integer or half-integer, and $K \subset \Gamma_0^n(4)$ in the latter case. Then each of the subspaces*

$$\mathfrak{N}_w(K, \chi, \psi) \subset \mathfrak{N}_w(K, \chi),$$

where ψ is a character of the group $(\mathbf{Z}/q\mathbf{Z})^*$—and hence also the entire space $\mathfrak{N}_w(K, \chi)$ of cusp-forms—has an orthogonal basis consisting of common eigenfunctions of all of the Hecke operators $|_{w,\chi}\tau$ for $\tau \in L(K)$ or $\tau \in \widehat{E}(K, \varkappa)$ in the cases $w = k$ and $w = k/2$, respectively.

PROOF. By Theorem 4.5 of Chapter 2, $\mathfrak{M}_w(K, \chi)$—and hence also each of the subspaces $V = \mathfrak{N}_w(K, \chi, \psi)$—is finite dimensional. Thus, the ring of Hecke operators on V can be regarded as a subring of the ring of matrices of a certain finite size. This implies that the ring of Hecke operators on V is finite dimensional (over $\mathbf{C}$). Thus, there exists a finite set of generators of the ring $L(K)$ or $\widehat{E}(K, \varkappa)$ of the form $\tau_i = (M_i)_K$ or $(\widehat{M}_i)_{\widehat{K}}$ $(i = 1, \ldots, d)$ such that every operator $|\tau = |_{w,\chi}\tau$ on V is a polynomial in the operators $|\tau_i$. We note that these operators, and in fact any Hecke operators on V, commute with one another, since, by Theorem 3.3 and Lemma 3.5 of Chapter 3 and Proposition 1.4, the rings $L(K)$ and $\widehat{E}(K, \varkappa)$ are isomorphic to $L^n(q)$ and $\widehat{E}^n(q, \varkappa)$, respectively. The latter rings are commutative, by Theorems 3.7, 4.6, and 4.13 of Chapter 3.

LEMMA 1.10. *Let V be a nonzero finite dimensional vector space over an algebraically closed field, and let $S_1, \ldots, S_d$ be a finite set of pairwise commuting linear operators on V. Then V contains a nonzero common eigenvector of $S_1, \ldots, S_d$.*

PROOF. The case $d = 1$ is obvious. Suppose that $d > 1$, and the lemma holds for sets of $d - 1$ operators. Let λ_1 denote an eigenvalue of S_1 on V, and let $V' = \{v \in V;\ v|S_1 = \lambda_1 v\}$ denote the corresponding eigenspace. Then V' is invariant relative to $S_2, \ldots, S_d$ because those operators commute with S_1. By the induction assumption, V' contains a nonzero eigenvector of all of these operators. $\square$

Returning to the proof of the theorem, we see that V contains a nonzero eigenfunction F_1 of all of the operators $|\tau_i$ $(i = 1, \ldots, d)$ (provided, of course, that $V \neq \{0\}$). We set $V_1 = \{\alpha F_1;\ \alpha \in \mathbf{C}\}$ and $V_2 = \{G \in V;\ (F_1, G) = 0\}$. Since the scalar product on V is hermitian and nondegenerate, V splits into the orthogonal direct sum of V_1 and V_2: $V = V_1 \oplus V_2$. By construction, V_1 is invariant relative to the operators $|\tau_i$. Then (1.26) implies that V_2 is also invariant relative to these operators; hence, if $V_2 \neq \{0\}$, then V_2 contains a nonzero common eigenfunction F_2. Repeating the same argument for V_2 and F_2 in place of V and F_1, and continuing in this way, after a finite number of steps we obtain an orthogonal basis for V consisting of common eigenfunctions for all of the operators $|\tau_i$. $\square$

PROBLEM 1.11. Suppose that (K, χ) is a q-regular pair of degree n, $w = k$ or $k/2$ is an integer or half-integer, and $K \subset \Gamma_0^n(4)$ in the latter case. Show that the subspace $\mathfrak{E}_w(K, \chi) \subset \mathfrak{M}_w(K, \chi)$ (see (5.10) of Chapter 2) is invariant relative to all of the Hecke operators $|_{w,\chi}\tau$, where $\tau \in L(K)$ or $\widehat{E}(K, \varkappa)$ in the case $w = k$ or $w = k/2$, respectively.

§2. Action of the Hecke operators

We now consider the problem of computing the action of the Hecke operators on modular forms and their Fourier coefficients. Since the details are different for different congruence subgroups $K \subset \Gamma^n$, as in Chapter 3 we shall focus our attention on the case $K = \Gamma_0^n(q)$, which is of the greatest arithmetic importance.

1. Hecke operators for $\Gamma_0^n(q)$. Let χ be a Dirichlet character modulo q, and let $[\chi]$ denote the *one-dimensional character* of the group $\Gamma_0^n(q)$ that corresponds to χ, i.e., the character given by

$$[\chi](M) = \chi(\det D) \quad \text{for } M = \begin{pmatrix} A & B \\ C & D \end{pmatrix} \in \Gamma_0^n(q).$$

LEMMA 2.1. *The pair $(\Gamma_0^n(q), [\chi])$, where χ is a Dirichlet character modulo q, is q-regular. For any integer $w = k$ or half-integer $w = k/2$ the normalized extension $X_{w,[\chi]}$ of $[\chi]$ to the group $S^n(q) = S(\Gamma_0^n(q))$ is given by the formula*

$$(2.1) \qquad X_{w,[\chi]}(M) = r(M)^{-wn + \langle n \rangle} \overline{\chi}(\det A) \quad \left(M = \begin{pmatrix} A & B \\ C & D \end{pmatrix} \in S^n(q) \right).$$

PROOF. By Lemma 3.5 of Chapter 3, the group $\Gamma_0^n(q)$ satisfies the q-symmetry condition. Hence, to prove the lemma it suffices to verify that the formula (2.1) gives a group homomorphism from $S^n(q)$ to $\mathbf{C}^*$, that the restriction to $\Gamma_0^n(q)$ of this homomorphism coincides with $[\chi]$, and that its restriction to $S(\Gamma^n(q))$ is the map ρ_w

given by (1.16). The first claim follows immediately from the description of $S^n(q)$ in Lemma 3.5 of Chapter 3. The second and third claims follow from the definition of the character $[\chi]$ and the map ρ_w. $\qquad\square$

In arithmetic applications, for the most part one encounters one-dimensional characters of groups of the type $\Gamma_0^n(q)$. Hence, we shall give explicit computations only for such characters. If $[\chi]$ is the one-dimensional character of $\Gamma_0^n(q)$ that corresponds to the Dirichlet character χ, and if F is a function on $\mathbf{H}_n$, then, by (1.3), (1.8), and (2.1), we have

$$(2.2) \qquad F|_{w,\chi}\xi = F|_{w,[\chi]}\xi = r(M)^{wn-\langle n\rangle}\chi(\det A)F|_w\xi,$$

where $\xi = M = \begin{pmatrix} A & B \\ C & D \end{pmatrix} \in S^n(q)$ or $\xi = \widehat{M} \in \widehat{S}^n(q)$ for $w = k$ or $w = k/2$, respectively. We set

$$\mathfrak{M}_w^n(q,\chi) = \mathfrak{M}_w(\Gamma_0^n(q),[\chi])$$

and for $\tau \in L^n(q)$ or $\tau \in \widehat{E}^n(q,\varkappa)$ we let

$$(2.3) \qquad |\tau = |_{w,\chi}\tau = |_{w,[\chi]}\tau$$

denote the *normalized Hecke operators on* $\mathfrak{M}_w^n(q,\chi)$. According to the definition of modular forms, we have the inclusion

$$(2.4) \qquad \mathfrak{M}_w^n(q,\chi) \subset \mathfrak{M}(\Gamma_0^n,[\chi_w]),$$

where $[\chi_w]$ is the one-dimensional character of Γ_0^n corresponding to the character

$$(2.5) \qquad \chi_w(m) = (\operatorname{sign} m)^{w_0}\chi(m)$$

in which $w_0 = k$ or 0 depending on whether $w = k$ or $w = k/2$, respectively. By Theorem 3.1 of Chapter 2, any modular form $F \in \mathfrak{M}(\Gamma_0^n,[\chi_w])$—and, in particular, any modular form $F \in \mathfrak{M}_w^n(q,\chi)$—has a Fourier expansion of the form

$$(2.6) \qquad F(Z) = \sum_{R\in\mathbf{A}_n} f(R)e\{RZ\},$$

in which the Fourier coefficients $f(R)$ satisfy the relations

$$(2.7) \qquad f(R[V]) = s(\det V)f(R) \quad \text{for } V \in GL_n(\mathbf{Z}),$$

where s is the character of the group $\{\pm 1\}$ given by setting

$$(2.8) \qquad s(-1) = \chi_w(-1).$$

Thus, the Fourier coefficients $f(R)$ of the modular form $F(Z)$ may be regarded as the values of a function $f: \mathbf{A}_n \to \mathbf{C}$ that satisfies (2.7). We let $\mathfrak{F}_w^n(q,\chi)$ denote the vector space of such functions. If $f \in \mathfrak{F}_w^n(q,\chi)$ and $\tau \in L^n(q)$ or $\tau \in \widehat{E}^n(q,\varkappa)$ for $w = k$ or for $w = k/2$, respectively, then

$$F = \sum_{R\in\mathbf{A}_n} f(R)e\{RZ\} \quad \text{and} \quad F|\tau \in \mathfrak{M}_w^n(q,\chi).$$

In particular, the function $F|\tau$ has a Fourier expansion of the same form as F. We write

$$(2.9) \qquad F|\tau = \sum_{R\in\mathbf{A}_n} (f|\tau)(R)e\{RZ\},$$

where $f|\tau$ is another function in $\mathfrak{F}_w^n(q, \chi)$. From Proposition 1.1 we immediately obtain

LEMMA 2.2. *The map*

$$\tau \to |\tau: f \to f|\tau \quad (f \in \mathfrak{F}_w^n(q, \chi))$$

is a linear representation of the Hecke ring $L^n(q)$ if $w = k$, and a linear representation of the Hecke ring $\widehat{E}^n(q, \varkappa)$ if $w = k/2$.

2. Hecke operators for Γ_0^n. In Chapter 3 we obtained expressions for elements of $L^n(q)$ and $\widehat{E}^n(q, \varkappa)$ in terms of components belonging to the Hecke ring of the triangular subgroup $\Gamma_0^n \subset \Gamma^n$. The Hecke operators corresponding to these components do not, in general, stay within the confines of the original spaces of modular forms. Thus, in order to apply these results of Chapter 3 when computing the action of the Hecke operators, we must first define suitable extensions of the spaces $\mathfrak{M}_w^n(q, \chi)$.

To each character s of the group $\{\pm 1\}$ we associate the character $\delta_s : \Gamma_0^n \to \{\pm 1\}$ defined by setting

$$(2.10) \qquad \delta_s(M) = s(\det D) \quad \text{for } M = \begin{pmatrix} A & B \\ 0 & D \end{pmatrix} \in \Gamma_0^n.$$

Let $T = T^n$ denote the kernel of $\delta_- = \delta_{s^-}$, where $s^-(i) = \text{sign}(i)$ is the odd character of the group $\{\pm 1\}$, and let $\mathfrak{M} = \mathfrak{M}(T, 1)$ be the space of modular forms for T that was defined in §3.1 of Chapter 2. If we set $F|_0 M = F(M\langle Z\rangle)$, we obtain a representation of the group Γ_0^n in the space $\mathfrak{M}$; it obviously reduces to a representation of the group Γ_0^n/T of order two. Hence, $\mathfrak{M}$ is the direct sum of the subspaces $\mathfrak{M}_+ = \mathfrak{M}_{s^+}$ and $\mathfrak{M}_- = \mathfrak{M}_{s^-}$, where

$$(2.11) \qquad \mathfrak{M}_s^n = \{F \in \mathfrak{M}; F|_0 M = \delta_s(M)F \text{ for } M \in \Gamma_0^n\}$$

and s^+ is the unit character of the group $\{\pm 1\}$. From the definitions and (2.4) it follows that

$$(2.12) \qquad \mathfrak{M}_w^n(q, \chi) \subset \mathfrak{M}_s^n, \quad \text{where } s(-1) = \chi_w(-1).$$

Following the scheme in §1, as in the case of congruence subgroups, we define representations of the global Hecke rings $\overline{L}_0^n(q) = D_{\mathbf{C}}(\Gamma_0^n, S_0^n(q))$ of the group Γ_0^n in the spaces $\mathfrak{M}_s^n$. Here we want to obtain operators that are compatible with the imbeddings (2.12) and with the normalized Hecke operators on the spaces $\mathfrak{M}_w^n(q, \chi)$. In view of these conditions and the inclusion (2.4), it is natural to start with an automorphy factor of the group $S_0^n(q)$ having the form

$$(2.13) \qquad \varphi_{w,\chi}(M, Z) = r(M)^{-wn+\langle n\rangle}\overline{\chi}_w(\det A)|\det D|^w,$$

where $M = \begin{pmatrix} A & B \\ 0 & D \end{pmatrix} \in S_0^n(q)$ and $Z \in \mathbf{H}_n$. Then the action of the group $S_0^n(q)$ on functions $F : \mathbf{H}_n \to \mathbf{C}$ is written in the form

$$(2.14) \qquad F|_{w,\chi} M = r(M)^{wn-\langle n\rangle}\chi_w(\det A)|\det D|^w F(M\langle Z\rangle).$$

We let V denote the space of all functions $F : \mathbf{H}_n \to \mathbf{C}$ such that $F|_{w,\chi} M = F$ for all $M \in \Gamma_0^n$. If for each $T = \sum_i a_i(\Gamma_0^n M_i)$ in $\overline{L}_0^n(q)$ we set

$$(2.15) \qquad F|_{w,\chi} T = \sum_i a_i F|_{w,\chi} M_i \quad (F \in V),$$

then by Proposition 1.14 of Chapter 3 we obtain a linear representation of the ring $\overline{\mathbf{L}}_0^n(q)$ in the endomorphism ring of V.

We now define the representation $|_{k/2,\chi}$ of the ring $\widehat{\mathbf{L}}_0^n(q) = D_{\mathbf{Q}}(\widehat{\Gamma}_0^n, \widehat{S}_0^n(q))$ (see §4 of Chapter 3) in the space V with $w = k/2$. To do this, we set $X = X_{k/2,[\chi]}$ in (1.8), and then give the action of $\widehat{S}_0^n(q)$ on a function $F: \mathbf{H}_n \to \mathbf{C}$ as follows:

$$(2.16) \qquad F|_{k/2,\chi}\widehat{M} = r(M)^{nk/2-\langle n \rangle}\chi(\det A)\varphi(Z)^{-k}F(M\langle Z \rangle),$$

where $\widehat{M} = (M, \varphi(Z)) \in \widehat{S}_0^n(q)$. From (4.2) of Chapter 3 it follows that the space V can also be defined by the condition that $F|_{k/2,\chi}\widehat{M} = F$ for all $\widehat{M} \in \widehat{\Gamma}_0^n$. Hence, if for $\widehat{T} = \sum_i a_i(\widehat{\Gamma}_0^n\widehat{M}_i) \in \widehat{\mathbf{L}}_0^n(q)$ we define the operator $|_{k/2,\chi}\widehat{T}$ on V by the formula

$$(2.17) \qquad F|_{k/2,\chi}\widehat{T} = \sum_i a_i F|_{k/2,\chi}\widehat{M}_i,$$

then we obtain a linear representation of the ring $\widehat{\mathbf{L}}_0^n(q)$ in V.

In the general case, the condition $F \in V$ is equivalent to requiring that

$$F|_0 M = F(M\langle Z \rangle) = \delta_s(M)F \quad \text{for } M \in \Gamma_0^n,$$

where $s(-1) = \chi_w(-1)$; hence, V contains $\mathfrak{M}_s^n$ for this s. Since the operators in (2.15) and (2.17) obviously do not destroy the analytic properties of functions in $\mathfrak{M}$, it follows that the subspace $\mathfrak{M}_s^n \subset V$ is invariant relative to all of these operators. We thereby have the following

LEMMA 2.3. *Let $w = k$ or $k/2$ be an integer or half-integer, let χ be a Dirichlet character modulo q, and let s be the character of the group $\{\pm 1\}$ that is determined by the condition $s(-1) = \chi_w(-1)$. Then every operator*

$$(2.18) \qquad |_{w,\chi}T \quad (T \in \overline{\mathbf{L}}_0^n(q)) \quad or \quad |_{k/2,\chi}\widehat{T} \quad (\widehat{T} \in \widehat{\mathbf{L}}_0^n(q))$$

maps the corresponding space $\mathfrak{M}_s^n$ to itself. The maps $T \to |_{w,\chi}T$ and $\widehat{T} \to |_{k/2,\chi}\widehat{T}$ are linear representations of the rings $\overline{\mathbf{L}}_0^n(q)$ and $\widehat{\mathbf{L}}_0^n(q)$ in the space $\mathfrak{M}_s^n$.

The operators (2.18), which we shall also call *Hecke operators*, are compatible with the imbeddings of Hecke rings in (4.101) and (5.3) of Chapter 3 and with the imbedding (2.12). More precisely, we have

LEMMA 2.4. *Under the assumptions in the previous lemma, one has:*

$$F|_{k,\chi}T = F|_{k,\chi}\varepsilon_q(T) \quad for \ F \in \mathfrak{M}_k^n(q,\chi) \ and \ T \in L^n(q),$$

$$F|_{k/2,\chi}\widehat{T} = F|_{k/2,\chi}\widehat{\varepsilon}_{q,k}(\widehat{T}) \quad for \ F \in \mathfrak{M}_{k/2}^n(q,\chi) \ and \ \widehat{T} \in \widehat{E}^n(q,\varkappa),$$

where $\varepsilon_q: L^n(q) \to \mathbf{L}_0^n(q)$ and $\widehat{\varepsilon}_{q,k}: \widehat{E}^n(q,\varkappa) \to \overline{\mathbf{L}}_0^n(q)$ are the imbeddings (5.3) and (5.5) of Chapter 3. In particular, the subspace $\mathfrak{M}_w^n(q,\varkappa) \subset \mathfrak{M}_s^n$ is invariant relative to all of the Hecke operators $|_{w,\chi}\tau$, where $\tau \in \mathbf{L}^n(q) = \varepsilon_q(L^n(q))$ or $\tau \in \widehat{\mathbf{E}}^n(q,\varkappa) = \widehat{\varepsilon}_{q,k}(\widehat{E}^n(q,\varkappa))$ in the cases $w = k$ and $w = k/2$, respectively.

PROOF. In the case $w = k$, the lemma follows immediately from the definitions. Suppose that $w = k/2$. We first note that $\widehat{\varepsilon}_{q,k} = \widehat{\varepsilon}_q \cdot P_k$ is the composition of the homomorphisms (4.101) and (5.3) of Chapter 3. According to the definitions of the operators in (2.3), (2.17), and (2.15), they satisfy the relations

$$(2.19) \qquad \begin{aligned} F|_{k/2,\chi}\widehat{T} &= F|_{k/2,\chi}\widehat{\varepsilon}_q(\widehat{T}) \quad \text{for } F \in \mathfrak{M}_{k/2}^n(q,\chi), \widehat{T} \in \widehat{E}^n(q,\varkappa), \\ F|_{k/2,\chi}\widehat{T} &= F|_{k/2,\chi}P_k(\widehat{T}) \quad \text{for } F \in \mathfrak{M}_k^s, \widehat{T} \in \widehat{\mathbf{L}}_0^n(q), \end{aligned}$$

where $s(-1) = \chi(-1)$, and this implies the lemma for $w = k/2$. $\qquad \square$

As already mentioned, every modular form $F \in \mathfrak{M}_s^n = \mathfrak{M}(\Gamma_0^n, [\chi_w])$, where $s(-1) = \chi_w(-1)$, has a Fourier expansion (2.6), and its Fourier coefficients $f(R)$ satisfy the relations (2.7). We let $\mathfrak{F}_s^n$ denote the set of such functions $f : \mathbf{A}_n \to \mathbf{C}$. Clearly,

$$(2.20) \qquad \mathfrak{F}_w^n(q,\chi) \subset \mathfrak{F}_s^n, \quad \text{if } s(-1) = \chi_w(-1).$$

If $f \in \mathfrak{F}_s^n$, if $F \in \mathfrak{M}_s^n$ is a function with Fourier coefficients $f(R)$, and if $T \in \overline{\mathbf{L}}_0^n(q)$, then we let $(f|_{w,\chi}T)(R)$, where $\chi_w(-1) = s(-1)$, denote the Fourier coefficients of the function

$$(2.21) \qquad F|_{w,\chi}T = \sum_{R \in \mathbf{A}_n} (f|_{w,\chi}T)(R)e\{RZ\}.$$

From Lemmas 2.3 and 2.4 we have (in the notation of those lemmas):

LEMMA 2.5. *The map $T \to |_{w,\chi}T$ is a linear representation of the ring $\overline{\mathbf{L}}_0^n(q)$ in the space $\mathfrak{F}_s^n$, and it satisfies the following relations*:

$$(2.22) \qquad \begin{aligned} f|_{k,\chi}T &= f|_{k,\chi}\varepsilon_q(T) \quad \text{for } f \in \mathfrak{F}_k^n(q,\chi), T \in L^n(q), \\ f|_{k/2,\chi}\widehat{T} &= f|_{k/2,\chi}\widehat{\varepsilon}_{q,k}(\widehat{T}) \quad \text{for } f \in \mathfrak{F}_{k/2}^n(q,\chi), \widehat{T} \in \widehat{E}^n(q,\varkappa), \end{aligned}$$

where ε_q and $\widehat{\varepsilon}_{q,k}$ are the imbeddings (5.3) and (5.5) of Chapter 3. In particular, the subspace $\mathfrak{F}_w^n(q,\chi) \subset \mathfrak{F}_s^n$ is invariant relative to all of the operators $|_{w,\chi}\tau$, where $\tau \in L^n(q)$ or $\tau \in \widehat{E}^n(q,\varkappa)$ in the cases $w = k$ and $w = k/2$, respectively.

The above constructions make it possible to apply the decompositions in Chapter 3 when computing the action of concrete Hecke operators on modular forms and their Fourier coefficients. Here we shall prove three general lemmas concerning the action of Hecke operators in $\mathbf{L}_0^n$. From now on we shall assume, without further mention, that $w = k$ or $k/2$ is an integer or half-integer, χ is a Dirichlet character modulo $q \in \mathbf{N}$, and s is the character of the group $\{\pm 1\}$ such that $s(-1) = \chi_w(-1)$.

LEMMA 2.6. *Let F be a function in $\mathfrak{M}_s^n$ with Fourier coefficients $f(R)$. Then for any matrix $M_0 = \begin{pmatrix} rD_0^* & B_0 \\ 0 & D_0 \end{pmatrix}$ in $S_0^n(q)$ the action of the Hecke operator for $(M_0)_{\Gamma_0^n}$ on F*

and f is given by the formulas

$$(F|_{w,\chi}(M_0)_{\Gamma_0^n})(Z)$$

(2.23)

$$= r^{wn-\langle n\rangle}\chi(r)^n \sum_{D,B} \overline{\chi}_w(\det D)|\det D|^{-w} F(rZ[D^{-1}] + BD^{-1}),$$

$$(f|_{w,\chi}(M_0)_{\Gamma_0^n})(R) = r^{wn-\langle n\rangle}\chi(r)^n {\sum_D}' \overline{\chi}_w(\det D)|\det D|^{-w}$$

(2.24)

$$\times l_{M_0}(r^{-1}R[{}^tD]; D) f(r^{-1}R[{}^tD]);$$

here $D \in \Lambda^n \setminus \Lambda^n D_0 \Lambda^n$, $B \in B_{M_0}(D)/\mathrm{mod}\,D$, and in (2.24) $R[{}^tD] \in r\mathbf{A}_n$, where $Z \in \mathbf{H}_n$, $R \in \mathbf{A}_n$, $B_{M_0}(D)$ is the set in Lemma 3.25 of Chapter 3, and for $R' \in \mathbf{A}_n$ (see (3.25) of Chapter 3)

(2.25)

$$l_{M_0}(R', D) = \sum_{B \in B_{M_0}(D)/\mathrm{mod}\,D} e\{R'BD^{-1}\}.$$

PROOF. The formula (2.23) follows immediately from (2.15), (2.14), and the left coset decomposition for $\Gamma_0^n M_0 \Gamma_0^n$ in (3.44) of Chapter 3. If we substitute the Fourier expansion of F in the right side of (2.23), we obtain the series

$$r^{wn-\langle n\rangle}\chi(r)^n \sum_{D,B,R'} \overline{\chi}_w(\det D)|\det D|^{-w} f(R')e\{R'rZ[D^{-1}] + R'BD^{-1}\}.$$

Since $e\{R'rZ[D^{-1}]\} = e\{rR'[D^*]Z\}$ and the resulting series still lies in $\mathfrak{M}_s^n$, we see that if we group together terms with fixed product $rR'[D^*] = R$, then the only terms that remain are those for which $R \in X = rD^{-1}\mathbf{A}_n D^* \cap \mathbf{A}_n$. Thus, if we set $r^{wn-\langle n\rangle}\chi(r)^n = \gamma$ and $\chi_w(\det D)|\det D|^{-w} = \psi(D)$, we find that our series is equal to

$$\gamma \sum_{D,B;R \in X} \psi(D)e\{r^{-1}DR\,{}^tDBD^{-1}\}f(r^{-1}R[{}^tD])e\{RZ\}$$

$$= \sum_{R \in \mathbf{A}_n} \left(\sum_{D,R[{}^tD]\in r\mathbf{A}_n} \psi(D)l_{M_0}(r^{-1}R[{}^tD]; D)f(r^{-1}R[{}^tD]) \right)e\{RZ\},$$

and this proves (2.24).   □

LEMMA 2.7. *Suppose that F is a function in $\mathfrak{M}_s^n$ with Fourier coefficients $f(R)$, and $M_0 = \begin{pmatrix} pD_0^* & 0 \\ 0 & D_0 \end{pmatrix} \in S_0^n(q)$ satisfies the condition $d_n(D_0)^2|r$ (respectively, $r|d_1(D_0)^2$), where $d_i(D)$ denotes the ith elementary divisor of D. Then we have*

(2.26)

$$(F|_{w,\chi}(M_0)_{\Gamma_0^n})(Z) = r^{wn-\langle n\rangle}\chi(r)^n \sum_D \overline{\chi}_w(\det D)|\det D|^{-w} F(rZ[D^{-1}]),$$

(2.27)

$$(f|_{w,\chi}(M_0)_{\Gamma_0^n})(R) = r^{wn-\langle n\rangle}\chi(r)^n {\sum_D}' \overline{\chi}_w(\det D)|\det D|^{-w} f(r^{-1}R[{}^tD]),$$

where $D \in \Lambda \setminus \Lambda D_0 \Lambda$ $(\Lambda = \Lambda^n)$, and $R['D] \in r\mathbf{A}_n$ in (2.27) (respectively, we have

(2.28)

$$(F|_{w,\chi}(M_0)_{\Gamma_0^n})(Z) = r^{wn-n(n+1)}\chi(r)^n|\det D_0|^{n+1}$$
$$\times \sum_D \overline{\chi}_w(\det D)|\det D|^{-w}\sum_{R'} f(R')e\{rR'|D^*|Z\},$$

(2.29)

$$(f|_{w,\chi}(M_0)_{\Gamma_0^n})(R) = r^{wn-n(n+1)}\chi(r)^n|\det D_0|^{n+1}$$
$$\times \sum_D \overline{\chi}_w(\det D)|\det D|^{-w} f(r^{-1}R|{}'D|),$$

where $D \in \Lambda \setminus \Lambda D_0 \Lambda$ and $R' \in \mathbf{A}_n \cap r^{-1}D\mathbf{A}_n{}'D).$

PROOF. The formulas (2.26) and (2.27) follow immediately from (2.23) and (2.24), since, by (5.22) of Chapter 3, in this case we can take $B_{M_0}(D) = \{0\}$ for all $D \in \Lambda D_0 \Lambda$. From (5.23) of Chapter 3 it follows that when $r | d_1(D_0)^2$ we can take

$$B_{M_0}(D) = \{rD^*S; S \in S_n/r^{-1} \cdot {}'DS_nD\},$$

where $S_n = S_n(\mathbf{Z})$. From this we deduce that in this case the sum (2.25) for $R \in \mathbf{A}_n$ and $D \in \Lambda D_0 \Lambda$ can be computed according to the formulas

$$(2.30) \qquad l_{M_0}(R, D) = \begin{cases} r^{-\langle n \rangle}|\det D_0|^{n+1}, & \text{if } R \in \mathbf{A}_n \cap r^{-1}D\mathbf{A}_n{}'D, \\ 0, & \text{if } R \notin \mathbf{A}_n \cap r^{-1}D\mathbf{A}_n{}'D. \end{cases}$$

In fact, by definition we have

$$l_{M_0}(R, D) = \sum_{S \in S_n/r^{-1} \cdot {}'DS_nD} e\{rR|D^*|S\}.$$

Under our assumptions, the function $S \to e\{rR[D^*]S\}$ is obviously a character of the finite additive quotient group $S_n/r^{-1} \cdot {}'DS_nD$. Hence, the above sum is equal to the order of this quotient group if the character is trivial, and is equal to zero otherwise. Clearly, the character is trivial if and only if $rR[D^*]$ is an integer matrix with even entries on the main diagonal, i.e. (since $R \in \mathbf{A}_n$), if and only if $R \in \mathbf{A}_n \cap r^{-1}D\mathbf{A}_n{}'D$. When computing the order of the quotient group, obviously we can replace D by its matrix of elementary divisors $\mathrm{ed}(D) = \mathrm{diag}(d_1, \ldots, d_n)$; we find that this order is

$$\prod_{1 \leqslant i \leqslant j \leqslant n} (r^{-1}d_id_j) = r^{-\langle n \rangle}(d_1 \cdots d_n)^{n+1} = r^{-\langle n \rangle}|\det D_0|^{n+1}.$$

This argument proves (2.30). The formulas (2.28) and (2.39) follow from (2.23), (2.24), and (2.30). $\qquad \square$

In the expansions in Chapter 3 we frequently encountered elements of $\mathbf{L}_0^n$ of the form

$$(2.31) \qquad \Pi_a^n(d) = (M_a^n(d))_{\Gamma_0^n} \quad (0 \leqslant a \leqslant n, d \in \mathbf{N}),$$

where

$$M_a^n(d) = U(d, D_a^n(d)) \quad \text{and} \quad D_a^n(d) = \begin{pmatrix} E_{n-a} & 0 \\ 0 & dE_a \end{pmatrix},$$

and elements of the form $\Delta_n(d) = (dE_{2n})_{\Gamma_0} = (d\Gamma_0^n)$ for $d \in \mathbf{N}$. The next lemma gives formulas for the action of the Hecke operators that correspond to these elements.

LEMMA 2.8. *Let F be a function in $\mathfrak{M}_s^n$ with Fourier coefficients $f(R)$. Then:*

$$(2.32) \quad (F|_{w,\chi}\Pi_a^n(d))(Z) = d^{wn-\langle n\rangle + \langle a\rangle}\chi(d)^n\sum_D\overline{\chi}_w(\det D)|\det D|^{-w}$$
$$\times \sum_R f(R)e\{dR[D^*]Z\},$$

where $D \in \Lambda \setminus \Lambda D_a^n(d)\Lambda$ $(\Lambda = \Lambda^n)$ and $R \in \mathbf{A}_n \cap d^{-1}D\mathbf{A}_n{}'D$;

$$(2.33) \quad (f|_{w,\chi}\Pi_a^n(d))(R) = d^{wn-\langle n\rangle + \langle a\rangle}\chi(d)^n{\sum_D}'\overline{\chi}_w(\det D)|\det D|^{-w}f(d^{-1}R[{}'D]),$$

where $D \in \Lambda \setminus \Lambda D_a^n(d)\Lambda$ and $R[{}'D] \in d\mathbf{A}_n$;

$$(2.34) \quad F|_{w,\chi}\Delta_n(d) = d^{n(w-n-1)}\chi(d)^nF, \quad f|_{w,\chi}\Delta_n(d) = d^{n(w-n-1)}\chi(d)^nf.$$

PROOF. We first show that

$$B_{M_0}(D) = B_0(D), \quad \text{if } M_0 = M_a^n(d) \text{ and } D \in \Lambda D_a^n(d)\Lambda,$$

where $B_0(D)$ is the set (3.60) of Chapter 3. Since the double coset $\Gamma_0^n M_0\Gamma_0^n$ consists of integer matrices, the left side of this equality is contained in the right side. Conversely, suppose that $B \in B_0(D)$ and $D = \alpha D_0\beta$, where $\alpha, \beta \in \Lambda$ and $D_0 = D_a^n(d)$. Then

$$M = \begin{pmatrix} dD^* & B \\ 0 & D \end{pmatrix} = U(\alpha)\begin{pmatrix} dD_0^{-1} & B_0 \\ 0 & D_0 \end{pmatrix}U(\beta),$$

where $B_0 = {}'\alpha B\beta^{-1}$. By Lemma 3.33(1) of Chapter 3, B_0 is contained in $B_0(D_0)$, and hence it can be written in the form $B_0 = \begin{pmatrix} S_1 & dS_2 \\ {}'S_2 & S_3 \end{pmatrix}$, where $S_1 \in S_{n-a}(\mathbf{Z})$, $S_3 \in S_a(\mathbf{Z})$, and $S_2 \in M_{n-a,a}$. Thus, the matrix

$$\begin{pmatrix} dD_0^{-1} & B_0 \\ 0 & D_0 \end{pmatrix} = T\left(\begin{pmatrix} S_1 & S_2 \\ {}'S_2 & 0 \end{pmatrix}\right)\begin{pmatrix} dD_0^{-1} & 0 \\ 0 & D_0 \end{pmatrix}T\left(\begin{pmatrix} 0 & 0 \\ 0 & S_3 \end{pmatrix}\right),$$

and hence M as well, is contained in the double coset $\Gamma_0^n M_a^n(d)\Gamma_0^n$. This means that $B \in B_{M_0}(D)$. We now show that the trigonometric sum (2.25) with $R \in \mathbf{A}_n$, $M_0 = M_a^n(d)$, and $D \in \Lambda D_a^n(d)\Lambda$ is given by the formulas

$$(2.35) \quad l_{M_0}(R, D) = \begin{cases} d^{\langle a\rangle}, & \text{if } R \in \mathbf{A}_n \cap d^{-1}D\mathbf{A}_n{}'D, \\ 0, & \text{if } R \notin \mathbf{A}_n \cap d^{-1}D\mathbf{A}_n{}'D. \end{cases}$$

In fact, if we write D in the form $D = \alpha D_0\beta$, where $\alpha, \beta \in \Lambda$ and $D_0 = D_a^n(d)$, and use Lemma 3.33 of Chapter 3, we obtain

$$l_{M_0}(R, D) = \sum_{B' \in B_0(D_0)/\bmod D_0} e\{R\alpha^*B'\beta(\alpha D_0\beta)^{-1}\}$$
$$= \sum_{B' \in B_0(D_0)/\bmod D_0} e\{R[\alpha^*]B'D_0^{-1}\}.$$

From the description of $B_0(D_0)/\bmod D_0$ in Lemma 3.33(2) of Chapter 3 it easily follows that the last sum is equal to the number of elements in this set, i.e., it is equal to $d^{\langle a\rangle}$ if the lower-right $a \times a$-block in the matrix $R[\alpha^*]$, when divided by d, gives an integer matrix with even entries on the main diagonal, and it is zero otherwise.

Since $R \in \mathbf{A}_n$, this condition for the trigonometric sum to be nonzero is equivalent to the condition $dR[\alpha^* D_0^{-1}] \in \mathbf{A}_n$; and this, in turn, is equivalent to the condition $dR[D^*] \in \mathbf{A}_n$. This proves (2.35). The formulas (2.32) and (2.33) follow from (2.23), (2.24), and (2.35). The formulas in (2.34) are a direct consequence of (2.14).   □

PROBLEM 2.9. Let $R \in \mathbf{A}_n$ and $D \in M_n$, $\det D \neq 0$. Set

$$S_D(R) = \sum_{B \in B_0(D)/\operatorname{mod} D} e\{RBD^{-1}\},$$

and let $\mathbf{A}_n(D) = \{R \in \mathbf{A}_n;\ S_D(R) \neq 0\}$. Prove that:

(1) If $\alpha, \beta \in \Lambda^n$, then $S_{\alpha D \beta}(R) = S_D(R[\alpha^*])$ and $\mathbf{A}_n(\alpha D \beta) = \alpha \mathbf{A}_n(D) {}'\alpha$.

(2) If $\operatorname{ed} D = \operatorname{diag}(d_1, \ldots, d_n)$, then

$$S_D(R) = \begin{cases} d_1^n d_2^{n-1} \cdots d_n, & \text{if } R \in \mathbf{A}_n(D), \\ 0, & \text{if } R \notin \mathbf{A}_n(D). \end{cases}$$

(3) If $d = \operatorname{ed} D = \operatorname{diag}(d_1, \ldots, d_n)$ and $R = ((1 + e_{\alpha\beta}) r_{\alpha\beta}) \in \mathbf{A}_n$, where $(e_{\alpha\beta}) = E_n$, then $R \in \mathbf{A}_n(D)$ if and only if $r_{\alpha\beta} \equiv 0(\operatorname{mod} d_\alpha)$ for $1 \leqslant \alpha \leqslant \beta \leqslant n$.

(4) Let $n = 2$ and $D = d_1 D_0$, where D_0 is an integer matrix with relatively prime entries. Then $\mathbf{A}_2(D) = d_1(d_0^{-1} D_0 \mathbf{A}_2 {}'D_0 \cap \mathbf{A}_2)$, where $d_0 = |\det D_0|$.

[**Hint:** See Lemma 3.33 of Chapter 3.]

PROBLEM 2.10. As before, let k, χ, and s be connected by the relation $(-1)^k \chi(-1) = s(-1)$. Show that for $(m, q) = 1$ the images $\mathbf{T}(m)$ of the elements (3.22) of Chapter 3 in $\mathbf{L}_0^n(q)$ act on the space $\mathfrak{F}_s^n$ according to the formula

$$(f|_{k,\chi} \mathbf{T}(m))(R) = m^{nk - \langle n \rangle} \chi(m)^n \sum_{d_1|d_2|\cdots|d_n|m} d_1^n d_2^{n-1} \cdots d_n$$
$$\times \sum_D \overline{\chi}(\det D)(\det D)^{-k} f(m^{-1} R['D]),$$

where $d_i \in \mathbf{N}$, $D \in \Lambda^n \setminus \Lambda^n \operatorname{diag}(d_1, \ldots, d_n)\Lambda^n$, $m^{-1} R['D] \in \mathbf{A}_n(D)$, $f \in \mathfrak{F}_s^n$, and $R \in \mathbf{A}_n$. In particular, if $n = 1$, then

$$(f|_{k,\chi} \mathbf{T}(m))(2a) = \sum_{\delta|m,a} \delta^{k-1} \chi(\delta) f(2ma/\delta^2),$$

and if $n = 2$, then

$$(f|_{k,\chi} \mathbf{T}(m))(R) = \sum_{\delta_1, \delta_2, \delta_3} \delta_2^{2k-2} \delta_3^{2k-3} \chi(\delta_2 \delta_3)^2$$
$$\times \sum_D \overline{\chi}(\det D)(\det D)^{-k} f((\delta_1/\delta_2\delta_3)R['D]),$$

where $\delta_1 \delta_2 \delta_3 = m$, $R \in \delta_3 \mathbf{A}_2$, and $D \in \Lambda^2 \setminus \Lambda^2 \begin{pmatrix} 1 & 0 \\ 0 & \delta_2 \end{pmatrix} \Lambda^2$, $R['D] \in \delta_2 \delta_3 \mathbf{A}_2$.

[**Hint:** For the first part use Problem 3.14 of Chapter 3, the formulas (2.24) for the double cosets in $\mathbf{T}(m)$, and the previous problem; for the second part set $m/d = \delta$; and for the third part set $d_1 = \delta_1$, $d_2 = \delta_1 \delta_2$, and $m = \delta_1 \delta_2 \delta_3$, and use part (4) of the previous problem.]

3. Hecke operators and the Siegel operator. We now study the relations between Hecke operators on the spaces $\mathfrak{M}_w^n(q, \chi)$ and $\mathfrak{M}_s^n$ and the Siegel operator Φ that was defined in §3.4 of Chapter 2. These relations make it possible to reduce certain questions in the theory of Hecke operators for groups of degree n to the analogous questions for groups of lower degree.

In §3.4 of Chapter 2 the Siegel operator Φ was originally defined on the space $\mathfrak{F}_q^n$ of Fourier series of the form (3.5) of Chapter 2 that converge absolutely on all of $\mathbf{H}_n$ and uniformly on subsets of the form $\mathbf{H}_n(\varepsilon)$ with $\varepsilon > 0$. From Theorem 3.1 of Chapter 2 it follows that $\mathfrak{M}_s^n \subset \mathfrak{F}_1^n$ for any character s. Thus, the operator Φ is defined on the spaces $\mathfrak{M}_s^n$, and hence also on the subspaces $\mathfrak{M}_w^n(q, \chi)$. From the definitions and Theorem 3.1 of Chapter 2 it follows that the subspace $\mathfrak{M}_s^n \subset \mathfrak{F}_1^n$ can be characterized as the subset of all $F \in \mathfrak{F}_1^n$ whose Fourier coefficients satisfy (2.7). If we consider these relations for matrices V in Λ^n of the form $\begin{pmatrix} V_1 & 0 \\ 0 & 1 \end{pmatrix}$, where $V_1 \in \Lambda^{n-1}$, and apply (3.50) of Chapter 2, we find that for any $F \in \mathfrak{M}_s^n$ the Fourier coefficients of the function $F|\Phi$ satisfy the conditions (2.7) for $n - 1$. Thus,

$$(2.36) \qquad \Phi \colon \mathfrak{M}_s^n \to \mathfrak{M}_s^{n-1} \quad (n \geqslant 1),$$

where we set

$$(2.37) \qquad \mathfrak{M}_s^0 = \begin{cases} \mathbf{C}, & \text{if } s(-1) = 1, \\ \{0\}, & \text{if } s(-1) = -1 \end{cases}$$

(in the case $s(-1) = -1$, the constant term in the Fourier expansion of any function of $\mathfrak{M}_s^n$ is obviously zero). Now let $q \in \mathbf{N}$, and let χ be a Dirichlet character modulo q. From the definitions it easily follows that if $K = \Gamma_0^n(q)$, then the group $K^{[n-1]}$ (see (3.55) of Chapter 2) is $\Gamma_0^{n-1}(q)$, and the character $[\chi]^{[n-1]}$ of this group (see (3.56) of Chapter 2) is the one-dimensional character corresponding to χ. Thus, by Proposition 3.12 of Chapter 2 we see that for any integer or half-integer w

$$(2.38) \qquad \Phi \colon \mathfrak{M}_w^n(q, \chi) \to \mathfrak{M}_w^{n-1}(q, \chi) \quad (n \geqslant 1),$$

where, in accordance with (2.20) and (2.37), we have set

$$(2.39) \qquad \mathfrak{M}_w^0(q, \chi) = \begin{cases} \mathbf{C}, & \text{if } \chi_w(-1) = 1, \\ \{0\}, & \text{if } \chi_w(-1) = -1. \end{cases}$$

We now turn to the Hecke operators. We show that the Hecke operators on the spaces $\mathfrak{M}_s^n$ and $\mathfrak{M}_w^n(q, \chi)$ are compatible with Φ in the sense that there exists a homomorphism $X \to X'$ of Hecke rings from degree n to degree $n - 1$ such that for every function F in the space under consideration one has $(F|_{w,\chi} X)|\Phi = (F|\Phi)|_{w,\chi} X'$. It is convenient to describe these homomorphisms in terms of the polynomial realizations of the Hecke rings by means of the spherical maps. Hence, we shall limit ourselves to the local Hecke rings. The global Hecke rings could be treated in an analogous manner; however, we do not need to do this, since all of the elements that interest us in the global Hecke ring are generated by local components.

Suppose that $n \in \mathbf{N}$, p is a prime, and

$$(2.40) \qquad X = \sum_i a_i \left(\Gamma_0^n \begin{pmatrix} p^{\delta_i} D_i^* & B_i \\ 0 & D_i \end{pmatrix} \right)$$

is an arbitrary element of the Hecke ring $\mathbf{L}_{0,p}^n$ (see (3.45) of Chapter 3). By choosing different Γ_0^n-left coset representatives, we can replace each matrix $D_i \in G_p^n$ by any matrix in the left coset $\Lambda^n D_i$. From Lemma 2.7 of Chapter 3 it then follows that all of the D_i may be assumed to have been chosen in the form

$$(2.41) \qquad D_i = \begin{pmatrix} D_i' & * \\ 0 & p^{d_i} \end{pmatrix}, \quad \text{where } D_i' \in G_p^{n-1} \text{ and } d_i \in \mathbf{Z}.$$

With this choice of representatives, we set

$$(2.42) \qquad \Psi(X, u) = \Psi_p^n(X, u) = \sum_i a_i u^{-\delta_i} (u p^{-n})^{d_i} \left(\Gamma_0^{n-1} \begin{pmatrix} p^{\delta_i} (D_i')^* & B_i' \\ 0 & D_i' \end{pmatrix} \right),$$

where u is an independent variable, and for every $n \times n$-matrix A we let A' denote the $(n-1) \times (n-1)$-matrix in the upper-left corner of A. If $n = 1$, then $D_i = p^{d_i}$, and we set

$$(2.43) \qquad \Psi_p^1(X, u) = \sum_i a_i u^{-\delta_i} (u p^{-1})^{d_i}.$$

It is easy to see that $\Psi(X, u)$ does not depend on the choice of representatives with these properties. We thus obtain a linear map

$$\Psi = \Psi_p^n : \mathbf{L}_{0,p}^n \to L(\Gamma_0^{n-1}, S_{0,p}^{n-1}) \otimes_{\mathbf{Z}} \mathbf{Q}[u^{\pm 1}]$$

from the ring $\mathbf{L}_{0,p}^n$ to the left coset module of the pair $(\Gamma_0^{n-1}, S_{0,p}^{n-1})$ over the ring of polynomials in $u^{\pm 1}$ with rational coefficients; in the case $n = 1$ we obtain a map

$$\Psi_p^1 : \mathbf{L}_{0,p}^1 \to \mathbf{Q}[u^{\pm 1}].$$

PROPOSITION 2.11. *Let $n \in \mathbf{N}$, and let p be a prime. Then:*
(1) *For any $X \in \mathbf{L}_{0,p}^n$ the element $\Psi(X, u)$ lies in the ring*

$$\mathbf{L}_{0,p}^{n-1}[u^{\pm 1}] = D_{\mathbf{Q}}(\Gamma_0^{n-1}, S_{0,p}^{n-1})[u^{\pm 1}]$$

of polynomials in $u^{\pm 1}$ over the Hecke ring $\mathbf{L}_{0,p}^{n-1}$ (we set $\mathbf{L}_{0,p}^0 = \mathbf{Q}$); the map

$$(2.44) \qquad \Psi = \Psi_p^n : \mathbf{L}_{0,p}^n \to \mathbf{L}_{0,p}^{n-1}[u^{\pm 1}]$$

is a ring homomorphism.
  (2) *The image of the restriction of Ψ_p^n to the subring C_{-p}^n, C_{+p}^n or $\mathbf{L}_p^n$ of the ring $\mathbf{L}_{0,p}^n$ is contained in $C_{-p}^{n-1}[u^{\pm 1}]$, $C_{+p}^{n-1}[u^{\pm 1}]$, $\mathbf{L}_p^{n-1}[u^{\pm 1}]$, respectively (we set $C_{\pm p}^0 = \mathbf{L}_p^0 = \mathbf{Q}$).*
  (3) *The following diagram commutes:*

$$(2.45) \qquad
\begin{array}{ccc}
\mathbf{L}_{0,p}^n & \xrightarrow{\;\Omega_p^n\;} & \mathbf{Q}[x_0^{\pm 1}, \ldots, x_n^{\pm 1}] \\
\Big\downarrow{\scriptstyle \Psi_p^n} & & \Big\downarrow{\scriptstyle \Xi_n} \\
\mathbf{L}_{0,p}^{n-1}[u^{\pm 1}] & \xrightarrow{\;\Omega_p^{n-1}\;} & \mathbf{Q}[x_0^{\pm 1}, \ldots, x_{n-1}^{\pm 1}, u^{\pm 1}],
\end{array}$$

where Ω_p^n is the spherical map (3.49) of Chapter 3, in the bottom row Ω_p^{n-1} is the homomorphism extending the spherical map on $\mathbf{L}_{0,p}^{n-1}$ that satisfies the condition $\Omega_p^{n-1}(u^{\pm 1}) = u^{\pm 1}$ (we define the spherical map on $\mathbf{L}_{0,p}^0 = \mathbf{Q}$ to be the identity map), Ψ_p^n is the homomorphism (2.44), and Ξ_n is the homomorphism of polynomial rings given on generators by:

$\Xi(x_0) = x_0 u^{-1}$, $\Xi(x_n) = u$, and $\Xi(x_i) = x_i$ for $1 \leqslant i \leqslant n-1$ (we take $\Xi_1(x_0) = u^{-1}$, $\Xi_1(x_1) = u$).

PROOF. If γ_1 is an arbitrary matrix in Γ_0^{n-1} and γ is the image of γ_1 under the map (3.53) of Chapter 2, then $\gamma \in \Gamma_0^n$, and from the condition $X \cdot \gamma = X$ for $X \in \mathbf{L}_{0,p}^n$ and the definition of $\Psi(X, u)$ it easily follows that $\Psi(X, u)\gamma_1 = \Psi(X \cdot \gamma, u) = \Psi(X, u)$, where γ_1 acts only on the left cosets and not on the coefficients. Hence, $\Psi(X, u) \in \mathbf{L}_{0,p}^{n-1}[u^{\pm 1}]$. From the definition of multiplication in Hecke rings it immediately follows that the map (2.44) is a homomorphism.

Next, using the definition of Ψ and the expansions (5.12) of Chapter 3, we obtain: $\Psi(\Pi_{\pm}^n(p), u) = u^{-1}\Pi_{\pm}^{n-1}(p)$. This, along with (6.1) of Chapter 3, implies the claim in the lemma concerning the Ψ-images of $C_{\pm p}^n$. As for L_p^n, if we define the map $\Psi: L_p^n \to L_p^{n-1}[u^{\pm 1}]$ by analogy with (2.42), then it is not hard to verify the commutativity of the diagram

$$
\begin{array}{ccc}
L_p^n & \xrightarrow{\ \varepsilon\ } & \mathbf{L}_{0,p}^n \\
\downarrow{\scriptstyle\Psi} & & \downarrow{\scriptstyle\Psi} \\
L_p^{n-1}[u^{\pm 1}] & \xrightarrow{\ \varepsilon\ } & \mathbf{L}_{0,p}^{n-1}[u^{\pm 1}],
\end{array}
$$

where ε denotes the imbeddings in Lemma 3.26 of Chapter 3; hence

$$\Psi(\mathbf{L}_p^n) = \Psi(\varepsilon(L_p^n)) = \varepsilon(\Psi(L_p^n)) \in \mathbf{L}_p^{n-1}[u^{\pm 1}].$$

To prove the third part of the proposition, we suppose that each matrix D_i in the expansion (2.40) of some $X \in \mathbf{L}_{0,p}^n$ has been chosen in the form (2.36) of Chapter 3 with diagonal entries $p^{d_{i1}}, \ldots, p^{d_{in}}$. Then, by the definition of the maps, we have

$$
\begin{aligned}
\Xi_n(\Omega_p^n(X)) &= \Xi_n\left(\sum_i a_i x_0^{\delta_i} \prod_{j=1}^n (x_j p^{-j})^{d_{ij}}\right) \\
&= \sum_i a_i u^{-\delta_i} (up^{-n})^{d_{in}} x_0^{\delta_i} \prod_{j=1}^{n-1} (x_j p^{-j})^{d_{ij}} \\
&= \sum_i a_i u^{-\delta_i} (up^{-n})^{d_{in}} \omega_p^{n-1}\left(\Phi_p^{n-1}\left(\Gamma_0^{n-1}\begin{pmatrix} p^{\delta_i}(D_i')^* & B_i' \\ 0 & D_i' \end{pmatrix}\right)\right) \\
&= \Omega_p^{n-1}(\Psi(X, u)). \qquad \square
\end{aligned}
$$

In what follows, to avoid worrying about the different fields of definition of the Hecke rings, we shall suppose that all of our maps of Hecke rings have been extended by linearity to the complexifications.

THEOREM 2.12 (the Zharkovskaia commutation relations). *Suppose that $q \in \mathbf{N}$, p is a prime not dividing q, χ is a Dirichlet character modulo q, and s is the character of the group $\{\pm 1\}$ that satisfies the condition $s(-1) = \chi_w(-1)$, where χ_w is the character (2.5). Then the following relation holds for any $F \in \mathfrak{M}_s^n$ and any $X \in \overline{\mathbf{L}}_{0,p}^n = \mathbf{L}_{0,p}^n \otimes_{\mathbf{Q}} \mathbf{C}$:*

$$(2.46) \qquad (F|_{w,\chi} X)|\Phi = (F|\Phi)_{w,\chi} \Psi(X, p^{n-w}\overline{\chi}(p)),$$

where Φ is the Siegel operator, $\Psi(X, p^{n-w}\overline{\chi}(p)) \in \overline{\mathbf{L}}_{0,p}^{n-1}$ is the element (2.42), and in the case $n = 1$ the operator $|_{w,\chi}\Psi$ acts on the right as multiplication by the complex number Ψ.

PROOF. Let $f(R)$ $(R \in \mathbf{A}_n)$ be the coefficients in the Fourier expansion (2.6) of F, and let X be written in the form (2.40), where $a_i \in \mathbf{C}$ and each D_i has the form (2.41). Using the definitions (see (2.14)), we have

$$F|_{w,\chi}X = \sum_i a_i \left(\sum_{R \in \mathbf{A}_n} f(R)e\{RZ\} \right)\bigg|_{w,\chi} \begin{pmatrix} p^{\delta_i}D_i^* & B_i \\ 0 & D_i \end{pmatrix}$$

$$= \sum_{i,R} a_i\alpha_i f(R)e\{p^{\delta_i}R[D_i^*]Z + RB_iD_i^{-1}\},$$

where

$$\alpha_i = p^{\delta_i(wn-\langle n\rangle)}\chi_w(\det(p^{\delta_i}D_i^*))|\det D_i|^{-w}.$$

We note that for $R = \begin{pmatrix} R' & * \\ * & r \end{pmatrix} \in \mathbf{A}_n$ the entry in the lower-right corner of the matrix $p^{\delta_i}R[D_i^*]$ is equal to $p^{\delta_i-2d_i}r$ (see (2.41)). Thus, if we set $Z = \begin{pmatrix} Z' & 0 \\ 0 & i\lambda \end{pmatrix}$ in the last expression, where $Z' \in \mathbf{H}_{n-1}$, and if we let λ approach $+\infty$, then all of the terms corresponding to matrices R with $r > 0$ will approach zero. Since $r \geqslant 0$, we have $r = 0$, and since $R \geqslant 0$ it follows that $R = \begin{pmatrix} R' & 0 \\ 0 & 0 \end{pmatrix}$. Finally, we note that for $R = \begin{pmatrix} R' & 0 \\ 0 & 0 \end{pmatrix}$, $Z = \begin{pmatrix} Z' & 0 \\ 0 & i\lambda \end{pmatrix}$, and D_i of the form (2.41) we have the relations

$$p^{\delta_i}R[D_i^*] = p^{\delta_i}\begin{pmatrix} R'[(D_i')^*]Z' & 0 \\ 0 & 0 \end{pmatrix} \quad \text{and} \quad RB_iD_i^{-1} = \begin{pmatrix} R'B_i'(D_i')^{-1} & 0 \\ 0 & 0 \end{pmatrix},$$

and we use the uniform convergence of the series on the subsets $\mathbf{H}_n(\varepsilon) \subset \mathbf{H}_n$. We obtain

$$(F|_{w,\chi}X)|\varphi = \lim_{\lambda \to +\infty} (F|_{w,\chi}X)\begin{pmatrix} Z' & 0 \\ 0 & i\lambda \end{pmatrix}$$

$$= \sum_{i,R'\in\mathbf{A}_{n-1}} a_i\alpha_i f\left(\begin{pmatrix} R' & 0 \\ 0 & 0 \end{pmatrix}\right)e\{p^{\delta_i}R'[(D_i')^*]Z' + R'B_i'(D_i')^{-1}\}$$

$$= \sum_i \beta_i a_i \left(\sum_{R'\in\mathbf{A}_{n-1}} f\left(\begin{pmatrix} R' & 0 \\ 0 & 0 \end{pmatrix}\right)e\{R'Z'\} \right)\bigg|_{w,\chi} \begin{pmatrix} p^{\delta_i}(D_i')^* & B_i' \\ 0 & D_i' \end{pmatrix}$$

$$= (F|\Phi)|_{w,\chi}\Psi(X, p^{n-w}\overline{\chi}(p)),$$

where $\beta_i = p^{\delta_i(w-n)}\chi(p^{\delta_i-d_i})p^{-wd_i}$, since, according to (3.50) of Chapter 2, the sum in the large parentheses is equal to $(F|\Phi)(Z')$. $\qquad\square$

We now consider the restriction of the map $\Psi(\cdot, p^{n-k}\overline{\chi}(p))$ to the complexification $\overline{\mathbf{L}}_p^n = \mathbf{L}_p^n \otimes_{\mathbf{Q}} \mathbf{C}$ of the ring (3.46) in Chapter 3.

PROPOSITION 2.13. Suppose that $n, q \in \mathbf{N}$, $k \in \mathbf{Z}$, and χ is a character modulo q. Then for any prime p not dividing q:

(1) *If* $X \in \overline{\mathbf{L}}_p^n$ *(respectively, if X belongs to the even subring*

(2.47)
$$\overline{\mathbf{E}}_p^n = \left\{ \sum_i a_i (\Gamma_0^n M_i) \in \overline{\mathbf{L}}_p^n ;\ r(M_i) = p^{\delta_i},\ \delta_i \in 2\mathbf{Z} \right\}$$

of $\overline{\mathbf{L}}_p^n$*), then* $\Psi(X, p^{n-k}\overline{\chi}(p)) \in \overline{\mathbf{L}}_p^{n-1}$ *(respectively,* $\overline{\mathbf{E}}_p^{n-1}$*), where we set* $\overline{\mathbf{L}}_p^0 = \overline{\mathbf{E}}_p^0 = \mathbf{C}$.
(2) *The map*

(2.48)
$$\Psi(\cdot, p^{n-k}\overline{\chi}(p)) \colon \overline{\mathbf{L}}_p^n \to \overline{\mathbf{L}}_p^{n-1}$$

is an epimorphism in all cases except when

(2.49)
$$k = n > 1 \quad \text{and} \quad \chi(p) = -1.$$

In that case the image of $\overline{\mathbf{L}}_p^n$ *is the even subring* $\overline{\mathbf{E}}_p^{n-1} \subset \overline{\mathbf{L}}_p^{n-1}$.
(3) *The map* (2.48) *gives an epimorphism of* $\overline{\mathbf{E}}_p^n$ *onto the ring* $\overline{\mathbf{E}}_p^{n-1}$.

PROOF. We form the complexifications of all of the rings in the diagram (2.45), i.e., we tensor with $\mathbf{C}$ over $\mathbf{Q}$, and we extend the maps by linearity to the complexifications. Obviously, the resulting diagram still commutes. Then, instead of the action of $\Psi_p^n(\cdot, p^{n-k}\overline{\chi}(p))$ on $\overline{\mathbf{L}}_p^n$ we can consider the action of Ξ_n with $u = p^{n-k}\overline{\chi}(p)$ on the image $\Omega_p^n(\overline{\mathbf{L}}_p^n)$ under the extended spherical map. From Theorem 3.30 of Chapter 3 it follows that this image is the polynomial ring

(2.50)
$$\Omega_p^n(\overline{\mathbf{L}}_p^n) = \mathbf{C}[t, \rho_0^{\pm 1}, \rho_1, \dots, \rho_{n-1}]$$
$$= \mathbf{C}[t, r_1, \dots, r_{n-1}, \rho_0^{\pm 1}] \quad (n \geqslant 1),$$

where the polynomials $r_a = r_a^n$, $\rho_a = \rho_a^n$, and $t = t^n$ are defined by (3.52)–(3.54) of Chapter 3, respectively. From the definition of the map Ω_p^n and the fact that it is a monomorphism on $\overline{\mathbf{L}}_p^n$ it follows that the image of the subring (2.47) under this map is the subset of polynomials in $\Omega_p^n(\overline{\mathbf{L}}_p^n)$ having even degree in the variable x_0. Hence, if we take into account the relation

(2.51)
$$(t)^2 = x_0^2 x_1 \cdots x_n \prod_{i=1}^n (1 + x_i)(1 + x_i^{-1})$$
$$= \rho_0 \sum_{a=0}^{2n} r_a = \rho_0 (2 + 2(r_1 + \cdots + r_{n-1}) + r_n) \quad (n \geqslant 1),$$

we obtain

(2.52)
$$\Omega_p^n(\overline{\mathbf{E}}_p^n) = \mathbf{C}[(t)^2, r_1, \dots, r_{n-1}, \rho_0^{\pm 1}] = \mathbf{C}[r_1, \dots, r_n, \rho_0^{\pm 1}].$$

We now examine how the map $\Xi = \Xi_n$ acts on the generators of these polynomial rings. By definition, we have

(2.53)
$$\Xi(\rho_0^n) = u^{-1} x_0^2 x_1 \cdots x_{n-1} = u^{-1} \rho_0^{n-1}, \quad \Xi(t^n) = u^{-1}(1 + u)t^{n-1}.$$

Using (3.52) of Chapter 3, we obtain

$$\sum_{a=0}^{2n}(-1)^a\Xi(r_a^n)v^a = (1-uv)(1-u^{-1}v)\prod_{i=1}^{n-1}(1-x_iv)(1-x_i^{-1}v)$$

(2.54)

$$= (1-(u+u^{-1})v+v^2)\sum_{a=0}^{2(n-1)}(-1)^a r_a^{n-1}(x_1,\dots,x_{n-1})v^a,$$

and hence

(2.55) $$\Xi(r_a^n) = r_a^{n-1} + (u+u^{-1})r_{a-1}^{n-1} + r_{a-2}^{n-1}.$$

These formulas imply that for any $u \in \mathbf{C}$—in particular, for $u = p^{n-k}\overline{\chi}(p)$—the map Ξ takes the ring $\Omega_p^n(\overline{\mathbf{L}}_p^n)$ to $\Omega_p^{n-1}(\overline{\mathbf{L}}_p^{n-1})$, and takes the ring $\Omega_p^n(\overline{\mathbf{E}}_p^n)$ to

$$\mathbf{C}[r_1^{n-1},\dots,r_n^{n-1},(\rho_0^{n-1})^{\pm 1}] = \mathbf{C}[r_1^{n-1},\dots,r_{n-1}^{n-1},(\rho_0^{n-1})^{\pm 1}] = \Omega_p^{n-1}(\overline{\mathbf{E}}_p^{n-1})$$

(obviously $r_n^{n-1} = r_{n-2}^{n-1}$). This proves part (1) (for $\overline{\mathbf{L}}_p^n$ this part also follows from Proposition 2.11(2)).

To find the images of these maps we use (2.53) and (2.54) to express the generators r_a^{n-1} and t^{n-1} in terms of their preimages. From (2.53) we have

(2.56) $$\rho_0^{n-1} = u\Xi_n(\rho_0^n) \quad\text{and}\quad t^{n-1} = u(1+u)^{-1}\Xi_n(t^n).$$

Furthermore, from (2.54) we obtain

(2.57) $$r_a^{n-1} = \sum_{0\leqslant i\leqslant a}(-1)^i c_i(u)\Xi_n(r_{a-i}^n),$$

where the $c_i(u)$ are determined by the condition

$$\sum_{i=0}^{\infty}c_i(u)v^i = \{(1-uv)(1-u^{-1}v)\}^{-1}.$$

If we set $u = p^{n-k}\overline{\chi}(p)$ in these formulas, we see that the image of $\Omega_p^n(\overline{\mathbf{L}}_p^n)$ contains all of the generators of the ring $\Omega_p^{n-1}(\overline{\mathbf{L}}_p^{n-1})$, except in the case when $n > 1$ and $p^{n-k}\overline{\chi}(p) = -1$, i.e., the case (2.49). In the latter case, the image contains all of the elements r_a^{n-1} and $(\rho_0^{n-1})^{\pm 1}$, but it does not contain t^{n-1}; hence, by (2.52) for $n-1$, this image is $\Omega_p^{n-1}(\overline{\mathbf{E}}_p^{n-1})$. The same formulas also imply that the image of $\Omega_p^n(\overline{\mathbf{E}}_p^n)$ is always $\Omega_p^{n-1}(\overline{\mathbf{E}}_p^{n-1})$.   □

The formulas (2.56) and (2.57) give us a practical way to compute the preimages of the generators of $\overline{\mathbf{L}}_p^{n-1}$ under the map (2.48).

Let $\widehat{\mathbf{L}}_{0,p}^n$ be the Hecke ring (4.99) of Chapter 3, and let P_k^n be the homomorphism (4.101) of Chapter 3. We define the homomorphism $\widehat{\Psi} = \widehat{\Psi}_p^n$ for the ring $\widehat{\mathbf{L}}_{0,p}^n$ in such a way that the following diagram commutes:

$$(2.58) \qquad \begin{array}{ccc} \widehat{\mathbf{L}}^n_{0,p} & \xrightarrow{\ P^n_k\ } & \overline{\mathbf{L}}^n_{0,p} \\ {\scriptstyle \widehat{\Psi}(\,,u)}\big\downarrow & & \big\downarrow{\scriptstyle \Psi(\,,u)} \\ \widehat{\mathbf{L}}^{n-1}_{0,p}[u^{\pm1}] & \xrightarrow{\ P^{n-1}_k\times 1\ } & \overline{\mathbf{L}}^{n-1}_{0,p}[u^{\pm1}], \end{array}$$

where $P^{n-1}_k \times 1$ is the homomorphism extending P^{n-1}_k and taking $u^{\pm1} \to u^{\pm1}$, and $\widehat{\mathbf{L}}^0_{0,p} = \mathbf{C}$. If

$$(2.59) \qquad \widehat{X} = \sum_i a_i \left(\widehat{\Gamma}^n_0 \begin{pmatrix} p^{\delta_i}(D'_i)^* & B'_i \\ 0 & D'_i \end{pmatrix}, t_i \right)$$

is an arbitrary element of $\widehat{\mathbf{L}}^n_{0,p}$ and the matrices D_i have the form (2.41), then we easily see that the map $\widehat{\Psi}$ that takes $\widehat{X}$ to

$$(2.60) \qquad \begin{aligned} \widehat{\Psi}(\widehat{X}, u) &= \sum_i a_i u^{-\delta_i} (up^{-n})^{d_i} \left(\widehat{\Gamma}^{n-1}_0 \left(\begin{pmatrix} p^{\delta_i}(D'_i)^* & B'_i \\ 0 & D'_i \end{pmatrix}, t_i p^{-d_i/2} \right) \right), \\ \widehat{\Psi}(\widehat{X}, u) &= \sum_i a_i u^{\delta_i} (up^{-1})^{d_i} t_i^{-k} \end{aligned}$$

for $n > 1$ and $n = 1$, respectively, has the above property. Moreover, if we use (2.19), (2.46), and the commutativity of the diagram (2.58), then for any $F \in \mathfrak{M}^n_s$, where $s(-1) = \chi(-1)$, and for any $\widehat{X} \in \widehat{\mathbf{L}}^n_{0,p}$, we obtain the relation

$$(2.61) \qquad (F|_{k/2,\chi}\widehat{X})|\Phi = (F|\Phi)|_{k/2,\chi}\widehat{\Psi}(\widehat{X}, p^{n-k/2}\overline{\chi}(p)).$$

LEMMA 2.14. *Let $\widehat{E}^n_p(q)$ be the even Hecke ring (4.37) of Chapter 3, and let $\widehat{\varepsilon}_q$ be the imbedding of this ring in $\widehat{\mathbf{L}}^n_{0,p}$ given in (4.100) of Chapter 3. Then the restriction of the map (2.60) to the subring $\widehat{\varepsilon}_q(\widehat{E}^n_p(q))$ gives a homomorphism*

$$(2.62) \qquad \widehat{\Psi}(\,,u)\colon \widehat{\varepsilon}_q(\widehat{E}^n_p(q)) \to \widehat{\varepsilon}_q(\widehat{E}^{n-1}_p(q))[u^{\pm1}],$$

where the ring on the right is $\mathbf{C}[u^{\pm1}]$ in the case $n = 1$.

PROOF. Since the lemma is obvious for $n = 1$, we shall suppose that $n > 1$. It was shown in §4.3 of Chapter 3 that the Hecke pairs $(\widehat{\Gamma}^m_0(q), \widehat{S}^m_p(q))$ and $(\widehat{\Gamma}^m_0, \widehat{S}^m_{0,p})$ satisfy the condition (1.26) of Chapter 3. Using the diagram (1.30) of Chapter 3, we define an action of the group $\widehat{\Gamma}^m_0(q)$ on the space $L = L_{\mathbf{Q}}(\widehat{\Gamma}^m_0, \widehat{S}^m_{0,p})$ and on its extension $L[u^{\pm1}]$, where $\widehat{\Gamma}^m_0$ acts only on the coefficients in L. Then, by Lemma 1.10 of Chapter 3, to prove (2.62) it suffices to show that

$$\widehat{\Psi}(\widehat{X}, u) \cdot \widehat{\gamma}' = \widehat{\Psi}(\widehat{X} \cdot \widehat{\gamma}, u) \quad \text{for } \widehat{X} \in \widehat{\varepsilon}_q(\widehat{E}^n_p(q)), \quad \gamma' \in \Gamma^{n-1}_0(q),$$

where γ is the image of γ' in the group $\Gamma^n_0(q)$ under the map (3.53) of Chapter 2. This relation, in turn, is a consequence of the following claim.

Let $\widehat{M_i} = (M_i, t_i) \in \widehat{S}^n_{0,p}$ with $M_i = \begin{pmatrix} p^{2\delta_i} D_i^* & B_i \\ 0 & D_i \end{pmatrix}$, where $i = 1, 2$, the matrices D_i have the form (2.41), and

$$(2.63) \qquad \widehat{M_1}\widehat{\gamma} = \widehat{\delta}\widehat{M_2} \quad \text{with } \delta = \begin{pmatrix} a & b \\ c & d \end{pmatrix} \in \Gamma^n_0(q).$$

Then

$$(2.64) \qquad \widehat{M'_1}\widehat{\gamma}' = \widehat{\delta}'\widehat{M'_2} \quad \text{for } \widehat{M'_i} = (M'_i, t_i p^{-d_i/2}) \in \widehat{S}^{n-1}_{0,p},$$

where $M'_i = \begin{pmatrix} p^{2\delta_i}(D'_i)^* & B'_i \\ 0 & D'_i \end{pmatrix}$ and $\delta' = \begin{pmatrix} a' & b' \\ c' & d' \end{pmatrix} \in \Gamma^{n-1}_0(q)$, and for any $n \times n$-matrix A we let A' denote the $(n-1) \times (n-1)$-matrix in its upper-left corner.

To prove this claim, we first note that $M_1\gamma = \delta M_2$. From this and from (2.7)–(2.8) of Chapter 1 we obtain $M'_1\gamma' = \delta' M'_2$, where $\delta' \in \Gamma^{n-1}_0(q)$, $d_1 = d_2$, and the matrix δ has blocks of the form $a = \begin{pmatrix} a' & 0 \\ * & 1 \end{pmatrix}$, $c = \begin{pmatrix} c' & 0 \\ 0 & 0 \end{pmatrix}$, and $d = \begin{pmatrix} d' & * \\ 0 & 1 \end{pmatrix}$. If we now compare the second components of the elements in (2.63), by (3.19) and (3.23) of Chapter 2 we obtain

$$t_1 j^n_{(2)}(\gamma, Z) = j^n_{(2)}(\delta, M_2\langle Z\rangle)t_2.$$

Setting $Z = \begin{pmatrix} Z' & 0 \\ 0 & i\lambda \end{pmatrix} \in \mathbf{H}_n$, we conclude from (3.61) of Chapter 2 and the last equality that we have

$$t_1 j^{n-1}_{(2)}(\gamma', Z') = j^{n-1}_{(2)}(\delta', M'_2\langle Z'\rangle)t_2.$$

This, together with the equality $M'_1\gamma' = \delta' M'_2$, gives us (2.64).   □

This lemma enables us to describe the action of Ψ on the subring $\widehat{\varepsilon}_{q,k}(\widehat{E}^n_p(q,\varkappa))$ of the ring $\overline{\mathbf{L}}^n_{0,p}$, and to define a map $\widehat{\Psi}$ for $\widehat{E}^n_p(q,\varkappa)$ that commutes with the Siegel operator Φ. Namely, we have

PROPOSITION 2.15. (1) *Let $\widehat{\mathbf{E}}^n_p(q,\varkappa)$ be the ring (4.104) of Chapter 3, let χ be a Dirichlet character modulo q, and let k be an odd integer. Then the map Ψ (see (2.42) and (2.43)) gives an epimorphism*

$$(2.65) \qquad \Psi(\,, p^{n-k/2}\overline{\chi}(p)): \widehat{\mathbf{E}}^n_p(q,\varkappa) \otimes_{\mathbf{Q}} \mathbf{C} \to \widehat{\mathbf{E}}^{n-1}_p(q,\varkappa) \otimes_{\mathbf{Q}} \mathbf{C},$$

where the ring on the right is $\mathbf{C}$ in the case $n = 1$.

(2) *Let $\widehat{\Psi} = \widehat{\Psi}(\cdot, p^{n-k/2}\overline{\chi}(p))$ be the map defined by the commutative diagram*

$$(2.66)$$

$$\begin{array}{ccc}
\widehat{E}^n_p(q,\varkappa) \otimes_{\mathbf{Q}} \mathbf{C} & \xrightarrow{\widehat{\varepsilon}^n_{q,k}} & \overline{\mathbf{L}}^n_{0,p} \\
\downarrow{\widehat{\Psi}(\,,p^{n-k/2}\overline{\chi}(p))} & & \downarrow{\Psi(\,,p^{n-k/2}\overline{\chi}(p))} \\
\widehat{E}^{n-1}_p(q,\varkappa) \otimes_{\mathbf{Q}} \mathbf{C} & \xrightarrow{\widehat{\varepsilon}^{n-1}_{q,k}} & \overline{\mathbf{L}}^{n-1}_{0,p},
\end{array}$$

where $\widehat{E}^m_p(q,\varkappa)$ is the ring (4.83) of Chapter 3 and $\widehat{\varepsilon}_{q,k}$ is the monomorphism (4.102) of Chapter 3 extended by linearity if $m > 0$, and is the identity map from $\mathbf{C}$ to $\mathbf{C}$ if $m = 0$. Then $\widehat{\Psi}$ is an epimorphism.

(3) *For any $F \in \mathfrak{M}^n_{k/2}(q, \chi)$ and $\widehat{X} \in \widehat{E}^n_p(q, \varkappa)$ one has*

(2.67) $$(F|_{k/2,\chi}\widehat{X})|\Phi = (F|\Phi)|_{k/2,\chi}\widehat{\Psi}(\widehat{X}, p^{n-k/2}\widehat{\chi}(p)),$$

where Φ is the Siegel operator, $|_{k/2,\chi}\widehat{X}$ is the operator (2.3), and in the case $n = 1$ the operator $|_{k/2,\chi}\widehat{\Psi}$ acts as multiplication by the complex number $\widehat{\Psi}$.

PROOF. Since part (1) is obvious for $n = 1$, we shall suppose that $n \geqslant 2$. Let $\widehat{T}^n(p^2)$ be one of the generators of $\widehat{E}^n_p(q, \varkappa)$ in (4.82) of Chapter 3, and let $\widehat{\varepsilon}_q$ be the imbedding (4.100) of Chapter 3. From (2.60) and Lemma 2.14 it follows that

$$\widehat{\Psi}(\widehat{\varepsilon}_q(\widehat{T}^n(p^2)), u) = \sum_s a_s u^{\alpha_s} \widehat{\varepsilon}_q((\xi_s)_{\widehat{\Gamma}})$$

(a finite sum), where $\xi_s = (K^{n-1}_{s'}, *) \in \mathfrak{G}$ (see (4.82) of Chapter 3), $a_s \in \mathbf{Q}$, $\alpha_s \in \mathbf{Z}$, and $\widehat{\Gamma} = \widehat{\Gamma}^{n-1}_0(q)$. But since $(\xi_s)_{\widehat{\Gamma}} = \widehat{T}^{n-1}_{s'}(p^2) \cdot ((E_{2n-2}, t_s))_{\widehat{\Gamma}}$, where $t_s \in \mathbf{C}_1$, it follows from the commutativity of the diagram (2.58) and the relation $P_k(\widehat{\varepsilon}_q((E_{2n-2}, t_s)_{\widehat{\Gamma}})) = t_s^{-k}$ that the map (2.65) is a homomorphism. That it is an epimorphism follows from the commutativity of the diagram

(2.68)

$$\begin{array}{ccc} \widehat{E}^n_p(q, \varkappa) \otimes_{\mathbf{Q}} \mathbf{C} & \xrightarrow{\Omega^n_p} & \mathbf{C}[r^n_1, \dots, r^n_n, (\rho^n_0)^{\pm 1}] \\ \Big\downarrow {\scriptstyle \Psi(\,, p^{n-k/2}\overline{\chi}(p))} & & \Big\downarrow {\scriptstyle \Xi_n} \\ \widehat{E}^{n-1}_p(q, \varkappa) \otimes_{\mathbf{Q}} \mathbf{C} & \xrightarrow{\Omega^{n-1}_p} & \mathbf{C}[r^{n-1}_1, \dots, r^{n-1}_{n-1}, (\rho^{n-1}_0)^{\pm q}], \end{array}$$

where the maps Ω^m_p are isomorphisms by Theorem 4.19 of Chapter 3 and (2.52), and from the fact that Ξ_n is an isomorphism (see Proposition 2.13).

According to Theorem 4.21(4) of Chapter 3, the map $\widehat{\varepsilon}_{q,k}$ gives a ring isomorphism between $\widehat{E}^m_p(q, \varkappa)$ and $\widehat{\mathbf{E}}^m_p(q, \varkappa)$. From this and part (1) we conclude that the map $\widehat{\Psi}$ in (2.66) is well defined and is an epimorphism.

Finally, the third part of the proposition follows from part (2), the commutation relation (2.46), and Lemma 2.4. $\qquad\qquad\qquad\qquad\qquad\qquad\qquad\qquad\qquad\qquad\square$

The Zharkovskaia relations are often used when one wants to answer certain questions concerning the action of Hecke operators on modular forms not in the kernel of the Siegel operator by reducing them to analogous questions for forms of lower degree. One such question is the existence of a basis of eigenfunctions of the Hecke operators. Theorem 1.9 gives a positive answer to this question in the case of cusp-forms. In many cases the Zharkovskaia relations enable one to carry this result over to the entire space of modular forms. Since the general case has not yet been sufficiently investigated, we shall limit ourselves to the simplest nontrivial case, that of the space

(2.69) $$\mathfrak{M}^n_k = \mathfrak{M}_k(\Gamma^n, 1) = \mathfrak{M}^n_k(1, 1)$$

of modular forms of integer weight and unit character for the modular group Γ^n.

THEOREM 2.16. *Any subspace V of $\mathfrak{M}^n_k$, where $n \in \mathbf{N}$ and $k \in \mathbf{Z}$, that is invariant relative to all of the Hecke operators $|_k T = |_{k,1} T$ for $T \in L(\Gamma^n) = L^n(1) = L^n$ has a basis consisting of eigenfunctions of all of these operators.*

PROOF. In the case under consideration, (1.26) obviously implies that for any forms $F, G \in \mathfrak{M}_k^n$, at least one of which is a cusp-form, and for any $T \in L^n$, one has

$$(2.70) \qquad (F|_k T, G) = (F, G|_k T).$$

Then the argument used to prove Theorem 1.9 can be applied to any invariant subspace V contained in $\mathfrak{N}_k^n$, so that our theorem is proved in that case. If V is an arbitrary invariant subspace, we set $V_2 = V \cap \mathfrak{N}_k^n$ and $V_1 = \{F \in V; (F, G) = 0 \text{ for all } G \in V_2\}$. Using the properties (3)–(5) of the scalar product in Theorem 5.3 of Chapter 2 and standard linear algebra, we see that V is the direct sum of the subspaces V_1 and V_2:

$$(2.71) \qquad V = V_1 \oplus V_2.$$

Since V_2 is the intersection of two invariant subspaces (see Proposition 1.5), it is invariant relative to all of the Hecke operators. It then follows from (2.70) that the subspace V_1 has the same property. In the case of $\mathfrak{M}_k^n$ the set of cusp-forms coincides with the kernel of Φ. The subspace V_1 does not contain cusp-forms, and so Φ gives an isomorphism of this space onto the image $V' = \Phi(V_1) \subset \mathfrak{M}_k^{n-1}$. Using the invariance of V_1, the Zharkovskaia relations (2.46), the surjectivity of the maps (2.48) in the present situation, and Lemma 2.4, we conclude that the space V' is invariant relative to all Hecke operators in L_p^{n-1} for each prime p, and hence, by Theorem 3.12 of Chapter 3, relative to all Hecke operators in L^{n-1}. Now suppose that the theorem has already been proved for subspaces of $\mathfrak{M}_k^{n-1}$. Then V' has a basis $F_1', \ldots, F_d'$ of eigenfunctions of all of the Hecke operators in L^{n-1}. Let $F_1 = \Phi^{-1}(F_1'), \ldots, F_d = \Phi^{-1}(F_d')$ denote the preimages of these eigenfunctions in V_1. Then $F_1, \ldots, F_d$ obviously form a basis of V_1. In addition, each of the functions F_i is an eigenfunction for all of the operators in L^n. In fact, by Theorem 3.12 of Chapter 3, it suffices to verify this for the operators corresponding to elements $T \in L_p^n$ for all primes p. For such T it follows from our assumptions and Lemma 2.4 that

$$F_i'|_k \Psi(\varepsilon(T), p^{n-k}) = \lambda_i(T) F_i',$$

where $\varepsilon: L_p^n \to \mathbf{L}_{0,p}^n$ is the imbedding (1.27) of Chapter 3 and $\lambda_i(T)$ is a scalar; hence, by (2.46), we have

$$(F_i|_k T - \lambda_i(T) F_i)|\Phi = F_i|_k \varepsilon(T)|\Phi - \lambda_i(T) F_i'$$
$$= F_i'|_k \Psi(\varepsilon(T), p^{n-k}) - \lambda_i(T) F_i' = 0,$$

and so $F_i|_k T = \lambda_i(T) F_i$. As noted before, the space V_2 has a basis of eigenfunctions of all of the Hecke operators. If we combine this basis with the basis $F_1, \ldots, F_d$ of V_1, we obtain the desired basis for $V = V_1 \oplus V_2$. To complete the induction it remains to prove the theorem in the case $n = 1$. We again represent the invariant subspace $V \subset \mathfrak{M}_k^1$ in the form (2.71), and note that in this case $\dim V_1 = 0$ or 1, since $\Phi(V_1) \subset \mathfrak{M}_k^0 = \mathbf{C}$. If $\dim V_1 = 0$, then $V \subset \mathfrak{N}_k^1$, and our claim has already been proved; if $\dim V_1 = 1$ and F is a function that spans the invariant subspace V_1, then F is automatically an eigenfunction of all of the Hecke operators. This F, together with the basis of eigenfunctions for V_2, form the desired basis for V. $\qquad \square$

Another application of the Zharkovskaia relations can be found in the next subsection.

PROBLEM 2.17. Let λ be a nonzero $\mathbf{Q}$-linear homomorphism from $\mathbf{L}_p^{n-1}$ to $\mathbf{C}$, where $n > 1$ and p is a prime; let $\alpha_0, \alpha_1, \ldots, \alpha_{n-1}$ be the parameters of λ (see Proposition 3.36 of Chapter 3), and let $\bar{\lambda}$ be the linear extension of λ to the complexification $\overline{\mathbf{L}}_p^{n-1}$. Show that $\alpha_0 u^{-1}, \alpha_1, \ldots, \alpha_{n-1}, u$ can be taken as parameters of the homomorphism $T \to \bar{\lambda}(\Psi_p^n(T, u))$ of the ring $\mathbf{L}_p^n$, where u is a nonzero complex number.

PROBLEM 2.18. Show that all of the eigenvalues of Hecke operators on $\mathfrak{M}_k^n$ are real.

4. Action of the middle factor in the symmetric factorization of Rankin polynomials. As mentioned before, the main purpose of this section is to study the action of the Hecke operators in $L^n(q)$ or $\widehat{E}(q, \varkappa)$ on modular forms and their Fourier coefficients. Our general philosophy is to replace the global rings by their local subrings $L_p^n(q)$ and $\widehat{E}_p^n(q, \varkappa)$, and then to replace these by the isomorphic subrings $\mathbf{L}_p^n$ and $\widehat{\mathbf{E}}_p^n(q, \varkappa)$ of the Hecke ring of the group Γ_0^n. The imbeddings of the local Hecke rings in the Hecke rings of the triangular subgroup Γ_0^n make it possible to decompose the elements of the local rings—in particular, their generators—into components having a simpler action on the modular forms and their Fourier coefficients. Among the decompositions of this sort, the most important are the ones in the symmetric factorization of the Rankin polynomials $\mathbf{R}_p^n(v)$ and $\widehat{\mathbf{R}}_p^n(v)$ (see (6.99) and (6.100) of Chapter 3), whose coefficients include all of the generators of the even local subrings $\mathbf{E}_p^n$ and $\widehat{\mathbf{E}}_p^n(q, \varkappa)$ (from (2.52) and (2.68) it follows that the coefficients of the Rankin polynomials precisely generate the even local subrings). Lemma 2.7 contains, in particular, formulas for the action of the first and third factors in the factorizations (6.99) and (6.100) of Chapter 3. Below we shall derive formulas for the action of the middle factors, regarded as polynomials with operator coefficients.

As before, $w = k$ or $k/2$ is an integer or half-integer, $n, q \in \mathbf{N}$, and q is divisible by 4 if $w = k/2$. In addition, p is a prime not dividing q, χ is a Dirichlet character modulo q, and s is the character of the group $\{\pm 1\}$ such that $s(-1) = \chi_w(-1)$ (see (2.5)).

We introduce some useful notation. If $P(v) = \sum_{i \geqslant 0} p_i v^i$ ($p_i \in \overline{\mathbf{L}}_{0,p}^n$) is a polynomial or formal power series over the ring $\overline{\mathbf{L}}_{0,p}^n$, then for any $F \in \mathfrak{M}_s^n$ (respectively, for any $f \in \mathfrak{F}_s^n$) we set

$$(2.72) \qquad F|_{w,\chi} P(v) = \sum_{i \geqslant 0} (F|_{w,\chi} p_i)(Z) v^i$$

(respectively,

$$(2.73) \qquad f|_{w,\chi} P(v) = \sum_{i \geqslant 0} (f|_{w,\chi} p_i)(R) v^i),$$

where the right side is understood as a formal sum in the case when P is a formal power series. If the values of f coincide with the Fourier coefficients of F, then obviously

$$(2.74) \qquad F|_{w,\chi} P(v) = \sum_{R \in \mathbf{A}_n} (f|_{w,\chi} P(v))(R) e\{RZ\}.$$

It also follows from the definitions that the product of polynomials or series corresponds to the product of the corresponding operators.

Thus, let

$$(2.75) \qquad B_p^n(v) = \sum_{i=0}^{n} (-1)^i b_i v^i, \quad \widehat{B}_p^n(v) = \sum_{i=0}^{n} (-1)^i \widehat{b}_i v^i$$

be the middle factors in (6.99) and (6.100) of Chapter 3, where the coefficients b_i and $\widehat{b}_i$ are linear combinations of elements of the form $\Pi_{n,0}^{(r)}$ and $\Pi_{n,0}^{(r)}(k)$. The next lemma reduces the study of the action of these elements on $\mathfrak{M}_s^n$ and $\mathfrak{F}_s^n$ to the computation of certain trigonometric sums.

LEMMA 2.19. *Under the above assumptions and notation, any $f \in \mathfrak{F}_s^n$ satisfies the relations*

$$f|_{k,\chi} \Pi_{n,0}^{(r)} = p^{n(k-n-1)} \chi(p)^n l_p(r, n; R) f(R),$$

$$f|_{k/2,\chi} \Pi_{n,0}^{(r)}(k) = p^{n(k/2-n-1)} \chi(p)^n l_p^k(r, n; R) f(R),$$

where

$$(2.76) \qquad l_p^k(r, n; R) = \sum_{A \in L_p(r,n)} \varkappa(A)^{-k} e\{p^{-1} R A\},$$

$L_p(r, n)$ is the set of symmetric $n \times n$-matrices of rank r over the field $\mathbf{F}_p = \mathbf{Z}/p\mathbf{Z}$, $\varkappa$ is the function (4.70) of Chapter 3, and

$$(2.77) \qquad l_p(r, n; R) = l_p^0(r, n; R).$$

PROOF. Let $F \in \mathfrak{M}_s^n$ have Fourier coefficients $f(R)$. Then from (2.14), (2.15), and (4.106) of Chapter 3 it follows that

$$F|_{w,\chi} \Pi_{n,0}^{(r)}(k) = p^{n(w-n-1)} \chi(p)^n \sum_{R, B_0 \in L_p(r,n)} f(R) \varkappa(B_0)^{-k} e\{R(Z + p^{-1} B_0)\}.$$

Combining similar terms containing $e\{RZ\}$ and setting $\varkappa \equiv 1$ when $w = k$, we obtain the lemma. $\qquad\square$

This lemma enables us to write the action of the polynomials $B_p^n(v)$ and $\widehat{B}_p^n(v)$ (in the sense of (2.73)) in terms of the trigonometric sums (2.76) and (2.77).

PROPOSITION 2.20. *Under the above assumptions and notation, any $f \in \mathfrak{F}_s^n$ satisfies the relations*

$$(2.78) \qquad \begin{aligned} (f|_{k,\chi} B_p^n(v))(R) &= B_p^n(v, R) f(R), \\ (f|_{k/2,\chi} \widehat{B}_p^n(v))(R) &= \widehat{B}_{p,k}^n(v, R) f(R), \end{aligned}$$

where $R \in \mathbf{A}_n$, $B_p^n(v, R)$ and $\widehat{B}_{p,k}^n(v, R)$ are the polynomials defined by setting

$$(2.79) \qquad \begin{aligned} B_p^n(v, R) &= \sum_{i=0}^{n} (-1)^i p^{-\langle i \rangle - i(n-i)} \left\{ \sum_{j=0}^{i} \alpha_{ij} l_p(i - j, n; R) \right\} v^i, \\ \widehat{B}_{p,k}^n(v, R) &= \sum_{i=0}^{n} (-1)^i p^{-\langle i \rangle - i(n-i)} \left\{ \sum_{j=0}^{n} \widehat{\alpha}_{ij} l_p^k(i - j, n; R) \right\} v^i, \end{aligned}$$

and α_{ij} and $\widehat{\alpha}_{ij}$ are the coefficients (6.70) and (6.86) of Chapter 3.

PROOF. The proposition follows directly from (2.73), (2.75), Theorem 6.24 of Chapter 3, and Lemma 2.19, if we take into account that $\Delta = \Pi_{n,0}^{(0)}$. $\square$

Thus, our task reduces to the computation of the polynomials (2.79), to which the rest of this subsection is devoted. We begin by computing the trigonometric sums (2.77). For $0 < b \leqslant n$ we set

$$(2.80) \qquad Pr_p(b,n) = \{M \in M_{b,n}(\mathbf{Z}/p\mathbf{Z}); r_p(M) = b\},$$

where, as before, $r_p(M)$ denotes the rank of the matrix M over the field of p elements; and for

$$(2.81) \qquad Q \in \mathbf{E}_n = \{(s_{ij}) \in S_n(\mathbf{Z}); s_{ii} \in 2\mathbf{Z}\}$$

(the set of matrices of integral quadratic forms in n variables) we define the set

$$(2.82) \qquad Pr_p(b,n;Q) = \{M \in Pr_p(b,n); Q['M] \equiv 0(\mathrm{mod}\, p)\},$$

where congruence modulo $d \in \mathbf{N}$ for $S, S' \in \mathbf{E}_m$ is understood in the following sense:

$$(2.83) \qquad S \equiv S'(\mathrm{mod}\, d) \quad \text{means that } S - S' \in d\mathbf{E}_m.$$

We shall let $\rho_p(b,n)$ and $\rho_p(b,n;Q)$ denote the number of elements in the sets (2.80) and (2.82), respectively:

$$\rho_p(b,n) = |Pr_p(b,n)|, \quad \rho_p(b,n;Q) = |Pr_p(b,n;Q)|,$$

and we set $\rho_p(0,n) = \rho_p(0,n;Q) = 1$.

LEMMA 2.21. *Let* $0 \leqslant r \leqslant n$, $R \in \mathbf{E}_n$, *and* p *be a prime number. Then the trigonometric sum* (2.77) *can be written in the form*

$$l_p(r,n;R) = \varphi_{n-r}^{-1} \sum_{a,b \geqslant 0, a+b=r} (-1)^a p^{b+\langle a-1 \rangle} \frac{\varphi_{n-b}}{\varphi_a \varphi_b} \rho_p(b,n;R),$$

where $\varphi_i = \varphi_i(p)$ *is the function* (2.29) *of Chapter* 3.

PROOF. In the case $r = 0$ the formula is obvious. If we apply Lemma 6.18 of Chapter 3 and use the notation of that lemma, we have

$$l_p(r,n;R) = \sum_{i \in I_{r,n}} \sum_{M \in L'_p(r,n;i)} e\{p^{-1}RM\}$$

$$= \sum_{i \in I_{r,n}} \sum_{A,V} \exp(\pi i \sigma(R['X]A)/p),$$

where $A \in L_p(r,r)$, $V \in V(\mathbf{i})$, and $X = (E_r, V)M(\mathbf{i})$.

To transform the last expression we note that the set that X runs through may be regarded as a set of representatives of the orbits of the group $G = GL_r(\mathbf{Z}/p\mathbf{Z})$ acting by left multiplication on the set $Pr_p(r,n)$:

$$(2.84) \qquad G \setminus Pr_p(r,n) = \bigcup_{i \in I_{r,n}} \{X = (E_r, V)M(i); V \in V(i)\}.$$

Namely, for every matrix T in $Pr_p(r, n)$ we let $\mathbf{i} = \mathbf{i}(T) \in I_{r,n}$ denote the first (in lexicographical order) set of indices $1 \leqslant i_1 < \cdots < i_r \leqslant n$ such that the columns of T indexed by $i_1, \ldots, i_r$ are linearly independent modulo p. Since obviously

$$(2.85) \qquad\qquad \mathbf{i}(gT) = \mathbf{i}(T), \quad \text{if } g \in G,$$

it follows by taking $g = T_1^{-1}$, where T_1 is the matrix made up of the i_1th, $\ldots$, i_rth columns of T, that the matrix $T' = T_1^{-1}T$ has the same index set as T, and the columns corresponding to these indices are equal to the corresponding columns of the identity matrix E_r. Since $\mathbf{i}$ is a minimal set, it follows that the entries t'_{α, j_β} in the columns of T' with indices $(j_1, \ldots, j_{n-r}) = \hat{\mathbf{i}}$ (the complement of $\mathbf{i} = (i_1, \ldots, i_r)$ in $(1, 2, \ldots, n)$) satisfy the condition $t'_{\alpha, j_\beta} \equiv 0 \pmod{p}$ if $i_\alpha > j_\beta$. If we then replace all of the entries in T' by their least nonnegative residues modulo p and use (6.51) of Chapter 3, we see that the matrix $T'M(\mathbf{i})^{-1} = T_1^{-1}TM(\mathbf{i})^{-1}$ has the form (E_r, V), where $V \in V(\mathbf{i})$ (see (6.52) of Chapter 3). Hence, the right side of (2.84) contains representatives of all of the orbits. If two matrices $X = (E_r, V)M(\mathbf{i})$ and $X' = (E_r, V')M(\mathbf{i}')$, where $V \in V(\mathbf{i})$ and $V' \in V(\mathbf{i}')$, belong to the same orbit, i.e., $X' \equiv gX \pmod{p}$, then from (2.85) and the obvious equalities $\mathbf{i}(X) = \mathbf{i}$, $\mathbf{i}(X') = \mathbf{i}'$ it follows that $\mathbf{i}' = \mathbf{i}$, and hence $g \equiv E_r \pmod{p}$ and $X = X'$. This proves (2.84). We now note that, because of the summation over A, the right side of the last expression for $l_p(r, n; R)$ does not change if A is replaced by $A[g]$ with $g \in G$. Hence, applying (2.84), we obtain

$$(2.86) \qquad \begin{aligned} l_p(r, n; R) &= |G|^{-1} \sum_{A \in L_p(r,r), g \in G} \sum_{X \in G \backslash Pr_p(r,n)} \exp\left(\frac{\pi i}{p} \sigma(gXR\,^t X\,^t gA) \right) \\ &= \rho_p(r, r)^{-1} \sum_{A \in L_p(r,r)} G_p^*(A, R), \end{aligned}$$

where $G_p^*(A, R)$ for $A \in S_r$, $R \in \mathbf{E}_n$ denotes the *reduced Gauss sum*

$$G_p^*(A, R) = \sum_{X \in Pr_p(r,n)} \exp\left(\frac{\pi i}{p} \sigma(RA[X]) \right).$$

On the other hand, the number of elements in a set of the form $Pr_p(b, n; R)$ can also be expressed in terms of reduced Gauss sums. Namely, from the obvious relations

$$\sum_{A \in S_b(\mathbf{Z}/p\mathbf{Z})} \exp\left(\frac{\pi i}{p} \sigma(QA) \right) = \begin{cases} p^{\langle b \rangle}, & \text{if } Q \in p\mathbf{E}_b, \\ 0, & \text{if } Q \notin p\mathbf{E}_b, \end{cases}$$

where $Q \in \mathbf{E}_b$, we have

$$\begin{aligned} p^{-\langle b \rangle} \sum_{A \in S_b(\mathbf{Z}/p\mathbf{Z})} G_p^*(A, R) &= p^{-\langle b \rangle} \sum_{X \in Pr_p(\mathbf{b},n)} \sum_{A \in S_b(\mathbf{Z}/p\mathbf{Z})} \exp\left(\frac{\pi i}{p} \sigma(R[^tX]A) \right) \\ &= \rho_p(b, n; R), \end{aligned}$$

where $0 \leqslant b \leqslant n$ and $R \in \mathbf{E}_n$; hence,

$$\rho_p(b, n; R) = p^{-\langle b \rangle} \sum_{s=0}^{b} \sum_{A \in L_p(s,b)} G_p^*(A, R).$$

If we note that none of the sums $G_p^*(A, R)$ is affected by any substitution of the form $A \to gA\,{}^tg$ with $g \in GL_b(\mathbf{Z}/p\mathbf{Z})$, and if we use Lemma 6.18 of Chapter 3, we can rewrite the inner sums in the last expression in the form

$$\sum_{A \in L_p(s,b)} G_p^*(A, R) = \frac{\varphi_b}{\varphi_s \varphi_{b-s}} \sum_{A = \left(\begin{smallmatrix} A' & 0 \\ 0 & 0 \end{smallmatrix}\right) \in L_p(s,b)} G_p^*(A, R),$$

where $A' = A^{(s)}$ is an $s \times s$-block. Every matrix $X \in Pr_p(b, n)$ can be written in the form $X = \begin{pmatrix} X_1 \\ X_2 \end{pmatrix}$, where $X_1 \in Pr_p(s, n)$. The number of such X with fixed X_1 clearly does not depend on X_1, and so this number is $\rho_p(b,n)/\rho_p(s,n)$. Thus, each sum G_p^* in the last expression can be rewritten in the form

$$G_p^*\left(\begin{pmatrix} A' & 0 \\ 0 & 0 \end{pmatrix}, R\right) = \sum_{\left(\begin{smallmatrix} X_1 \\ X_2 \end{smallmatrix}\right) \in Pr_p(b,n)} e\left\{p^{-1}R[({}^tX_1, {}^tX_2)]\begin{pmatrix} A' & 0 \\ 0 & 0 \end{pmatrix}\right\}$$

$$= \frac{\rho_p(b,n)}{\rho_p(s,n)} \sum_{X_1 \in Pr_p(s,n)} e\{p^{-1}R[{}^tX_1]A'\}$$

$$= \frac{\rho_p(b,n)}{\rho_p(s,n)} G_p^*(A', R).$$

If we substitute the resulting expressions into the formula for $\rho_p(b, n; R)$ and use the formulas for ρ_p in Lemma 6.16 of Chapter 3, after obvious cancellations we obtain the formula

$$\rho_p(b, n; R) = \sum_{\substack{s,t \geqslant 0 \\ s+t=b}} p^{-\langle s-1 \rangle} \frac{\varphi_b \varphi_{n-s}}{\varphi_s \varphi_t \pi_{n-b}} \sum_{A \in L_p(s,s)} G_p^*(A, R).$$

Using these expressions and the relations (6.82) of Chapter 3, we find that the right side of the equality in the lemma is equal to

$$\varphi_{n-r}^{-1} \sum_{s=0}^{r} \left(\sum_{\substack{a+t=r-s \\ (a,t \geqslant 0)}} (-1)^a \frac{p^{\langle a-1 \rangle}}{\varphi_a \varphi_t} \right) p^{-\langle s-1 \rangle} \frac{\varphi_{n-s}}{\varphi_s} \sum_{A \in L_p(s,s)} G_p^*(A, R)$$

$$= p^{-\langle r-1 \rangle} \varphi_r^{-1} \sum_{A \in L_p(r,r)} G_p^*(A, R).$$

Since, by Lemma 6.16 of Chapter 3, the factor in front of the last sum is equal to $\rho_p(r, r)^{-1}$, it follows that the last expression is equal to the expression (2.86) for $l_p(r, n; R)$. □

We use the theory of quadratic spaces to compute $\rho_p(b, n; Q)$. The properties of quadratic spaces that we shall need are given in Appendix 2.

LEMMA 2.22. *Suppose that* $n, b \in \mathbf{N}$, $0 < b \leqslant n$, p *is a prime,* $Q \in \mathbf{E}_n$, *and* $q = q(x_1, \ldots, x_n)$ *is the quadratic form* (1.4) *of Chapter 1 having matrix* Q. *Then the number* $\rho_p(b, n; Q)$ *of elements in the set* (2.82) *is equal to the number* $i(V_p, f_p; b)$ *of isotropic sets of* b *vectors in any quadratic space* (V_p, f_p) *of type* $\{q\}$ *over* $\mathbf{F}_p$.

PROOF. Let $e_1, \ldots, e_n$ be a basis of V_p in which the quadratic form of the space (V_p, f_p) is equal to q modulo p. It follows from the definitions that a set of vectors $\mathbf{m}_1, \ldots, \mathbf{m}_b \in V_p$, where $\mathbf{m}_i = \sum_{j=1}^{n} m_{ij} e_j$, is isotropic if and only if the matrix $M = (m_{ij})$ is contained in the set $Pr_p(b, n; Q)$. $\square$

We say that two matrices Q and Q_1 in $\mathbf{E}_n$ are *equivalent* modulo some $d \in \mathbf{N}$ and write $Q \sim Q_1 (\mathrm{mod}\, d)$ if there exists $M \in M_n(\mathbf{Z})$ with $(\det M, d) = 1$ such that

$$Q_1 \equiv Q[M](\mathrm{mod}\, d),$$

where the congruence is understood in the sense of (2.83). If Q is equivalent modulo d to a matrix of the form $\begin{pmatrix} Q_1 & 0 \\ 0 & 0 \end{pmatrix}$, where $Q_1 \in \mathbf{E}_{n-1}$, then we say that Q is *degenerate* modulo d. Otherwise, we say that Q is *nondegenerate* modulo d. If $d = p$ is a prime, then the relation $Q \sim Q_1(\mathrm{mod}\, p)$ is obviously equivalent to the relation $q \sim q_1$ over $\mathbf{F}_p$ between the quadratic forms corresponding to the two matrices (see (1.4) of Chapter 1), and Q is nondegenerate modulo p if and only if the quadratic space over $\mathbf{F}_p$ of type $\{q\}$ is nondegenerate (see Appendix 2).

We can now use the results of Appendix 2.4 to finish the computation of the polynomials $B_p^n(v, R)$.

THEOREM 2.23. *Let $n \in \mathbf{N}$, p be a prime, $R \in \mathbf{E}_n$, and $B_p^n(v, R)$ be the polynomial (2.79). Then:*

(1) $B_p^n(v, R) = B_p^n(v, R_1)$ *if $R \sim R_1 \ (\mathrm{mod}\ p)$;*

(2) *if the matrix R is degenerate modulo p, i.e.,*

$$R \sim \begin{pmatrix} R_1 & 0 \\ 0 & 0 \end{pmatrix} (\mathrm{mod}\, p), \quad \text{where } R_1 \in \mathbf{E}_{n-1},$$

then $B_p^n(v, R) = B_p^{n-1}(v, R_1)$;

(3) *if R is a nondegenerate matrix modulo p, then*

$$B_p^n(v, R) = \begin{cases} (1+v)(1 - \chi_R(p)\frac{v}{p^m})\prod_{i=1}^{m-1}\left(1 - \frac{v^2}{p^{2i}}\right), & \text{if } n = 2m, \\ (1+v)\prod_{i=1}^{m}\left(1 - \frac{v^2}{p^{2i}}\right), & \text{if } n = 2m+1, \end{cases}$$

where for n even $\chi_R(p)$ denotes the sign of the quadratic space over $\mathbf{F}_p$ (see (2.21) of Appendix 2) whose quadratic form has matrix (in some basis) congruent to R modulo p.

PROOF. Part (1) follows from (2.79), since the sums $l_p(r, n; R)$ clearly depend only on the class of R modulo p.

To prove part (2), we apply the Siegel operator and the Zharkovskaia relations. In what follows we use the notation (2.73), and, when we apply maps defined on Hecke rings to polynomials over these rings, we let them act only on the coefficients. We first show that the following equality holds identically in v and u:

(2.87) $\Psi(B_p^n(v), u) = B_p^{n-1}(v),$

where $B_p^n(v)$ is the polynomial (2.75) over $\mathbf{L}_{0,p}^n$ and Ψ is the map from $\mathbf{L}_{0,p}^n$ to $\mathbf{L}_{0,p}^{n-1}[u^{\pm 1}]$ defined by (2.42)–(2.43). Using the definition of the polynomials $\mathbf{R}^n(v) = \mathbf{R}_p^n(v)$ (see (6.28) of Chapter 3) and the commutativity of the diagram (2.45), we obtain

$$\Omega^{n-1}(\Psi(\mathbf{R}^n(v), u)) = \Xi(\Omega^n \mathbf{R}^n)(v) = (1 - u^{-1}v)(1 - uv)(\Omega^{n-1}\mathbf{R}^{n-1})(v).$$

By Proposition 2.11(2), all of the coefficients of the polynomial $\Psi(\mathbf{R}^n(v), u)$ lie in the ring $\mathbf{L}_p^{n-1}$, which also contains all of the coefficients of $\mathbf{R}^{n-1}(v)$. Since, by Theorem 3.30(3) of Chapter 3, Ω is a monomorphism on $\mathbf{L}_p^{n-1}$, the last relation implies that

$$\Psi(\mathbf{R}^n(v), u) = (1 - u^{-1}v)(1 - uv)\mathbf{R}^{n-1}(v).$$

Now let $X_-^n(v)$ and $X_+^n(v)$ be the first and third factors in (6.99) of Chapter 3. Using the commutativity of the diagram (2.45) and Lemma 3.34 of Chapter 3, we obtain

$$\Omega^{n-1}(\Psi(X_-^n(v), u)) = \Xi(\Omega^n X_-^n)(v) = \Xi\left(\sum_{i=0}^n (-1)^i s_i(x_1^{-1}, \ldots, x_n^{-1})v^i\right)$$

$$= (1 - u^{-1}v)(\Omega^{n-1} X_-^{n-1})(v).$$

Similarly,

$$\Omega^{n-1}(\Psi(X_+^n(v), u)) = (1 - uv)(\Omega^{n-1} X_+^{n-1})(v).$$

According to (6.34) of Chapter 3, the coefficients of the polynomial X_-^n lie in the subring C_-^n of $\mathbf{L}_{0,p}^n$. From the duality relations (6.2) of Chapter 3 and (6.33) of Chapter 3 it then follows that the coefficients of X_+^n lie in C_+^n. If we take Proposition 2.11(2) into account and use the fact that Ω is a monomorphism on $C_\pm^{n-1}$, we arrive at the relations

$$\Psi(X_-^n(v), u) = (1 - u^{-1}v)X_-^{n-1}(v), \quad \Psi(X_+^n(v), u) = (1 - uv)X_+^{n-1}(v).$$

Applying Ψ to (6.99) of Chapter 3 and using the above formulas, we obtain

$$(1 - u^{-1}v)(1 - uv)\mathbf{R}^{n-1}(v) = \Psi(\mathbf{R}^n(v), u)$$
$$= (1 - u^{-1}v)X_-^{n-1}(v)\Psi(B_p^n(v), u)(1 - uv)X_+^{n-1}(v),$$

so that

$$\mathbf{R}^{n-1}(v) = X_-^{n-1}(v)\Psi(B_p^n(v), u)X_+^{n-1}(v).$$

On the other hand,

$$\mathbf{R}^{n-1}(v) = X_-^{n-1}(v)B_p^{n-1}(v)X_+^{n-1}(v).$$

The relation (2.87) follows from these factorizations, since the polynomials $X_\pm^{n-1}$ have constant term 1, and so are invertible in the ring of formal power series over $\mathbf{L}_{0,p}^n$.

We now proceed directly to part (2). By part (1), we may assume that

$$R = \begin{pmatrix} R_1 & 0 \\ 0 & 0 \end{pmatrix}.$$

We can take R_1 to be an arbitrary matrix in its residue class modulo $p\mathbf{E}_{n-1}$. Hence, without loss of generality we may assume that $R_1 > 0$ (for example, we can arrange this by choosing sufficiently large representatives modulo $2p$ of the diagonal entries in R_1—see Theorem 1.5 of Appendix 1). We then set $Q = \begin{pmatrix} R_1 & 0 \\ 0 & 2 \end{pmatrix} \in \mathbf{A}_n^+$, and we consider the theta-series $\theta^n(Z, Q)$ of degree n for the matrix Q (see (1.9) of Chapter 1). Since the theta-series is obviously invariant relative to the transformations $Z \to M\langle Z\rangle$ for $M \in \Gamma_0^n$, it follows from Proposition 1.3 of Chapter 1 that $\theta^n(Z, Q) \in \mathfrak{M}_s^n$, where s is the unit character of the group $\{\pm 1\}$. We take $k = 0$ and $\chi = 1$, and in two different ways we compute the R_1-Fourier coefficient of $(F|_{0,1} B_p^n(v))|\Phi$, where $F = \theta^n(Z, Q)$

and Φ is the Siegel operator (see (2.74)). On the one hand, from Proposition 2.20 and (3.50) of Chapter 2 we see that this coefficient is (see §1.2 of Chapter 1)

$$B_p^n\left(v, \begin{pmatrix} R_1 & 0 \\ 0 & 0 \end{pmatrix}\right) r\left(Q, \begin{pmatrix} R_1 & 0 \\ 0 & 0 \end{pmatrix}\right) = B_p^n(v, R) r(Q, R_1).$$

On the other hand, if we use (2.46), (3.50) of Chapter 2, (2.87), and Proposition 2.20 for the function $F|\Phi$, we conclude that this coefficient is equal to

$$B_p^{n-1}(v, R_1) r\left(Q, \begin{pmatrix} R_1 & 0 \\ 0 & 0 \end{pmatrix}\right) = B_p^{n-1}(v, R_1) r(Q, R_1).$$

Since obviously $r(Q, R_1) \geqslant 1$, we obtain part (2) by equating the last two expressions.

Now suppose that the matrix R is nondegenerate modulo p. Let $r(x_1, \ldots, x_n)$ denote the quadratic form with matrix R, and let (V_p, f_p) denote the quadratic space of type $\{r\}$ over the field $\mathbf{F}_p = \mathbf{Z}/p\mathbf{Z}$. As noted before, the nondegeneracy of R modulo p implies nondegeneracy of the quadratic space (V_p, f_p). If we apply Lemma 2.22, we can rewrite the expressions for $l_p(r, n; R)$ in Lemma 2.21 in the form

$$l_p(r, n; R) = \varphi_{n-r}^{-1} \sum_{a,b \geqslant 0; a+b=r} (-1)^a p^{b+\langle a-1 \rangle} \frac{\varphi_{n-b}}{\varphi_a \varphi_b} i(b),$$

where $i(b) = i(V_p, f_p, b)$ is the number of isotropic sets of b vectors in the space (V_p, f_p). If we substitute these expressions into (2.79) and use the formulas for the α_{ij} in (6.70) of Chapter 3, we obtain

$$B_p^n(v, R) = \sum_{i=0}^{n} (-1)^i p^{-\langle i \rangle - i(n-i)} \frac{1}{\varphi_{n-i}} B_i v^i,$$

where

$$\begin{aligned}
B_i &= \sum_{j+r=i} \varphi_{n-r} \sum_{c+d=j} \frac{(-p)^c}{\varphi_c \varphi_d} \cdot \varphi_{n-r}^{-1} \sum_{a+b=r} (-1)^a p^{\langle a-1 \rangle + b} \frac{\varphi_{n-b}}{\varphi_a \varphi_b} i(b) \\
&= \sum_{c+d+a+b=i} \frac{(-p)^c}{\varphi_c} \cdot \frac{(-1)^a p^{\langle a-1 \rangle}}{\varphi_d \varphi_a} \cdot \frac{p^b \varphi_{n-b}}{\varphi_b} i(b).
\end{aligned}$$

For fixed c and b we sum the terms in this expression over all nonnegative integers d and a such that $d + a = i - c - b$, and we use (6.82) of Chapter 3; this gives us

$$B_i = p^i \sum_{c+b=i} \frac{(-1)^c \varphi_{n-b}}{\varphi_c \varphi_b} i(b).$$

We let λ denote the dimension of a maximal isotropic subspace of (V_p, f_p). By Corollary 2.15 of Appendix 2, we have $\lambda = m$ if $n = 2m$ and $\chi_R(p) = 1$, $\lambda = m - 1$ if $n = 2m$ and $\chi_R(p) = -1$, and $\lambda = m$ if $n = 2m + 1$. This implies that in all cases the desired expression for the polynomial $B_p^n(v, R)$ can be written in the form

(2.88)
$$\left\{ \prod_{i=0}^{n-\lambda-1} (1 + p^{-i} v) \right\} \left\{ \prod_{i=1}^{\lambda} (1 - p^{-i} v) \right\}.$$

Using the above formulas and taking into account that $i(b) = 0$ if $b > \lambda$, we obtain

$$B_p^n(v, R) = \sum_{i=0}^{n}(-1)^i p^{\langle i \rangle - in} \varphi_{n-i}^{-1}\left(\sum_{c+b=i}\frac{(-1)^c \varphi_{n-b}}{\varphi_c \varphi_b}i(b)\right)v^i$$

$$\overset{(n-i=a)}{=}\sum_{a+b+c=n}(-1)^{b+2c}p^{\langle b \rangle + \langle c \rangle + bc - (b+c)n}i(b)\frac{\varphi_{a+c}}{\varphi_a\varphi_b\varphi_c}v^{b+c}$$

$$= \sum_{b=0}^{\lambda}(-1)^b p^{\langle b \rangle bn}i(b)\varphi_b^{-1}v^b\sum_{a+c=n-b}\frac{p^{\langle c \rangle}\varphi_{a+c}}{\varphi_a\varphi_c}(vp^{b-n})^c.$$

By (2.34) of Chapter 3, we can represent the inner sum as a product

$$\prod_{i=1}^{n-b}(1 + vp^{i+b-n}) = \left\{\prod_{j=0}^{n-\lambda-1}(1+p^{-j}v)\right\}\left\{\prod_{i=1}^{\lambda-b}(1+vp^{i+b-n})\right\},$$

where the second product is taken to be 1 in the case $b = \lambda$. If we substitute this expression in the last formula for B_p^n, we find that

$$B_p^n(v, R) = \left\{\prod_{j=0}^{n-\lambda-1}\left(1+\frac{v}{p^j}\right)\right\}G(v),$$

where

$$G(v) = \sum_{b=0}^{\lambda}(-1)^b p^{\langle b \rangle - bn}i(b)\varphi_b^{-1}v^b\prod_{i=1}^{\lambda-b}(1+vp^{i+b-n}),$$

from which it follows that to complete the proof of part (3) it suffices to verify that the polynomial $G(v)$ is equal to the second factor in (2.88). Since $G(0) = 1$ and the degree of $G(v)$ is at most λ, we see that it is sufficient to show that $G(p^\mu) = 0$ for $\mu = 1, 2, \ldots, \lambda$. For integers $j \geqslant -1$ we define the numbers $\varkappa_j = \varkappa_j(p)$ by setting $\varkappa_j = \varphi_j(p^2)\varphi_j(p)^{-1}$ if $j \geqslant 0$, and $\varkappa_{-1} = 1/2$. It is not hard to see that

$$(2.89) \qquad \frac{\varkappa_{g-1}}{\varkappa_{h-1}} = \begin{cases} 1, & \text{if } g = h \geqslant 0, \\ \prod_{h \leqslant j \leqslant g-1}(p^j + 1), & \text{if } g > h \geqslant 0. \end{cases}$$

In this notation, if we substitute $v = p^\mu$, where $1 \leqslant \mu \leqslant \lambda$, into one of the products in the expression for $G(v)$, then we obtain

$$\prod_{i=1}^{\lambda-b}(1 + p^{i+b+\mu-n}) = \prod_{i=1}^{\lambda-b}p^{i+b+\mu-n}(p^{n-(i+b+\mu)} + 1)$$

$$= p^{\langle \lambda-b \rangle + (\lambda-b)(b+\mu-n)}\prod_{j=n-\lambda-\mu}^{n-b-\mu-1}(p^j + 1)$$

$$= p^{\langle \lambda \rangle - \lambda\mu - \lambda n - b\mu - \langle b \rangle + bn}\varkappa_{n-b-\mu-1}\varkappa_{n-\lambda-\mu-1}^{-1}$$

(recall that this product is assumed to be 1 in the case $b = \lambda$). Hence,

$$G(p^\mu) = p^{\langle \lambda \rangle + \lambda\mu - \lambda n}\varkappa_{n-\lambda-\mu-1}^{-1}\sum_{b=0}^{\lambda}(-1)^b i(b)\varphi_b^{-1}\varkappa_{n-b-\mu-1}.$$

We now use the formulas in Appendix 2.4 for the numbers $i(b)$, but first we rewrite these formulas in a more convenient form. From (2.24)–(2.25) of Appendix 2 and the definitions we obtain

$$i(b) = p^{\langle b-1 \rangle}(p^\lambda - 1)(p^{\lambda-b} - 1)\varphi_{\lambda-b}(p^2)^{-1}$$
$$= \frac{(p^\lambda - 1)\varphi_{\lambda-1}(p^2)}{\varphi_\lambda(p)} \cdot \frac{1}{\varkappa_{\lambda-b-1}} \cdot \frac{p^{\langle b-1 \rangle}\varphi_\lambda(p)}{\varphi_{\lambda-b}(p)},$$

if $n = 2m$ and $\chi_R(p) = 1$;

$$i(b) = p^{\langle b-1 \rangle}(p^{\lambda+1} + 1)(p^{\lambda+1-b} - 1)\varphi_\lambda(p^2)\varphi_{\lambda+1-b}(p^2)^{-1}$$
$$= \frac{(p^{\lambda+1} + 1)\varphi_\lambda(p^2)}{\varphi_\lambda(p)} \cdot \frac{1}{\varkappa_{\lambda-b+1}} \cdot \frac{p^{\langle b-1 \rangle}\varphi_\lambda(p)}{\varphi_{\lambda-b}(p)},$$

if $n = 2m$ and $\chi_R(p) = -1$; and in the case $n = 2m + 1$

$$i(b) = \frac{\varphi_\lambda(p^2)}{\varphi_\lambda(p)} \cdot \frac{1}{\varkappa_{\lambda-b}} \cdot \frac{p^{\langle b-1 \rangle}\varphi_\lambda(p)}{\varphi_{\lambda-b}(p)}.$$

These formulas imply that in all cases $i(b)$ can be written in the form

$$i(b) = c\,\varkappa_{n-b-\mu-1}^{-1}\,p^{\langle b-1 \rangle}\varphi_\lambda\varphi_{\lambda-b}^{-1},$$

where c does not depend on b. Substituting these expressions into the formula for $G(p^\mu)$, we find that

$$G(p^\mu) = c' \sum_{b=0}^{\lambda} \frac{\varkappa_{n-b-\mu-1}}{\varkappa_{n-b-\lambda-1}} \cdot \frac{(-1)^b p^{\langle b-1 \rangle}\varphi_\lambda}{\varphi_b\varphi_{\lambda-b}}.$$

We now note that

$$\varkappa_{n-b-\mu-1}\varkappa_{n-b-\lambda-1}^{-1} = \sum_{i=0}^{\lambda-\mu} c_i\,p^{-bi},$$

where the coefficients c_i do not depend on b. This is clear in the case $\mu = \lambda$, while if $\mu < \lambda$, then by (2.89) we obtain the expression

$$\varkappa_{n-b-\mu-1}\varkappa_{n-b-\lambda-1}^{-1} = \prod_{i=1}^{\lambda-\mu}(p^{i+n-\lambda-1}p^{-b} + 1),$$

which, after we expand the parentheses and combine similar terms, reduces to the form indicated. If we substitute these expressions in $G(p^\mu)$ and change the order of summation, we find that

$$G(p^\mu) = c' \prod_{i=0}^{\lambda-\mu} c_i \sum_{b=0}^{\lambda}(-p^{-i-1})^b \frac{p^{\langle b \rangle}\varphi_\lambda}{\varphi_b\varphi_{\lambda-b}} = c' \sum_{i=0}^{\lambda-\mu} c_i \prod_{t=1}^{\lambda}(1 - p^{t-i-1}),$$

where the inner sum can be transformed to a product using the identity (2.43) of Chapter 3. Since $\mu \geqslant 1$, it follows that $1 \leqslant i+1 \leqslant \lambda$ for each $i = 0,\ldots,\lambda - \mu$. Hence, each product in the last expression is zero, and so also $G(p^\mu) = 0$. $\square$

We now compute the polynomial $\widehat{B}_{p,k}^n(v, R)$ (see (2.79)), where $p \neq 2$ and k is odd. In this case Theorems 1.2 and 1.3 of Appendix 1 imply that any matrix $R \in \mathbf{E}_n$ of rank $r_p(R) = r$ is equivalent to a diagonal matrix:

$$(2.90) \qquad R \sim \begin{pmatrix} R_1 & 0 \\ 0 & 0 \end{pmatrix} \pmod{p},$$

where R_1 is a diagonal matrix in $\mathbf{E}_r$ and $r_p(R_1) = r$. In addition, if $r_p(R) = n$ and $n = 2m$ is an even number, then, by (2.21) of Appendix 2, the sign of the quadratic space over $\mathbf{F}_p = \mathbf{Z}/p\mathbf{Z}$ whose quadratic form has matrix Q is

$$(2.91) \qquad \chi_R(p) = \left(\frac{(-1)^m \det R}{p} \right).$$

LEMMA 2.24. *Let R be a matrix in $\mathbf{E}_n$ that is nondegenerate modulo the prime p, let*

$$R \sim \begin{pmatrix} R_1 & 0 \\ 0 & R_2 \end{pmatrix} \pmod{p}, \qquad where \; R_1 \in \mathbf{E}_{n-2}, R_2 \in \mathbf{E}_2,$$

and let $\lambda(R_2) = \chi_{R_2}(p)p$. Then for $n > 2$ the trigonometric sums (2.76) satisfy the relation

$$l_p^k(r, n; R) = -p\lambda(R_2)^{r-2} l_p^k(r-2, n-2; R_1) + \lambda(R_2)^r l_p^k(r, n-2; R_1).$$

PROOF. From (6.91) of Chapter 3 and (2.76) it follows that the sum $l_p^k(r, n; R)$ depends only on the equivalence class modulo p of the matrix R. Hence, we may suppose that $R = \begin{pmatrix} R_1 & 0 \\ 0 & R_2 \end{pmatrix} \in \mathbf{E}_n$ and $R_2 = \begin{pmatrix} \lambda_1 & 0 \\ 0 & \lambda_2 \end{pmatrix}$. We rewrite the sum (2.76) in the form

$$(2.92) \qquad l_p^k(r, n; R) = \sum_{d \geqslant 0} \sum_{Z \in L_p(d, n-2)} \sigma(d; Z, R),$$

where

$$\sigma(d; Z, R) = \sum_{A \in L_p(r,n), A^{(n-2)}=Z} \varkappa(A)^{-k} e\{p^{-1}RA\},$$

and we let $A^{(n-2)}$ denote the $(n-2) \times (n-2)$-matrix in the upper-left corner of A. Let $Z = \begin{pmatrix} Z_1 & 0 \\ 0 & 0 \end{pmatrix}[U_1]$, where $U_1 \in GL_{n-2}(\mathbf{Z}/p\mathbf{Z})$ and $Z_1 = \operatorname{diag}(z_1, \ldots, z_d)$ is a matrix that is nondegenerate modulo p. If we then replace A by $A[U]$ in the last sum, with $U = \begin{pmatrix} U_1 & 0 \\ 0 & E_2 \end{pmatrix}$, and use (6.91) of Chapter 3, we obtain

$$\sigma(d; Z, R) = \sigma\left(d; \begin{pmatrix} Z_1 & 0 \\ 0 & 0 \end{pmatrix}, R[{}^tU]\right).$$

Any matrix $A \in S_n(\mathbf{Z}/p\mathbf{Z})$ with $A^{(n-2)} = \begin{pmatrix} Z_1 & 0 \\ 0 & 0 \end{pmatrix}$ can be represented in the form

$$A = \begin{pmatrix} Z_1 & 0 & X_1 \\ 0 & 0 & X_2 \\ {}^tX_1 & {}^tX_2 & Y \end{pmatrix} = \begin{pmatrix} Z_1 & 0 & & 0 \\ 0 & & & \\ & & X & \\ 0 & & & \end{pmatrix} \left[\begin{pmatrix} E & 0 & Z_1^{-1} & X_1 \\ 0 & E & & 0 \\ 0 & 0 & & E \end{pmatrix} \right],$$

where $X = \begin{pmatrix} 0 & X_2 \\ {}^tX_2 & Y_1 \end{pmatrix}$ and $Y_1 = Y - Z_1^{-1}[X_1]$, and in all of the matrices the matrices on the main diagonal are of size $d \times d$, $(n-d-2) \times (n-d-2)$, and 2×2, respectively. From this, (4.70) of Chapter 3, and the last equality for $\sigma(d; Z, R)$ we find that

$$\sigma(d; Z, R) = \sum_{\substack{A \in L_p(r,n), \\ A^{(n-2)} = \begin{pmatrix} Z_1 & 0 \\ 0 & 0 \end{pmatrix}}} \varkappa(A)^{-k} e\{p^{-1} R[{}^tU]A\}$$

$$= \varkappa(Z_1)^{-k} e\{p^{-1} R_1 Z\} \sum_{X_1, X_2, Y; r_p(X) = r - d} \varkappa(X)^{-k} e\{p^{-1} R_2 Y\}$$

$$= \varkappa(Z)^{-k} e\{p^{-1} R_1 Z\} \sigma(r - d)\sigma,$$

where $\sigma(\rho)$ and σ denote the sums

$$\sigma(\rho) = \sum_{X_2, Y; r_p(X) = \rho} \varkappa(X)^{-k} e\{p^{-1} R_2 Y_1\},$$

$$\sigma = \sum_{X_1} e\{p^{-1} R_2 Z_1^{-1}[X_1]\},$$

and X_1, X_2, and $Y = {}^tY$ run through the set of all matrices over $\mathbf{Z}/p\mathbf{Z}$ of size $d \times 2$, $(n - d - 2) \times 2$, and 2×2, respectively. We first find a value for σ. Using the form of the matrices R_2 and Z_1 and Lemma 4.14 of Chapter 1, we have

$$\sigma = \prod_{i=1}^{2} \prod_{j=1}^{d} \left(\varepsilon_p \left(\frac{\lambda_i z_j^{-1}}{p} \right) p^{1/2} \right) = (\chi_{R_2}(p)p)^d = \lambda(R_2)^d.$$

To compute $\sigma(\rho)$ we fix the following notation:

$$U = \begin{pmatrix} U_1 & 0 \\ 0 & E_2 \end{pmatrix}, \qquad V = \begin{pmatrix} E_{n-d-2} & 0 \\ 0 & V_1 \end{pmatrix}, \qquad X = \begin{pmatrix} 0 & X_2 \\ {}^tX_2 & Y \end{pmatrix},$$

where $U_1 \in GL_{n-d-2}(\mathbf{Z}/p\mathbf{Z})$, $V_1 \in GL_2(\mathbf{Z}/p\mathbf{Z})$, and X_2 and Y are the same as in the matrix A. If $r_p(X_2) = 2$, then there exists a matrix U_1 such that $U_1 X_2 = \begin{pmatrix} X_2' \\ 0 \end{pmatrix}$ and $X_2' \in GL_2(\mathbf{Z}/p\mathbf{Z})$. Consequently,

$$X[{}^tU] \sim \begin{pmatrix} 0 & 0 & 0 \\ 0 & 0 & X_2' \\ 0 & {}^tX_2' & Y \end{pmatrix},$$

where $\sim$ denotes equivalence of matrices over $\mathbf{Z}/p\mathbf{Z}$ in the sense of §1 of Appendix 1. This implies that $r_p(X) = 4$ and $\varkappa(X) = 1$. If, on the other hand, $r_p(X_2) = 1$, then there exist matrices U_1 and V_1 such that $U_1 X_2 V_1 = \begin{pmatrix} 0 & 1 \\ 0 & 0 \end{pmatrix}$, and hence

$$X[{}^t(UV)] = \begin{pmatrix} 0 & 0 & 0 & 0 \\ 0 & 0 & 0 & 1 \\ 0 & 0 & & \\ 0 & 1 & & Y_1 \end{pmatrix}, \qquad \text{where } Y_1 = Y[{}^tV_1] = \begin{pmatrix} y_1 & y_2 \\ y_2 & y_3 \end{pmatrix}.$$

From this we find that $r_p(X) = 3$ and $\varkappa(X) = \varepsilon_p\left(\frac{y_1}{p}\right)$ if $y_1 \not\equiv 0 \pmod{p}$, and $r_p(X) = 2$ and $\varkappa(X) = 1$ if $y_1 \equiv 0 \pmod{p}$. Finally, if $r_p(X_2) = 0$, then obviously $r_p(X) = r_p(Y)$ and $\varkappa(X) = \varkappa(Y)$.

Thus, we obtain the following values directly from the above observations: $\sigma(0) = 1$, $\sigma(2) = l_p^k(2, 2; R_2)$, and $\sigma(\rho) = 0$ for $\rho \geqslant 3$. As for the sum $\sigma(1)$, we note that if $Y \in S_2(\mathbf{Z}/p\mathbf{Z})$ and $r_p(Y) = 1$, then Y has the form

$$Y = a \begin{pmatrix} 1 & v \\ v & v^2 \end{pmatrix} \quad \text{or} \quad \begin{pmatrix} 0 & 0 \\ 0 & a \end{pmatrix},$$

where $a \in (\mathbf{Z}/p\mathbf{Z})^*$, $v \in \mathbf{Z}/p\mathbf{Z}$. Since the summation in $\sigma(1)$ is taken over such Y, we have

$$\sigma(1) = \sum_{a \in \mathbf{Z}/p\mathbf{Z}} \left(\varepsilon_p^{-1}\left(\frac{-a}{p}\right)\right)^{-k} e\left\{\frac{\lambda_1 a}{p}\right\} \sum_{v \in \mathbf{Z}/p\mathbf{Z}} e\left\{\frac{\lambda_2 a v^2}{p}\right\}$$

$$+ \sum_{a \in \mathbf{Z}/p\mathbf{Z}} \left(\varepsilon_p^{-1}\left(\frac{-a}{p}\right)\right)^{-k} e\left\{\frac{\lambda_2 a}{p}\right\}.$$

By assumption, k is odd, and λ_1 and λ_2 are even numbers prime to p. Hence, if we apply the formulas for Gauss sums (see (4.28) and Lemma 4.14 of Chapter 1) to the second and third sums, we find that $\sigma(1) = 0$. We now substitute these values for σ and $\sigma(\rho)$ into the expression for $\sigma(d; Z, R)$ and use (2.92); we find that $l_p^k(r, n; R)$ is equal to the sum

$$\lambda(R_2)^{r-2} l_p^k(2, 2; R_2) l_p^k(r - 2, n - 2; R) + \lambda(R_2)^r l_p^k(r, n - 2; R_1).$$

Thus, to prove the lemma it remains to evaluate $l_p^k(2, 2; R_2)$. By the definition (2.76), this sum can be divided into two parts as follows:

$$\left(\sum_{A = \begin{pmatrix} 0 & x \\ x & y \end{pmatrix} \in L_p(2,2)} + \sum_{z \in (\mathbf{Z}/p\mathbf{Z})^*} \sum_{A = \begin{pmatrix} z & x \\ x & y \end{pmatrix} \in L_p(2,2)} \right) \varkappa(A)^{-k} e\{p^{-1} R_2 A\}.$$

But since $\begin{pmatrix} z & x \\ x & y \end{pmatrix} \sim \begin{pmatrix} z & 0 \\ 0 & y - x^2 z^{-1} \end{pmatrix}$ over $\mathbf{F}_p = \mathbf{Z}/p\mathbf{Z}$ (see (1.2) of Appendix 1) and $R_2 = \begin{pmatrix} \lambda_1 & 0 \\ 0 & \lambda_2 \end{pmatrix}$, we have

$$l_p^k(2, 2; R_2) = \sum_{x \neq 0, y \in \mathbf{Z}/p\mathbf{Z}} e\left\{\frac{\lambda_2 y}{p}\right\}$$

$$+ \sum_{z \in (\mathbf{Z}/p\mathbf{Z})^*} \varkappa(z)^{-k} e\left\{\frac{\lambda_1 z}{p}\right\} \sum_{\substack{x, y \in \mathbf{Z}/p\mathbf{Z} \\ (y \not\equiv x^2 z^{-1} \pmod{p})}} \varkappa(y - x^2 z^{-1}) e\left\{\frac{\lambda_2 y}{p}\right\}.$$

Hence, if we make the change of variables $y \to y' + x^2 z^{-1}$ and use (4.28) and Lemma 4.14 of Chapter 1, we conclude that $l_p^k(2, 2; R_2) = -p$. $\qquad \square$

The recursive relations in this lemma allow us to obtain explicit formulas for the sums $l_p^k(r, n; R)$.

LEMMA 2.25. *Let $l_p^k(r, n; R)$ be the trigonometric sum (2.76), where k is an odd number and R is a matrix in $\mathbf{E}_n$ that is nondegenerate modulo the prime $p \neq 2$, and let $\varphi_m^+ = \varphi_m^+(p)$ be the function in (6.83) of Chapter 3. Then:*

$$l_p^k(2r, 2m; R) = l_p^k(2r, 2m + 1; R) = \varphi_{m,r},$$

$$l_p^k(2r + 1, 2m; R) = 0,$$

$$l_p^p(2r + 1, 2m + 1; R) = \chi_{R,k}(p) p^{m+1/2} \varphi_{m,r},$$

where r and m are integers, $0 \leqslant r \leqslant m$, and

$$(2.93) \qquad \varphi_{m,r} = \frac{\varphi_{2m}^+ (-1)^r p^{r^2}}{\varphi_{2m-2r}^+ \varphi_{2r}^+}, \quad \chi_{R,k}(p) = \left(\frac{-1}{p}\right)^{(k-1)/2} \left(\frac{(-1)^m \det R}{p}\right).$$

PROOF. For $v = 0$ or 1 we define the generating polynomials

$$F_v^n(z; R) = \sum_{0 \leqslant r \leqslant n; r \equiv v \,(\mathrm{mod}\, 2)} l_p^k(r, n; R) z^r$$

and show that the sum $F^n(z; R) = F_0^n(z; R) + F_1^n(z; R)$ is equal to the product

$$(1 + c_n \chi_{R,k}(p) p^{m+1/2} z) \prod_{j=0}^{m-1} (1 - p^{2j+1} z^2),$$

where $c_n = 0$ or 1 depending on whether $n = 2m$ or $n = 2m + 1$, respectively.

According to formula (2.90), from the very beginning we may assume that $R = \begin{pmatrix} R_1 & 0 \\ 0 & R_2 \end{pmatrix}$, just as in Lemma 2.24. If we apply the recursive relation in Lemma 2.24 to the coefficients of the polynomial $F_v^n(z; R)$, we see that for $n > 2$

$$F_v^n(z; R) = (1 - pz^2) F_v^{n-2}(\lambda(R_2)z; R_1).$$

On the other hand, if $n \leqslant 2$, then the proof of Lemma 2.24 and the formulas for the Gauss sums immediately imply that

$$l_p^k(1, 2; R) = 0, \quad l_p^k(2, 2; R) = -p,$$

$$l_p^k(1, 1; R) = \left(\frac{-1}{p}\right)^{(k-1)/2} \left(\frac{\det R}{p}\right) p^{1/2};$$

and, by the definition (2.76), $l_p^k(0, n; R) = 1$. These equalities, along with the formula for decreasing the size of the matrix R, give the required product formula for $F^n(z; R)$.

On the other hand, we can obtain a similar product by making the change of variables $v \to \sqrt{-1} \cdot z$ in the identity in Lemma 6.22 of Chapter 3 and using (6.98) of Chapter 3. We find that

$$(2.94) \qquad \sum_{r=0}^m \frac{\varphi_{2m}^+}{\varphi_{2m-2r}^+} \cdot \frac{(-1)^r p^{r^2}}{\varphi_{2r}^+} z^{2r} = \prod_{j=0}^{m-1} (1 - p^{2j+1} z^2).$$

It is easy to see that this gives all of the equalities for $l_p^k(r, n; R)$ in the lemma. □

We use Lemma 2.25 to compute the second polynomial in (2.79).

THEOREM 2.26. *Let $\widehat{B}^n_{p,k}(v, R)$ be the polynomial in (2.79), where $p \neq 2$, k is odd, and $R \in \mathbf{E}_n$. Then:*

(1) $\widehat{B}^n_{p,k}(v, R) = \widehat{B}^n_{p,k}(v, R_1)$,   if $R \sim R_1 (\bmod p)$;

(2) if R is a degenerate matrix modulo p, i.e.,

$$R \sim \begin{pmatrix} R_1 & 0 \\ 0 & 0 \end{pmatrix} (\bmod p), \quad \text{where } R_1 \in \mathbf{E}_{n-1},$$

then $\widehat{B}^n_{p,k}(v, R) = \widehat{B}^{n-1}_{p,k}(v, R_1)$;

(3) if R is a nondegenerate matrix modulo p, then

$$\widehat{B}^n_{p,k}(v, R) = \left(1 - c_n \chi_{R,k}(p) \frac{v}{p^{m+1/2}}\right) \prod_{j=0}^{m-1} \left(1 - \frac{v^2}{p^{2j+1}}\right),$$

where $\chi_{R,k}$ is the character in Lemma 2.25 and $c_n = 0$ or 1 depending on whether $n = 2m$ or $n = 2m + 1$, respectively.

PROOF. The first two parts are proved in exactly the same way as in Theorem 2.23. We now prove part (3). We rewrite the polynomial $\widehat{B}^n_{p,k}(v, R)$ in the form

$$(2.95) \qquad \widehat{B}^n_{p,k}(v, R) = \sum_{i=0}^{n} (-1)^i \widehat{B}^n_i v^i,$$

where $\widehat{B}^n_i$ denotes the sum

$$\widehat{B}^n_i = p^{\langle n-i \rangle - \langle n \rangle} \sum_{j=0}^{i} \widehat{\alpha}_{ij} l^k_p(i - j, n; R),$$

and we first suppose that $n = 2m$. Then from (6.86) of Chapter 3 and Lemma 2.25 we have $\widehat{B}^{2m}_{2i+1} = 0$ for $0 \leqslant i \leqslant m - 1$, and

$$(2.96)
\begin{aligned}
\widehat{B}^{2m}_{2i} &= p^{\langle n-2i \rangle - \langle n \rangle} \sum_{j=0}^{i} \frac{(-p)^j \varphi_{n-2i+2j}}{\varphi^+_{2j} \varphi_{n-2i}} \cdot \frac{\varphi^+_n (-1)^{i-j} p^{(i-j)^2}}{\varphi^+_{n-2i+2j} \varphi^+_{2i-2j}} \\
&\overset{(s=i-j)}{=} p^{\langle n-2i \rangle - \langle n \rangle} (-p)^i \frac{\varphi^+_n}{\varphi_{n-2i} \varphi^+_{2i}} \sum_{s=0}^{i} \frac{\varphi^-_{n-2s}}{\varphi^-_{n-2i}} \cdot \frac{\varphi^+_{2i} p^{s^2}}{\varphi^+_{2i-2s} \varphi^+_{2s}} (p^{-1/2})^{2s},
\end{aligned}$$

where $\varphi^\pm_s = \varphi^\pm_s(p)$ are the functions (6.83) of Chapter 3. From the definition of these functions it follows that for $n = 2m$ and $0 \leqslant s \leqslant i$

$$\frac{\varphi^-_{n-2s}}{\varphi^-_{n-2i}} = \prod_{m-i \leqslant j \leqslant m-s-1} (p^{2j+1} - 1)$$

$$= (-1)^{i-s} \prod_{0 \leqslant t \leqslant i-s-1} (1 - p^{2t+1}(p^{m-i})^2) = (-1)^{i-s} \Pi(i - s, \sqrt{-1} \cdot p^{m-i}),$$

where we made the substitution $j = m - i + t$ in the second step, and where

$$\Pi(a, v) = \prod_{0 \leqslant j \leqslant a-1} (1 + p^{2j+1} v^2) \quad \text{or} \quad 1$$

depending on whether $a \geqslant 1$ or $a = 0$, respectively. Similarly, if we make the change of variables $z = \sqrt{-1} \cdot p^{-1/2}v$ in the identity (2.94), under the same conditions we obtain

$$(2.97) \qquad \frac{\varphi_{2i}^{+} p^{s^2}}{\varphi_{2i-2s}^{+} \varphi_{2s}^{+}} (p^{-1/2})^{2s} = \Pi_s(i, p^{-1/2}v),$$

where $\Pi_s(i, av)$ is the coefficient of v^{2s} in the polynomial $\Pi(i, av)$. If we substitute these expressions in the last equality for $\widehat{B}_{2i}^{2m}$ and introduce the new notation

$$s(i, m) = \sum_{s=0}^{i} (-1)^{i-s} \Pi(i - s, \sqrt{-1} \cdot p^{m-i}) \Pi_s(i, p^{-1/2}v),$$

then we can write

$$\widehat{B}_{2i}^{2m} = (-1)^i p^{-2i(n-i)} \frac{\varphi_n^{+}}{\varphi_{n-2i}^{+} \varphi_{2i}^{+}} s(i, m).$$

We arrange the pairs (i, m) in lexicographic order, and prove by induction on (i, m) that

$$(2.98) \qquad s(i, m) = p^{i(2m-i)} \quad (0 \leqslant i \leqslant m).$$

For pairs $(0, m)$ the equality is obvious. Suppose that it holds for all pairs less than (i, m). The rest of the proof is based on the following obvious properties of the polynomials $\Pi(a, v)$ for $a \geqslant 1$:

$$\Pi(a, v) = (1 + p^{2a-1}v^2)\Pi(a - 1, v) = (1 + pv^2)\Pi(a - 1, pv),$$

which implies the relation

$$(2.99) \qquad \Pi_s(1, p^{-1/2}v) = \Pi_s(i - 1, p^{-1/2}v) + \Pi_{s-1}(i - 1, p^{-1/2}v)p^{2i-2}$$

for $1 \leqslant s \leqslant i$ and the relation

$$(2.100) \qquad \Pi(i - s, \sqrt{-1} \cdot p^{m-i}) = (1 - p^{2m-2i+1})\Pi((i - 1) - s, \sqrt{-1} \cdot p^{m-(i-1)})$$

for $0 \leqslant s \leqslant i - 1$. If we now separate the extreme terms in $s(i, m)$ and use (2.99), we obtain

$$\begin{aligned} s(i, m) =&(-1)^i \Pi(i, \sqrt{-1} \cdot p^{m-i}) + \Pi_i(i, p^{-1/2}v) \\ &+ \sum_{0 < s < i} (-1)^{i-s} \Pi(i - s, \sqrt{-1} \cdot p^{m-i}) \Pi_s(i - 1, p^{-1/2}v) \\ &+ p^{2i-2} \sum_{0 < s < i} (-1)^{i-s} \Pi(i - s, \sqrt{-1} \cdot p^{m-i}) \Pi_{s-1}(i - 1, p^{-1/2}v). \end{aligned}$$

Since (2.100) and (2.99) for $1 \leqslant i \leqslant m$ imply that

$$\Pi(i, \sqrt{-1} \cdot p^{m-i}) = (1 - p^{2m-2i+1})\Pi(i - 1, \sqrt{-1} \cdot p^{m-(i-1)}),$$
$$\Pi_i(i, p^{-1/2}v) = p^{2i-2}\Pi_{(i-1)}(i - 1, p^{-1/2}v),$$

it follows that, if we again apply (2.100) to the first sum on the right in the equality for $s(i, m)$, we find that $s(i, m)$ satisfies the following recursive relation:

$$s(i, m) = (p^{2m-2i+1} - 1)s(i - 1, m) + p^{2i-2}s(i - 1, m - 1).$$

We now use (2.98) for $(i-1, m)$ and $(i-1, m-1)$, and make the corresponding transformations on the right side of the resulting relation; we obtain the formula (2.98) for (i, m).

We now finish the computation of $\widehat{B}_{p,k}^n(v, R)$ for $n = 2m$. Since $\widehat{B}_{2i+1}^{2m} = 0$, and, by (2.98),

$$(2.101) \qquad \widehat{B}_{2i}^{2m} = (-1)^i \frac{\varphi_{2m}^+ p^{i^2}}{\varphi_{2m-2i}^+ \varphi_{2i}^+} p^{-2im} \qquad (0 \leqslant 1 \leqslant m),$$

it follows that, substituting these expressions in place of the coefficients in (2.95) and using (2.94), we obtain

$$\widehat{B}_{p,k}^n(v, R) = \sum_{i=0}^{m} (-1)^i \frac{\varphi_{2m}^+ p^{i^2}}{\varphi_{2m-2i}^+ \varphi_{2i}^+} (p^{-m} v)^{2i}$$

$$= \prod_{j=0}^{m-1} (1 - p^{2j+1} p^{-2m} v^2) = \prod_{j=0}^{m-1} \left(1 - \frac{v^2}{p^{2j+1}}\right).$$

Thus, to prove the theorem it remains to consider the polynomial $\widehat{B}_{p,k}^n(v, R)$ for $n = 2m + 1$. By Lemma 2.25 and (6.86) of Chapter 3, we can write the coefficients of this polynomial in the form

$$p^{\langle n-2i \rangle - \langle n \rangle} \sum_{j=0}^{i} \frac{(-p)^j \varphi_{2m+1-2i+2j}}{\varphi_{2j}^+ \varphi_{2m+1-2i}} \cdot \frac{\varphi_{2m}^+ (-1)^{i-j} p^{(i-j)^2}}{\varphi_{2m-2i+2j}^+ \varphi_{2i-2j}^+}$$

$$\overset{(i-j=s)}{=} p^{\langle n-2i \rangle - \langle n \rangle} (-p)^i \frac{\varphi_{2m}^+}{\varphi_{n-2i}^+ \varphi_{2i}^+} \sum_{s=0}^{i} \frac{\varphi_{n-2s}^-}{\varphi_{n-2i}^-} \cdot \frac{\varphi_{2i}^+ p^{s^2}}{\varphi_{2i-2s}^+ \varphi_{2s}^+} (p^{-1/2})^{2s}$$

or, if we use (2.97), the relation

$$\frac{\varphi_{n-2s}^-}{\varphi_{n-2i}^-} = (-1)^{i-s} \Pi(i - s, \sqrt{-1} \cdot p^{m-i+1})$$

and (2.98), in the form

$$(-1)^i p^{-2i(n-i)} \frac{\varphi_{2m}^+}{\varphi_{2m-2i}^+ \varphi_{2i}^+} \sum_{s=0}^{i} (-1)^{i-s} \Pi(i - s, \sqrt{-1} \cdot p^{m-i+1}) \Pi_s(i, p^{-1/2} v)$$

$$= (-1)^i p^{-i(n-1-i)} \frac{\varphi_{2m}^+}{\varphi_{2m-2i}^+ \varphi_{2i}^+},$$

and this, by (2.101), is equivalent to the equality

$$\widehat{B}_{2i}^{2m+1} = \widehat{B}_{2i}^{2m} \qquad (0 \leqslant i \leqslant m).$$

We similarly have the following formula for the coefficients $\widehat{B}_{2i+1}^n$:

$$\widehat{B}_{2i+1}^n = p^{\langle n-2i-1 \rangle - \langle n \rangle} \sum_{j=0}^{i} \frac{(-p)^j \varphi_{2m-2i+2j}}{\varphi_{2j}^+ \varphi_{2m-2i}} \cdot \frac{\varphi_{2m}^+ (-1)^{i-j} p^{(i-j)^2}}{\varphi_{2m-2i+2j}^+ \varphi_{2i-2j}^+} \chi_{R,k}(p) p^{m+1/2},$$

which, together with the first equality in (2.96), implies that

$$\widehat{B}_{2i+1}^{2m+1} = \chi_{R,k}(p)p^{-(m+1/2)}\widehat{B}_{2i}^{2m} \quad (0 \leqslant i \leqslant m).$$

To complete the proof, we substitute these values for the coefficients in (2.95) and use the formula we proved for $\widehat{B}_{p,k}^{2m}(v, R)$. We obtain part (3) for $n = 2m + 1$:

$$\widehat{B}_{p,k}^{n}(v, R) = \sum_{i=0}^{m} \widehat{B}_{2i}^{2m}v^{2i} - v\sum_{i=0}^{m}(\chi_{R,k}(p)p^{-(m+1/2)}\widehat{B}_{2i}^{2m})v^{2i}$$

$$= \left(1 - \chi_{R,k}(p)\frac{v}{p^{m+1/2}}\right)\prod_{j=0}^{m-1}\left(1 - \frac{v^2}{p^{2j+1}}\right). \qquad \square$$

§3. Multiplicative properties of the Fourier coefficients

In this section we apply the above technique to study the multiplicative properties of the Fourier coefficients of modular forms. The plan is as follows. First, the matrices of the Hecke operators in a fixed basis of some invariant space of modular forms satisfy the same relations as the corresponding elements of the original Hecke ring. On the other hand, the Hecke operators, while acting on the modular forms, also act on their Fourier coefficients; and the matrices of the Hecke operators appear in relations that reflect this action. In many cases it is possible to use these relations to express the Fourier coefficients of the basis forms (or certain combinations of the coefficients) in terms of the entries in the matrices of suitable Hecke operators; in that way one can investigate how the multiplicative properties of the Hecke operators are reflected in the Fourier coefficients. In the case of modular forms in several variables it is usually not possible to express individual Fourier coefficients in terms of the matrix entries of Hecke operators; rather, it is convenient to establish the relationships between them in the form of identities that express suitable Dirichlet series constructed from the Fourier coefficients of the modular forms in terms of Dirichlet series that are constructed from the matrices of the Hecke operators (the latter Dirichlet series are called "zeta-functions"). Because of the multiplicative properties of Hecke operators, these zeta-functions have a special type of Euler product expansion, and this gives the desired multiplicative relations for the Fourier coefficients. In the analytic theory of zeta-functions, which we hardly touch upon in this book, the same identities serve another purpose. Namely, they make it possible to express the zeta-functions as suitable integral transforms of the original modular forms. In many cases this enables one to prove that the zeta-functions have analytic continuations onto the entire complex plane, and to find functional equations that they satisfy.

In this section our basic object of study will be the case of one-dimensional invariant subspaces, i.e., eigenfunctions of the Hecke operators. This does not really represent much loss of generality, since the invariant subspaces are usually spanned by such eigenfunctions. For modular forms of degree 1 and 2 we consider first only the case of integer weight. This limitation is related to the fact that their multiplicative properties and their zeta-functions are connected in this case with the full Hecke ring. However, as we showed in Chapter 3, in the case of modular forms of half-integer weight, i.e., in the case of Hecke rings for the symplectic covering group, the full Hecke ring and its even subring induce the same ring of Hecke operators on the spaces of modular forms of half-integer weight. This is one of the main differences between modular forms of integer and half-integer weight. On the other hand, if we consider the multiplicative properties connected with the even Hecke rings, then here the theories for integer and half-integer weight are parallel—and, moreover, they can be developed for modular forms of arbitrary degree.

1. Modular forms in one variable. We consider a modular form

$$F(z) = \sum_{a=0}^{\infty} f(2a)e^{2\pi i a z} \in \mathfrak{M}_k(q, \chi) = \mathfrak{M}_k^1(q, \chi)$$

of weight k and character χ for the group

$$\Gamma_0(q) = \Gamma_0^1(q).$$

We suppose that it is an eigenfunction for all of the Hecke operators $|T = |_{k,\chi} T$ for $T \in L(q) = L^1(q)$. In particular, for the elements $T(m)$ of the form (3.19) of Chapter 3 we have the relations

$$(3.1) \qquad\qquad F|T(m) = \lambda(m)F \quad (m \in \mathbf{N}_{(q)}),$$

where $\lambda(m)$ is the corresponding eigenvalue. When we compute the Fourier coefficients $(f|T(m))(2a)$ of the function $F|T(m)$, using (5.48) of Chapter 3 and Lemma 2.7 (or 2.8), we find that

$$
\begin{aligned}
(f|T(m))(2a) &= \sum_{d_1, d_2 \in \mathbf{N},\, d_1 d_2 = m} ((f|\Pi_+^1(d_1))|\Pi_-^1(d_2))(2a) \\
&= \sum_{d|m,a} d^{k-1}\chi(d)(f|\Pi_+^1(m/d))(2a/d) \\
&= \sum_{d|m,a} d^{k-1}\chi(d)f(2ma/d^2).
\end{aligned}
$$

If we equate the Fourier coefficients with the same indices on both sides of (3.1), we obtain

$$(3.2) \qquad \sum_{d|m,a} d^{k-1}\chi(d)f(2ma/d^2) = \lambda(m)f(2a) \quad (m \in \mathbf{N}_{(q)}, a = 0, 1, \dots).$$

After replacing m and a by m/δ and a/δ, where $\delta \in \mathbf{N}$ is a common divisor of m and a, we obtain

$$\sum_{d|m/\delta, a/\delta} d^{k-1}\chi(d)f(2ma/(\delta d)^2) = \lambda(m/\delta)f(2a/\delta),$$

from which, if we multiply both sides by $\delta^{k-1}\chi(\delta)\mu(\delta)$, where μ is the Möbius function, and sum over all common divisors $\delta \in \mathbf{N}$ of a and m, we obtain

$$\sum_{\delta, d \in \mathbf{N},\, \delta d|m,a} (\delta d)^{k-1}\chi(\delta d)\mu(\delta)f(2ma/(\delta d)^2) = \sum_{\delta|m,a} \delta^{k-1}\chi(\delta)\lambda(m/\delta)f(2a/\delta).$$

Since for $b \in \mathbf{N}$

$$(3.3) \qquad\qquad \sum_{\delta|b} \mu(\delta) = \begin{cases} 1, & \text{if } b = 1, \\ 0, & \text{if } b > 1, \end{cases}$$

it follows that the left side of the last formula is

$$\sum_{b|m,a} b^{k-1}\chi(b)f(2ma/b^2)\sum_{\delta|b}\mu(\delta) = f(2ma),$$

and we arrive at the following multiplicative identity for f:

(3.4) $$f(2ma) = \sum_{d\in\mathbf{N},\,d|m,a} d^{k-1}\chi(d)\mu(d)\lambda(m/d)f(2a/d),$$

where $m \in \mathbf{N}_{(q)}$ and $a = 0, 1, \ldots$. This series of identities is actually equivalent to (3.2).

These identities show, in particular, that for any $a \in \mathbf{N}$

(3.5) $$f(2a) = \lambda(m)f(2a/m),$$

where $m = m(a)$ is the greatest divisor of a that is in $\mathbf{N}_{(q)}$. Thus, the question of the dependence of f on divisors of the argument that are prime to q reduces to the study of the corresponding eigenvalues $\lambda(m)$. The next theorem, which is also based on these identities, gives a complete description of the multiplicative properties of the eigenvalues $\lambda(m)$.

THEOREM 3.1. *The eigenvalues* $\lambda(m) = \lambda(m, F)$ *of* $T(m)$ *that correspond to a nonzero eigenfunction* $F \in \mathfrak{M}_k(q, \chi)$ *satisfy the following multiplicative relations for* $m, m_1 \in \mathbf{N}_{(q)}$:

(3.6) $$\lambda(m)\lambda(m_1) = \sum_{d|m,m_1} d^{k-1}\chi(d)\lambda(mm_1/d^2),$$

(3.7) $$\lambda(mm_1) = \sum_{d|m,m_1} d^{k-1}\chi(d)\mu(d)\lambda(m/d)\lambda(m_1/d).$$

If $k > 0$, *then the eigenvalues satisfy the inequalities*

(3.8) $$|\lambda(m)| \leqslant c_F m^k \quad (m \in \mathbf{N}_{(q)})$$

(*they satisfy the inequalities*

(3.9) $$|\lambda(m)| \leqslant c_F m^{k/2} \quad (m \in \mathbf{N}_{(q)}),$$

if F *is a cusp-form*), *where* c_F *depends only on* F.

PROOF. As before, let $f(2a)$, $a = 0, 1, 2, \ldots$, denote the Fourier coefficients of F. First suppose that $f(0) \neq 0$. If we write the relations (3.2) for $a = 0$ and divide both sides by $f(0)$, we obtain

(3.10) $$\lambda(m) = \sum_{d|m} d^{k-1}\chi(d) \quad (m \in \mathbf{N}_{(q)}),$$

from which (3.6) and (3.7) easily follow by an elementary number-theoretic argument. We leave the details to the reader as an exercise. Now suppose that there exist nonzero a for which $f(2a) \neq 0$ (this is always the case if $k > 0$), and let $\varkappa = \varkappa(F)$ be the smallest such a. Let d denote the largest divisor of $\varkappa$ that is in $\mathbf{N}_{(q)}$. Since d and $\varkappa/d$ are relatively prime, the relation (3.2) gives $f(2\varkappa) = \lambda(d)f(2\varkappa/d)$, and hence $f(2\varkappa/d) \neq 0$. Thus, $d = 1$, i.e.,

(3.11) $$\varkappa(F)|q^\infty \quad (\text{in particular}, \varkappa(F) = 1, \text{ if } q = 1).$$

From this and from (3.2) for $a = \varkappa$ we have the formulas

(3.12) $$\lambda(m) = f(2m\varkappa)/f(2\varkappa) \quad (m \in \mathbf{N}_{(q)}).$$

These formulas together with (3.2) imply that

$$\begin{aligned}
\lambda(m)\lambda(m) &= \lambda(m)f(2m_1\varkappa)/f(2\varkappa) \\
&= \sum_{d|m,m_1\varkappa} d^{k-1}\chi(d)f(2mm_1\varkappa/d^2)/f(2\varkappa) \\
&= \sum_{d|m,m_1} d^{k-1}\chi(d)\lambda(mm_1/d^2),
\end{aligned}$$

since, by (3.11), the common divisors of m and $m_1\varkappa$ for $m, m_1 \in \mathbf{N}_{(q)}$ must divide m_1; this proves (3.6). Similarly, from (3.12), (3.4), and (3.11) we obtain

$$\begin{aligned}
\lambda(mm_1) &= f(2mm_1\varkappa)/f(2\varkappa) \\
&= \sum_{d|m,m_1\varkappa} d^{k-1}\chi(d)\mu(d)\lambda(m/d)f(2m_1\varkappa/d)/f(2\varkappa) \\
&= \sum_{d|m,m_1} d^{k-1}\chi(d)\mu(d)\lambda(m/d)\lambda(m_1/d);
\end{aligned}$$

and this proves (3.7).

If $k > 0$, then F is not a constant, and so some of its Fourier coefficients with nonzero indices must be nonzero. Hence, $\varkappa = \varkappa(F)$ exists, and we can represent the eigenvalues $\lambda(m)$ in the form (3.12). If we apply the inequalities (3.35) of Chapter 2 ((3.70) of Chapter 2 if F is a cusp-form) to the Fourier coefficients $f(2m\varkappa)$, we obtain (3.5) ((3.6) if F is a cusp-form). □

The identities (3.7) show, in particular, that the eigenvalue $\lambda(m)$ for any $m \in \mathbf{N}_{(q)}$ can be explicitly written as a polynomial in the eigenvalues $\lambda(p)$, where p runs through the prime divisors of m.

The identities (3.4) and the identities in Theorem 3.1 have the following elegant reformulation in the language of Dirichlet series.

THEOREM 3.2. *Let*

$$F = \sum_{a=0}^{\infty} f(2a)e^{2\pi i a z} \in \mathfrak{M}_k(q, \chi)$$

be a nonzero modular form of weight $k \in \mathbf{N}$ and character χ for the group $\Gamma_0^1(q)$. Suppose that F is an eigenfunction of all of the Hecke operators on $\mathfrak{M}_k(q, \chi)$ of the form $|_{k,\chi}T(m)$ for $m \in \mathbf{N}_{(q)}$, and let $\lambda(m) = \lambda(m, F)$ be the corresponding eigenvalues. Then for any $a \in \mathbf{N}$ the Dirichlet series

(3.13) $$D(s, a; F) = \sum_{m \in \mathbf{N}_{(q)}} \frac{f(2am)}{m^s}$$

converges absolutely and uniformly in any right half-plane of the complex variable s of the form $\mathrm{Re}\, s \geqslant k + 1 + \varepsilon$ (of the form $\mathrm{Re}\, s \geqslant k/2 + 1 + \varepsilon$ if F is a cusp-form) with

$\varepsilon > 0$. *In that region it factors as follows*:

$$(3.14) \qquad D(s, a; F) = \left(\sum_{d \in \mathbf{N}_{(q)}, d \mid a} \chi(d)\mu(d)f(2a/d)d^{k-1-s} \right) \zeta(s, F),$$

where μ is the Möbius function and

$$(3.15) \qquad \zeta(s, F) = \zeta(s, F; q) = \sum_{m \in \mathbf{N}_{(q)}} \frac{\lambda(m)}{m^s}$$

is the zeta-function corresponding to the eigenfunction F. The zeta-function $\zeta(s, F)$ converges absolutely and uniformly for s in any of the half-planes indicated above, and in that region it has an absolutely and uniformly convergent Euler product of the form

$$(3.16) \qquad \zeta(s, F) = \prod_{p \in \mathbf{P}_{(q)}} (1 - \lambda(p)p^{-s} + \chi(p)p^{k-1-2s})^{-1}.$$

PROOF. The absolute and uniform convergence of the series (3.13) in the indicated regions follows in the usual way from the estimates (3.35) of Chapter 2 (from (3.70) of Chapter 2 if F is a cusp-form). The convergence of the series (3.15) and the product (3.16) follows from the estimates in Theorem 3.1.

Using (3.4), we have

$$D(s, a; F) = \sum_{m \in \mathbf{N}_{(q)}} \frac{1}{m^s} \sum_{d \mid m, a} d^{k-1} \chi(d)\mu(d)\lambda(m/d)f(2a/d),$$

from which, replacing m by dm, we obtain

$$\sum_{d, m \in \mathbf{N}_{(q)}, d \mid a} \frac{1}{(dm)^s} d^{k-1} \chi(d)\mu(d)f(2a/d)\lambda(m)$$

$$= \left(\sum_{d \in \mathbf{N}_{(q)}, d \mid a} \chi(d)\mu(d)f(2a/d)d^{k-1-s} \right) \sum_{m \in \mathbf{N}_{(q)}} \frac{\lambda(m)}{m^s},$$

and this proves (3.14).

To prove the Euler expansion (3.16) we first note that, by (3.6),

$$(3.17) \qquad \lambda(m)\lambda(m_1) = \lambda(mm_1), \quad \text{if } (m, m_1) = 1,$$

from which, by the unique factorization of integers, we obtain

$$(3.18) \qquad \sum_{m \in \mathbf{N}_{(q)}} \frac{\lambda(m)}{m^s} = \prod_{p \in \mathbf{P}_{(q)}} \sum_{v=0}^{\infty} \frac{\lambda(p^v)}{p^{vs}} = \prod_{p \in \mathbf{P}_{(q)}} \zeta_p(s),$$

where

$$(3.19) \qquad \zeta_p(s) = \zeta_p(s, F) = \sum_{v=0}^{\infty} \lambda(p^v)v^v \quad (v = p^{-s})$$

are the so-called *local zeta-functions of the modular form F*. To sum the power series (3.19) we make use of a special case of (3.6):

$$\lambda(p)\lambda(p^v) = \lambda(p^{v+1}) + p^{k-1}\chi(p)\lambda(p^{v-1}) \quad (v \geqslant 1).$$

Using this relation, we have

$$\lambda(p)\zeta_p(s) = \lambda(p) + \sum_{\nu=1}^{\infty} \lambda(p)\lambda(p^{\nu})v^{\nu}$$

$$= \lambda(p) + \sum_{\nu=1}^{\infty} \lambda(p^{\nu+1})v^{\nu} + p^{k-1}\chi(p)\sum_{\nu=1}^{\infty} \lambda(p^{\nu-1})v^{\nu}$$

$$= (\zeta_p(s) - 1)v^{-1} + p^{k-1}\chi(p)v\zeta_p(s),$$

from which, if we solve for $\zeta_p(s)$, we obtain

(3.20) $\zeta_p(s) = (1 - \lambda(p)v + p^{k-1}\chi(p)v^2)^{-1}$ $(v = p^{-s})$,

and this, along with (3.18), proves (3.16).   □

REMARK. The Euler product expansion (3.16) also follows from the properties of the elements $T(m)$ that we studied in §3.3. That is, the relations (3.17), and hence (3.18), are a direct consequence of (3.20) of Chapter 3. The identity (3.20) can be obtained if the elements $T(p^{\delta})$ in the first identity of Proposition 3.35 of Chapter 3 are replaced by the corresponding eigenvalues and we use the fact that, by (2.34), $F|_{k,\chi}\Delta_1(p) = p^{k-2}\chi(p)F$.

The expansions in Theorem 3.2 do not give any new information about the multiplicative properties of the Fourier coefficients of eigenfunctions of Hecke operators that was not contained, for example, in the identities (3.4). But they make it possible to express the zeta-function $\zeta(s, F)$ in terms of the original modular form, and in many cases this enables one to investigate its analytic properties (see Problem 3.9). In the case of modular forms of degree $n > 1$ it seems that there are no universal identities that express individual eigenvalues in terms of the Fourier coefficients of an eigenfunction, or vice-versa. (Note that when $n > 1$ the Fourier coefficients and the eigenvalues are even indexed by sets that are not related to one another in any clear way: the set of integral equivalence classes of matrices in $\mathbf{A}_n$ in the first case, and a set of double cosets, i.e., diagonal matrices of a special type, in the second case.) As for the Dirichlet series, in the multivariable case one is able to express certain Dirichlet series constructed from the Fourier coefficients of eigenfunctions in terms of Euler products (zeta-functions) constructed from the eigenvalues, and vice-versa. On the one hand, the resulting identities reveal the multiplicative nature of the Fourier coefficients; on the other hand, as in the one-variable case, they enable us to investigate the analytic properties of the zeta-functions that appear. Unfortunately, the identity in Theorem 3.2 has thus far been generalized only to modular forms of degree $n = 2$. This generalization will be explained in the second part of this section.

PROBLEM 3.3. Prove that a modular form $F \in \mathfrak{M}_k^1(q, \chi)$ is an eigenfunction for all of the Hecke operators $|_{k,\chi}T = |T$ for $T \in L^1(q)$ if it is an eigenfunction for all T of the form $T(p)$ with $p \in \mathbf{P}_{(q)}$.

PROBLEM 3.4. Let $F_1, \ldots, F_h$ be a basis of a subspace of $\mathfrak{M}_k^1(q, \chi)$ that is invariant relative to all of the Hecke operators $|T$ $(T \in L^1(q))$, and let

$$F_i|T = \sum_{j=1}^{h} \lambda_{ij}(T)F_j \quad \text{for } i = 1, \ldots, h.$$

Let $\mathbf{f}(2a)$ for $a = 0, 1, \ldots$ denote the column made up of the Fourier coefficients $f_i(2a)$ of the form F_i; and for $m \in \mathbf{N}_{(q)}$ let $\Lambda(m)$ denote the matrix with entries $\lambda_{ij}(T(m))$. Prove the following relations:

$$\sum_{d \mid m,a} d^{k-1}\chi(d)\mathbf{f}(2ma/d^2) = \Lambda(m)\mathbf{f}(2a) \quad (m \in \mathbf{N}_{(q)});$$

$$\mathbf{f}(2ma) = \sum_{d \mid m,a} d^{k-1}\chi(d)\mu(d)\Lambda(m/d)\mathbf{f}(2a/d) \quad (m \in \mathbf{N}_{(q)});$$

$$\Lambda(m)\Lambda(m_1) = \sum_{d \mid m,m_1} d^{k-1}\chi(d)\Lambda(mm_1/d^2) \quad (m, m_1 \in \mathbf{N}_{(q)});$$

$$\Lambda(mm_1) = \sum_{d \mid m,m_1} d^{k-1}\chi(d)\mu(d)\Lambda(m/d)\Lambda(m_1/d) \quad (m, m_1 \in \mathbf{N}_{(q)});$$

$$\sum_{m \in \mathbf{N}_{(q)}} m^{-s}\mathbf{f}(2am) = \left(\sum_{m \in \mathbf{N}_{(q)}} m^{-s}\Lambda(m) \right)$$

$$\times \left(\sum_{d \in \mathbf{N}_{(q)}, d \mid a} \chi(d)\mu(d)d^{k-1-s}\mathbf{f}(2a/d) \right) \quad (a \in \mathbf{N}),$$

where the identity is understood in the formal sense;

$$\sum_{m \in \mathbf{N}_{(q)}} m^{-s}\Lambda(m) = \prod_{p \in \mathbf{P}_{(q)}} (E_h - p^{-s}\Lambda(p) + \chi(p)p^{k-1-2s}E_h)^{-1},$$

where this is again understood as a formal identity.

PROBLEM 3.5. In the notation of the previous problem, show that in the case $k > 0$ the entries $\lambda_{ij}(T(m))$ in the matrix $\Lambda(m)$ satisfy the inequalities

$$|\lambda_{ij}(T(m))| \leqslant cm^k \quad (m \in \mathbf{N}_{(q)})$$

(and satisfy the inequality

$$|\lambda_{ij}(T(m))| \leqslant cm^{k/2} \quad (m \in \mathbf{N}_{(q)}),$$

if $F_1, \ldots, F_h$ are cusp-forms), where c does not depend on m. Using these estimates, investigate the convergence of the matrix series and products in the previous problem.

[Hint: Let $a_1, \ldots, a_h \in \mathbf{N}$ be chosen so that the matrix $A = (f_i(2a_j))$ is invertible. Then $\Lambda(m) = B(m)A^{-1}$, where $B(m)$ is the matrix whose columns are $\sum_{d \mid m,a_j} d^{k-1}\chi(d)\mathbf{f}(2ma_j/d^2)$.]

PROBLEM 3.6. Let $q(X) = q(x_1, \ldots, x_m)$ be a positive definite integral quadratic form whose matrix has determinant 1. Show that there exists a finite set of functions $\lambda_1, \ldots, \lambda_h \colon \mathbf{N} \to \mathbf{C}$ satisfying the relations

$$\lambda_i(ab) = \sum_{d \mid a,b} d^{m/2-1}\lambda_i(ab/d^2) \quad \text{for } a, b \in \mathbf{N}$$

and having the property that the number $r(q, a)$ of integer solutions of the equation $q(X) = a$ can be represented in the form

$$r(q, a) = \sum_{i=1}^{h} \alpha_i \lambda_i(a) \quad (a \in \mathbf{N})$$

with constant coefficients α_i. Then deduce that for any $a \in \mathbf{N}$ and any prime p the power series

$$\sum_{v=0}^{\infty} r(q, ap^v) v^v$$

is a rational function in v with denominator

$$\prod_{i=1}^{h} (1 - \lambda_i(p) v + p^{m/2-1} v^2).$$

PROBLEM 3.7. Let

$$F(z) = \sum_{a=0}^{\infty} f(2a) e^{2\pi i a z} \in \mathfrak{M}_k^1(q, \chi).$$

Using the integral representation of the gamma-function

$$\Gamma(s) = \int_0^{\infty} t^{s-1} e^{-t} \, dt \quad (\text{Re } s > 0),$$

prove that the Dirichlet series with coefficients $f(2a)$ has the following integral representation in terms of the original modular form F:

$$\Psi(s; F) = (2\pi)^{-s} \Gamma(s) \sum_{a \in \mathbf{N}} \frac{f(2a)}{a^s} = \int_0^{\infty} t^{s-1} (F(it) - f(0)) \, dt \quad (\text{Re } s > k).$$

PROBLEM 3.8. In the notation of the previous problem, suppose that $q = 1$ and $\chi = 1$. Derive the identity

$$\Psi(s; F) = \int_1^{\infty} t^{s-1} (F(it) - f(0)) \, dt$$
$$+ i^k \int_1^{\infty} t^{k-s-1} (F(it) - f(0)) \, dt - f(0) \left(\frac{i^k}{k - s} + \frac{1}{s} \right) \quad (\text{Re } s > k).$$

From this deduce that the function $\Psi(s; F)$ has a meromorphic continuation to the entire s-plane, the function

$$\Psi(s; F) + f(0) \left(\frac{i^k}{k - s} + \frac{1}{s} \right)$$

is an entire function, and $\Psi(s; F)$ satisfies the functional equation

$$\Psi(k - s; F) = i^k \Psi(s; F).$$

[Hint: Divide the integral from 0 to ∞ in the previous problem into the integral from 1 to ∞ and the integral from 0 to 1. In the latter integral make the change of variable $t \to 1/t$ and use the relation $F(i/t) = F(-1/it) = (it)^k F(it)$.]

PROBLEM 3.9. Prove the following properties for the zeta-function $\zeta(s, F)$ corresponding to a nonzero eigenfunction $F \in \mathfrak{M}_k^1(1, 1)$ for all of the Hecke operators $|T(m)$:

(1) $\zeta(s, F)$ has a meromorphic continuation to the entire s-plane;

(2) the function

$$\varphi(s; F) + \frac{f(0)}{f(2)} \left(\frac{i^k}{k - s} + \frac{1}{s} \right),$$

where $\varphi(s; F) = (2\pi)^{-s} \Gamma(s) \zeta(s, F)$ and $f(2a)$ are the Fourier coefficients of F, is an entire function;

(3) $\zeta(s, F)$ satisfies the functional equation

$$\varphi(k - s; F) = i^k \varphi(s; F).$$

PROBLEM 3.10. Prove that:

(1) $H_q^{-1} \Gamma_0^1(q) H_q = \Gamma_0^1(q)$, where $H_q = \begin{pmatrix} 0 & -1 \\ q & 0 \end{pmatrix}$.

(2) The map $F \to F|_k H_q = (qz)^{-k} F(-1/qz)$ gives an isomorphism between $\mathfrak{M}_k^1(q, \chi)$ and $\mathfrak{M}_k^1(q, \overline{\chi})$, where $\overline{\chi}$ is the conjugate of χ.

(3) If χ is a real character modulo q, then

$$\mathfrak{M}_k^1(q, \chi) = \mathfrak{M}_k^+(q, \chi) \oplus \mathfrak{M}_k^-(q, \chi),$$

where $F \in \mathfrak{M}_k^\pm(q, \chi)$ if $F|_k H_q = \pm q^{-k/2} i^k F$.

(4) If $F \in \mathfrak{M}_k^1(q, \chi)$, then

$$(F|_k H_q)|_{k, \overline{\chi}} T(m) = \overline{\chi}(m) (F|_{k, \chi} T(m))|_k H_q \quad (m \in \mathbf{N}_{(q)}),$$

so that in the case of a real character χ the subspaces $\mathfrak{M}_k^\pm(q, \chi)$ are invariant relative to all of the Hecke operators $|T(m)$.

(5) Let F be a modular form in $\mathfrak{M}_k^\pm(q, \chi)$, where χ is a real character. In the notation of Problem 3.7, prove that one has the following integral representation:

$$\Psi(s; F) = \int_{q^{-1/2}}^\infty t^{s-1} (F(it) - f(0)) \, dt \pm q^{k/2-s} \int_{q^{-1/2}}^\infty t^{k-s-1} (F(it) - f(0)) \, dt$$

$$- f(0) q^{-s/2} \left(\frac{1}{s} \pm \frac{1}{k - s} \right) \quad (\mathrm{Re}\, s > k).$$

From this deduce that $\Psi(s; F)$ has a meromorphic continuation to the entire s-plane, the function

$$\Psi(s; F) + f(0) q^{-s/2} \left(\frac{1}{s} \pm \frac{1}{k - s} \right)$$

is an entire function, and $\Psi(s; F)$ satisfies the functional equation

$$\Psi(k - s; F) = \pm q^{s - k/2} \Psi(s; F).$$

(6) Let F be a nonzero modular form in $\mathfrak{M}_k^1(q, \chi)$ with Fourier coefficients $f(0), f(2), \ldots$. Suppose that F is an eigenfunction for all of the Hecke operators

$|T(m)$ $(m \in \mathbf{N}_{(q)})$, and let $\lambda(m)$ be the corresponding eigenvalues. Show that one has the factorization

$$\sum_{a \in \mathbf{N}} \frac{f(2a)}{a^s} = \left(\sum_{a \in \mathbf{N}, a | q^{\infty}} \frac{f(2a)}{a^s} \right) \left(\sum_{m \in \mathbf{N}_{(q)}} \frac{\lambda(m)}{m^s} \right) \quad (\mathrm{Re}\, s > k).$$

2. Modular forms of degree 2, Gaussian composition, and zeta-functions. As we mentioned before, in the case of modular forms of several variables it is natural to express the relations between the Fourier coefficients of eigenfunctions for the Hecke operators and the corresponding eigenvalues in the form of identities between suitable Dirichlet series. In this subsection we obtain analogues of the identities (3.14) for modular forms of degree 2. A special feature of degree 2 is that the Fourier coefficients are indexed by the matrices of binary quadratic forms, for which in many cases there exists a natural multiplication—the so-called Gaussian composition of quadratic forms. This composition turns out to be essential for understanding the multiplicative nature of the Fourier coefficients of eigenfunctions. At present there do not exist generalizations of the identities (3.14) to modular forms of degree greater than 2.

Consider a modular form

$$F(Z) = \sum_{A \in \mathbf{A}_2} f(A) e\{AZ\} \in \mathfrak{M}_k^2(q, \chi)$$

of integer weight k and one-dimensional character $[\chi]$ for the group $\Gamma_0^2(q)$, where $q \in \mathbf{N}$ and χ is a Dirichlet character modulo q. We suppose that F is an eigenfunction for all of the Hecke operators $|T = |_{k,\chi} T$ for $T \in L^2(q)$ with eigenvalues $\lambda(T) = \lambda(T; F)$:

$$F|T = \lambda(T)F.$$

The Dirichlet series constructed from the Fourier coefficients of F that we shall work with is the series

$$(3.21) \qquad D(s, A, \gamma; F) = \sum_{m \in \mathbf{N}_{(q)}} \frac{f(mA)\gamma(m)}{m^s},$$

where $A \in \mathbf{A}_2$ and $\gamma \colon \mathbf{N}_{(q)} \to \mathbf{C}$ is a completely multiplicative function, i.e., a function satisfying the relation

$$(3.22) \qquad \gamma(mm_1) = \gamma(m)\gamma(m_1) \quad \text{for all } m, m_1 \in \mathbf{N}_{(q)},$$

and we shall also consider certain linear combinations of these series. On the other hand, the Dirichlet series constructed from the eigenvalues of F that arise are Euler products of the form

$$(3.23) \qquad \zeta(s, \gamma; F) = \prod_{p \in \mathbf{P}_{(q)}} Q_p(\gamma(p)p^{-s}; F)^{-1},$$

where γ is the same as above, $Q_p(v; F)$ for $p \in \mathbf{P}_{(q)}$ denotes the polynomial

$$(3.24) \qquad Q_p(v; F) = \sum_{j=0}^{4} (-1)^j \lambda(q_j(p)) v^j$$

and $q_j(p) = q_j^2(p)$ is defined as the preimage in $L_p^2(q)$ of the element $\mathbf{q}_j^2(p) \in \mathbf{L}_p^2$ (see (3.77) of Chapter 3) under the isomorphism ε_q (see (3.45) of Chapter 3). We call (3.23) the *zeta-function of F with "character"* γ. Its p-factors

$$(3.25) \qquad \zeta_p(s, \gamma; F) = Q_p(\gamma(p)p^{-s}; F)^{-1}$$

are called the *local zeta-functions of F (with character γ)*.

We begin our search for "global" connections between the series (3.21) and the products (3.23) by finding local relations, i.e., relations between the power series

$$(3.26) \qquad D_p(s, A, \gamma; F) = \sum_{\delta=0}^{\infty} f(p^\delta A)v^\delta, \quad \text{where } v = \gamma(p)p^{-s},$$

and the local zeta-functions $\zeta_p(s, \gamma; F)$. In the computations below we use the technique of Chapter 3, based on extending Hecke rings of the symplectic group to Hecke rings of the triangular subgroup. According to the philosophy of §2.2, we consider $\mathfrak{M}_k^2(q, \chi)$ as an invariant subspace relative to all of the Hecke operators in $\mathbf{L}^2(q) = \varepsilon_q(L^2(q))$ inside the space $\mathfrak{M}_s^2$, where $s(-1) = (-1)^k \chi(-1)$; and we consider the space $\mathfrak{F}_k^2(q, \chi)$ of Fourier coefficients of functions in $\mathfrak{M}_k^2(q, \chi)$ as an invariant subspace of $\mathfrak{F}_s^2$ relative to the same Hecke operators. We shall systematically make use of the notation (2.72) and the relations (2.74). For most of what we do, what is important is simply that F lies in $\mathfrak{M}_s^2$ and is an eigenfunction for certain Hecke operators. Everywhere in what follows k, χ, and s are fixed, and are connected by the relation $s(-1) = (-1)^k \chi(-1)$.

LEMMA 3.11. *Let* $F \in \mathfrak{M}_s^2$ *be an eigenfunction for all of the Hecke operators* $|T = |_{k,\chi} T$ *for* $T \in \mathbf{L}_p^2$, *where* p *is a fixed prime, let* $\lambda(T)$ *be the corresponding eigenvalues, and let* $f(A)$ $(A \in \mathbf{A}_2)$ *be the Fourier coefficients of* F. *Then the following formal identity holds for every matrix* $B \in \mathbf{A}_2$:

$$(3.27) \qquad \begin{aligned} d_p(v, B) &= \sum_{\delta=0}^{\infty} f(p^\delta B)v^\delta \\ &= Q_p(v; F)^{-1}(f|_{k,\chi}(1 - \Pi_- v)(1 - \Pi_1 v + p(\Pi_{2,0}^{(1)} + \Pi_{2,0}^{(0)})v^2))(B), \end{aligned}$$

where, by analogy with (3.24), *we set*

$$Q_p(v; F) = \sum_{j=0}^{4} (-1)^j \lambda(\mathbf{q}_j^2(p))v^j,$$

$\Pi_- = \Pi_-^2(p) = \Pi_0^2(p)$ *and* $\Pi_1 = \Pi_1^2(p)$ *are the elements* (3.59) *of Chapter 3, and* $\Pi_{2,0}^{(r)} = \Pi_{2,0}^{(r)}(p)$ *are the elements* (3.62) *of Chapter 3 for* $n = 2$.

PROOF. First of all, by (2.21) the function $f \in \mathfrak{F}_s^2$ is an eigenfunction for all of the Hecke operators in $\mathbf{L}_p^2$ and has the same eigenvalues as F; this implies that for any $B' \in \mathbf{A}_2$

$$Q_p(v; F)f(B') = \sum_{j=0}^{4} (-1)^j (f|\mathbf{q}_j^2(p))(B')v^j = (f|\mathbf{Q}_p^2(v))(B'),$$

where $\mathbf{Q}_p^2(v)$ is the polynomial (3.78) of Chapter 3 over the ring $\mathbf{L}_p^2 \subset \mathbf{L}_{0,p}^2$. Furthermore, it follows from (2.33) that for any $g \in \mathfrak{F}_s^2$ and any $B \in \mathbf{A}_2$ we can write

$$g(p^\delta B) = (g|\Pi_2^2(p^\delta))(B) = (g|\Pi_+^\delta)(B) \quad (\delta \geqslant 0),$$

where $\Pi_+ = \Pi_+^2(p) = \Pi_2^2(p)$. Using these formulas and the factorization of the polynomial $\mathbf{Q}_p^2(v)$ in Proposition 6.13 of Chapter 3, we obtain

$$Q_p(v;F)d_p(v,B) = \sum_{j,\delta}(-1)^j(f|\mathbf{q}_j^2(p))(p^\delta B)v^{j+\delta}$$

$$= \sum_{j,\delta}(-1)^j(f|\mathbf{q}_j^2(p)|\Pi_+^\delta)(B)v^{j+\delta}$$

$$= \left(f|\mathbf{Q}_p^2(v)|\sum_{\delta=0}^{\infty}\Pi_+^\delta v^\delta\right)(B)$$

$$= (f|(1-\Pi_- v)(1-\Pi_1 v + p(\Pi_{2,0}^{(1)}+\Pi_{2,0}^{(0)})v^2))(B). \qquad \square$$

The identity (3.27) shows that each series $d_p(v,B)$ is a rational function with denominator $Q_p(v;F)$. To compute the numerator we need formulas for the action of the operators on functions $f \in \mathfrak{F}_s^2$. First, if we specialize the formula (2.33) to the case $n = 2$ and $a = 0,1$, we obtain the formulas

(3.28)
$$(f|\Pi_-)(B) = \begin{cases} p^{2k-3}\chi(p)^2 f(p^{-1}B), & \text{if } B \in p\mathbf{A}_2, \\ 0, & \text{if } B \notin p\mathbf{A}_2, \end{cases}$$

(3.29)
$$(f|\Pi_1)(B) = p^{k-2}\chi(p)\sum_{\substack{D\in\Lambda^+\setminus\Lambda^+D_1^2(p)\Lambda^+ \\ B['D]\in p\mathbf{A}_2}}f(p^{-1}B['D]),$$

where $\Lambda^+ = SL_2(\mathbf{Z})$. Next, by Lemma 2.19 we have

(3.30)
$$(f|\Pi_{2,0}^{(1)})(B) = p^{2(k-3)}\chi(p)^2 l_p(1,2;B)f(B)$$

and

(3.31)
$$(f|\Pi_{2,0}^{(0)})(B) = (f|\Delta_2(p))(B) = p^{2(k-3)}\chi(p)^2 f(B).$$

By Lemma 2.21,

$$l_p(1,2;B) = (p-1)^{-1}p\rho_p(1,2;B) - (p+1),$$

where for $B = \begin{pmatrix} 2b_1 & b_2 \\ b_2 & 2b_3 \end{pmatrix}$ we let $\rho_p(1,2;B)$ denote the number of nontrivial solutions of the congruence

(3.32)
$$b(x,y) = b_1 x^2 + b_2 xy + b_3 y^2 \equiv 0 (\bmod\, p).$$

According to Lemma 2.22, this number is equal to the number $i(V_p,b_p,1)$ of nonzero isotropic vectors in the two-dimensional quadratic space (V_p,b_p) over $\mathbf{F}_p$ with quadratic form $b_p = b(x,y)/\bmod p$. If (V_p,b_p) is a nondegenerate space, then, by Proposition 2.14 of Appendix 2, we have $i(V_p,b_p,1) = (p-\varepsilon)(1+\varepsilon)$, where $\varepsilon = \varepsilon(V_p,b_p)$ is the sign of the space (V_p,b_p) (see (2.21) of Appendix 2). Since $\varepsilon = \pm 1$, the last formula can be rewritten in the form

(3.33)
$$i(V_p,b_p,1) = (\varepsilon+1)(p-1).$$

This formula is actually valid for any two-dimensional quadratic space (V_p, b_p), if we set

$$(3.34) \qquad \varepsilon = \varepsilon(V_p, b_p) = \begin{cases} 0, & \text{if } (V_p, b_p) \text{ is degenerate but } b_p \neq 0, \\ p, & \text{if } b_p = 0. \end{cases}$$

In fact, in the first case the space (V_p, b_p) is the sum of two one-dimensional subspaces, one null and one not null, and every nonzero isotropic vector must belong to the null one-dimensional subspace; thus, the number of isotropic vectors is $p - 1$. In the second case, every nonzero vector is isotropic, so there are $p^2 - 1$ of them.

For every matrix $B \in \mathbf{A}_2$ we define the p-sign $\varepsilon_p(B)$ by setting

$$(3.35) \qquad \varepsilon_p(B) = \varepsilon_p(V_p, b_p(x, y)),$$

where $b_p(x, y)$ is the quadratic form with matrix B, considered over $\mathbf{F}_p$. Here the right side of (3.35) is understood in the sense of (2.21) of Appendix 2 if the two-dimensional quadratic space (V_p, b_p) is nondegenerate, and in the sense of (3.34) if this space is degenerate. Then, by the above observations, we can write

$$l_p(1, 2; B) = p(\varepsilon_p(B) + 1) - (p + 1) = p\varepsilon_p(B) - 1,$$

and from this and (3.30)–(3.31) we finally obtain a formula for the action of the operator $\Pi_{2,0}^{(1)} + \Pi_{2,0}^{(0)}$ on $\mathfrak{F}_s^2$:

$$(3.36) \qquad (f | \Pi_{2,0}^{(1)} + \Pi_{2,0}^{(0)})(B) = p^{2k-5} \chi(p)^2 \varepsilon_p(B) f(B).$$

In order to apply the identities (3.27), it remains to interpret the right side of (3.29). If $\Delta = \det B \neq 0$, then any of the matrices $p^{-1} B[{}^t D]$ on the right in (3.29) is the matrix of a positive definite integral quadratic form of discriminant $-\Delta$, and hence it is naturally associated with a module of the imaginary quadratic field $\mathbf{Q}(\sqrt{-\Delta})$. This enables us to interpret the right side of (3.29) in terms of the composition of quadratic forms. We shall use the language, notation, and results of Appendix 3.

We first look at the conditions under the summation in (3.29). According to Lemma 1.2 of Chapter 3, we can take our set of representatives of the left cosets $\Lambda^+ \backslash \Lambda^+ D_1^2(p)\Lambda^+$ to be matrices of the form $D_1^2(p) \begin{pmatrix} u_1 & u_2 \\ v_1 & v_2 \end{pmatrix} = \begin{pmatrix} u_1 & u_2 \\ pv_1 & pv_2 \end{pmatrix}$,

where $\begin{pmatrix} u_1 & u_2 \\ v_1 & v_2 \end{pmatrix}$ runs through a set of left coset representatives of $\Lambda^+ = \Gamma^1$ modulo the subgroup

$$(3.37) \qquad \Gamma^1 \cap D_1^2(p)^{-1} \Gamma^1 D_1^2(p) = \left\{ \begin{pmatrix} \alpha & \beta \\ \gamma & \delta \end{pmatrix} \in \Gamma^1; \beta \equiv 0 (\bmod p) \right\}.$$

It is clear that two matrices in Γ^1 with first rows (u_1, u_2) and (u_1', u_2') lie in the same left coset modulo the above group if and only if $u_1' u_2 \equiv u_2' u_1 (\bmod p)$, i.e., if and only if the pairs (u_1, u_2) and (u_1', u_2') are proportional modulo p:

$$(3.38) \qquad (u_1', u_2') \equiv \alpha(u_1, u_2)(\bmod p), \quad \text{where } \alpha \not\equiv 0(\bmod p).$$

Thus, we can take

(3.39)
$$\Delta^+ \setminus \Delta^+ D_1^2(p)\Delta^+ = \left\{ \begin{pmatrix} 1 & 0 \\ 0 & p \end{pmatrix} U; \right.$$
$$\left. U = \begin{pmatrix} u_1 & u_2 \\ * & * \end{pmatrix} \in \Lambda^+, (u_1, u_2) \in P^1(\mathbf{Z}/p\mathbf{Z}) \right\},$$

where $P^1(\mathbf{Z}/p\mathbf{Z})$ denotes an arbitrary set of representatives of the equivalence classes of pairs of relatively prime integers under the equivalence (3.38) (this is the projective line over $\mathbf{Z}/p\mathbf{Z}$). Let $B = \begin{pmatrix} 2b_1 & b_2 \\ b_2 & 2b_3 \end{pmatrix} \in \mathbf{A}_2$, and let $b(x, y) = b_1 x^2 + b_2 xy + b_3 y^2$ be the quadratic form with matrix B. If $D = \begin{pmatrix} 1 & 0 \\ 0 & p \end{pmatrix} U = \begin{pmatrix} u_1 & u_2 \\ pv_1 & pv_2 \end{pmatrix}$, then, multiplying out the matrices, we find that

$$B['D] = \begin{pmatrix} 2b(u_1, u_2) & pb_2(U) \\ pb_2(U) & 2p^2 b(v_1, v_2) \end{pmatrix},$$

where

(3.40)
$$b_2(U) = 2b_1 u_1 v_1 + b_2(u_1 v_2 + u_2 v_1) + 2b_3 u_2 v_2.$$

In particular, if U is an integer matrix, then the condition $B['D] \in p\mathbf{A}_2$ is equivalent to the congruence $b(u_1, u_2) \equiv 0 \pmod{p}$. Thus, if we use the set of representatives (3.39), we can rewrite (3.29) in the form

(3.41) $(f|\Pi_1)(B) = p^{k-2}\chi(p) \sum_{\substack{(u_1,u_2)\in P^1(\mathbf{Z}/p\mathbf{Z}), \\ b(u_1,u_2)\equiv 0 (\mathrm{mod}\, p)}} f\left(\begin{pmatrix} 2b(u_1, u_2)/p & b_2(U) \\ b_2(U) & 2pb(v_1, v_2) \end{pmatrix} \right),$

where for each pair of relatively prime integers (u_1, u_2) we take (v_1, v_2) to be an arbitrary pair of integers for which $u_1 v_2 - u_2 v_1 = 1$, and where $U = \begin{pmatrix} u_1 & u_2 \\ v_1 & v_2 \end{pmatrix}$.

The next proposition, which plays a central role in this discussion, interprets the right side of (3.41) for positive definite matrices B not divisible by p in terms of the composition of matrices of quadratic forms and modules of the corresponding quadratic field.

Let $\mathfrak{O}'$ be an order of the algebraic number field K. For brevity, we shall use the term *regular ideal* (of the ring $\mathfrak{O}'$) to refer to a full submodule of K that is contained in $\mathfrak{O}'$ and has the property that its ring of multipliers is $\mathfrak{O}'$.

PROPOSITION 3.12. *Let* $A = \begin{pmatrix} 2a & b \\ b & 2c \end{pmatrix}$ *be a positive definite even matrix with relatively prime* a, b, c, *and let* $D = b^2 - 4ac$ *be its discriminant. Let* d *denote the discriminant of the imaginary quadratic field* $K = \mathbf{Q}(\sqrt{D})$, *and let* $\mathfrak{O}_l$ *be the subring of index* $l = \sqrt{D/d}$ *in the ring of integers of* K. *Then the following results hold for any prime* p, *any natural number* m *not divisible by* p, *and any function* $f \in \mathfrak{F}_s^2$:

(1) *If the* p-*sign* $\varepsilon_p(A)$ *is 1, then the ring* $\mathfrak{O}_l$ *contains exactly two regular ideals of norm* p, *say* $\mathfrak{p}$ *and* $\mathfrak{p}'$; *these ideals are conjugate, i.e.,* $\mathfrak{p}' = \bar{\mathfrak{p}}$; *and one has the formula*

$$(f|_{k,\chi}\Pi_1^2(p))(mA) = p^{k-2}\chi(p)(f(m(A \times \mathfrak{p})) + f(m(A \times \bar{\mathfrak{p}}))).$$

(2) *If, on the other hand,* $\varepsilon_p(A) = -1$, *then* $\mathfrak{O}_l$ *contains no regular ideals of norm* p, *and one has:* $(f|_{k,\chi}\Pi_1^2(p))(mA) = 0$.

(3) *If* $\varepsilon_p(A) = 0$ *and* $p \nmid l$, *then there exists a unique regular ideal* $\mathfrak{p}$ *of norm* p *in the ring* $\mathfrak{O}_l$, *we have* $\bar{\mathfrak{p}} = \mathfrak{p}$, *and*

$$(f|_{k,\chi}\Pi_1^2(p))(mA) = p^{k-2}\chi(p)f(m(A \times \mathfrak{p})).$$

(4) *If* $\varepsilon_p(A) = 0$ *and* $p|l$, *then*

$$(f|_{k,\chi}\Pi_1^2(p))(mA) = p^{k-2}\chi(p)f(mp(A \times \mathfrak{O}_{l/p})),$$

where $\mathfrak{O}_{l/p}$ *is the subring of index* l/p *in the ring of integers of* K.

In all cases the composition of matrices and modules is understood in the sense of Appendix 3.3.

PROOF. Let $q(x, y) = ax^2 + bxy + cy^2$ be the quadratic form with matrix A, and let $V = V_p = (V_p, q \bmod p)$ be the two-dimensional quadratic space over $\mathbf{F}_p$ with quadratic form q, regarded modulo p. By the definition of the p-sign of the matrix of a quadratic form that is not divisible by p, we find that if $p \neq 2$, then

$$(3.42) \qquad \varepsilon_p(A) = \varepsilon_p(V) = \left(\frac{-\det A}{p}\right) = \left(\frac{D}{p}\right),$$

and so the equality $\varepsilon_p(A) = 1$ means that $p \nmid D$ and the congruence $x^2 \equiv D = dl^2(\bmod p)$ is solvable. Since $d \equiv 0, 1(\bmod 4)$, this implies that the congruence $x^2 \equiv d(\bmod 4p)$ is also solvable. The equality $\varepsilon_p(A) = -1$ means that $p \nmid D$ and the congruence $x^2 \equiv D(\bmod p)$ has no solution, in which case neither does the congruence $x^2 \equiv d(\bmod 4p)$. The equality $\varepsilon_p(A) = 0$ is equivalent to the condition $p|D$, so that either $p|l$, or else $p \nmid l$ but $p|d$, and in the latter case the congruence $x^2 \equiv d(\bmod 4p)$ is obviously solvable. Finally, if $p = 2$, then

$$(3.43) \qquad \varepsilon_p(A) = \begin{cases} 1, & \text{if } \det A \equiv -1(\bmod 8), \\ -1, & \text{if } \det A \equiv 3(\bmod 8), \\ 0, & \text{if } \det A \equiv 0(\bmod 2), \end{cases}$$

since the square of any odd integer is $\equiv 1(\bmod 8)$, so that in the first case $d \equiv -l^2 \det A \equiv 1(\bmod 8)$ and the congruence $x^2 \equiv d(\bmod 8)$ is solvable. In the second case $d \equiv -l^{-2} \det A \equiv 5(\bmod 8)$ and the congruence $x^2 \equiv d(\bmod 8)$ has no solutions; and in the third case, if l is odd, then d is congruent to 0 or 4 modulo 8, and the congruence $x^2 \equiv d(\bmod 8)$ is solvable.

These observations and §2 of Appendix 3 give us the statements in the first three parts of the proposition about the existence and properties of ideals of norm p in $\mathfrak{O}$. In particular, in these cases there are exactly $\varepsilon_p(A) + 1$ such ideals.

On the other hand, the expression $\varepsilon_p(A) + 1$ appears in the formula (3.33) for the number of nonzero isotropic vectors in V_p, which can obviously be interpreted as the number of distinct pairs (modulo p) of relatively prime integers (u_1, u_2) satisfying the congruence $q(u_1, u_2) \equiv 0(\bmod p)$. If we divide these pairs into classes of pairs that are proportional to one another modulo p in the sense of (3.38), and if we take into account that each such class contains exactly $p - 1$ pairs, we conclude that the number of terms in the sum (3.41) for $B = mA$ with $m \not\equiv 0(\text{mod } p)$ is equal to

(3.44)
$$|\{(u_1, u_2) \in P^1(\mathbf{Z}/p\mathbf{Z}); q(u_1, u_2) \equiv 0(\text{mod } p)\}|$$
$$= (p - 1)^{-1}i(V_p, q \text{ mod } p) = \varepsilon_p(A) + 1.$$

Hence, in cases (1), (2), and (3) of the proposition it is natural to expect that the terms in (3.41) are directly connected with the regular ideals in $\mathfrak{O}_l$ of norm p.

Suppose that $\varepsilon_p(A) = 1$. In this case, by (3.44), there exist exactly two pairs of relatively prime integers (u_1, u_2) and (u_1', u_2') that are not proportional modulo p and that satisfy the congruence $q(x, y) \equiv 0(\text{mod } p)$. Then $q(u_1, u_2) = pa_1$ and $q(u_1', u_2') = pa_2$, where $a_1, a_2 \in \mathbf{N}$. We choose integers v_1, v_2, v_1', v_2' such that

$$U_1 = \begin{pmatrix} u_1 & u_2 \\ v_1 & v_2 \end{pmatrix}, \qquad U_2 = \begin{pmatrix} u_1' & u_2' \\ v_1' & v_2' \end{pmatrix} \in SL_2(\mathbf{Z}),$$

and for $i = 1, 2$ we set

$$A_i = A[{}^tU_i] = \begin{pmatrix} 2pa_i & b_i \\ b_i & 2c_i \end{pmatrix}, \qquad A_i' = \begin{pmatrix} 2a_i & b_i \\ b_i & 2pc_i \end{pmatrix}.$$

In this notation, by (3.41) we can write

(3.45)
$$(f|\Pi_1)(mA) = p^{k-2}\chi(p)(f(mA_1') + f(mA_2')).$$

We consider the following full modules of $K = \mathbf{Q}(\sqrt{D})$:

$$M_1 = \left\{p, \frac{b_1 - \sqrt{D}}{2}\right\} \quad \text{and} \quad M_2 = \left\{p, \frac{b_2 - \sqrt{D}}{2}\right\}.$$

The number $\gamma_1 = (b_1 - \sqrt{D})/2p$ is obviously a root of the polynomial $pv^2 - b_1 v + a_1 c_1$. Since $D = b^2 - 4ac = b_1^2 - 4pa_1 c_1$ is not divisible by p, it follows that b_1 is not divisible by p, and the coefficients of this polynomial are relatively prime. Then, by §2 of Appendix 3, the ring of multipliers $\mathfrak{O}_{M_1}$ of the module $M_1 = p\{1, \gamma_1\}$ is $\mathfrak{O}_l$, and $N(M_1) = N(p)/p = p$. In addition, obviously $M_1 \subset \mathfrak{O}_l$. Thus, the module M_1 is a regular ideal of $\mathfrak{O}_l$ of norm p. By the same argument this is also true of the module M_2. We now claim that $M_1 \neq M_2$. It suffices to show that $M_1 + M_2 = \mathfrak{O}_l$; and for this it is enough to see that the numbers $(b_2 - b_1)/2$ and p, both of which lie in $M_1 + M_2$, are relatively prime, since in that case 1 is an integer linear combination of these numbers and hence lies in $M_1 + M_2$. We set

$$U_2 U_1^{-1} = \begin{pmatrix} u_1' & u_2' \\ v_1' & v_2' \end{pmatrix}\begin{pmatrix} v_2 & -u_2 \\ -v_1 & u_1 \end{pmatrix} = T = \begin{pmatrix} t_1 & t_2 \\ t_3 & t_4 \end{pmatrix} \in SL_2(\mathbf{Z}).$$

Since the pairs (u_1, u_2) and (u_1', u_2') are not proportional to one another modulo p, it follows that $t_2 = u_1 u_2' - u_2 u_1' \not\equiv 0(\text{mod } p)$. If we use the obvious relation $A_1[{}^tT] = A_2$ and equate coefficients, we find that $pa_1 t_1^2 + b_1 t_1 t_2 + c_1 t_2^2 = pa_2$ and $2pa_1 t_1 t_3 + b_1(t_1 t_4 + t_2 t_3) + 2c_1 t_2 t_4 = b_2$. Since $t_2 \not\equiv 0(\text{mod } p)$, the first congruence implies that $b_1 t_1 + c_1 t_2 \equiv 0(\text{mod } p)$. Since $t_1 t_4 = 1 + t_2 t_3$, the second congruence implies that $(b_2 - b_1)/2 \equiv t_2(b_1 t_3 + c_1 t_4)(\text{mod } p)$. Thus, if $(b_1 - b_2)/2$ were divisible by p, we would have the system of congruences $b_1 t_1 + c_1 t_2 \equiv b_1 t_3 + c_1 t_4 \equiv 0(\text{mod } p)$, from which it would follow that $b_1 \equiv c_1 \equiv 0(\text{mod } p)$. But this is impossible, since the discriminant $D = b_1^2 - 4pa_1 c_1$ is not divisible by p. This proves the claim, and implies that $M_1 = \mathfrak{p}$ and $M_2 = \bar{\mathfrak{p}}$ are the only regular ideals of $\mathfrak{O}_l$ of norm p. Let

q_1, q_2, q_1', and q_2' be the quadratic forms with matrices A_1, A_2, A_1', and A_2', respectively, and let $M(q_1)$, $M(q_2)$, $M(q_1')$, and $M(q_2')$ be the modules corresponding to these quadratic forms (see §3 of Appendix 3). We set $\delta_1 = (b_1 - \sqrt{D})/2$. Since obviously $\delta_1^2 = b_1\delta_1 - pa_1c_1$, and since b_1 is not divisible by p (see above), it follows that

$$M(q_1')\mathfrak{p} = \{a_1, \delta_1\}\{p, \delta_1\} = \{a_1p, a_1\delta_1, p\delta_1, b\delta_1 - pa_1c_1\} = \{pa_1, \delta_1\} = M(q_1),$$

from which, using §2 of Appendix 3, we obtain

$$M(q_1') = p^{-1}M(q_1')\mathfrak{p}\bar{\mathfrak{p}} = p^{-1}M(q_1)\bar{\mathfrak{p}}.$$

Since the quadratic form q_1 is properly equivalent to q, it follows that the last module is similar to the module $M(q)\bar{\mathfrak{p}}$, and so the matrix A_1' of the quadratic form q_1' is properly equivalent to the matrix $A \times \bar{\mathfrak{p}}$; hence, $f(mA_1') = f(m(A \times \bar{\mathfrak{p}}))$. Similarly, A_2' is properly equivalent to $A \times \mathfrak{p}$, and $f(mA_2') = f(m(A \times \mathfrak{p}))$. These relations, together with (3.45), prove the first part of the proposition.

The second part follows from the above arguments and from (3.41) and (3.44).

Now suppose that $\varepsilon_p(A) = 0$. Then, according to (3.44), all of the pairs of relatively prime integers satisfying $q(x, y) \equiv 0 \pmod{p}$ are proportional to one such pair, say, (u_1, u_2). Let $q(u_1, u_2) = pa_1$. We choose $v_1, v_2 \in \mathbf{Z}$ so that

$$U = \begin{pmatrix} u_1 & u_2 \\ v_1 & v_2 \end{pmatrix} \in SL_2(\mathbf{Z}),$$

and we set

$$A_1 = A[{}^tU] = \begin{pmatrix} 2pa_1 & b_1 \\ b_1 & 2c_1 \end{pmatrix} \quad \text{and} \quad A_1' = \begin{pmatrix} 2a_1 & b_1 \\ b_1 & 2pc_1 \end{pmatrix}.$$

Then, by (3.41), we have

$$(3.46) \qquad\qquad (f|\Pi_1)(mA) = p^{k-2}\chi(p)f(mA_1').$$

We have already observed that in the case under consideration $D = b^2 - 4ac = b_1^2 - 4pa_1c_1$ is divisible by p; in particular, this implies that

$$(3.47) \qquad\qquad b_1 \equiv 0 \pmod{p} \quad \text{and} \quad D \equiv -4pa_1c_1 \pmod{p^2}.$$

We consider two subcases. First suppose that $p \nmid l$. Then $p \mid d$. We show that a_1c_1 is not divisible by p. If $p \neq 2$, then this follows immediately from the second congruence in (3.47), since d, and hence D, are not divisible by p^2. If $p = 2$, then $d = 4d_0$, where $d_0 \equiv 2, 3 \pmod{4}$. If we set $b_1 = 2b_0$, then $D = 4d_0l^2 = 4(b_0^2 - 2a_1c_1)$, and hence

$$2a_1c_1 = b_0^2 - d_0l^2 = b_0^2 - d_0 \pmod{4}.$$

The last expression cannot be divisible by 4, since $b_0^2 \equiv 0, 1 \pmod{4}$ and $d_0 \equiv 2, 3 \pmod{4}$; hence a_1c_1 is odd. Since the quadratic form q_1 with matrix A_1 is primitive (because q is primitive), and since a_1 is not divisible by p, it follows that the quadratic form q_1' with matrix A_1' is also primitive, and so it is associated with a module $M(q_1')$ of K. We consider the module $M = \left\{p, \frac{b - \sqrt{D}}{2}\right\}$. Since the number $\gamma_1 = (b_1 - \sqrt{D})/2p$ is a root of the polynomial $pv^2 - b_1v + a_1c_1$, and since a_1c_1 is not divisible by p, it follows from §2 of Appendix 3 that the ring of multipliers of the module $M = p\{1, \gamma_1\}$ is $\mathfrak{O}_l$, and $N(M) = N(p)/p = p$. Since $M \subset \mathfrak{O}_l$, we see that $M = \mathfrak{p}$ is the unique regular ideal of $\mathfrak{O}_l$ of norm p. We set $\delta_1 = p\gamma_1 = (b_1 - \sqrt{D})/2$. Since $\delta_1^2 = b_1\delta_1 - pa_1c_1$ and a_1 is not divisible by p, it follows that

$$M(q_1')\mathfrak{p} = \{a_1, \delta_1\}\{p, \delta_1\} = \{a_1 p, a_1\delta_1, p\delta_1, b_1\delta_1 - pa_1c_1\}$$
$$= \{a_1 p, \delta_1\} = M(q_1),$$

where q_1 is the quadratic form with matrix A_1. From this, if we take into account that $\bar{\mathfrak{p}} = \mathfrak{p}$ and use §2 of Appendix 3, we obtain

$$M(q_1') = p^{-1} M(q_1')\mathfrak{p}\bar{\mathfrak{p}} = p^{-1} M(q_1)\bar{\mathfrak{p}} = p^{-1} M(q_1)\mathfrak{p}.$$

Since q_1 is properly equivalent to q, it follows that the last module is similar to the module $M(q)\mathfrak{p}$, and so the matrix A_1' is properly equivalent to the matrix $A \times \mathfrak{p}$. This means that $f(mA_1') = f(m(A \times \mathfrak{p}))$, and this, together with (3.46), completes the proof of part (3).

Finally, suppose that $p \mid l$. Since b_1 is divisible by p, and the quadratic form q_1 with matrix A_1 is primitive (because q is primitive), it follows that c_1 is not divisible by p. We show that p divides a_1. If $p \neq 2$, then this follows immediately from the second congruence in (3.47). If $p = 2$, then the congruence $d \equiv 0, 1 \pmod 4$ implies that $D = dl^2 = b_1^2 - 8a_1c_1 \equiv 0, 4 \pmod{16}$. Then $(b_1/2)^2 - 2a_1c_1 \equiv 0, 1 \pmod 4$, and so $2a_1c_1 \equiv 0 \pmod 4$ and $a_1 \equiv 0 \pmod 2$. We set $A_1' = pA_2$. From the above observations we see that A_2 is an integer matrix and is even. The primitivity of the quadratic form q_1 implies that the quadratic form q_2 with matrix A_2 is primitive. The number $\gamma_1 = (b_1 - \sqrt{D})/2p$ is a root of the polynomial $v^2 - (b_1/p)v + a_1c_1/p$, which has rational integer coefficients. Hence, from §2 of Appendix 3 it follows that the module $\{1, \gamma_1\}$ is an order of the field K, having discriminant $(b_1/p)^2 - 4pa_1c_1/p^2 = D(l/p)^2$. Hence, $\{1, \gamma_1\} = \mathfrak{O}_{l/p}$. We now obtain

$$M(q_1)\mathfrak{O}_{l/p} = \{pa_1, p\gamma_1\}\{1, \gamma_1\} = \{pa_1, p\gamma_1, pa_1\gamma_1, p(b_1\gamma_1/p - a_1c_1/p)\}$$
$$= \{a_1c_1, pa_1, p\gamma_1\} = \{a_1, p\gamma_1\} = pM(q_2).$$

As before, the module $M(q_1)$ is similar to the module $M(q)$, and so $M(q_2)$ is similar to $M(q) \cdot \mathfrak{O}_{l/p}$, and the matrix A_2 is properly equivalent to the matrix $A \times \mathfrak{O}_{l/p}$; hence, $f(mA_1') = f(pmA_2) = f(pm(A \times \mathfrak{O}_{l/p}))$. $\qquad\square$

We now return to the identities in Lemma 3.11, and apply the above formulas to compute the numerators of the rational functions on the right in these identities. If we set $B = mA$, where $m \in \mathbf{N}$ is not divisible by p and $A = \begin{pmatrix} 2a & b \\ b & 2c \end{pmatrix}$ is a matrix in $\mathbf{A}_2^+$ with relatively prime a, b, c, and if we use (3.36) and (3.28), then we obtain

$$\begin{aligned}
(3.48) \qquad Q_p(v; F)d_p(v, mA) &= (f \mid (1 - \Pi_1 v + p(\Pi_{2,0}^{(1)} + \Pi_{2,0}^{(0)})v^2 \\
&\quad - \Pi_- v + \Pi_- \Pi_1 v^2 - p\Pi_- (\Pi_{2,0}^{(1)} + \Pi_{2,0}^{(0)})v^3))(mA) \\
&= f(mA) - (f \mid \Pi_1)(mA)v + p^{2k-4}\chi(p)^2 \varepsilon_p(A)f(mA)v^2 \\
&\quad + (f \mid \Pi_- \Pi_1)(mA)v^2,
\end{aligned}$$

because $mA \notin p\mathbf{A}_2$, and obviously $\varepsilon_p(mA) = \varepsilon_p(A)$. The next transformations of the numerators of the series $d_p(v, mA)$ are based on the formulas for Π_1 in Proposition 3.12; however, rather than apply the formulas to the individual $d_p(v, mA)$, it is more convenient to apply them to suitable linear combinations of these series.

We fix an integer $D < 0$, and let $A_1, \ldots, A_{h(D)}$ be a set of representatives of the proper equivalence classes (see §3 of Appendix 3) of matrices $A = \begin{pmatrix} 2a & b \\ b & 2c \end{pmatrix} \in \mathbf{A}_2^+$ with g. c. d. $(a, b, c) = 1$ and $b^2 - 4ac = D$. These equivalence classes form a finite abelian group $H(D)$ under the composition in §3 of Appendix 3. We fix an arbitrary character ξ of the group $H(D)$. Given a prime p, a natural number m, and $f \in \mathfrak{F}_s^2$, we define the formal power series

$$(3.49) \qquad d_p(v, m, \xi, D) = \sum_{i=1}^{h(D)} \xi(A_i) d_p(v, m A_i) = \sum_{\delta=0}^{\infty} f(p^\delta m, \xi) v^\delta,$$

where

$$f(t, \xi) = \sum_{i=1}^{h(D)} \xi(A_i) f(t A_i).$$

Before we sum the series (3.49), notice that, by (3.42) and (3.43), the p-sign $\varepsilon_p(A_i)$ does not depend on i, but rather depends only on D. Hence, we set

$$(3.50) \qquad \varepsilon_p(D) = \varepsilon_p(A_i) = \left(\frac{D}{p} \right), \quad \text{if } p \neq 2,$$

and

$$(3.51) \qquad \varepsilon_2(D) = \varepsilon_2(A_i) = \begin{cases} 1, & \text{if } D \equiv 1 \pmod 8, \\ -1, & \text{if } D \equiv 5 \pmod 8, \\ 0, & \text{if } D \equiv 0 \pmod 2. \end{cases}$$

LEMMA 3.13. *Suppose that the values of the function $f \in \mathfrak{F}_s^2$ are the Fourier coefficients of an eigenfunction $F \in \mathfrak{M}_s^2$ of all of the Hecke operators $|\mathbf{T} = |_{k,\chi} \mathbf{T}$ for $\mathbf{T} \in \mathbf{L}_p^2$, where p is a fixed prime. Let D be a negative integer, ξ be a character of the group $H(D)$, and m be a natural number prime to p. Then the power series $d_p(v) = d_p(v, m, \xi; D)$ is formally equal to a rational function*

$$d_p(v) = Q_p(v; F)^{-1} P_p(v),$$

whose denominator is $Q_p(v, F)$ and whose numerator $P_p(v)$ is a polynomial in v of degree at most 2. The numerator is given by the following formulas (where, as before, $\mathfrak{O}_l$ denotes the subring of index $l = \sqrt{D/d}$ in the ring of integers of $K = \mathbf{Q}(\sqrt{D})$, and d is the discriminant of the field K):
  (1) *If $\varepsilon_p(D) = 1$, then*

$$P_p(v) = (1 - (N\mathfrak{p})^{k-2} \chi(N\mathfrak{p}) \xi(\mathfrak{p}) v)(1 - (N\overline{\mathfrak{p}})^{k-2} \chi(N\overline{\mathfrak{p}}) \xi(\overline{\mathfrak{p}}) v) f(m, \xi),$$

where $\mathfrak{p}$ and $\overline{\mathfrak{p}}$ are the unique regular ideals of $\mathfrak{O}_l$ of norm p.
  (2) *If $\varepsilon_p(D) = -1$, then*

$$P_p(v) = (1 - (N\mathfrak{p})^{k-2} \chi(N\mathfrak{p}) \xi(\mathfrak{p}) v^2) f(m, \xi),$$

where $\mathfrak{p} = p\mathfrak{O}_l$ is the unique regular ideal of $\mathfrak{O}_l$ of norm p^2.
  (3) *If $\varepsilon_p(D) = 0$ and $p \nmid l$, then*

$$P_p(v) = (1 - (N\mathfrak{p})^{k-2} \chi(N\mathfrak{p}) \xi(\mathfrak{p}) v) f(m, \xi),$$

where $\mathfrak{p} = \overline{\mathfrak{p}}$ is the unique regular ideal of $\mathfrak{O}_l$ of norm p.

(4) If $\varepsilon_p(D) = 0$ and $p|l$, then

$$P_p(v) = (f|(1 - \Pi_- v)(1 - \Pi_1 v))(m, \xi) = \sum_{i=1}^{h(D)} \xi(A_i)(f|(1 - \Pi_- v)(1 - \Pi_1 v))(mA_i).$$

PROOF. First suppose that p does not divide l. Then from Proposition 3.12 and the formula (3.28) it follows that $(f|\Pi_-\Pi_1)(mA_i) = 0$ for any $i = 1, \ldots, h(D)$. Then from (3.48) and (3.50)–(3.51) we obtain

$$P_p(v) = Q_p(v; F) d_p(v) = f(m, \xi) - (f|\Pi_1)(m, \xi)v + p^{2k-4}\chi(p^2)\varepsilon_p(D)f(m, \xi)v^2.$$

If $\varepsilon_p(D) = 1$, then, using the first part of Proposition 3.12, we have

$$\begin{aligned}
(f|\Pi_1)(m, \xi) &= \sum_{i=1}^{h(D)} \xi(A_i)(f|\Pi_1)(mA_i) \\
&= p^{k-2}\chi(p) \sum_i \xi(A_i)(f(m(A_i \times \mathfrak{p})) + f(m(A_i \times \overline{\mathfrak{p}}))) \\
&= p^{k-2}\chi(p)\bigg\{ \xi(\overline{\mathfrak{p}}) \sum_i \xi(A_i \times \mathfrak{p}) f(m(A_i \times \mathfrak{p})) \\
&\qquad\qquad + \xi(\mathfrak{p}) \sum_i \xi(A_i \times \overline{\mathfrak{p}}) f(m(A_i \times \overline{\mathfrak{p}})) \bigg\} \\
&= p^{k-2}\chi(p)(\xi(\mathfrak{p}) + \xi(\overline{\mathfrak{p}}))f(m, \xi),
\end{aligned}$$

since $\xi(\mathfrak{p})\xi(\overline{\mathfrak{p}}) = \xi(\mathfrak{p}\overline{\mathfrak{p}}) = \xi(p\mathfrak{O}_l) = 1$, and the matrices $A_i \times \mathfrak{p}$ and $A_i \times \overline{\mathfrak{p}}$ each run through a complete set of representatives of the equivalence classes in $H(D)$ whenever the matrices A_i do. This implies that in the case under consideration $P_p(v)$ is equal to

$$\begin{aligned}
&(1 - p^{k-2}\chi(p)(\xi(\mathfrak{p}) + \xi(\overline{\mathfrak{p}}))v + p^{2k-4}\chi(p^2)v^2)f(m, \xi) \\
&= (1 - p^{k-2}\chi(p)\xi(\mathfrak{p})v)(1 - p^{k-2}\chi(p)\xi(\overline{\mathfrak{p}})v)f(m, \xi).
\end{aligned}$$

Now let $\varepsilon_p(D) = -1$. Then, using the second part of Proposition 3.12, we have

$$(f|\Pi_1)(m, \xi) = \sum_i \xi(A_i)(f|\Pi_-)(mA_i) = 0,$$

and so

$$P_p(v) = (1 - p^{2k-4}\chi(p^2)v^2)f(m, \xi).$$

Finally, if $\varepsilon_p(D) = 0$ (and $p\nmid l$), then, by Proposition 3.12(3), we have

$$\begin{aligned}
(f|\Pi_1)(m, \xi) &= p^{k-2}\chi(p) \sum_{i=1}^{h(D)} \xi(A_i) f(m(A_i \times \mathfrak{p})) \\
&= p^{k-2}\chi(p)\xi(\overline{\mathfrak{p}}) \sum_i \xi(A_i \times \mathfrak{p}) f(m(A_i \times \mathfrak{p})) = p^{k-2}\chi(p)\xi(\mathfrak{p})f(m, \xi),
\end{aligned}$$

since $\overline{\mathfrak{p}} = \mathfrak{p}$; thus,

$$P_p(v) = (1 - p^{k-2}\chi(p)\xi(\mathfrak{p})v)f(m, \xi).$$

Now suppose that $\varepsilon_p(D) = 0$ and $p|l$. Since $(f|\Pi_-)(mA_i) = 0$, the formula (3.48) for $A = A_i$ can be rewritten in the form

$$Q_p(v; F)d_p(v, mA_i) = f(mA_i) - (f|\Pi_1)(mA_i)v + (f|\Pi_-\Pi_1)(mA_i)v^2$$
$$= (f|(1 - \Pi_-v)(1 - \Pi_1v))(mA_i).$$

If we multiply this relation by $\xi(A_i)$ and sum over i, we obtain the last formula in the lemma. $\square$

We are near the end of our examination of the multiplicative properties of the Fourier coefficients of modular forms of degree 2. It remains for us to bring together the local information into a single global picture.

THEOREM 3.14. *Let*

$$F(Z) = \sum_{A \in \mathbf{A}_2} f(A)e\{AZ\} \in \mathfrak{M}_k^2(q, \chi)$$

be a nonzero modular form of integer weight k and one-dimensional character $[\chi]$ for the group $\Gamma_0^2(q)$. Suppose that F is an eigenfunction for all of the Hecke operators $|T = |_{k,\chi}T$ for $T \in L^2(q)$. Further suppose that D is an arbitrary negative integer, $l \in \mathbf{N}$ is determined from the condition $D = dl^2$, where d is the discriminant of the field $K = \mathbf{Q}(\sqrt{D})$, and ξ is an arbitrary character of the class group $H(D)$, regarded as an abstract group.

Then the following formal identity holds for any natural number a such that $a|q^\infty$ and for any completely multiplicative function $\gamma\colon \mathbf{N}_{(q)} \to \mathbf{C}$:

(3.52)
$$\sum_{i=1}^{h(D)} \xi(A_i) \sum_{m \in \mathbf{N}_{(q)}} \frac{f(maA_i)\gamma(m)}{m^s}$$
$$= \rho(s) \left\{ \prod_{\mathfrak{p}, N\mathfrak{p} \not\mid (ql)^2} \left(1 - \frac{\chi(N\mathfrak{p})\gamma(N\mathfrak{p})\xi(\mathfrak{p})}{(N\mathfrak{p})^{s-k+2}} \right) \right\} \zeta(s, \gamma; F),$$

where $A_1, \ldots, A_{h(D)}$ is an arbitrary set of representatives of the elements of $H(D)$, regarded as proper equivalence classes of the matrices of positive definite primitive integral binary quadratic forms of discriminant D; where $\rho(s) = \rho(s, a, D, \xi, \gamma; F)$ is a finite sum given by one of the expressions

(3.53)
$$\rho(s) = \sum_{i=1}^{h(D)} \xi(A_i) \left(f \left| \prod_{\substack{p \in \mathbf{P}_{(q)} \\ p|l}} \left(1 - \frac{\Pi_0^2(p)\gamma(p)}{p^s} \right) \right. \right.$$
$$\left. \left. \times \left(1 - \frac{\Pi_1^2(p)\gamma(p)}{p^s} \right) \right) (aA_i) \right.$$
$$= \sum_{i=1}^{h(D)} \xi(A_i) \sum_{\substack{\delta, \delta_1 \in \mathbf{N}_{(q)} \\ \delta|\delta_1|l}} \frac{\chi(\delta^2\delta_1)\gamma(\delta\delta_1)\mu(\delta)\mu(\delta_1)}{\delta^{s-2k+3}\delta_1^{s-k+2}} f\left(\frac{a\delta_1}{\delta}(A_i \times \mathfrak{O}_{1/\delta_1}) \right),$$

in which μ is the Möbius function, $\mathfrak{O}_t$ denotes the subring of index t in the ring of integers of K, and the composition $\times$ is understood in the sense of §3 of Appendix 3; where $\mathfrak{p}$ in the product in (3.52) runs through all regular prime ideals of $\mathfrak{O}_l$ whose norm $N\mathfrak{p}$ is prime *to ql; and where $\zeta(s, \gamma; F)$ is the Euler product (3.23) corresponding to the eigenfunction F.*

Further suppose that $k \geqslant 0$, $f(A) \neq 0$ for some nondegenerate matrix $A \in \mathbf{A}_2^+$, and $|\gamma(m)| \leqslant cm^\sigma$ for all $m \in \mathbf{N}_{(q)}$, where c and σ are real numbers that do not depend on m. Then the Dirichlet series on the left in (3.52) and the infinite products on the right in this identity converge absolutely and uniformly in any right half-plane of the complex variable s of the form $\operatorname{Re} s > 2\rho k + \sigma + 1 + \varepsilon$ with $\varepsilon > 0$, where $\rho = 1$ in the general case and $\rho = 1/2$ if F is a cusp-form; and the resulting holomorphic functions on the indicated half-planes are connected by the identity (3.52).

PROOF. Let $p_1, \ldots, p_b$ be all of the distinct prime divisors of l that are prime to q, and let $p_{b+1}, p_{b+2}, \ldots$ be the sequence of all primes in $\mathbf{P}_{(ql)}$ arranged in increasing order. We first prove by induction on c that for all $c \in \mathbf{N}$ one has the formal identity

(3.54)
$$
\sum_{m \in \mathbf{N}_{(q)}} \frac{f(am, \xi)\gamma(m)}{m^s} = \left\{ \sum_{m \in \mathbf{N}_{(q \cdot q(c))}} \frac{f(am, \xi)\gamma(m)}{m^s} \right\}
$$
$$
\times \left\{ \prod_{\mathfrak{p}, N\mathfrak{p}|q(c)^2} \left(1 - \frac{\chi(N\mathfrak{p})\gamma(N\mathfrak{p})\xi(\mathfrak{p})}{(N\mathfrak{p})^{s-k+2}} \right) \right\} \prod_{p|q(c)} Q_p\left(\frac{\gamma(p)}{p^s}; F \right)^{-1},
$$

where

$$
q(c) = p_{b+1} \cdots p_{b+c}, \quad f(am, \xi) = \sum_i \xi(A_i) f(amA_i),
$$

and $\mathfrak{p}$ in the middle term on the right runs through all regular prime ideals of $\mathfrak{O}_l$ whose norm satisfies the condition given. If $c = 1$, then, setting $p_{b+1} = p$ and $\gamma(p)p^{-s} = v$ and using the identity in Lemma 3.13 that corresponds to the value of $\varepsilon_p(D)$, we obtain

(3.55)
$$
Q_p(\gamma(p)p^{-s}; F) \sum_{m \in \mathbf{N}_{(q)}} \frac{f(am, \xi)\gamma(m)}{m^s}
$$
$$
= \sum_{m \in \mathbf{N}_{(qp)}} \frac{\gamma(m)}{m^s} Q_p(\gamma(p)p^{-s}; F) \sum_{\delta=0}^{\infty} f(amp^\delta, \xi)(\gamma(p)p^{-s})^\delta
$$
$$
= \left\{ \sum_{m \in \mathbf{N}_{(qp)}} \frac{f(am, \xi)\gamma(m)}{m^s} \right\} \prod_{\mathfrak{p}, N\mathfrak{p}|p^2} \left(1 - (N\mathfrak{p})^{k-2}\chi(N\mathfrak{p})\xi(\mathfrak{p})\frac{\gamma(N\mathfrak{p})}{(N\mathfrak{p})^s} \right),
$$

where, according to §2 of Appendix 3 and Lemma 3.13, $\mathfrak{p}$ runs through all regular prime ideals of $\mathfrak{O}_l$ of norm p or p^2; this proves (3.54) in the case under consideration. Suppose that (3.54) has been proved for some value of c. If we set $p_{b+c+1} = p$ and use the identity (3.55) with $q \cdot q(c)$ in place of q, we obtain

$$Q_p(\gamma(p)p^{-s}; F) \sum_{m \in \mathbf{N}_{(q)}} \frac{f(am, \xi)\gamma(m)}{m^s}$$

$$= Q_p(\gamma(p)p^{-s}; F) \left\{ \sum_{m \in \mathbf{N}_{(q \cdot q(c))}} \frac{f(am, \xi)\gamma(m)}{m^s} \right\}$$

$$\times \left\{ \prod_{\mathfrak{p}, N\mathfrak{p}|q(c)^2} \left(1 - \frac{\chi(N\mathfrak{p})\gamma(N\mathfrak{p})\xi(\mathfrak{p})}{(N\mathfrak{p})^{s-k+2}} \right) \right\} \prod_{p|q(c)} Q_p(\gamma(p)p^{-s}; F)^{-1}$$

$$= \left\{ \sum_{m \in \mathbf{N}_{(q \cdot q(c+1))}} \frac{f(am, \xi)\gamma(m)}{m^s} \right\} \left\{ \prod_{\mathfrak{p}, N\mathfrak{p}|q(c+1)^2} \left(1 - \frac{\chi(N\mathfrak{p})\gamma(N\mathfrak{p})\xi(\mathfrak{p})}{(N\mathfrak{p})^{s-k+2}} \right) \right\}$$

$$\times \prod_{p|q(c)} Q_p(\gamma(p)p^{-s}; F)^{-1},$$

which implies that (3.54) holds for $c + 1$.

If we now take the limit as $c \to \infty$ in (3.54) coefficient by coefficient, we obtain the identity

$$\sum_{m \in \mathbf{N}_{(q)}} \frac{f(am, \xi)\gamma(m)}{m^s} = \left\{ \sum_{m|(p_1 \dots p_b)^\infty} \frac{f(am, \xi)\gamma(m)}{m^s} \right\}$$

$$\times \left\{ \prod_{\mathfrak{p}, N\mathfrak{p} \nmid (ql)^2} \left(1 - \frac{\chi(N\mathfrak{p})\gamma(N\mathfrak{p})\xi(\mathfrak{p})}{(N\mathfrak{p})^{s-k+2}} \right) \right\} \prod_{p \in \mathbf{P}_{(ql)}} Q_p(\gamma(p)p^{-s}; F)^{-1}.$$

Thus, to prove (3.52) with the factor ρ given by the first formula in (3.53), it remains to verify that

$$(3.56) \qquad \left\{ \prod_{p|p_1 \dots p_r} Q_p(\gamma(p)p^{-s}; F) \right\} \left\{ \sum_{m|(p_1 \dots p_r)^\infty} \frac{f(am, \xi)\gamma(m)}{m^s} \right\} = \rho_r(s),$$

where

$$\rho_r(s) = \sum_{i=1}^{h(D)} \xi(A_i) \left(f | \prod_{p|p_1 \dots p_r} \left(1 - \frac{\Pi_0^2(p)\gamma(p)}{p^s} \right) \left(1 - \frac{\Pi_0^2(p)\gamma(p)}{p^s} \right) \right) (A_i)$$

and $p_1, \dots, p_r$ are the distinct primes in $\mathbf{P}_{(q)}$ that divide l. We use induction on r. If $r = 1$, then (3.56) holds by Lemma 3.13(4). Now suppose that (3.56) has been proved for r primes in $\mathbf{P}_{(q)}$ that divide l, and let $p_{r+1} = p$ be another such prime. Using the induction assumption, we obtain

$$\left\{ \prod_{p|p_1 \dots p_r} Q_p\left(\frac{\gamma(p)}{p^s}; F \right) \right\} Q_p\left(\frac{\gamma(p)}{p^s}; F \right) \sum_{m|(p_1 \dots p_r p)^\infty} \frac{f(am, \xi)\gamma(m)}{m^s}$$

$$= Q_p\left(\frac{\gamma(p)}{p^s}; F \right) \sum_{\delta=0}^{\infty} \left\{ \prod_{p|p_1 \dots p_r} Q_p\left(\frac{\gamma(p)}{p^s}; F \right) \right.$$

$$(3.57) \qquad \qquad \left. \times \sum_{m|(p_1 \dots p_r)^\infty} \frac{f(amp^\delta, \xi)\gamma(m)}{m^s} \right\} \left(\frac{\gamma(p)}{p^s} \right)$$

$$= Q_p\left(\frac{\gamma(p)}{p^s}; F\right) \sum_{\delta=0}^{\infty} f'(ap^\delta, \xi)\left(\frac{\gamma(p)}{p^s}\right)^\delta,$$

where

$$f'(ap^\delta, \xi) = \sum_{i=1}^{h(D)} \xi(A_i)f'(ap^\delta A_i)$$

and

$$f'(A) = \left(f\left|\prod_{p|p_1\ldots p_r}\left(1 - \frac{\Pi_0^2(p)\gamma(p)}{p^s}\right)\left(1 - \frac{\Pi_1^2(p)\gamma(p)}{p^s}\right)\right)\right.(A).$$

It is clear that for any fixed value of s the function f' is contained in the space $\mathfrak{F}_s^2$. By Proposition 5.12 of Chapter 3, in the Hecke ring $\mathbf{L}_0^2$ every element of $\mathbf{L}_p^2$ commutes with any of the elements $\Pi_0^2(p_j) \in C_{-p_j}^2$ $(j = 1, \ldots, r)$. Since $\mathbf{L}^2$ is a commutative ring, the elements of $\mathbf{L}_p^2$ commute with all elements of $\mathbf{L}_{p_j}^2$, and in particular with $\mathbf{T}^2(p_j) = \Pi_0^2(p_j) + \Pi_1^2(p_j) + \Pi_2^2(p_j)$ (see Proposition 5.14 of Chapter 3); from this, if we again use Proposition 5.12 of Chapter 3, we conclude that every element of $\mathbf{L}_p$ commutes with any of the elements $\Pi_1^2(p_j) = \mathbf{T}^2(p_j) - \Pi_0^2(p_j) - \Pi_2^2(p_j)$. From these observations it follows that the function $f' \in \mathbf{F}_s^2$, along with f, is an eigenfunction for all of the Hecke operators corresponding to elements of $\mathbf{L}_p^2$, and it has the same eigenvalues as f. Thus, we can compute the expression in (3.57) using Lemma 3.13(4), according to which it is equal to

$$\left(f'\left|\left(1 - \frac{\Pi_0^2(p)\gamma(p)}{p^s}\right)\left(1 - \frac{\Pi_1^2(p)\gamma(p)}{p^s}\right)\right)\right.(a, \xi),$$

and this proves (3.56) for $r + 1$ primes.

The above argument proves the formal identity (3.52) with factor $\rho(s) = \rho_b(s)$ given by the first formula in (3.53). To prove the second formula for $\rho(s)$ we note that, by Proposition 5.12 of Chapter 3, the elements $\Pi_0^2(p_i)$ and $\Pi_1^2(p_j) = \mathbf{T}^2(p_j) - \Pi_0^2(p_j) - \Pi_2^2(p_j)$ (see (5.49) of Chapter 3) commute with one another if $p_i \neq p_j$. Hence,

$$\prod_{i=1}^{b}\left(1 - \frac{\Pi_0^2(p_i)\gamma(p_i)}{p_i^s}\right)\left(1 - \frac{\Pi_1^2(p_i)\gamma(p_i)}{p_i^s}\right)$$

$$= \prod_{i=1}^{b}\left(1 - \frac{\Pi_0^2(p_i)\gamma(p_i)}{p_i^s}\right)\prod_{i=1}^{b}\left(1 - \frac{\Pi_1^2(p_i)\gamma(p_i)}{p_i^s}\right)$$

$$= \left(\sum_{\delta|p_1\ldots p_b}\frac{\mu(\delta)\Pi_0(\delta)\gamma(\delta)}{\delta^s}\right)\left(\sum_{\delta_1|p_1\ldots p_b}\frac{\mu(\delta_1)\Pi_1(\delta_1)\gamma(\delta_1)}{\delta_1^s}\right)$$

$$= \sum_{\delta,\delta_1|p_1\ldots p_b}\frac{\mu(\delta)\mu(\delta_1)\gamma(\delta\delta_1)}{(\delta\delta_1)^s}\Pi_0(\delta)\Pi_1(\delta_1),$$

where for squarefree δ and δ_1 we set $\Pi_0(\delta) = \prod_{p|\delta}\Pi_0^2(p)$ and $\Pi_1(\delta_1) = \prod_{p|\delta_1}\Pi_1^2(p)$ (it follows from (5.13), (5.49), and Proposition 5.12 of Chapter 3 that the order of the primes makes no difference). Using Proposition 3.12(4) and induction on the number of prime divisors of the squarefree number δ_1 dividing l, we easily derive the formulas (see §2 of Appendix 3)

$$(f|_{k,\chi}\Pi_1(\delta_1))(aA_i) = \delta_1^{k-2}\chi(\delta_1)f(a\delta_1(A_i \times \mathfrak{D}_{l/\delta_1})),$$

and from this and (3.28) we obtain the following relations for $i = 1, \ldots, h(D)$:

$$(f|_{k,\chi}\Pi_0(\delta)\Pi_1(\delta_1))(aA_i) = (f|\Pi_0(\delta)|\Pi_1(\delta_1))(aA_i)$$
$$= \begin{cases} \delta^{2k-3}\delta_1^{k-2}\chi(\delta^2\delta_1)f\left(\frac{a\delta_1}{\delta}(A_i \times \mathfrak{D}_{l/\delta_1})\right), & \text{if } \delta|\delta_1, \\ 0, & \text{if } \delta \nmid \delta_1. \end{cases}$$

The second expression for $\rho(s)$ follows from the first expression and the above formulas. The identity (3.52) is proved.   □

It remains to examine the convergence of the series and products in (3.52). According to (3.35) of Chapter 2 ((3.70) of Chapter 2 if F is a cusp-form) we have the inequalities $|f(mA_i)| \leqslant \gamma_F D^{k\rho}m^{2k\rho}$ ($m \in \mathbf{N}$), where γ_F depends only on F. This implies that the Dirichlet series on the left in (3.52) converges absolutely and uniformly in any of the half-planes indicated in the theorem. The infinite product over $\mathfrak{p}$ on the right of (3.52) converges absolutely and uniformly in any right half-plane of the variable s of the form $\operatorname{Re} s \geqslant k + \sigma - 1 + \varepsilon$ with $\varepsilon > 0$, since the following estimate is a consequence of the description in Appendix 3.2 of the regular prime ideals of $\mathfrak{D}_l$ whose norm does not divide l:

$$\sum_{\mathfrak{p},\, N\mathfrak{p} \nmid (ql)^2} \left|\frac{\chi(N\mathfrak{p})\gamma(N\mathfrak{p})\xi(\mathfrak{p})}{(N\mathfrak{p})^{s-k+2}}\right| \leqslant \sum_{\mathfrak{p},\, N\mathfrak{p}=p} \frac{p^\sigma}{p^{\operatorname{Re} s - k + 2}} + \sum_{\mathfrak{p},\, N\mathfrak{p}=p^2} \frac{p^{2\sigma}}{p^{2(\operatorname{Re} s - k + 2)}}$$

$$\leqslant 2\sum_{p\in\mathbf{P}} \frac{1}{p^{\operatorname{Re} s - (k+\sigma-1)+1}} + \sum_{p\in\mathbf{P}} \frac{1}{p^{2(\operatorname{Re} s - (k+\sigma-1))+2}}$$

(the norm of any prime ideal of $\mathfrak{D}_l$ is obviously either p or p^2, where $p \in \mathbf{P}$). In addition, in any of the indicated half-planes the modulus of the product is clearly bounded from below by a positive constant. From these observations and from the formal identity

$$(3.58) \quad \rho(s)\zeta(s,\gamma;F) = \left\{\prod_{\mathfrak{p}}\left(1 - \frac{\chi(N\mathfrak{p})\gamma(N\mathfrak{p})\xi(\mathfrak{p})}{(N\mathfrak{p})^{s-k+1}}\right)\right\}^{-1} \sum_{m\in\mathbf{N}_{(q)}} \frac{f(am,\xi)\gamma(m)}{m^s}$$

it follows that the product $\rho(s)\zeta(s;\gamma,F)$, regarded as a Dirichlet series, converges absolutely and uniformly in any of the half-planes indicated in the theorem. From this we cannot immediately conclude anything about the convergence of the Dirichlet series for $\zeta(s,\gamma;F)$, since the factor $\rho(s)$ could turn out to be identically zero. However, since the product $\zeta(s,\gamma;F)$ does not depend on D, ξ, or a, we might try to choose these parameters in such a way that $\rho(s) = \rho(s,a,D,\xi,\gamma,F) \neq 0$. From the conditions in the theorem it obviously follows that there exist an integer $D < 0$, a character ξ of the group $H(D)$, $m \in \mathbf{N}_{(q)}$, and $a \in \mathbf{N}$ with $a|q^\infty$ such that (in the notation of the theorem) we have

$$(3.59) \qquad\qquad f(am,\xi) = \sum_{i=1}^{h(D)} \xi(A_i)f(amA_i) \neq 0.$$

Let D_0 denote the smallest such D in absolute value, and let ξ_0 and a_0 denote the corresponding ξ and a. Using the second formula in (3.53) and the minimality of D_0, we find that

$$\rho_0 = \rho(s, a_0, D_0, \xi_0, \gamma, F) = \sum_{i=1}^{h(D_0)} \xi_0(A_i) f(a_0 A_i) = f(a_0, \xi_0),$$

since, according to (3.6) and §2 of Appendix 3, the quadratic form with matrix $A_i \times \mathfrak{D}_{l/\delta_1}$ has discriminant equal to D_0/δ_1^2. If we write out the identity (3.52) for $D = D_0, \xi = \xi_0$, $a = a_0$, and $\gamma = 1$, and if we take into account that, by (3.59), the series on the left is not formally equal to zero, then we conclude that $f(a_0, \xi_0) = \rho_0 \neq 0$. Thus, the factor $\rho(s)$ in the identity (3.52) for $D = D_0, \xi = \xi_0$, and $a = a_0$ is equal to a nonzero constant. Hence, the Dirichlet series for $\zeta(s, \gamma; F)$, along with the corresponding Euler product, converges absolutely and uniformly in the same regions as the Dirichlet series on the right. $\qquad\square$

The identities in (3.52) are analogous to those in (3.14). We call $\zeta(s, \gamma; F)$ the *zeta-function with character γ that is associated to the eigenfunction F*.

PROBLEM 3.15. Let $F \in \mathfrak{M}_s^n$, and let $f(A)$ for $A \in \mathbf{A}_n$ be the Fourier coefficients of F. Suppose that for some prime p the modular form F is an eigenfunction of all of the Hecke operators $|\mathbf{T} = |_{k,\chi}\mathbf{T}$ for $\mathbf{T} \in \mathbf{L}_p^n$, where $(-1)^k \chi(-1) = s(-1)$, with eigenvalues $\lambda(\mathbf{T}) = \lambda(\mathbf{T}, F)$. Set

$$Q_p(v; F) = \sum_{j=0}^{2^n} (-1)^j \gamma(\mathbf{q}_j^n(p)) v^j,$$

where $\mathbf{q}_j^n(p) \in \mathbf{L}_p^n$ are the coefficients of the polynomial (3.78) of Chapter 3, and let $\alpha_i = \alpha_i(p, F)$ for $i = 0, 1, \ldots, n$ be the parameters of the homomorphism $\mathbf{T} \to \lambda(\mathbf{T})$ in the sense of Proposition 3.36 of Chapter 3. Prove the following:

(1) (Zharkovskaia) The following formal identity holds for any fixed matrix $A \in \mathbf{A}_n$:

$$\sum_{\delta=0}^{\infty} f(p^\delta A) v^\delta = Q_p(v; F)^{-1} P(v),$$

where $P(v) = P_p(v; A, F)$ is a polynomial in v of degree at most $2^n - 1$.

[**Hint:** Use Proposition 6.10 of Chapter 3.]

(2) The polynomial $Q_p(v; F)$ decomposes into linear factors as follows:

$$Q_p(v; F) = (1 - \alpha_0 v) \prod_{r=1}^{n} \prod_{1 \leqslant i_1 < \cdots < i_r \leqslant n} (1 - \alpha_0 \alpha_{i_1} \cdots \alpha_{i_r} v),$$

where $\alpha_i = \alpha_i(p, F)$; the parameters α_i satisfy the relation

$$\alpha_0^2 \alpha_1 \cdots \alpha_n = \chi(p)^n p^{nk - \langle n \rangle};$$

in particular, if all of the roots of $Q_p(v; F)$ have the same absolute value, then this absolute value is equal to $p^{-(nk - \langle n \rangle)/2}$, and the α_i satisfy the conditions

$$|\alpha_0| = p^{(nk - \langle n \rangle)/2}, \quad |\alpha_1| = \cdots = |\alpha_n| = 1.$$

(3) (Maass–Zharkovskaia) Suppose that $n > 1$ and either $k \neq n$, or else $k = n$ and $\chi(p) \neq -1$. Then the modular form $F|\Phi \in \mathfrak{M}_s^{n-1}$, where Φ is the Siegel operator,

is an eigenfunction for all of the Hecke operators $|_{k,\chi}\mathbf{T}'$ for $\mathbf{T}' \in \mathbf{L}_p^{n-1}$. Supposing that $F|\Phi \neq 0$, we let $\lambda(\mathbf{T}', F|\Phi)$ denote the corresponding eigenvalue, and we let $\alpha_i = \alpha_i(p, F|\Phi)$ for $i = 0, 1, \ldots, n-1$ denote the parameters of the homomorphism $\mathbf{T}' \to \lambda(\mathbf{T}', F|\Phi)$. Then we can take

$$(\alpha_0(p, F), \alpha_1(p, F), \ldots, \alpha_n(p, F)) = (\alpha_0 p^{k-n}\chi(p), \alpha_1, \ldots, \alpha_{n-1}, p^{n-k}\overline{\chi}(p)).$$

In particular, we have the relation

$$Q_p(v; F) = Q_p(v; F|\Phi)Q_p(p^{k-n}\chi(p)v; F|\Phi).$$

[**Hint:** Apply Theorem 2.12, Proposition 2.13, and Problem 2.17.]

PROBLEM 3.16. Let $F \in \mathfrak{M}_k^2(q, \chi)$, $F \neq 0$, where $k, q \in \mathbf{N}$ and χ is a Dirichlet character modulo q. Suppose that F is an eigenfunction for all of the Hecke operators $|T = |_{k,\chi}T$ for $T \in L^2(q)$ with eigenvalues $\lambda(T) = \lambda(T, F)$. For $p \in \mathbf{P}_{(q)}$ let $\alpha_0(p, F)$, $\alpha_1(p, F)$, $\alpha_2(p, F)$ be the parameters of the homomorphism $T \to \lambda(T, F)$ of the ring $L_p^2 \simeq \mathbf{L}_p^2$. Prove that:

(1) One has the inequalities

$$\max(|\alpha_0(p, F)|, |\alpha_0(p, F)\alpha_1(p, F)|, |\alpha_0(p, F)\alpha_2(p, F)|,$$
$$|\alpha_0(p, F)\alpha_1(p, F)\alpha_2(p, F)|) \leqslant p^{2\rho k},$$

where ρ is the same as in Theorem 3.14.

(2) If $F|\Phi \neq 0$, where Φ is the Siegel operator, and if either $k \neq 2$ or else $k = 2$ and $\chi(p) \neq -1$ for all $p \in \mathbf{P}_{(q)}$, then the modular form $F|\Phi \in \mathfrak{M}_k^1(q, \chi)$ is an eigenfunction for all of the Hecke operators $|T$ for $T \in L^1(q)$, and the zeta-function (3.23) associated to F can be expressed in terms of the zeta-function associated to $F|\Phi$

$$\zeta(s, \gamma; F|\Phi) = \sum_{m \in \mathbf{N}_{(q)}} \frac{\lambda(T^1(m), F|\Phi)\gamma(m)}{m^s}$$

$$= \prod_{p \in \mathbf{P}_{(q)}} \left(1 - \frac{\lambda(T^1(p), F|\Phi)\gamma(p)}{p^s} + \frac{\chi(p)\gamma(p^2)}{p^{2s-k+1}}\right)^{-1},$$

in the following way:

$$\zeta(s, \gamma; F) = \zeta(s, \gamma; F|\Phi)\zeta(s - k + 2, \chi\gamma; F|\Phi).$$

(3) Suppose that $q = 1$, $\chi = 1$, and F is not a cusp-form. Then the zeta-function $\zeta(s, F) = \zeta(s, 1; F)$ has a meromorphic continuation to the entire s-plane, it is holomorphic except possibly at $s = 0, k - 2, k, 2k - 2$, where it can have at most simple poles, and it satisfies the functional equation

$$\Psi(2k - 2 - s, F) = (-1)^k \Psi(s, F),$$

where

$$\Psi(s, F) = (2\pi)^{-2s}\Gamma(s)\Gamma(s - k + 2)\zeta(s, F)$$

and $\Gamma(s)$ is the gamma-function.

3. Modular forms of arbitrary degree and even zeta-functions. The point of departure for the theory presented below is the symmetric factorization of the polynomials $\mathbf{R}_p^n(v)$ and $\widehat{\mathbf{R}}_p^n(v)$ in §6.5 of Chapter 3. Since the coefficients of these polynomials lie in the even subrings of the corresponding Hecke rings, we begin by making some remarks about the even subrings.

Let

$$(3.60) \qquad E^n(q) = D_q(\Gamma_0^n(q), S^n(q)^+),$$

where $S^n(q)^+$ is the subgroup of $S^n(q)$ consisting of matrices M for which $r(M)$ is the square of a rational number, be the *even subring* of $L^n(q)$. By analogy with Theorem 3.12 of Chapter 3, one easily verifies that the ring $E^n(q)$ is generated by the even subrings (see (4.37) of Chapter 3)

$$(3.61) \qquad E_p^n(q) = D_Q(\Gamma_0^n(q), S_p^n(q)^+) = L_p^n(q) \cap E^n(q)$$

of the rings $L_p^n(q)$ for $p \in \mathbf{P}_{(q)}$. The imbedding ε_q in (5.3) of Chapter 3 maps each $E_p^n(q)$ isomorphically onto the even subring $\mathbf{E}_p^n$ of $\mathbf{L}_p^n \subset \mathbf{L}_{0,p}^n$. Using the definition of the spherical map $\Omega = \Omega_p^n$ and the fact that it is a monomorphism on $\mathbf{L}_p^n$, we see that the image of $\mathbf{E}_p^n$ under this map is the subset of polynomials in $\Omega(\mathbf{L}_p^n)$ having even degree in the variable x_0. In particular, all of the coefficients of the polynomial (3.52) of Chapter 3 lie in $\Omega(\mathbf{E}_p^n)$, and hence their preimages in $\mathbf{L}_p^n$—i.e., the coefficients $\mathbf{r}_a^n(p)$ of the polynomial $\mathbf{R}_p^n(v)$—lie in $\mathbf{E}_p^n$:

$$(3.62) \qquad \mathbf{R}_p^n(v) \in \mathbf{E}_p^n[v].$$

On the other hand, from (3.56)–(3.57) and Theorem 3.30 of Chapter 3 it follows that

$$\Omega(\mathbf{E}_p^n) = \mathbf{Q}[(t)^2, r_0^{\pm 1}, r_1, \dots, r_{n-1}],$$

and hence

$$\Omega(\mathbf{L}_p^n) = \mathbf{Q}[t, r_0^{\pm 1}, r_1, \dots, r_{n-1}] = \Omega(\mathbf{E}_p^n)[t] \quad \text{with } t^2 \in \Omega(\mathbf{E}_p^n).$$

Returning to the preimages, we conclude that $\mathbf{L}_p^n$ is an extension of degree two of the ring $\mathbf{E}_p^n$, or more precisely:

$$\mathbf{L}_p^n = \mathbf{E}_p^n \oplus \mathbf{T}\mathbf{E}_p^n, \quad \text{where } \mathbf{T} = \Omega^{-1}(t) \in \mathbf{L}_p^n, \mathbf{T}^2 \in \mathbf{E}_p^n.$$

This decomposition implies that any nonzero $\mathbf{Q}$-linear homomorphism λ from $\mathbf{E}_p^n$ to $\mathbf{C}$ extends to a nonzero $\mathbf{Q}$-linear homomorphism $\lambda' \colon \mathbf{L}_p^n \to \mathbf{C}$ if we set $\lambda'(\mathbf{T}) = \pm\lambda(\mathbf{T}^2)^{1/2}$, and this extension is unique except for the choice of sign in front of the square root. Let $\alpha_0, \alpha_1, \dots, \alpha_n$ be the parameters of λ' (see Proposition 3.36 of Chapter 3). Since the choice of sign in $\lambda'(\mathbf{T})$ affects only the sign on the right in the last equation in the system of equations (3.83) of Chapter 3 that determines the parameters $\alpha_0, \alpha_1 \dots, \alpha_n$, it follows that replacing $\lambda'(\mathbf{T})$ by $-\lambda'(\mathbf{T})$ leads to a change from α_0 to $-\alpha_0$ and otherwise leaves the parameters fixed. The numbers $\pm\alpha_0, \alpha_1, \dots, \alpha_n$ are called the *parameters of the homomorphism* $\lambda \colon \mathbf{E}_p^n \to \mathbf{C}$.

In the case of the Hecke ring $\widehat{L}^n(q)$, from the very beginning we singled out the even subring $\widehat{E}^n(q, \varkappa)$ that is generated by the local subrings $\widehat{E}_p^n(q, \varkappa)$ for all $p \in \mathbf{P}_{(q)}$ (see (4.83) and (5.4) of Chapter 3). These subrings and their isomorphic images $\widehat{\mathbf{E}}_p^n(q, \varkappa)$ in the ring $\overline{\mathbf{L}}_{0,p}^n$ have all of the same properties as the rings $E_p(q)$ and $\mathbf{E}_p^n$. For example, by Proposition 4.22 of Chapter 3, any **Q**-linear homomorphism λ from $\widehat{\mathbf{E}}_p^n(q, \varkappa)$ to **C** is also determined by the parameters $\pm\alpha_0, \alpha_1, \ldots, \alpha_n$, and the polynomial $\widehat{\mathbf{R}}^n(v)$ (see (6.27) of Chapter 3) has the property that

$$(3.63) \qquad\qquad \widehat{\mathbf{R}}_p^n(v) \in \widehat{\mathbf{E}}_p^n(q, \varkappa)[v].$$

We introduce some definitions and notation. Suppose that the character s of the group $\{\pm 1\}$, the integer or half-integer $w = k$ or $k/2$, and the Dirichlet character χ are connected by the relation $s(-1) = \chi_w(-1)$, where χ_w is the character (2.5). If $F \in \mathfrak{M}_s^n$ is an eigenfunction for all of the Hecke operators $|\tau = |_{w,\chi}\tau$, where $\tau \in \mathbf{E}_p^n$ for $w = k$ and $\tau \in \widehat{\mathbf{E}}_p^n(q, \varkappa)$ for $w = k/2$, with eigenvalues $\lambda(\tau, F)$, then by (3.62)–(3.63) we can associate to F the polynomial with complex coefficients

$$F_p^n(v, F) = \sum_{a=0}^{2n} (-1)^a \lambda(\mathbf{r}_a^n(p), F)v^a$$

$$(3.64) \qquad\qquad\qquad\qquad \text{or}$$

$$\widehat{R}_p^n(v, R) = \sum_{a=0}^{2n} (-1)^a \lambda(\widehat{\mathbf{r}}_a^n(p), F)v^a.$$

Similarly, for any eigenfunction $f \in \mathfrak{F}_s^n$ we define the polynomial

$$R_p^n(v, f) = \sum_{a=0}^{2n} (-1)^a \lambda(\mathbf{r}_a^n(p), f)v^a$$

$$(3.65) \qquad\qquad\qquad\qquad \text{or}$$

$$\widehat{R}_p^n(v, f) = \sum_{a=0}^{2n} (-1)^a \lambda(\widehat{\mathbf{r}}_a^n(p), f)v^a.$$

If the values of f are the Fourier coefficients of a modular form F, and if $\pm\alpha_0, \alpha_1, \ldots, \alpha_n$ are the parameters of the homomorphism $\tau \to \lambda(\tau, F) = \lambda(\tau, f)$, then from the definitions we easily find that

$$R_p^n(v, F) = R_p^n(v, f) = \prod_{i=1}^n (1 - \alpha_i(p)^{-1}v)(1 - \alpha_i(p)v),$$

$$(3.66)$$

$$\widehat{R}_p^n(v, F) = \widehat{R}_p^n(v, f) = \prod_{i=1}^n (1 - \alpha_i(p)^{-1}v)(1 - \alpha_i(p)v).$$

Now let $F \in \mathfrak{M}_w^n(q, \chi)$ be a modular form of weight w and one-dimensional character $[\chi]$ for $\Gamma_0^n(q)$, and suppose that F is an eigenfunction for all of the even Hecke operators $|\tau = |_{w,\chi}\tau$, where $\tau \in E^n(q)$ for $w = k$ and $\tau \in \widehat{E}^n(a, \chi)$ for $w = k/2$. If we regard F as an element of $\mathfrak{M}_s^n$ with $s(-1) = \chi_w(-1)$, we see that F is an eigenfunction for all of the operators $|\tau$ for $\tau \in \mathbf{E}_p^n$ or $\tau \in \widehat{\mathbf{E}}_p^n(q, \varkappa)$ for $p \in \mathbf{P}_{(q)}$. Thus, for each such p we have a corresponding polynomial (3.64). If $\gamma: \mathbf{N}_{(q)} \to \mathbf{C}$ is an arbitrary completely multiplicative function, then we call the function

$$(3.67) \qquad \zeta_p^+(s, \gamma; F) = R_p\left(\frac{\gamma(p)}{p^s}, F\right)^{-1} \quad \text{or} \quad \widehat{R}_p\left(\frac{\gamma(p)}{p^s}, F\right)^{-1}$$

(depending on whether $w = k$ or $w = k/2$, respectively) and the Euler product

$$(3.68) \qquad \zeta^+(s, \gamma; F) = \prod_{p \in \mathbf{P}_{(q)}} \zeta_p^+(s, \gamma; F)$$

the *local even zeta-function* and the (*global*) *even zeta-function of the modular form F with character* γ.

Finally, in the above notation, given $A \in \mathbf{A}_n$, $f \in \mathfrak{F}_s^n$, and an arbitrary subset $\Delta \subset \mathbf{N}_{(q)}$, we define the formal Dirichlet series

$$(3.69) \qquad D_w(s, A, f, \Delta) = \sum_{\substack{M \in \Lambda \backslash M_n \\ |\det M| \in \Delta}} \frac{\gamma(|\det M|)\overline{\chi}_w(\det M) f(A['M])}{|\det M|^{s+w-1}},$$

where $\Lambda = \Lambda^n = GL_n(\mathbf{Z})$ and $\overline{\chi}_w = \chi_w^{-1}$ (see (2.5)).

We are now ready to give the fundamental result of this subsection.

THEOREM 3.17. *Let*

$$F(Z) = \sum_{A \in \mathbf{A}_n} f(A)e\{AZ\} \in \mathfrak{M}_w^n(q, \chi)$$

be a nonzero modular form of integer or half-integer weight $w = k$ *or* $k/2$ *and one-dimensional character* $[\chi]$ *for the group* $\Gamma_0^n(q)$, *where* $n, q \in \mathbf{N}$, χ *is a Dirichlet character modulo q, and q is divisible by* 4 *if* $w = k/2$. *Suppose that F is an eigenfunction for all of the Hecke operators* $|\tau = |_{w,\chi}\tau$ *for* τ *in the even Hecke ring* $E^n(q)$ *for* $w = k$ *or* $\widehat{E}^n(q, \varkappa)$ *for* $w = k/2$. *Then the following formal identity holds for any completely multiplicative function* $\gamma \colon \mathbf{N}_{(q)} \to \mathbf{C}$ *and any fixed matrix* $A_0 \in \mathbf{A}_n^+$:

$$(3.70) \qquad D_w(s, A_0, f, \mathbf{N}_{(q)}) = X(s, A_0, f)B_w\left(\frac{\gamma(p)}{p^s}, A_0\right)\zeta^+(s, \gamma; F),$$

where $\zeta^+(s, \gamma; F)$ *is the even zeta-function* (3.68) *of the modular form F with character* γ;

$$B_k(v, A_0) = \prod_{p \in \mathbf{P}_{(q)}} B_p^n(v, A_0)$$

and

$$B_{k/2}(v, A_0) = \prod_{p \in \mathbf{P}_{(q)}} \widehat{B}_{p,k}^n(v, A_0),$$

in which the polynomials on the right are the ones defined in Theorems 2.23 *and* 2.26; *and the finite sum* $X(s, A_0, f)$ *is given by setting*

$$(3.71) \qquad X(s, A_0, f) = \left(f|_{w,\chi} \prod_{\substack{p \in \mathbf{P}_{(q)} \\ p \mid \det A_0}} \left(\sum_{i=0}^n (-1)^i x_{-i}^n(p)\left(\frac{\gamma(p)}{p^s}\right)^i\right)\right)(A_0),$$

in which the function $f \colon \mathbf{A}_n \to \mathbf{C}$ *is regarded as an element of* $\mathfrak{F}_s^n$ *with* $s(-1) = \chi_w(-1)$, $x_{-i}^n(p) \in \mathbf{L}_{0,p}^n$ *are the coefficients of the polynomial* (6.101) *of Chapter* 3, *and the right side is understood in the sense of* (2.72).

Further suppose that $f(A) \neq 0$ *for some* $A \in \mathbf{A}_n^+$ *and* $|\gamma(m)| \leqslant cm^\sigma$ *for all* $m \in \mathbf{N}_{(q)}$, *where c and* σ *are real numbers that do not depend on m. Then the Dirichlet series on the left in* (3.70) *and the infinite products on the right converge absolutely and uniformly*

in any right half-plane of the variable s of the form $\operatorname{Re} s > (2\rho - 1)w + \sigma + n + 1 + \varepsilon$
with $\varepsilon > 0$, *where* $\rho = 1$ *in the general case and* $\rho = 1/2$ *if* F *is a cusp-form; and these*
holomorphic functions on the half-planes indicated are connected by the identity (3.70).

REMARKS. (1) From (6.34) of Chapter 3 it follows that $x^n_{-i}(p) \in C^n_{-p} \subset C^n_{-}$. By
Theorem 5.8 of Chapter 3, $C^n_{-} = C^n_{-}(1)$ is a commutative ring. Thus, it makes no
difference in what order the primes appear on the right in (3.71).

(2) The action of the operators $|_{w,\chi} x^n_{-i}(p)$ on the space $\mathfrak{F}^n_s$ can be computed from
the formulas (3.74).

(3) At the end of the proof of this theorem we show that, under the conditions of
the theorem, there exist matrices $A_0 \in \mathbf{A}^+_n$ for which $X(s, A_0, f) = f(A_0) \neq 0$.

We first prove three lemmas.

LEMMA 3.18. *For* $a \in \mathbf{N}$ *set*

$$t^+(a) = t^+_n(a) = \sum_{\substack{D \in \Lambda \backslash \Lambda M_n \Lambda \\ \det D = \pm a}} (U(D))_{\Gamma^n_0} \in \mathbf{L}^n_0.$$

Then:

(1) *the elements* $t^+(a)$ *belong to the subring* $C^n_+ = C^n_+(1)$ *of* $\mathbf{L}^n_{0,p}$, *and in particular*
commute with one another;

(2) *if* a *and* b *are relatively prime, then*

$$(3.72) \qquad\qquad t^+(a)t^+(b) = t^+(ab);$$

(3) *for any* $g \in \mathfrak{F}^n_s$ *one has the formula*

$$(3.73) \qquad a \sum_{\substack{M \in \Lambda \backslash M_n \\ \det M = \pm a}} \overline{\chi}_w(\det M)|\det M|^{-w} g(A[{}^t M])$$

$$= (g|_{w,\chi} a^{-n} t^+(a))(A) \quad (a \in \mathbf{N}, A \in \mathbf{A}_n),$$

where we assume that $s(-1) = \chi_w(-1)$ *and* $\chi(a) \neq 0$.

PROOF. Part (1) follows from Proposition 5.5 and Theorem 5.8 of Chapter 3.

If $D \in M_n$ and $a = |\det D| > 0$, then from (5.23) and (6.109) of Chapter 3 and
from the definition of the map Φ^n in (5.29) of Chapter 3 we obtain

$$\Phi^n((U(D))_{\Gamma_0}) = a^{n+1}(D)_\Lambda,$$

and hence $\Phi^n(t^+(a)) = a^{n+1} t(a)$, where $t(a) = t_n(a) \in H^n$ are the elements (2.10) of
Chapter 3. Part (2) of the lemma follows from these formulas and the relations (2.14)
of Chapter 3 for the elements $t(a)$, since, by Theorem 5.8 of Chapter 3, the restriction
of Φ^n to C^n_+ is a monomorphism.

By (2.29) we have

$$(g|_{w,\chi}(U(D))_{\Gamma_0})(A) = a^{n+1} \sum_{M \in \Lambda \backslash \Lambda D \Lambda} \overline{\chi}_w(\det M)|\det M|^{-w} g(A[{}^t M]).$$

If we divide both sides by a^n and sum over all double cosets $\Lambda D \Lambda \subset M_n$ with
$\det M = \pm a$, we obtain (3.73). $\square$

LEMMA 3.19. *Suppose that the function* $g \in \mathfrak{F}_s^n$ *is an eigenfunction for all of the Hecke operators* $|\tau = |_{w,\chi}\tau$, *where* $\tau \in \mathbf{E}_p^n$ *for* $w = k$ *and* $\tau \in \widehat{\mathbf{E}}_p^n(q, \varkappa)$ *for* $w = k/2$, *and* $s(-1) = \chi_w(-1)$. *Then in the notation* (3.65) *the following formal identity holds for any matrix* $A \in \mathbf{A}_n$:

$$R_p^n(v, g) \sum_{\delta=0}^{\infty} (g|p^{-\delta n}t^+(p^\delta))(A)v^\delta = B_p^n(v, A) \sum_{i=0}^{n} (-1)^i (g|x_{-i}^n(p))(A)v^i,$$

if $w = k$, *and*

$$\widehat{R}_p^n(v, g) \sum_{\delta=0}^{\infty} (g|p^{-\delta n}t^+(p^\delta))(A)v^\delta = \widehat{B}_{p,k}^n(v, A) \sum_{i=0}^{n} (-1)^i (g|x_{-i}^n(p))(A)v^i,$$

if $w = k/2$, *where* $B_p^n(v, A)$ *and* $\widehat{B}_{p,k}^n(v, A)$ *are the polynomials in Theorems 2.23 and 2.26, and* $x_{-i}^n(p) \in \mathbf{L}_{0,p}^n$ *are the coefficients of the polynomials* (6.101) *of Chapter 3.*

PROOF. Since $\lambda(\mathbf{r}_a^n(p), g)g = g|\mathbf{r}_a^n(p)$, it follows that, using the notation (2.73), we can rewrite the left side of the identity in the lemma in the form

$$\sum_{\delta=0}^{\infty} \sum_{a=0}^{2n} (-1)^a (g|\mathbf{r}_a^n(p)|p^{-\delta n}t^+(p^\delta))(A)v^a v^\delta$$

$$= \left(g \middle| \left(\sum_{a=0}^{2n} (-1)^a \mathbf{r}_a^n(p)v^a \right) \middle| \sum_{\delta=0}^{\infty} p^{-\delta n}t^+(p^\delta)v^\delta \right)(A)$$

$$= \left(g|\mathbf{R}_p^n(v) \sum_{\delta=0}^{\infty} p^{-\delta n}t^+(p^\delta)v^\delta \right)(A).$$

Using the factorization of $\mathbf{R}_p^n(v)$ in (6.99) of Chapter 3 and the identity (6.105) of Chapter 3, we see that the last expression is equal to

$$\left(g \middle| \left(\sum_{i=0}^{n} (-1)^i x_{-i}v^i \right) \left(\sum_{i=0}^{n} (-1)^i b_i v^i \right) \right)(A)$$

$$= \sum_{i=0}^{n} (-1)^i ((g|x_{-i})|B(v))(A)v^i \quad (x_{-i} = x_{-i}^n(p)),$$

where $B(v) = B_p^n(v) = \sum_{i=0}^{n} (-1)^i b_i v^i$. Since each of the functions $g|x_{-i}$ also lies in $\mathfrak{F}_s^n$, it follows by Proposition 2.20 that

$$(g|x_{-i}|B(v))(A) = B_p^n(v, A)(g|x_{-i})(A),$$

where $B_p^n(v, A)$ is the polynomial in (2.79). If we substitute these expressions into the last formula and take Theorem 2.23 into account, we obtain the desired identity for $w = k$.

In the case $w = k/2$, instead of (6.99) one must use (6.100) of Chapter 3, and instead of Theorem 2.23 one uses Theorem 2.26. $\qquad\square$

LEMMA 3.20. *The elements* $x^n_{-i}(p)$ $(i = 0, 1, \ldots, n)$ *act on the space* $\mathfrak{F}^n_s$ *according to the formulas*

(3.74)
$$
\begin{aligned}
(g|_{w,\chi} x^n_{-i}(p))(A) = {}&p^{wn - \langle n \rangle + \langle n - i \rangle} \chi(p)^n \\
&\times \sum_{D \in \Lambda \backslash \Lambda D_{n-i} \Lambda} \overline{\chi}_w(\det D) |\det d|^{-w} g(p^{-2} A[{}^t D]),
\end{aligned}
$$

where, as usual, $s(-1) = \chi_w(-1)$ *and* $D_a = D^n_a(p)$ *are the matrices* (2.28) *of Chapter 3. In particular,*

(3.75)
$$
(g|_{w,\chi} x^n_{-i}(p))(A) = 0, \quad \text{if } p^{2i} \nmid \det A.
$$

PROOF. The formulas in (3.74) follow directly from the formulas in (6.101) of Chapter 3 for the elements $x^n_{-i}(p)$ and from the formulas in Lemma 2.8 for the action of $\Delta_n(p)$, $\Pi_- = \Pi^n_0(p)$, and $\Pi^n_{n-i}(p)$ on the space $\mathfrak{F}^n_s$. (3.75) is a consequence of (3.74), since in the case $p^{2i} \nmid \det A$ the matrix $A[{}^t D]$ cannot be divisible by p^2 for any D with $\det D = \pm p^{n-i}$.   □

We are now ready to prove Theorem 3.17.

PROOF OF THE THEOREM. Using (3.73), we can rewrite each series $D_w(s, A, g, \Delta)$ for arbitrary $g \in \mathfrak{F}^n_s$ in the form

$$
\begin{aligned}
\sum_{a \in \Delta} \frac{\gamma(a)}{a^s} a \sum_{\substack{M \in \Lambda \backslash M_n \\ \det M = \pm a}} \overline{\chi}_w(\det M) |\det M|^{-w} g(A[{}^t M]) \\
= \sum_{a \in \Lambda} \frac{\gamma(a)(g|_{w,\chi} a^{-n} t^+(a))(A)}{a^s}.
\end{aligned}
$$

We now suppose that for some prime $p \in \mathbf{P}_{(q)}$ the function g is an eigenfunction for all of the Hecke operators $|\tau = |_{w,\chi}\tau$ for $\tau \in \mathbf{E}^n_p$ if $w = k$ and $\tau \in \widehat{\mathbf{E}}^n_p(q, \varkappa)$ if $w = k/2$, with the same eigenvalues as f, and we suppose that the set Δ can be represented in the form

(3.76)
$$
\Delta = \bigcup_{\delta=0}^{\infty} \Delta_1 p^\delta, \quad \text{where } \Delta_1 \subset \mathbf{N}_{(pq)}.
$$

Then, using (3.66) and (3.72), we obtain

$$
\begin{aligned}
R^n_p(v_p, F) &D_k(s, A, g, \Delta) \\
&= R^n_p(v_p, g) \sum_{a \in \Delta_1} \sum_{\delta=0}^{\infty} \frac{\gamma(ap^\delta)(g|(ap^\delta)^{-n} t^+(ap^\delta))(A)}{(ap^\delta)^s} \\
&= \sum_{a \in \Delta_1} \frac{\gamma(a)}{a^s} R^n_p(v_p, g) \sum_{\delta=0}^{\infty} ((g|a^{-n} t^+(a)) | p^{-\delta n} t^+(p^\delta))(A) v^\delta_p,
\end{aligned}
$$

where $v_p = \gamma(p)p^{-s}$. Let $a \in \Delta_1$. We write a as a product of prime powers: $a = p^{\alpha_1}_1 \cdots p^{\alpha_t}_t$. Then, by (3.72), we have $t^+(a) = t^+(p^{\alpha_1}_1) \cdots t^+(p^{\alpha_t}_t)$. Since $t^+(p^{\alpha_i}_i)$ lies in $C^n_{p_i}$ and $p_i \neq p$ for every $i = 1, \ldots, t$, it follows by Proposition 5.12(1) of Chapter 3 that any element $\mathbf{T} \in \mathbf{E}^n_p \subset \mathbf{L}^n_p$ commutes with each of the $t^+(p^{\alpha_i}_i)$, and hence commutes with $t^+(a)$. Thus, for every $a \in \Delta_1$ the function $g_a = g|a^{-n}t^+(a)$, along with $g = g_1$, is an eigenfunction for all of the operators $|_{k,\chi}\mathbf{T}$ for $\mathbf{T} \in \mathbf{E}_p^n$ and has the same eigenvalues as g. Hence, using Lemma 3.19, we obtain

$$R_p^n(v_p, g) \sum_{\delta=0}^{\infty}(g|a^{-n}t^+(a)|p^{-\delta n}t^+(p^\delta))(A)v_p^\delta$$

$$= B_p^n(v_p, g_a) \sum_{i=0}^{n}(-1)^i (g_a|x_{-i}^n(p))(A)v_p^i.$$

If we again write $t^+(a)$ as a product of elements of the form $t^+(p_i^{\alpha_i})$ with primes $p_i \neq p$ and take into account that $x_{-j}^n(p) \in C_{-p}^n$, we conclude from Proposition 5.12(2) of Chapter 3 that the elements $t^+(a)$ for $a \in \Delta_1$ commute with the elements $x_{-j}^n(p)$. Hence,

$$(3.77) \qquad g_a|x_{-j}^n(p) = g|a^{-n}t^+(a)|x_{-j}^n(p) = g|x_{-j}^n(p)|a^{-n}t^+(a).$$

Combining these formulas, we arrive at the formal identity

$$(3.78) \qquad \begin{aligned} R_p^n&(v_p, F)D_k(s, A, g, \Delta) \\ &= B_p^n(v_p, A) \sum_{a \in \Delta_1} \frac{\gamma(a)}{a^s} \sum_{i=0}^{n}(-1)^i (g|x_{-i}^n(p)|a^{-n}t^+(a))(A)v_p^i \\ &= B_p^n(v_p, A) \sum_{i=0}^{n}(-1)^i D_k(s, A, g|x_{-i}^n(p), \Delta_1)v_p^i. \end{aligned}$$

According to Proposition 5.12(1) of Chapter 3, the elements $\widehat{\mathbf{T}} \in \widehat{\mathbf{E}}_p^n(q, \varkappa)$ and $t^+(a)$ with $a \in \Delta_1$ commute with one another; thus, by analogy with (3.78), we obtain

$$(3.79) \quad \widehat{R}_p^n(v_p, F)D_{k/2}(s, A, g, \Delta) = \widehat{B}_{p,k}^n(v_p, A) \sum_{i=0}^{n}(-1)^i D_{k/2}(s, A, g|x_{-i}^n(p), \Delta_1)v_p^i.$$

In particular, if $(p, \det A) = 1$, then (3.77) and (3.75) imply the identities

$$D_w(s, A, g|x_{-i}^n(p), \Delta_1) = \sum_{a \in \Delta_1} \frac{\gamma(a)}{a^s}(g|a^{-n}t^+(a)|x_{-i}^n(p))(A),$$

where the right side is equal to 0 for $i = 1, \ldots, n$ and is equal to $D_w(s, A, g, \Delta_1)$ for $i = 0$. Thus, in this case (3.78) and (3.79) are transformed, respectively, to

$$(3.80) \qquad \begin{aligned} R_p^n(v_p, F)D_k(s, A, g, \Delta) &= B_p^n(v_p, A)D_k(s, A, g, \Delta_1), \\ \widehat{R}_p^n(v_p, F)D_{k/2}(s, A, g, \Delta) &= \widehat{B}_{p,k}^n(v_p, A)D_{k/2}(s, A, g, \Delta_1). \end{aligned}$$

To prove (3.70) we follow the same plan as in the proof of (3.52). Let $p_1, \ldots, p_b$ denote all of the distinct prime divisors of $\det A_0$ that are in $\mathbf{N}_{(q)}$, let $p_{b+1}, p_{b+2}, \ldots$ denote all of the primes in $\mathbf{P}_{(q \det A_0)}$ listed in increasing order, and set $q(c) = p_{b+1} \cdots p_{b+c}$.

Using the first equation in (3.80) and an obvious induction on $c = 1, 2, \ldots$, we obtain the identity

$$D_k\big(s, A_0, f, \mathbf{N}_{(q)}\big)$$
$$= D_k\big(s, A_0, f, \mathbf{N}_{(q \cdot q(c))}\big)\bigg\{ \prod_{p \mid q(c)} B_p^n(v_p, A_0) \bigg\} \prod_{p \mid q(c)} \big(R_p^n(v_p, F)\big)^{-1},$$

which, if we take the limit as $c \to \infty$ coefficient by coefficient, gives us the identity

$$D_k\big(s, A_0, f, \mathbf{N}_{(q)}\big)$$
(3.81)
$$= D_k\big(s, A_0, f, \Delta\big)\bigg\{ \prod_{p \in \mathbf{P}_{(q \det A_0)}} B_p^n(\gamma(p)p^{-s}, A_0) \bigg\} \prod_{p \in \mathbf{P}_{(q \det A_0)}} \zeta_p^+(s, \gamma, F),$$

where $\Delta = \Delta(p_1, \ldots, p_b) = \{a \in \mathbf{N}; \ a \mid (p_1 \cdots p_b)^\infty\}$. To prove (3.70) with $w = k$ it remains to verify the identities

(3.82)
$$\bigg\{ \prod_{p \mid p_1 \cdots p_d} R_p^n(v_p, F) \bigg\} D_k\big(s, A, g, \Delta(p_1, \ldots, p_d)\big)$$
$$= X_d(s, A, g)\bigg\{ \prod_{p \mid p_1 \cdots p_d} B_p^n(v_p, A) \bigg\}$$

for $p_1, \ldots, p_d \in \mathbf{P}_{(q)}$, $A \in \mathbf{A}_n$, and any $g \in \mathfrak{F}_s^n$ that is an eigenfunction for all of the operators $|\mathbf{T}$ for $\mathbf{T} \in \mathbf{E}_{p_1}^n, \ldots, \mathbf{E}_{p_d}^n$ with the same eigenvalues as f, where

$$X_d(s, A, g) = \bigg(g \big| \prod_{p \mid p_1 \cdots p_d} \bigg(\sum_{i=0}^n (-1)^i x_{-i}^n(p) v_p^i \bigg) \bigg)(A).$$

We use induction on d. For $d = 1$ the identity (3.82) is the same as (3.78) for $\Delta = \Delta(p_1)$ and $\Delta_1 = \{1\}$. Suppose that (3.82) has already been proved for some $d \in \mathbf{N}$, and let $p_{d+1} = p \in \mathbf{P}_{(q)}$ be distinct from $p_1, \ldots, p_d$. By (3.78), we have

$$R_p^n(v_p, F) D_k\big(s, A, g, \Delta(p_1, \ldots, p_d, p)\big)$$
$$= \bigg(\sum_{i=0}^n (-1)^i D_k\big(s, A, g \big| x_{-i}^n(p), \Delta(p_1, \ldots, p_d)\big) v_p^i \bigg) B_p^n(v_p, A).$$

Since $x_{-i}^n(p) \in C_{-p}^n$, it follows from Proposition 5.12(1) of Chapter 3 that the functions $g | x_{-i}^n(p)$, along with g, are eigenfunctions for all of the operators $|\mathbf{T}$ for $\mathbf{T} \in \mathbf{E}_{p_1}^n, \ldots, \mathbf{E}_{p_d}^n$ with the same eigenvalues as f. Hence, using the last relation and the induction assumption, we obtain

$$\bigg\{ \prod_{p \mid p_1 \cdots p_d} R_p^n(v_p, F) \bigg\} R_p^n(v_p, F) D_k\big(s, A, g, \Delta(p_1, \ldots, p_d, p)\big)$$
$$= \bigg(\sum_{i=0}^n (-1)^i X_d\big(s, A, g \big| x_{-i}^n(p)\big) v_p^i \bigg) = \bigg\{ \prod_{p \mid p_1 \cdots p_d} B_p^n(v_p, A) \bigg\} B_p^n(v_p, A).$$

Since all elements of the form $x^n_{-i}(p)$ lie in the commutative ring C^n_-, it follows that

$$\sum_{i=0}^{n}(-1)^i X_d(s, A, g | x^n_{-i}(p)) v_p^i$$

$$= \sum_{i=0}^{n}\left(g | x^n_{-i}(p) | \prod_{p | p_1 \cdots p_d}\left(\sum_{j=0}^{n}(-1)^j x^n_{-j}(p) v_p^j\right)\right)(A) v_p^i$$

$$= \left(g | \prod_{p | p_1 \cdots p_{d+1}}\left(\sum_{i=0}^{n}(-1)^i x^n_{-i}(p) v_p^i\right)\right)(A) = X_{d+1}(s, A, g),$$

which proves (3.82), and hence also (3.70) for $w = k$. We note that the second equation in (3.80) implies (3.81) (with B_p^n replaced by $\widehat{B}_{p,k}^n$ on the right) for the Dirichlet series $D_{k/2}(s, A_0, f, \mathbf{N}_{(q)})$. Furthermore, by Proposition 5.12 of Chapter 3, elements of the rings C^n_{-p} and $\widehat{\mathbf{E}}_{p_1}^n(q, \varkappa), \ldots, \widehat{\mathbf{E}}_{p_d}^n(q, \varkappa)$ commute in pairs. Thus, under the same conditions as above, from (3.79) we obtain the identity

$$\left\{\prod_{p | p_1 \cdots p_d}\widehat{R}_p^n(v_p, F)\right\}D_{k/2}(s, A, g, \Delta(p_1, \ldots, p_d))$$

$$= X_d(s, A, g)\prod_{p | p_1 \cdots p_d}\widehat{B}_{p,k}^n(v_p, A),$$

which implies (3.70) for $w = k/2$.

It remains to examine the convergence of the series and products in (3.70). If we apply the bound (3.35) in the general case and (3.70) of Chapter 2 in the case when F is a cusp-form, we find that the Dirichlet series on the left in (3.70) is majorized by the series

$$c' \sum_{M \in \Lambda \backslash M_n, | \det M | \in \mathbf{N}_{(q)}} | \det M |^{-\operatorname{Re} s + (2\rho - 1)w + \sigma + 1}$$

with constant $c' \geqslant 0$. If we take Λ-left coset representatives of the form (2.15) of Chapter 3 and note that the number of representatives with fixed diagonal $c_{11}, c_{22}, \ldots, c_{nn}$ is obviously equal to $c_{22}c_{33}^2 \cdots c_{nn}^{n-1}$, we conclude that the last series is majorized by the series

$$c' \sum_{c_{11}, c_{22}, \ldots, c_{nn}}\left(\prod_{i=1}^{n} c_{ii}^{i-1}\right)\bigg/\left(\prod_{i=1}^{n} c_{ii}\right)^t$$

$$= c'\prod_{i=1}^{n}\sum_{c \in \mathbf{N}_{(q)}} c^{-t+i-1}, \quad \text{where } t = \operatorname{Re} s - (2\rho - 1)w - \sigma - 1,$$

which has positive terms and is uniformly convergent for $t > n + \varepsilon$, i.e., for $\operatorname{Re} s > (2\rho - 1)w + \sigma + n + 1 + \varepsilon$, where $\varepsilon > 0$. The formulas in Theorems 2.23 and 2.26 imply that the product of the polynomials $B_p^n(\gamma(p)p^{-s}, A_0)$ on the right of the identity converges absolutely and uniformly for $\operatorname{Re} s > 1 + \sigma + \varepsilon$, while the product of $\widehat{B}_{p,k}^n(\gamma(p)p^{-s}, A_0)$ converges absolutely and uniformly for $\operatorname{Re} s > 1/2 + \sigma + \varepsilon$; and in any of these half-planes the modulus is bounded from below by a positive constant. From these observations and the identity (3.70), we conclude that the product $X(s, A_0, f)\zeta^+(s, \gamma, F)$, regarded as a Dirichlet series, converges absolutely and uniformly in any of the half-planes indicated in the theorem. Since $\zeta^+(s, \gamma, F)$

does not depend on the choice of A_0, to complete the proof of the theorem it suffices to show that the sum $X(s, A_0, f)$ becomes a nonzero constant for some $A_0 \in \mathbf{A}_n^+$. By assumption, there exist matrices $A \in \mathbf{A}_n^+$ for which $f(A) \neq 0$. Among these matrices we choose a matrix A_0 of minimal determinant. Then (3.71) and (3.74) imply that $X(s, A_0, f) = f(A_0) \neq 0$. $\qquad\qquad\qquad\qquad\qquad\qquad\qquad\qquad\qquad\qquad\qquad\quad\square$

PROBLEM 3.21. With the notation and assumptions of Theorem 3.17, show that the roots $\alpha_i(p)^{\pm 1}$ of the polynomials in (3.66) satisfy the inequalities

$$|\alpha_i(p)^{\pm 1}| \leqslant p^{(2\rho-1)w+n} \quad (i = 1, \ldots, n; p \in \mathbf{P}_{(q)}).$$

PROBLEM 3.22. Let $F \in \mathfrak{M}_s^n$ be an eigenfunction for all of the Hecke operators $|_{w,\chi}\tau$ for $\tau \in \mathbf{E}_p^n$ if $w = k$ and for $\tau \in \widehat{\mathbf{E}}_p^n(q, \varkappa)$ if $w = k/2$, where $\chi_w(-1) = s(-1)$. Suppose that $F|\Phi \neq 0$, where Φ is the Siegel operator. Prove that the modular form $F|\Phi \in \mathfrak{M}_s^{n-1}$ is an eigenfunction for all of the operators $|_{w,\chi}\tau'$ for $\tau' \in \mathbf{E}_p^{n-1}$ (for $\tau' \in \widehat{\mathbf{E}}_p^{n-1}(q, \varkappa)$ in the case $w = k/2$), and the polynomials (3.64) corresponding to F and $F|\Phi$ are connected by the relation

$$R_p^n(v, F) = (1 - p^{n-k}\overline{\chi}(p)v)(1 - p^{k-n}\chi(p)v)R_p^{n-1}(v, F|\Phi)$$

(in the case $w = k/2$ by the relation

$$\widehat{R}_p^n(v, F) = (1 - p^{n-k/2}\overline{\chi}(p)v)(1 - p^{k/2-n}\chi(p)v)\widehat{R}_p^{n-1}(v, F|\Phi)).$$

[**Hint:** See Problem 3.15(3).]

PROBLEM 3.23. Using the previous problem and Theorem 3.17, investigate the convergence of the zeta-function $\zeta^+(s, \gamma, F)$ in the case when the Fourier coefficients f of the eigenfunction F have the property that $f(A) = 0$ for any nondegenerate matrix A in $\mathbf{A}_n$.

Symmetric Matrices Over a Field

1. Arbitrary fields. We let $S_n(K)$ denote the set of all symmetric $n \times n$-matrices over the field K. Two matrices $A, A' \in S_n(K)$ are said to be *equivalent* over K if

$$(1.1) \qquad A' = {}^tCAC = A[C], \quad \text{where } C \in GL_n(K).$$

The following identity, which is easy to verify, is often useful if one wants to simplify a matrix by replacing it with an equivalent one: if the upper-left $r \times r$-block $A_1 = A^{(r)}$ in the matrix $A = \begin{pmatrix} A_1 & A_2 \\ {}^tA_2 & A_4 \end{pmatrix} \in S_n(K)$ is nonsingular, where $0 < r < n$, then

$$(1.2) \qquad A = \begin{pmatrix} A_1 & A_2 \\ {}^tA_2 & A_4 \end{pmatrix} = \begin{pmatrix} A_1 & 0 \\ 0 & A_4 - A_1^{-1}[A_2] \end{pmatrix} \left[\begin{pmatrix} E_r & A_1^{-1}A_2 \\ 0 & E_{n-r} \end{pmatrix} \right].$$

THEOREM 1.1. *Let $A \in S_n(K)$. Suppose that there exists a column $c \in M_{n,1}(K)$ such that*
$$a = A[c] \neq 0.$$
Then A is equivalent over K to a matrix of the form $\begin{pmatrix} a & 0 \\ 0 & A_1 \end{pmatrix}$, where $A_1 \in S_{n-1}(K)$.

PROOF. If we replace A by the matrix $A[C]$, where C is a nonsingular matrix with first column c, we may suppose that $A = \begin{pmatrix} a & * \\ * & * \end{pmatrix}$. The theorem then follows from (1.2). $\qquad\square$

If the characteristic of the field K is not 2, then the assumption in Theorem 1.1 obviously holds for any nonzero matrix $A \in S_n(K)$. Hence, using Theorem 1.1 and induction on n, we have

THEOREM 1.2. *If the characteristic of the field K is not 2, then any matrix in $S_n(K)$ is equivalent over K to a diagonal matrix.*

THEOREM 1.3. *Suppose that the rank of $A \in S_n(K)$ is equal to r, where $0 < r < n$. Then A is equivalent over K to a matrix of the form*

$$(1.3) \qquad \begin{pmatrix} A_1 & 0 \\ 0 & 0 \end{pmatrix}, \quad \text{where } A_1 \in S_r(K) \text{ and } \det A_1 \neq 0.$$

PROOF. Since rank $A = r$, there exist $n - r$ linearly independent columns $c_{r+1}, \ldots, c_n \in M_{n,1}(K)$ satisfying the equation $Ax = 0$. Let C be a nonsingular matrix whose last columns are $c_{r+1}, \ldots, c_n$. Then the matrix $A[C]$ has the required form. $\qquad\square$

2. The field of real numbers.

THEOREM 1.4. *For any matrix $A \in S_n(\mathbf{R})$ there exists an orthogonal real matrix C such that the matrix $A[C]$ is diagonal:*

$$(1.4) \qquad\qquad A[C] = \mathrm{diag}(\lambda_1, \ldots, \lambda_n).$$

Here $\lambda_1, \ldots, \lambda_n$ are the eigenvalues of A.

The proof of this theorem is well known, and can be found in virtually any linear algebra textbook.

A matrix $A \in S_n(\mathbf{R})$ is said to be *semidefinite* and we write $A \geqslant 0$ if $A[l] \geqslant 0$ for any column-vector $l \in M_{n,1}(\mathbf{R})$. If $A \geqslant 0$ and the equality $A[l] = 0$ holds only for $l = 0$, then A is said to be *positive definite* and we write $A > 0$. From the definition it follows that if A is semidefinite, then so is any equivalent matrix; and similarly if A is positive definite. Clearly a diagonal matrix is semidefinite (or positive definite) if and only if all of the diagonal entries are nonnegative (respectively, positive). Thus, Theorem 1.4 implies that a matrix is semidefinite if and only if all of its eigenvalues are nonnegative, and it is positive definite if and only if all of its eigenvalues are positive. If $A = (a_{ij}) \geqslant 0$, then, since

$$A[e_\alpha \pm e_\beta] = a_{\alpha\alpha} \pm 2a_{\alpha\beta} + a_{\beta\beta} \geq 0,$$

where $e_1, \ldots, e_n$ are the columns of the identity matrix E_n, it follows that $|a_{\alpha\beta}| \leqslant (a_{\alpha\alpha} + a_{\beta\beta})/2$. Hence,

$$(1.5) \qquad\qquad |a_{\alpha\beta}| \leq \sigma(A) \quad (A = (a_{ij}) \geq 0, \, \alpha, \, \beta = 1, \ldots, n).$$

Furthermore, let $A - B \geqslant 0$ and $R \geqslant 0$, in which case $R = \Lambda[C]$, where C is a nonsingular matrix and Λ is a diagonal matrix with nonnegative entries. Then

$$\sigma((A - B)R) = \sigma((A - B)[{}^t C]\Lambda) \geq 0,$$

since $(A - B)[{}^t C]$ is a semidefinite matrix, and so all of its diagonal entries are nonnegative. Thus, we have

$$(1.6) \qquad\qquad \sigma(AR) \geq \sigma(BR) \quad \text{for } A - B \geq 0 \text{ and } R \geq 0.$$

Finally, we obviously have

$$(1.7) \qquad \begin{array}{l} A + B \geq 0, \quad \text{if } A \geq 0, \quad B \geq 0, \\ A + B > 0, \quad \text{if } A > 0, \quad B \geq 0. \end{array}$$

We write $A \geqslant B \ (A > B)$ if $A - B \geqslant 0 \ (A - B > 0)$.

THEOREM 1.5. *The following three conditions on a matrix $A = (a_{ij}) \in S_n(\mathbf{R})$ are equivalent:*

(1) $A > 0$;

(2) $\det A^{(r)} = \det \begin{pmatrix} a_{11} & \cdots & a_{1r} \\ \cdot & \cdot & \cdot \\ a_{r1} & \cdots & a_{rr} \end{pmatrix} > 0$ *for $r = 1, \ldots, n$;*

(3) $A = {}^t LL$, *where* $L = \begin{pmatrix} l_{11} & l_{12} & \cdots & l_{1n} \\ 0 & l_{22} & \cdots & l_{2n} \\ \cdot & \cdot & \cdot & \cdot \\ 0 & 0 & \cdots & l_{nn} \end{pmatrix} \in GL_n(\mathbf{R}).$

PROOF. (1) → (2). If $A > 0$, then all of the matrices $A^{(1)} = (a_{11})$, $A^{(2)} = \begin{pmatrix} a_{11} & a_{12} \\ a_{21} & a_{22} \end{pmatrix}, \ldots, A^{(n)} = A$ are also obviously positive definite. This implies that all of the determinants of these matrices—which are products of eigenvalues—are positive.

(2) → (3). We use induction on n. The case $n = 1$ is obvious. Suppose that the implication holds for $(n-1) \times (n-1)$-matrices, $n \geqslant 2$. If $A \in S_n(\mathbf{R})$ satisfies (2), then, by the induction assumption,

$$A^{(n-1)} = {}^tL'L', \quad \text{where } L' = \begin{pmatrix} l'_{11} & l'_{12} & \cdots & l'_{1,n-1} \\ 0 & l'_{22} & \cdots & l'_{2,n-1} \\ \cdot & & \cdot & \cdot \\ \cdot & & & \cdot \\ 0 & 0 & \cdots & l'_{n-1,n-1} \end{pmatrix} \in GL_{n-1}(\mathbf{R}).$$

Since $\det A^{(n-1)} \neq 0$, it follows from (1.2) that A can be written in the form

$$A = \begin{pmatrix} A^{(n-1)} & 0 \\ 0 & d_n \end{pmatrix} \left[\begin{pmatrix} E_{n-1} & h \\ 0 & 1 \end{pmatrix} \right] = \begin{pmatrix} A^{(n-1)} & 0 \\ 0 & 1 \end{pmatrix} \left[\begin{pmatrix} E_{n-1} & h \\ 0 & \sqrt{d_n} \end{pmatrix} \right],$$

since $d_n = \det A / \det A^{(n-1)} > 0$. Then

$$A = {}^tLL, \quad \text{where } L = \begin{pmatrix} L' & 0 \\ 0 & 1 \end{pmatrix} \begin{pmatrix} E_{n-1} & h \\ 0 & \sqrt{d_n} \end{pmatrix} = \begin{pmatrix} L' & L'h \\ 0 & \sqrt{d_n} \end{pmatrix}.$$

(3) → (1). This is obvious, since (3) implies that A is equivalent to the identity matrix, which is positive definite.   □

In particular, Theorem 1.5 implies that, if P_n denotes the subset of all positive definite matrices in $S_n(\mathbf{R})$, then P_n is closed in $S_n(\mathbf{R})$.

THEOREM 1.6. *Let $A, B \in S_n(\mathbf{R})$. Suppose that $A > 0$. Then there exists a matrix $C \in GL_n(\mathbf{R})$ such that $A[C] = E_n$ and the matrix $B[C]$ is diagonal.*

PROOF. According to the previous theorem, A is equivalent to the identity matrix: $A[C_1] = E_n$. By Theorem 1.4, there exists an orthogonal matrix C_2 such that the matrix $B[C_1][C_2] = B[C_1C_2]$ is diagonal. Since $A[C_1C_2] = {}^tC_2C_2 = E_n$, it follows that $C = C_1C_2$ is the required matrix.   □

We conclude this section by proving some useful determinant inequalities.

THEOREM 1.7. *One has*:

(1.8) $\det A \leq a_{12}a_{22} \cdots a_{nn}$,   *if $A = (a_{ij}) \in S_n(\mathbf{R})$, $A \geq 0$*;

(1.9) $\det(A + B) \geq \det A + \det B$,   *if $A, B \in S_n(\mathbf{R})$, $A, B \geq 0$*;

(1.10) $|\det(A + iB)| \geq \det B$,   *if $A, B \in S_n(\mathbf{R})$, $B > 0$*.

PROOF. If A is semidefinite but not positive definite, then $\det A = 0$, and (1.8) is obvious. The inequality (1.8) holds for $n = 1$. Suppose that (1.8) has been proved for positive definite $(n-1) \times (n-1)$-matrices, $n \geqslant 2$. If A is a positive definite $n \times n$-matrix, then $A^{(n-1)} > 0$, and we can use (1.2) with $A_1 = A^{(n-1)}$ and the induction assumption to obtain

$$\det A = \det A^{(n-1)} \cdot (a_{nn} - (A^{(n-1)})^{-1}[A_2]) \leq a_{11} \cdots a_{n-1,n-1}a_{nn}.$$

This proves (1.8).

The inequality (1.9) holds if $\det A = \det B = \det(A + B) = 0$. But if, for example, $\det(A + B) \neq 0$, then $A + B > 0$ and, by Theorem 1.6, the two matrices $A + B$ and B—and hence the two matrices A and B—can be simultaneously reduced to diagonal form. This reduces the proof of (1.9) to the case of semidefinite diagonal matrices, where it is obvious.

Using Theorem 1.6, we reduce (1.10) to the case when $B = E_n$ and A is a diagonal matrix. In that case it is obvious. $\qquad\qquad\qquad\qquad\qquad\qquad\qquad\qquad\qquad\qquad\qquad\qquad$ $\square$

APPENDIX 2

Quadratic Spaces

It is often convenient to use geometrical language when solving problems in the theory of quadratic forms, especially in cases when the ground ring is a finite field. Here we give a basic description of the geometrical approach, along with some applications.

1. Geometrical language. Let K be a commutative ring, and let V be a free K-module having finite dimension over K. A function $f : V \to K$ is said to be *quadratic* if in some basis $e_1, \ldots, e_n$ of V over K it is given by a quadratic form in the coordinates:

$$(2.1) \qquad f\left(\sum_{i=1}^{n} u_i e_i \right) = q(u_1, \ldots, u_n) \quad (u_1, \ldots, u_n \in K),$$

where

$$q(x_1, \ldots, x_n) = \sum_{1 \leq i \leq j \leq n} q_{ij} x_i x_j \quad (q_{ij} \in K).$$

This definition clearly does not depend on the choice of basis. We use the term *quadratic space* (over K) to denote a pair (V, f) consisting of a free K-module V of finite dimension and a quadratic function f on V. The quadratic form q is called the *form of the space* (V, f) *in the basis* $e_1, \ldots, e_n$. A different choice of basis clearly leads to a form that is equivalent to q over K (see §1.1 of Chapter 1). Conversely, quadratic forms that are equivalent to one another over K may be regarded as the forms of a fixed quadratic space in different bases. Thus, the class $\{q\} = \{q\}_K$ of q over K is uniquely determined by the space (V, f). We call the class $\{q\}_K$ the *type of the space* (V, f).

By a *morphism* $\varphi : (V, f) \to (V_1, f_1)$ *between two quadratic spaces over a ring* K we mean a linear map $\varphi : V \to V_1$ that satisfies the condition $f_1(\varphi(v)) = f(v)$ for all $v \in V$. A one-to-one surjective morphism is called an *isomorphism*, and in that case we say that the two spaces are *isomorphic*. It is clear that two spaces are isomorphic if and only if they correspond to the same class of forms.

Let (V, f) be a quadratic space over K. Given a pair of vectors $u, v \in V$, we define the scalar product $u \cdot v \in K$ by setting

$$(2.2) \qquad u \cdot v = f(u + v) - f(u) - f(v).$$

We say that the vectors $u, v \in V$ are *orthogonal* if $u \cdot v = 0$. If we write f in the form (2.1), we obtain an expression for the scalar product in terms of the coordinates in the basis $e_1, \dots, e_n$:

$$(2.3) \quad \left(\sum_{i=1}^{n} u_i e_i \right) \cdot \left(\sum_{j=1}^{n} v_j e_j \right) = \sum_{1 \leq i \leq j \leq n} q_{ij} \{(u_i + v_i)(u_j + v_j) - u_i u_j - v_i v_j\}$$

$$= \sum_{1 \leq i \leq j \leq n} q_{ij}(u_i v_j + v_i u_j) = \sum_{i,j=1}^{n} Q_{ij} u_i v_j,$$

where $Q_{ij} = Q_{ji} = q_{ij}$ for $1 \leq i < j \leq n$, and $Q_{ii} = 2q_{ii}$ for $1 \leq i \leq n$. In particular, this implies that the scalar product is linear in each factor. The matrix

$$Q = (Q_{ij}) = (e_i \cdot e_j) \in S_n(K)$$

is called the matrix of f (or of the scalar product (2.2)) in the basis $e_1, \dots, e_n$; it is the same as the matrix of q in (1.3) of Chapter 1. It is easy to see that, if we make a change of basis from $e_1, \dots, e_n$ to $e_i' = \sum_{j=1}^{n} a_{ij} e_j$, the matrix Q is replaced by the matrix

$$(2.4) \qquad\qquad Q' = (a_{ij}) Q^t (a_{ij}).$$

This implies that the coset $\det Q \cdot (K^*)^2$ of the number $\det Q$ modulo the group of squares of units of the ring K is independent of the choice of basis. This coset is called the *determinant of the space* (V, f) and is denoted $d(V) = d(V, f)$. If $d(V)$ is a unit of the ring K, we say that the space (V, f) is *nonsingular*.

A quadratic space (V_1, f_1) is said to be a *subspace* of (V, f) if $V_1 \subset V$ and f_1 coincides with the restriction of f to V_1. We say that (V, f) splits into a *direct sum of* $(V_i, f_i) \subset (V, f)$ $(1 \leq i \leq t)$ and write

$$(V, f) = \bigoplus_{i=1}^{t} (V_i, f_i),$$

if every element $v \in V$ can be uniquely written in the form $v = v_1 + \cdots + v_t$, where $v_i \in V_i$ and $f(v) = f_1(v_1) + \cdots + f_t(v_t)$. In this case, if we write f in a basis of V obtained as a union of bases of $V_1, \dots, V_t$, then we easily see that

$$(2.5) \qquad\qquad d(V, f) = \prod_{i=1}^{t} d(V_i, f_i).$$

Suppose that K is a field. By the *orthogonal complement of the subspace* $(V_1, f_1) \subset (V, f)$ *(in the space (V, f))* we mean the space $(V_1, f_1)^\perp = (V_1^\perp, f_1^\perp)$, where

$$V_1^\perp = \{u \in V; u \cdot v = 0 \text{ for all } v \in V_1\},$$

and $f_1^\perp$ is the restriction of f to $V_1^\perp$. The set $R(V) = V^\perp$ is called the *radical of the space* (V, f).

LEMMA 2.1. *Let (V, f) be a quadratic space over a field. Then the conditions $R(V) = \{0\}$ and $d(V, f) \neq 0$ are equivalent.*

PROOF. Writing the scalar product on V in the form (2.3), we see that $u = \sum_i u_i e_i \in R(V)$ if and only if $\sum_i Q_{ij} u_i = 0$ for $j = 1, \dots, n$. The existence of a nonzero solution of this homogeneous system of linear equations is equivalent to the vanishing of its determinant. $\square$

LEMMA 2.2. *Let* (V_1, f_1) *be a subspace of the quadratic space* (V, f) *over a field. Suppose that* $V_1 \neq \{0\}$ *and* $V_1 \subset R(V)$. *Then there exists a subspace* $(V_2, f_2) \subset (V, f)$ *such that*

$$(2.6) \qquad\qquad (V, f) = (V_1, f_1) \oplus (V_2, f_2).$$

PROOF. Let $e_1, \ldots, e_r$ be a basis of V_1. We complete this basis to a basis $e_1, \ldots, e_r,$ $e_{r+1}, \ldots, e_n$ of V. We set $V_2 = \{Ke_{r+1} + \cdots + Ke_n\}$, and we let f_2 be the restriction of f to V_2. We then obviously have the direct sum decomposition (2.6). □

THEOREM 2.3. *Let* (V_1, f_1) *be a nonsingular subspace of the quadratic space* (V, f) *over a field. Then one has the direct sum decomposition*

$$(2.7) \qquad\qquad (V, f) = (V_1, f_1) \oplus (V_1^{\perp}, f_1^{\perp}).$$

PROOF. We choose a basis $e_1, \ldots, e_n$ of V in such a way that the first $d = \dim V_1$ basis vectors $e_1, \ldots, e_d$ form a basis of V_1. Then the condition $u = \sum_{j=1}^{n} u_j e_j \in V_1^{\perp}$ is equivalent to the system of equations

$$e_i \cdot u = \sum_{j=1}^{n} u_j e_i \cdot e_j = 0 \quad (i = 1, \ldots, d),$$

whose matrix has rank d, since the $d \times d$-minor made up of the first d columns is obviously equal to $d(V_1) \neq 0$. Hence $\dim V_1^{\perp} = n - d$. On the other hand, $V_1 \cap V_1^{\perp} = \{0\}$, since $V_1 \cap V_1^{\perp} \subset R(V_1)$, and $R(V_1) = \{0\}$ by Lemma 2.1. These facts obviously imply that every $u \in V$ can be uniquely written in the form $u = v_1 + v_2$ with $v_1 \in V_1$ and $v_2 \in V_1^{\perp}$. Since

$$f(u) = f(v_1 + v_2) = v_1 \cdot v_2 + f(v_1) + f(v_2) = f(v_1) + f(v_2),$$

the theorem is proved. □

A quadratic space is said to be *reducible* if it splits into a direct sum of proper subspaces; otherwise, we say that the quadratic space is *irreducible*. From the definition it follows that every quadratic space splits into a direct sum of irreducible subspaces.

THEOREM 2.4. *Let* (V, f) *be an irreducible quadratic space over a field* K. *Then* $\dim V \leqslant 2$. *If the characteristic of* K *is not 2, then* $\dim V \leqslant 1$.

PROOF. First suppose that the characteristic of K is not 2. If f is identically zero on V, then any one-dimensional subspace of (V, f) is a direct summand; hence, $\dim V \leqslant 1$. Now suppose that there exists $e_1 \in V$ with $f(e_1) \neq 0$. Then the determinant of the one-dimensional subspace $V_1 = \{Ke_1\} \subset V$ is equal to $e_1 \cdot e_1 = 2f(e_1)$, which is nonzero; hence, by the previous theorem, V splits into the direct sum of V_1 and $V_1^{\perp}$. Thus, $V = V_1$ and $\dim V = 1$.

We now consider the case of characteristic 2. If the scalar product is identically zero on V, then any one-dimensional subspace is a direct summand, and hence $\dim V \leqslant 1$. Now suppose that there exist vectors $u, v \in V$ with $u \cdot v = \alpha \neq 0$. Then obviously the vectors u and v are linearly independent, and the determinant of the two-dimensional space $V_1 = \{Ku + Kv\}$ is equal to $2f(u) \cdot f(v) - (u \cdot v)^2 = -\alpha^2$, which is nonzero. If we again apply the previous theorem, we see that $V = V_1$ and $\dim V = 2$. □

If (V, f) is a one-dimensional quadratic space, then obviously

(2.8)   type $(V, f) = \{ax^2\}$, $d(V, f) = 2a$,   where $a \in K$,

and two spaces of types $\{ax^2\}$ and $\{bx^2\}$ are isomorphic if and only if $b = at^2$ for some nonzero $t \in K$.

PROPOSITION 2.5. *Every irreducible two-dimensional quadratic space over the field* $K = \mathbf{Z}/2\mathbf{Z}$ *is of type* $\{q_+^2\}$ *or* $\{q_-^2\}$, *where*

(2.9) $q_+^2(x_1, x_2) = x_1 x_2$ *and* $q_-^2(x_1, x_2) = x_1^2 + x_1 x_2 + x_2^2$.

Two spaces of types $\{q_+^2\}$ *and* $\{q_-^2\}$ *are not isomorphic.*

PROOF. If (V, f) is an irreducible two-dimensional quadratic space over the field of two elements, then we saw at the end of the proof of Theorem 2.4 that V contains a basis e_1, e_2 with $e_1 \cdot e_2 = 1$. If $f(e_1) = f(e_2) = 1$, then V is of type $\{q_-^2\}$. If one of these values, say $f(e_1)$, is equal to 1 and the other is equal to zero, then the change of basis from e_1, e_2 to $e_1, e_1 + e_2$ reduces this case to the previous one. Finally, if $f(e_1) = f(e_2) = 0$, then V is of type $\{q_+^2\}$. Two spaces of types $\{q_+^2\}$ and $\{q_-^2\}$ are not isomorphic, since the quadratic function for the first space vanishes on all nonzero vectors except for one, while the quadratic function for the second space is nonzero on all nonzero vectors. Since the determinant of any one-dimensional space over the field $\mathbf{Z}/2\mathbf{Z}$ is zero, while the determinant of a space of type $\{q_+^2\}$ or $\{q_-^2\}$ is nonzero, it follows that our spaces are irreducible (see (2.5)).   □

If the number 2 is a unit of the ring K, then the quadratic function $f(u)$ can be recovered from the scalar product $u \cdot v$ using the formula $f(u) = (1/2)u \cdot u$; hence, the quadratic space may be regarded in the usual way simply as a vector space with bilinear scalar product. Otherwise, it may very well happen that the scalar product is identically zero while the quadratic function is nonzero. Thus, in the general case it is more convenient to start out with the quadratic function.

2. Nondegenerate spaces. A quadratic space (V, f) is said to be *degenerate* if it splits into a direct sum of the form

$$(V, f) = (V_1, 0) \oplus (V_2, f_2),$$

where $\dim V_1 \geqslant 1$ and 0 denotes the zero function on V_1. If there is no such direct sum decomposition, then the space is said to be *nondegenerate*.

LEMMA 2.6. *If the quadratic space* (V, f) *is nonsingular, then it is nondegenerate.*

PROOF. Since $d(V_1, 0) = 0$, the lemma follows from (2.5).   □

LEMMA 2.7. *A nondegenerate quadratic space* (V, f) *over a field of characteristic* $\neq 2$ *is nonsingular.*

PROOF. Suppose that $d(V, f) = 0$. Then $R(V) \neq \{0\}$ by Lemma 2.1. Let f_1 be the restriction of f to $R(V)$. By Lemma 2.2, the subspace $(R(V), f_1)$ is a direct summand in (V, f). On the other hand, if $e \in R(V)$, then $f(e) = (1/2)e \cdot e = 0$, so that $f_1 = 0$. Hence (V, f) is degenerate.   □

We now turn to the case of fields of characteristic 2.

LEMMA 2.8. *If (V, f) is any quadratic space of odd dimension over a field of characteristic 2, then the determinant $d(V, f)$ is zero.*

PROOF. Let $Q = (Q_{ij})$ be the matrix of f in some basis $e_1, \ldots, e_n$, and let $A_{\alpha\beta}$ be the cofactors of Q. If we expand $\det Q$ along the ith row and sum these expansions as i goes from 1 to n, we obtain

$$n \det Q = \sum_{i,j=1}^{n} Q_{ij} A_{ij} = \sum_{i=1}^{n} Q_{ii} A_{ii} = 0,$$

since $Q_{ij} = Q_{ji}$, $A_{ij} = A_{ji}$, and $Q_{ii} = 2f(e_i) = 0$; since n is odd, this implies that $\det Q = 0$. $\qquad \square$

THEOREM 2.9. *Let (V, f) be a nondegenerate quadratic space over the field $K = \mathbf{Z}/2\mathbf{Z}$. If $n = \dim V$ is even, then (V, f) is nonsingular and is of one of the types $\{q_+^n\}$, $\{q_-^n\}$, where*

(2.10) $$q_+^n(x_1, \ldots, x_n) = x_1 x_2 + x_3 x_4 + \cdots + x_{n-1} x_n,$$

(2.11) $$q_-^n(x_1, \ldots, x_n) = x_1 x_2 + \cdots + x_{n-3} x_{n-2} + x_{n-1}^2 + x_{n-1} x_n + x_n^2;$$

if n is odd, then the space (V, f) is of type $\{q^n\}$, where

(2.12) $$q^n(x_1, \ldots, x_n) = x_1 x_2 + \cdots + x_{n-2} x_{n-1} + x_n^2.$$

PROOF. First suppose that n is even. Suppose that $d(V, f) = 0$. Then by Lemma 2.1 there exists a nonzero vector e_1 such that $e_1 \cdot v = 0$ for all $v \in V$. We complete e_1 to a basis $e_1, e_2, \ldots, e_n$ of V, and we let $V_2 = \{Ke_2 + \cdots + Ke_n\}$. Since $n - 1$ is odd, it follows by Lemma 2.7 that $d(V_2, f_2) = 0$, where f_2 is the restriction of f to V_2. Again applying Lemma 2.1, we find that the space V_2 contains a nonzero vector e_1' that is orthogonal to V_2. From the choice of e_1 and e_1' it obviously follows that the space $V_1 = \{Ke_1 + Ke_1'\}$ is two-dimensional, V_1 is a direct summand in V, and $e_1 \cdot e_1' = 0$. In addition, the space (V_1, f_1), where f_1 is the restriction of f to V_1, is degenerate, since it contains a one-dimensional subspace $(V_0, 0)$ that is a direct summand. Namely, if, say, $f(e_1) = 0$, then we can take $V_0 = \{Ke_1\}$; while if $f(e_1) = f(e_1') = 1$, then we take $V_0 = \{K(e_1 + e_1')\}$. Thus, the entire space (V, f) is degenerate. We conclude that we must have had $d(V, f) \neq 0$. By Theorem 2.4, the space (V, f) can be decomposed into a direct sum of irreducible subspaces (V_i, f_i) $(i = 1, \ldots, k)$ of dimension $\dim V_i \leqslant 2$. If $\dim V_i = 1$ for some i, then obviously $d(V_i, f_i) = 0$, and hence $d(V, f) = \prod_i d(V_i, f_i) = 0$ (see (2.5)). Thus, $\dim V_i = 2$ for all i. Then by Proposition 2.5 each of the spaces (V_i, f_i) is of type $\{q_+^2\}$ or $\{q_-^2\}$, and to complete the proof of the theorem in the case of even n it suffices to verify that the direct sum of two spaces of type $\{q_-^2\}$ is isomorphic to the direct sum of two spaces of type $\{q_+^2\}$. If (V, f) is the direct sum of two spaces of type $\{q_+^2\}$, then V_1 has a basis e_1, e_2, e_1', e_2' such that $f(e_1) = f(e_2) = f(e_1') = f(e_2') = 1$, $e_1 \cdot e_2 = e_1' \cdot e_2' = 0$, and $e_i \cdot e_j' = 0$ for $i, j = 1, 2$. Then the subspace V_1 with basis $e_1 + e_1', e_2 + e_2'$ and the subspace V_2 with basis $e_1 + e_2 + e_1' + e_2', e_1 + e_2 + e_2'$ are both easily seen to be subspaces of type $\{q_+^2\}$, and V is equal to their direct sum.

Now suppose that n is odd. By Lemmas 2.8 and 2.1, there exists a nonzero vector $e = e_n \in R(V)$. We set $V_2 = \{Ke_n\}$, and we let f_2 be the restriction of f to V_2. Then, by Lemma 2.2, $(V, f) = (V_1, f_1) \oplus (V_2, f_2)$ for some $(V_1, f_1) \subset (V, f)$.

Since the space (V, f) is nondegenerate, so are the subspaces (V_1, f_1) and (V_2, f_2). Then, by what was proved above, the space (V_1, f_1) is of type $\{q_+^{n-1}\}$ or $\{q_-^{n-1}\}$, and $f(e_n) = 1$. In the first case (V, f) is clearly of type $\{q_+^{n-1} + x_n^2\} = \{q^n\}$. In the second case (V_1, f_1) can be decomposed into the direct sum of subspaces (V_3, f_3) and (V_4, f_4) of types $\{q_+^{n-3}\}$ and $\{q_-^3\}$, respectively. Then the sum $(V_4, f_4) \oplus (V_2, f_2)$ has a basis v_1, v_2, e_n satisfying the following conditions: $f(v_1) = f(v_2) = f(e_n) = 1$, $v_1 \cdot e_n = v_2 \cdot e_n = 0$, and $v_1 \cdot v_2 = 1$. If we replace this basis by $v_1 + e_n, v_2 + e_n, e_n$, we see that our direct sum is of type $\{x_1 x_2 + x_3^2\} = \{q^3\}$. Hence, the full space $(V, f) = (V_3, f_3) \oplus (V_4, f_4) \oplus (V_2, f_2)$ is of type $\{q_+^{n-3} + x_{n-2} x_{n-1} + x_n^2\} = \{q^n\}$. $\square$

3. Gauss sums. By the *Gauss sum of the quadratic space* (V, f) *over* $K = \mathbf{F}_p = \mathbf{Z}/p\mathbf{Z}$ we mean the following sum of pth roots of unity:

$$(2.13) \qquad G_p(\alpha, f) = \sum_{v \in V} \exp\left(\frac{2\pi i \alpha f(v)}{p}\right),$$

where $\alpha \in K$ and each element $\alpha f(v) \in \mathbf{F}_p = \mathbf{Z}/p\mathbf{Z}$ is regarded as an integer of the corresponding residue class modulo p. If (V, f) is of type $\{q(x_1, \ldots, x_n)\}$, then the Gauss sum $G_p(\alpha, f)$ is obviously equal to the *Gauss sum of the form* αq:

$$G_p(\alpha, f) = G_p(\alpha q) = \sum_{\alpha_1, \ldots, \alpha_n \in \mathbf{Z}/p\mathbf{Z}} \exp(2\pi i \alpha q(\alpha_1, \ldots, \alpha_n)/p).$$

Since for fixed α the Gauss sum (2.13) depends only on the set of values of f, it follows that isomorphic spaces (and equivalent forms q) correspond to the same Gauss sum. Obviously,

$$(1.14) \qquad G_p(\alpha, f) = G_p(\alpha q) = p^{\dim V}, \qquad \text{if } \alpha = 0 \text{ or } f = 0.$$

Furthermore, if $(V, f) = (V_1, f_1) \oplus \cdots \oplus (V_t, f_t)$, then

$$
\begin{aligned}
G_p(\alpha, f) &= \sum_{v_j \in V_j (1 \leq j \leq t)} \exp(2\pi i \alpha (f_1(v_1) + \cdots + f_t(v_t))/p) \\
&= \prod_{j=1}^{t} \sum_{v_j \in V_j} \exp(2\pi i \alpha f_j(v_j)/p) = \prod_{j=1}^{t} G_p(\alpha, f_j),
\end{aligned}
$$

(2.15)

which reduces the calculation of Gauss sums to the case of irreducible spaces.

Before proceeding to the computations, we recall the definition and basic properties of the Legendre symbol. Suppose that p is an odd prime, $K = \mathbf{F}_p$, K^* is the multiplicative group of the field K, and $(K^*)^2 = \{d^2; d \in K^*\}$ is the subgroup of squares. Since the kernel of the homomorphism $d \to d^2$ from K^* to $(K^*)^2$ consists of 1 and $-1 \neq 1$, it follows that $(K^*)^2$ has index 2 in K^*. Let $a \to (a/p)$ denote the unique nontrivial character of the quotient group $K^*/(K^*)^2$, regarded as a function on K^*. In other words, $(a/p) = 1$ or -1 depending on whether or not a is a square. If a is an integer not divisible by p, then the *Legendre symbol* (a/p) is defined as $(\bar{a}/p)$, where $\bar{a}$ denotes the residue class of a in $\mathbf{Z}/p\mathbf{Z}$. From the definition it follows that the Legendre symbol has the multiplicative property

$$(2.16) \qquad \left(\frac{ab}{p}\right) = \left(\frac{a}{p}\right)\left(\frac{b}{p}\right), \qquad \text{if } (ab, p) = 1.$$

These relations still hold for all a and b if we agree to set $(a/p) = 0$ when $a \equiv 0 (\text{mod } p)$. Since the group K^* has the same number of squares as nonsquares, it follows that

$$(2.17) \qquad \sum_{a=1}^{p-1} \left(\frac{a}{p}\right) = \sum_{\bar{a} \in K^*} \left(\frac{\bar{a}}{p}\right) = 0.$$

We now return to the calculation of Gauss sums.

LEMMA 2.10. *One has the following formulas:*

$$(2.18) \qquad G_2(x^2) = 0, \quad G_2(q_+^2) = 2, \quad G_2(q_-^2) = -2,$$

where $q_\pm^2$ are the forms (2.9);

$$(2.19) \qquad G_p(bx^2) = \left(\frac{b}{p}\right) G_p(x^2), \quad G_p(x^2)^2 = \left(\frac{-1}{p}\right) p,$$

where p is an odd prime and $b \not\equiv 0 (\text{mod } p)$.

PROOF. (2.18) is easily verified by direct computation. The formulas in (2.19) were verified in the course of the proof of Proposition 4.9 of Chapter 1. □

PROPOSITION 2.11. *Let (V, f) be a nondegenerate quadratic space over $K = \mathbf{F}_p$, and let $\alpha \in K^*$. If $\dim V = 2k$, then*

$$(2.20) \qquad G_p(\alpha, f) = \varepsilon(V, f) p^k,$$

where

$$(2.21) \qquad \varepsilon(V, f) = \begin{cases} \left(\dfrac{(-1)^k d(V,f)}{p}\right), & \text{if } p \neq 2, \\ 1, & \text{if } p = 2 \text{ and } (V, f) \text{ is of type } \{q_+^n\}, \\ -1, & \text{if } p = 2 \text{ and } (V, f) \text{ is of type } \{q_-^n\}. \end{cases}$$

If $\dim V = 2k + 1$, then

$$(2.22) \qquad G_p(\alpha, f) = \left(\frac{\alpha}{p}\right) G_p(x^2) \left(\frac{(-1)^k 2 d(V,f)}{p}\right) p^k \quad \text{for } p \neq 2$$

and

$$G_2(1, f) = 0.$$

PROOF. In the case $p \neq 2$ it follows from Theorem 2.4 that the space (V, f) splits into the direct sum of n one-dimensional subspaces $(V_1, f_1), \ldots, (V_n, f_n)$, each of which is nondegenerate, because (V, f) is nondegenerate. Then (V_i, f_i) is of type $\{\alpha_i x^2\}$, where $\alpha_i \neq 0$, and we can use (2.15), (2.16), and (2.19) to obtain

$$G_p(\alpha, f) = \prod_{i=1}^n G_p(\alpha, f_i) = \prod_{i=1}^n G_p(\alpha a_i x^2) = \left\{\prod_{i=1}^n \left(\frac{\alpha a_i}{p}\right)\right\} G_p(x^2)^n$$

$$= \left(\frac{\alpha}{p}\right)^n \left(\frac{2}{p}\right)^n \left(\frac{2a_1 \cdots 2a_n}{p}\right) G_p(x^2)^n = \left(\frac{2\alpha}{p}\right)^n \left(\frac{d(V,f)}{p}\right) G_p(x^2)^n$$

(according to (2.5) we have $d(V, f) = d(V_1, f_1) \cdots d(V_n, f_n) = 2a_1 \cdots 2a_n$). Substituting $G_p(x^2)^2 = (-1/p)p$, we obtain the required formulas in the case of odd p.

The formulas for $G_2(1, f)$ follow from (2.15) and (2.18), since in the case of type $\{q_+^{2k}\}$ the space (V, f) is the direct sum of k spaces of type $\{q_+^2\}$, in the case of type $\{q_-^{2k}\}$ it is the direct sum of $k - 1$ spaces of type $\{q_+^2\}$ and one space of type $\{q_-^2\}$, and in the case of type $\{q^{2k+1}\}$ the direct summands include a space of type $\{x^2\}$ with zero Gauss sum. $\square$

The number $\varepsilon(V, f)$ is called the *sign of the nondegenerate even-dimensional quadratic space (V, f) over the field $\mathbf{F}_p$*. From (2.15) and (2.20) it follows that the sign is multiplicative in the sense that

$$(2.23) \qquad \varepsilon((V_1, f_1) \oplus (V_2, f_2)) = \varepsilon(V_1, f_1) \cdot \varepsilon(V_2, f_2),$$

if (V_i, f_i) are nondegenerate even-dimensional spaces.

The *Jacobi symbol* is a generalization of (2.16). For any odd $b = p_1 \cdots p_r$ with prime divisors $p_1, \ldots, p_r$ it is defined by setting

$$\left(\frac{a}{b} \right) = \left(\frac{a}{p_1} \right) \cdots \left(\frac{a}{p_r} \right), \qquad \text{where } (a, b) = 1.$$

4. Isotropy subspaces of nondegenerate spaces over residue fields. A nonzero quadratic space with zero quadratic form is called an *isotropy space*.

PROPOSITION 2.12. *Suppose that (V, f) is a nondegenerate quadratic space over the field $K = \mathbf{F}_p$, where p is a prime, $V_1 = (V_1, 0)$ is an isotropy subspace of (V, f), and $e_1, \ldots, e_r$ are an arbitrary basis of V_1. Then there exist vectors $e_1^*, \ldots, e_r^* \in V$ satisfying the conditions*

(1) $e_i \cdot e_i^* = 1$ *and* $f(e_i^*) = 0$ *for* $i = 1, \ldots, r$;
(2) *the subspaces* $(P_1, f_1), \ldots, (P_r, f_r)$, *where* $P_i = \{Ke_i + Ke_i^*\}$ *and* f_i *is the restriction of f to P_i, are pairwise orthogonal.*

We first prove a lemma.

LEMMA 2.13. *If (V, f) is an arbitrary quadratic space with $d(V, f) \neq 0$ and $(V_1, f_1) \subset (V, f)$ is any subspace, then*

$$\dim V_1^\perp = \dim V - \dim V_1, \quad (V_1^\perp)^\perp = V_1.$$

PROOF. We choose a basis $e_1, \ldots, e_n$ of V in such a way that the first r vectors form a basis of V_1. Since $d(V, f) = \det(e_i \cdot e_j) \neq 0$, the rows of the matrix $(e_i \cdot e_j)$ are linearly independent. In particular, the first r rows of this matrix are linearly independent. This implies that the system of r linear equations $u \cdot e_1 = 0, \ldots, u \cdot e_r = 0$ in the coordinates of the vector $u = \sum_i u_i e_i \in V_1^\perp$ has $n - r$ linearly independent solutions; this proves the dimension formula in the lemma. From this formula it follows that

$$\dim(V_1^\perp)^\perp = \dim V - (\dim V - \dim V_1) = \dim V_1.$$

On the other hand, we obviously have $V_1 \subset (V_1^\perp)^\perp$. $\square$

PROOF OF THE PROPOSITION. We first use induction on r to treat the case when $n = \dim V$ is even. Since $f(e_1) = 0$ and (V, f) is nondegenerate, there exists $v \in V$ with $e_1 \cdot v = \alpha \neq 0$ (see Lemma 2.2). We set $v_1 = \alpha^{-1} v$ and $e_1^* = v_1 - f(v_1)e_1$. Then $e_1 \cdot e_1^* = e_1 \cdot v_1 = 1$ and $f(e_1^*) = v_1 \cdot (-f(v_1)e_1) + f(v_1) = 0$, which gives the result in the case $r = 1$. Suppose that $r > 1$ and the proposition has already been proved for $(r - 1)$-dimensional subspaces. Set $V_0 = \{Ke_1 + \cdots + Ke_{r-1}\}$. From Lemma 2.7

and Theorem 2.9 it follows that in the case under consideration $d(V, f) \neq 0$. Hence, by Lemma 2.13, $(V_0^\perp)^\perp = V_0$. This implies that the vector e_r, which obviously lies in $V_0^\perp$ but not in V_0, is therefore not contained in $(V_0^\perp)^\perp$. Thus, there exists $u \in V_0^\perp$ with $e_r \cdot u = \beta \neq 0$. If we replace u by $e_r^* = u_1 - f(u_1)e_r$, where $u_1 = \beta^{-1}u$, we obtain a vector $e_r^* \in V_0^\perp$ that satisfies the conditions $e_r \cdot e_r^* = e_r \cdot u_1 = 1$ and $f(e_r^*) = u_1 \cdot (-f(u_1)e_r) + f(u_1) = 0$. We set $P_r = \{Ke_r + Ke_r^*\}$. Since $P_r \subset V_0^\perp$, it follows that $V_0 \subset P_r^\perp$. Since the plane P_r has determinant 1, it follows from Theorem 2.3, the relations (2.5), and Lemma 2.6 that $P_r^\perp$ is a nondegenerate space of dimension $n - 2$. By the induction assumption, there exist vectors $e_1^*, \ldots, e_{r-1}^* \in P_r^\perp$ that satisfy the required conditions with respect to the basis $e_1, \ldots, e_{r-1}$ of the isotropy subspace $V_0 \subset P_r^\perp$. Then the vectors $e_1^*, \ldots, e_{r-1}^*, e_r^*$ satisfy the conditions of the proposition.

Now suppose that $n = \dim V$ is odd. Since the matrix of the quadratic form (2.12) over $K = \mathbf{F}_2$ is clearly of rank $n - 1$, it follows that the radical $R(V) = V^\perp$ is one-dimensional, and, since V is nondegenerate, it contains a unique vector e_0 with $f(e_0) = 1$. Then $e_0 \notin V_1$. Hence, the vectors $e_0, e_1, \ldots, e_r$ are linearly independent, and they can be completed to a basis $e_0, e_1, \ldots, e_r, \ldots, e_{n-1}$ of the space V. We set $V' = \{Ke_1 + \cdots + Ke_{n-1}\}$. Then

$$(V, f) = (\{Ke_0\}, f_0) \oplus (V', f'),$$

where f_0 and f' are the restrictions of f to the corresponding spaces. From this decomposition and the nondegeneracy of (V, f) it follows that (V', f') is a nondegenerate even-dimensional space. Since $V_1 \subset V'$, the proposition now follows from the even-dimensional case considered above. $\square$

A set of vectors $e_1, \ldots, e_r$ in a quadratic space (V, f) is said to be *isotropic* if they are linearly independent and span an isotropy subspace.

PROPOSITION 2.14. *Suppose that (V, f) is a nondegenerate quadratic space over the field $K = \mathbf{F}_p$. Then the number $i(V, f; r)$ of isotropic sets of r vectors in (V, f) (where $r = 1, \ldots, \dim V$) is equal to*

$$(2.24) \qquad p^{\langle r-1 \rangle}(p^k - \varepsilon)(p^{k-r} + \varepsilon)(p^{2(k-1)} - 1) \cdots (p^{2(k-r+1)} - 1),$$

if $\dim V = 2k$, *where $\varepsilon = \varepsilon(V, f)$ is the sign* (2.21) *of (V, f), and is equal to*

$$(2.25) \qquad p^{\langle r-1 \rangle}(p^{2k} - 1)(p^{2(k-1)} - 1) \cdots (p^{2(k-r+1)} - 1),$$

if $\dim V = 2k + 1$.

PROOF. We use induction on r. By definition, $i(V, f; 1)$ is equal to the number of nonzero vectors $v \in V$ such that $f(v) = 0$. The total number of vectors (including the zero vector) satisfying this condition is obviously equal to

$$i(V, f; 1) + 1 = p^{-1} \sum_{\alpha=0}^{p-1} G_p(\alpha, f) = p^{n-1} + p^{-1} \sum_{\alpha=1}^{p-1} G_p(\alpha, f),$$

where $n = \dim V$. Hence, if $n = 2k$, then from (2.20) we obtain

$$i(V, f; 1) = p^{2k-1} - 1 + \varepsilon p^{k-1}(p - 1) = (p^k - \varepsilon)(p^{k-1} + \varepsilon),$$

while if $n = 2k + 1$, then from (2.22) and (2.17) we have

$$i(V, f; 1) = p^{n-1} - 1 = p^{2k} - 1,$$

which proves the proposition in the case $r = 1$. Now suppose that $r > 1$ and the proposition has already been proved for smaller isotropic sets of vectors. If $i(V, f; r - 1) = 0$, then $i(V, f; r) = 0$. In that case, by the induction assumption, the corresponding expression (2.24) or (2.25) for $r - 1$ is equal to zero; but then the expression for r is also clearly equal to zero, and the proposition is proved. So we suppose that $i(V, f; r - 1) \neq 0$, and we let $e_1, \ldots, e_{r-1}$ be one of the isotropic sets of $r - 1$ vectors. Since $e_1, \ldots, e_{r-1}$ span an isotropy subspace, it follows that for the basis $e_1, \ldots, e_{r-1}$ of this subspace there exists a set of vectors $e_1^*, \ldots, e_{r-1}^*$ with the properties in Proposition 2.12. Each of the pairwise orthogonal subspaces $(P_1, f_1), \ldots, (P_{r-1}, f_{r-1})$ in this proposition has determinant -1. It hence follows from Theorem 2.3 that the space (V, f) splits into the direct sum of subspaces of the form

$$(2.26) \qquad (V, f) = (P_1, f_1) \oplus \cdots \oplus (P_{r-1}, f_{r-1}) \oplus (V', f').$$

As a result of this decomposition, every vector $v \in V$ can be uniquely written in the form

$$v = \sum_{i=1}^{r-1} \alpha_i e_i + \sum_{j=1}^{r-1} \beta_j e_j^* + v',$$

where $\alpha_i, \beta_j \in K$, $v' \in V'$. Since $v \cdot e_i = \beta_i$, it follows that v is orthogonal to $e_1, \ldots, e_{r-1}$ if and only if $\beta_1 = \cdots = \beta_{r-1} = 0$. In that case, since the vectors $e_1, \ldots, e_{r-1}, v'$ are pairwise orthogonal, we find that $f(v) = f(\alpha_1 e_1) + \cdots + f(\alpha_{r-1} e_{r-1}) + f(v') = f(v')$, and the condition $f(v) = 0$ is equivalent to the condition $f(v') = 0$. Finally, the vectors $e_1, \ldots, e_{r-1}, v$ are linearly independent if and only if $v' \neq 0$. Thus, the set $e_1, \ldots, e_{r-1}$ can be completed to an isotropic set of r vectors $e_1, \ldots, e_{r-1}, v$ in exactly $p^{r-1} i(V', f'; 1)$ different ways, i.e.,

$$(2.27) \qquad i(V, f; r) = p^{r-1} i(V', f'; 1) i(V, f; r - 1).$$

From (2.26) it follows that the subspace (V', f') is nondegenerate, and $\dim V' = n - 2(r - 1)$. Since each of the subspaces (P_i, f_i) obviously has sign 1, it follows from (2.26) and (2.23) that if $n > 2(r - 1)$ is even, then the sign $\varepsilon(V', f')$ is the same as the sign $\varepsilon(V, f)$. If $n = 2(r - 1)$, then $i(V, f; r) = 0$ and the expression (2.24) is also zero. If $n > 2(r - 1)$, then, substituting into (2.27) the value of $i(V', f'; 1)$ found above and using the induction assumption, we obtain the required formulas.   □

COROLLARY 2.15. *Let (V, f) be a nondegenerate quadratic space over $K = \mathbf{F}_p$, and let $V_0 = (V_0, 0)$ be a maximal isotropy subspace in (V, f). Then*

$$\lambda = \dim V_0 = \begin{cases} k, & \text{if } \dim V = 2k \text{ and } \varepsilon(V, f) = 1, \\ k - 1, & \text{if } \dim V = 2k \text{ and } \varepsilon(V, f) = -1, \\ k, & \text{if } \dim V = 2k + 1. \end{cases}$$

Modules in Quadratic Fields
and Binary Quadratic Forms

The proofs of the facts given in Appendix 3 and more detailed information on this topic can be found in [13].

1. Modules in algebraic number fields. Let k be a subfield of the field K. If K has finite dimension $n = [K : k]$ as a vector space over k, then we say that K is a *finite extension of k* (of degree n). By the *matrix* (a_{ij}) *of an element* $\alpha \in K$ in the basis $\{\omega_i\}$ of K over k we mean the matrix whose rows are the coordinates of $\alpha\omega_i$ in the basis $\{\omega_i\}$. The trace $S(\alpha)$ and the determinant $N(\alpha)$ of this matrix (a_{ij}) are called, respectively, the *trace* and the *norm of α* (from K to k).

In particular, if $k = \mathbf{Q}$, then K is called an *algebraic number field*. In this case we say that an element $\alpha \in K$ is an *integer* if the coefficients of its characteristic polynomial $\mathrm{ch}(v, \alpha) = \det((a_{ij}) - vE_n)$ belong to the ring of rational integers $\mathbf{Z}$. By a *module in K* we mean any finitely generated $\mathbf{Z}$-submodule $M \subset K$. As a free abelian group M has basis (over $\mathbf{Z}$) $\omega_1, \ldots, \omega_m$. If the *rank m* of M is equal to $n = [K : \mathbf{Q}]$, we say that M is a *full module*. Suppose that M_1 and M_2 are modules with bases $\{\omega_i\}$ and $\{\eta_j\}$. Then the set of integer linear combinations of the products $\omega_i\eta_j$ is also a module, which is denoted $M_1 M_2$ and is called the *product* of M_1 and M_2. The product of two full modules is a full module.

A full module is called an *order of the field K* if it contains 1 and is a ring. Since the matrix entries for an element in any order $\mathfrak{O}'$ with respect to a basis of the same order are rational integers, it follows that $\mathfrak{O}'$ is contained in the set $\mathfrak{O} = \mathfrak{O}_K$ of all integers of K. A typical example of an order of K is the *ring of multipliers* of a full module M, defined as $\mathfrak{O}_M = \{\alpha \in K; \alpha M \subset M\}$. If M_1 and M_2 are *similar modules*, i.e., if $M_2 = \alpha M_1$ for some nonzero $\alpha \in K$, then obviously $\mathfrak{O}_{M_1} = \mathfrak{O}_{M_2}$. For every full module M there exists a similar module contained in $\mathfrak{O}_M$. The *norm $N(M)$ of a module M* is the absolute value of the determinant of the transition matrix from a basis of M to a basis of $\mathfrak{O}_M$. $N(M)$ does not depend on the choice of bases; if $M \subset \mathfrak{O}_M$, it is equal to the index of M in $\mathfrak{O}_M$. Let $\{\omega_i\}$ be a basis of the full module M. The number $d(M) = \det(S(\omega_i\omega_j))$, which is also independent of the choice of basis, is called the *discriminant of M*.

2. Modules and primes in quadratic fields. Any extension $K \supset \mathbf{Q}$ of degree two is of the form $K = \mathbf{Q} + \mathbf{Q}\sqrt{d_0} = \mathbf{Q}(\sqrt{d_0})$, where $d_0 \neq 0, 1$ is a squarefree rational integer. If we compute the matrix of the element $\alpha = a + b\sqrt{d_0}, a, b \in \mathbf{Q}$, in the basis $1, \sqrt{d_0}$, we find that $\mathrm{ch}(v, \alpha) = v^2 - 2av + (a^2 - d_0 b^2)$. Consequently, $\mathrm{ch}(v, \alpha) = (v - \alpha)(v - \alpha')$, where $\alpha' = a - b\sqrt{d_0}$ is the *conjugate* of α (over $\mathbf{Q}$). The element α has trace $S(\alpha) = \alpha + \alpha' = 2a$ and norm $N(\alpha) = \alpha\alpha' = a^2 - d_0 b^2$; it is an integer of K if $S(\alpha)$ and $N(\alpha)$ are both in $\mathbf{Z}$. This implies that the set $\mathfrak{O}$ of integers of the field

$K = \mathbf{Q}(\sqrt{d_0})$ is the order with basis $1, \omega$, where $\omega = (1 + \sqrt{d_0})/2$ or $\sqrt{d_0}$ depending on whether $d \equiv 1$ or $2, 3 \pmod 4$; the discriminant $d = d(\mathfrak{O})$, which is called the *discriminant of the field K*, is equal to d_0 or $4d_0$, respectively. Any order of K has the form $\mathfrak{O}_l = \mathbf{Z} + \mathbf{Z}l\omega$, where $l \in \mathbf{N}$ is the index of $\mathfrak{O}_l$ in $\mathfrak{O}$; the discriminant of the order is dl^2.

Every full module $M \subset K$ is generated by two elements α, β, where $\alpha \neq 0$ and $\gamma = \beta/\alpha \notin \mathbf{Q}$; in this case we write $M = \{\alpha, \beta\}$. If $\gamma \in K$, we let $\overline{\mathrm{ch}}(v, \gamma) = av^2 + bv + c$ denote the polynomial obtained by multiplying the characteristic polynomial $\mathrm{ch}(v, \gamma)$ by a rational number and such that $a > 0$ and a, b, c are relatively prime integers. The significance of the polynomial $\overline{\mathrm{ch}}(v, \gamma)$ is that for $M = \{1, \gamma\}$ with $\gamma \notin \mathbf{Q}$ we have

$$\mathfrak{O}_M = \{1, a\gamma\}, \quad d(\mathfrak{O}_M) = b^2 - 4ac, \quad \text{and} \quad N(M) = 1/a.$$

Suppose that M and M_1 are arbitrary full modules of K, $M' = \{\alpha'; \alpha \in M\}$ is the *module conjugate to M*, $\mathfrak{O}_M = \mathfrak{O}_l$, $\mathfrak{O}_{M_1} = \mathfrak{O}_{l_1}$, and m is the greatest common divisor of l and l_1. Then

$$\mathfrak{O}_{M'} = \mathfrak{O}_M, \quad MM' = N(M)\mathfrak{O}_M$$

and for the product module MM_1 we have

$$N(MM_1) = N(M)N(M_1) \quad \text{and} \quad \mathfrak{O}_{MM_1} = \mathfrak{O}_m.$$

This implies that the full modules of K with fixed ring of multipliers $\mathfrak{O}_l$ form a commutative group under multiplication with identity $\mathfrak{O}_l$; the inverse of M is the module $N(M)^{-1}M'$. The modules similar to $\mathfrak{O}_l$ form a subgroup of this group, and the quotient group is called the *class group of modules of the ring $\mathfrak{O}_l$* and is denoted

$$H(\mathfrak{O}_l) = H(d(\mathfrak{O}_l)),$$

where $d(\mathfrak{O}_l) = dl^2$ is the discriminant of the order $\mathfrak{O}_l$. If l_1 divides l, then the map $M \to \mathfrak{O}_{l_1}M$ induces an epimorphism of class groups

$$v(l, l_1): H(\mathfrak{O}_l) \to H(\mathfrak{O}_{l_1}).$$

Suppose that p is a prime number not dividing l. Then the field K of discriminant d has a full module M satisfying the conditions

$$\mathfrak{O}_M = \mathfrak{O}_l, \quad M \subset \mathfrak{O}_l, \quad \text{and} \quad N(M) = p$$

if and only if the congruence $x^2 \equiv d \pmod{4p}$ has a solution. If $p \nmid d$, then there exist exactly two such modules $\mathfrak{p}$ and $\mathfrak{q} = \mathfrak{p}'$; if $p \mid d$, then there exists exactly one module $\mathfrak{p}$, and $\mathfrak{p}' = \mathfrak{p}$. Finally, if this congruence does not have a solution, then K has a unique full module $M = \mathfrak{p} = p\mathfrak{O}_l$ satisfying the conditions

$$\mathfrak{O}_M = \mathfrak{O}_l, \quad M \subset \mathfrak{O}_l, \quad \text{and} \quad N(M) = p^2.$$

All of these modules are prime ideals of the ring $\mathfrak{O}_l$.

3. Modules in imaginary quadratic fields, and quadratic forms. We shall not need the case of indefinite forms and real quadratic fields, and so shall limit ourselves to positive definite forms and quadratic fields of negative discriminant (imaginary quadratic fields). Let

$$q(x, y) = ax^2 + bxy + cy^2$$

be a positive definite binary quadratic form with matrix $Q = \begin{pmatrix} 2a & b \\ b & 2c \end{pmatrix}$ and with discriminant $d(q) = -\det Q < 0$. We say that such a form is *integral* if $a, b, c \in \mathbf{Z}$, and is *primitive* if its *divisor* $\delta = \delta(q) = \text{g. c. d.}(a, b, c)$ is equal to 1. If the forms q and q_1 have matrices that are connected by the relation

$$Q_1 = Q[U] \quad \text{with } U \in SL_2(\mathbf{Z}),$$

then the forms are said to be *properly equivalent*; in that case they have the same discriminant and divisor. Theorem 1.12 of Chapter 2 implies that the number of proper equivalence classes of positive definite integral binary quadratic forms of fixed discriminant is finite.

We define the map

$$q \mapsto M(q) = \{a, a\gamma_q\},$$

where $M(q)$ is a full module of the imaginary quadratic field $K_q = \mathbf{Q}(\sqrt{d(q)})$ of discriminant d having ring of multipliers $\mathfrak{O}_l$ and

$$\gamma_q = (-b + \sqrt{d(q)})/2 \quad \text{and} \quad l = \sqrt{d(q)/d}.$$

For any fixed negative integer D this map gives a bijection between the set of proper equivalence classes of primitive positive definite forms q of discriminant $d(q) = D$ and the group $H(D)$ of similarity classes of modules of the field $\mathbf{Q}(\sqrt{D})$ with ring of multipliers of discriminant D. The inverse map takes the full module $M = \{\alpha, \beta\}$ to the quadratic form

$$q(M) = N(M)^{-1}(\alpha z + \beta y)(\overline{\alpha} x + \overline{\beta} y),$$

where the basis α, β is chosen so that $\text{Im}(-\beta/\alpha) > 0$.

Let q and q_1 be two primitive integral positive definite binary quadratic forms, let $Q = Q(q)$ and $Q_1 = Q(q_1)$ be the matrices of these forms, and let $D = -\det Q$, $D_1 = -\det Q_1$ be their discriminants. Suppose that the ratio D/D_1 is the square of a rational number. Then $\mathbf{Q}(\sqrt{D}) = \mathbf{Q}(\sqrt{D_1}) = K$. Thus, the modules $M(q)$ and $M(q_1)$ are full modules of the same imaginary quadratic field, and so can be multiplied by one another or by any full module of that field. We define the product $q \times q_1$ of the forms q and q_1 by setting

$$q \times q_1 = q(M(q)M(q_1)),$$

and we let $Q \times Q_1$ denote the matrix of the form $q \times q_1$:

$$Q \times Q_1 = Q(q \times q_1).$$

In addition, for any full module M of K we set

$$q \times M = M \times q = q(M(q)M)$$

and we let $Q \times M$ denote the matrix of the form $q \times M$:

$$Q \times M = M \times Q = Q(q \times M).$$

Notes

In 1987 the first author published the book *Quadratic Forms and Hecke Operators*, Grundlehren der Mathematischen Wissenschaften **286**, Springer-Verlag. It was devoted to the multiplicative properties of modular forms of integer weight and quadratic forms in an even number of variables. Meanwhile, the second author had carried over a large part of the theory to modular forms of half-integer weight and quadratic forms in an odd number of variables. Hence, when the question arose of preparing a Russian edition, it was decided that, rather than merely reproduce the original English version, we would expand it by including the multiplicative properties of modular forms of half-integer weight. In order not to increase the size of the book, it was necessary to omit sections on the action of Hecke operators on the theta-series of quadratic forms. The result was the present volume.

Notes for Chapter 1.

§§3.1, 3.2–The exposition of the properties of theta-functions and theta-groups follows [16], Appendix to Chapter 1.

§3.3–Eichler [16] was the first to have the idea of considering theta-series of degree 1 as specializations of suitable theta-functions and in this way finding the groups of automorphic transformations of theta-series; we follow [14], where this idea was extended to theta-series of arbitrary degree.

§4.2–The automorphic properties of theta-series of positive definite integral quadratic forms of level 1 were first studied by Witt in [48].

§§4.3–4.5–For theta-series of degree 1, the expression for the multipliers in terms of Gauss sums and the computation of those Gauss sums are classical (see, for example, [38, 34, 16]). For theta-series of arbitrary degree, the expression for the multipliers in terms of Gauss sums was found in [14], where a technique was also developed for reducing the multipliers for theta-series of arbitrary degree to the case of degree 1.

Notes for Chapter 2.

More details on the theory of Siegel modular forms can be found in the books [29, 31, 22]. Admirers of Bourbaki might want to consult [39]. Good expositions of modular forms in one variable can be found in [32] and [30]; for an initial exposure to the subject [40] is recommended; connections with algebraic geometry are featured in [42]. Our exposition has made essential use of [31] and [27].

§1.2–For more details on reduction theory see, for example, [26].

§2.3–Our definition of modular forms of half-integer weight for $\Gamma_0^n(q)$ in the case $n = 1$ is the same as that of Shimura [43].

§3.3–The notation (3.14) is credited to Petersson.

§4.1–The proof of Proposition 4.1 follows [40].

§5.1–The scalar product of modular forms of degree 1 was introduced by Petersson in [33]; the scalar product for arbitrary degree and both integer and half-integer weight is based on the same idea.

§5.2–Asymptotic formulas for the Fourier coefficients of modular forms and theta-series can be found in [25, 32, 35].

Notes for Chapter 3.

§1.1–Hecke operators for modular forms of degree 1 and integer weight were defined in [24]; the definition for forms of half-integer weight was essentially given in [43].

§1.2–It is hard to say who was the first to define abstract Hecke rings. In early works we used Shimura's definition in [41]; we later replaced it by the equivalent but more convenient definition that apparently appeared first in [8].

§1.4–The properties of the anti-automorphism j were examined in [8].

§2–Our exposition of the Hecke ring of the general linear group is in the spirit of [47].

§2.2–The formulas in Lemma 2.18 are a special case of the formulas in Lemmas 6 and 9 of [2].

§2.3–See [2].

§3–The structure of the Hecke rings for Γ^n was given in [37] and [41] (who was first?). The transition to congruence subgroups relies upon an idea of Hecke in [24]. But our exposition uses the analogy with the case of the general linear group and some common sense, rather than any particular sources in the literature.

§3.3–The theory of spherical mappings was developed in [37] in an equivalent form, namely, as the theory of zonal spherical functions on reductive algebraic groups over p-adic fields. We use an elementary approach, based on explicit computations. The summation of the series (3.71) is carried out in [41] for $n = 1, 2$, and for arbitrary n in [1] and [2].

§§4.1, 4.2–Hecke rings for a covering of the symplectic group were defined and their basic properties were proved in [52].

§4.3–See [53].

§5–The imbeddings of the local Hecke rings of the symplectic group in the Hecke rings of the triangular subgroup were first used in [6] in connection with the problem of factoring Hecke polynomials. The basic properties of the centralizers of elements $\Pi_+(p)$ were also examined in [6]. The case of the symplectic covering group was studied in [53].

§5.3–The idea of the expansion (5.48) goes back to Hecke [24]. The analogue of (5.49) for suitable operators on the Fourier coefficients of modular forms of degree 2 was studied and used in an essential way in [4].

§6–The factorization theory of Hecke polynomials arose from trying to understand and generalize the expansions of the operator polynomials corresponding to $Q_p^2(v)$ and $R_p^2(v)$ (see [4] and [5], respectively). The first version of the theory was given in [6]; the formulas (6.69) were derived in [7]; and in [8] duality considerations were brought into the theory, making it possible to obtain a symmetric factorization of the polynomial $R_p^n(v)$. A similar factorization of the polynomial $\widehat{R}_p^n(v)$ was obtained in [53].

Notes for Chapter 4.

§1.1–Hecke operators for $\Gamma^1 = SL_2(\mathbf{Z})$ and certain of its subgroups were defined in [24]. Hecke operators were first defined on Siegel modular forms in [45] and [46];

after a gap caused by the War, these operators were examined in [28]. For modular forms of degree 1 and half-integer weight, Hecke operators were introduced in [49]; however, the approach adopted by Shimura in [43] turned out to be more fruitful.

§1.2–The existence of a basis of eigenfunctions of the Hecke operators in the invariant spaces of cusp-forms of degree 1 was proved by Petersson in [33]. Petersson's idea was first used for Siegel modular forms by Maass in [28]. In [18] the author proved the existence of a basis of eigenfunctions in the entire space of modular forms of integer weight for a broad class of congruence subgroups; however, this paper had some errors, partially noted in [19], that make it difficult to use. In the present book we prove the existence of a basis of eigenfunctions only for spaces of cusp-forms relative to q-regular pairs (Theorem 1.9) and for invariant subspaces of the space of all modular forms for the full modular group Γ^n (Theorem 2.16). Here we do not even touch upon the important and extensively studied question of spaces spanned by theta-series that are invariant relative to the Hecke operators. In this connection see [20, 8–12, 21].

§2.3–The relations (2.46) were obtained in [50] for the full modular group and the trivial character. Our exposition follows the same idea. Zharkovskaya's work arose as a result of attempts to generalize the Maass commutation relations [28] for the Siegel operators and the Hecke operators corresponding to $T^2(m)$.

§2.4–The computations in this subsection were carried out in [7, 8, 53].

§3.1–The results in this subsection are essentially due to Hecke [24].

§3.2–The results in this subsection were obtained for the case $q = 1$ in [3] and in their final form in [4]. Here we follow the same ideas. Similar questions were examined for groups of the form $\Gamma^2(q)$ in [17]. In [4] in the case $q = 1$ it was proved that the function

$$\Psi(s, F) = (2\pi)^{-2s}\Gamma(s)\Gamma(s - k - 2)\xi(s, 1; F),$$

where $\Gamma(s)$ is the gamma-function, can be analytically continued to a meromorphic function on the entire s-plane, where it has at most four simple poles and satisfies the functional equation

$$\Psi(2k - 2 - s, F) = (-1)^k\Psi(s, F),$$

where k is the weight of the eigenfunction F.

§3.3–The results presented here were obtained for $n = 1$ in [43] and [44], for $n = 2$ and integer weight in [5], and for arbitrary degree and weight in [7, 8, 53]. Analogous series for the group $\Gamma^2(q)$ were studied in [23]. In [13] it was proved that, if F is a modular form of even weight and γ is a Dirichlet character, then the even zeta-function $\zeta^+(s, \gamma; F)$ extends meromorphically onto the entire s-plane; and in the case $q = 1$, $\gamma = 1$ (and under certain restrictions) a functional equation was found for the even zeta-function. The subsequent development of the theory of even zeta-functions is due to Böcherer, who, in particular, managed to remove the restriction alluded to above (S. Böcherer, *Über die Funktionalgleichung automorpher L-Funktionen zur Siegelschen Modulgruppe*, J. Reine Angew. Math. **362** (1985), 146–168).

References

1. A. N. Andrianov, *Rationality theorems for Hecke series and Zeta-functions of the groups GL_n and Sp_n over local fields*, Izv. Akad. Nauk SSSR Ser. Mat. **33** (1969), no. 3, 466–505; English transl. in Math. USSR-Izv. **3** (1969).

2. _____, *Spherical functions for GL_n over local fields, and the summation of Hecke series*, Mat. Sb. **83** (1970), no. 3, 429–451; English transl. in Math. USSR-Sb. **12** (1970).

3. _____, *Dirichlet series with Euler product in the theory of Siegel modular forms of genus 2*, Trudy Mat. Inst. Steklov. **112** (1971), 73–94; English transl. in Proc. Steklov Inst. Math. **112** (1973).

4. _____, *Euler products that correspond to Siegel's modular forms of genus 2*, Uspekhi Mat. Nauk **29** (1974), no. 3, 43–110; English transl. in Russian Math. Surveys **29** (1974).

5. _____, *Symmetric squares of zeta-functions of Siegel modular forms of genus 2*, Trudy Mat. Inst. Steklov. **142** (1976), 22–45; English transl. in Proc. Steklov Inst. Math. **1979**, no. 3.

6. _____, *The expansion of Hecke polynomials for the symplectic group of genus n*, Mat. Sb. **104** (1977), no. 3, 390–427; English transl. in Math. USSR-Sb. **33** (1977).

7. _____, *Euler expansions of the theta-transform of Siegel modular forms of genus n*, Mat. Sb. **105** (1978), no. 3, 291–341; English transl. in Math. USSR-Sb. **34** (1978).

8. _____, *Multiplicative arithmetic of Siegel's modular forms*, Uspekhi Mat. Nauk **34** (1979), no. 1, 67–135; English transl. in Russian Math. Surveys **34** (1979).

9. _____, *Action of Hecke operator $T(p)$ on theta series*, Math. Ann. **247** (1980), 245–254.

10. _____, *Integral representations of quadratic forms by quadratic forms: multiplicative properties*, Proceedings of the International Congress of Mathematicians, Vol. 1, 2 (Warsaw 1983), PWN, Warsaw, 1984, pp. 465–474.

11. _____, *Hecke operators and representations of binary quadratic forms*, Trudy Mat. Inst. Steklov. **165** (1984), 4–15; English transl. in Proc. Steklov Inst. Math. **1985**, no. 3.

12. _____, *Representations of an even zeta-function by theta-series*, Zap. Nauchn. Sem. Leningrad. Otdel. Mat. Inst. Steklov. (LOMI) **134** (1984), 5–14; English transl. in J. Soviet Math. **36** (1987).

13. A. N. Andrianov and V. L. Kalinin, *Analytic properties of standard zeta-functions of Siegel modular forms*, Mat. Sb. **106** (1978), no. 3, 323–339; English transl. in Math. USSR-Sb. **35** (1979).

14. A. N. Andrianov and G. N. Maloletkin, *Behavior of theta-series of genus n under modular substitutions*, Izv. Akad. Nauk SSSR Ser. Mat. **39** (1975), no. 2, 243–258; English transl. in Math. USSR-Izv. **9** (11975).

15. Z. I. Borevich and I. R. Shafarevich, *Number theory*, Pure Appl. Math., vol. 20, Academic Press, New York, 1966.

16. M. Eichler, *Introduction to the theory of algebraic numbers and functions*, Pure Appl. Math., vol. 23, Academic Press, New York, 1966.

17. S. A. Evdokimov, *Euler products for congruence subgroups of the Siegel group of genus 2*, Mat. Sb. **99** (1976), no. 4, 483–513; English transl. in Math. USSR-Sb. **28** (1976).

18. _____, *A basis composed of eigenfunctions of Hecke operators in the theory of modular forms of genus n*, Mat. Sb. **115** (1981), no. 3, 337–363; English transl. in Math. USSR-Sb. **43** (1982).

19. _____, *Letter to the editors*, Mat. Sb. **116** (1981), no. 4, 603; English transl. in Math. USSR-Sb. **44** (1983).

20. E. Freitag, *Die Invarianz gewisser von Thetareinen erzeugter Vektorräume unter Heckeoperatoren*, Math. Z. **156** (1977), no. 2, 141–155.

21. _____, *Eine Bemerkung zu Andrianovs expliziten Formeln für die Wirkung der Heckeoperatoren auf Thetareihen*, E. B. Christoffel (Aachen/Monschau, 1979), Birkhäuser, Basel, 1981, pp. 336–351.

22. _____, *Siegelsche Modulfunktionen*, Springer, New York, 1983.

23. V. A. Gritsenko, *Symmetric squares of the zeta functions for a principal congruence subgroup of the Siegel group of genus 2*, Mat. Sb. **104** (1977), no. 1, 22–41; English transl. in Math. USSR-Sb. **33** (1977).

24. E. Hecke, *Über Modulfunktionen und die Dirichletschen Reihen mit Eulerscher Produktentwicklung.* I, II, Math. Ann. **114** (1937), 1–28; 316–351.

25. _____, *Analytische Arithmetik der positiven quadratischen Formen*, Danske Vid. Selsk. Math.-Fys. Medd. **17** (1940), no. 12.

26. J. W. S. Cassels, *Rational quadratic forms*, Academic Press, New York, 1978.

27. M. Koecher, *Zur Theorie der Modulformen n-ten Grades.* I, II, Math. Z. **59** (1954), 399–416; **61** (1955), 455–466.

28. H. Maass, *Die Primzahlen in der Theorie der Siegelschen Modulfunktionen*, Math. Ann. **124** (1951), 87–122.

29. _____, *Lectures on Siegel's modular functions*, Tata Institute of Fundamental Research, Bombay, 1954–1955.

30. _____, *Lectures on modular functions of one complex variable*, Tata Institute of Fundamental Research, Bombay, 1964; revised 1983.

31. _____, *Siegel's modular forms and Dirichlet series*, Lecture Notes in Math., vol. 216, Springer, New York, 1971.

32. A. Ogg, *Modular forms and Dirichlet series*, Benjamin, New York, 1969.

33. H. Petersson, *Konstruktion der sämtlichen Lösungen einer Riemannschen Funktionalgleichung durch Dirichlet-Reihen mit Eulerscher Produktentwicklung.* I, II, II, Math. Ann. **116** (1936), 401–412; **117** (1939/40), 39–64; 277–300.

34. W. Pfetzer, *Die Wirkung der Modulsubstitutionen auf mehrafache Thetareihen zu quadratischen Formen ungerader Variablenzahl*, Arch. Math. **4** (1953), 448–454.

35. S. Raghavan, *Modular forms of degree n and representation by quadratic forms*, Ann. of Math. (2) **70** (1959), 446–477.

36. R. A. Rankin, *Contributions to the theory of Ramanujan's function $\tau(n)$ and similar arithmetical functions.* I, II, Proc. Cambridge Philos. Soc. **35** (1939), 351–372.

37. I. Satake, *Theory of spherical functions on reductive algebraic groups over p-adic fields*, Inst. Hautes Études Sci. Publ. Math. **1963**, no. 18, 5–69.

38. B. Schoeneberg, *Das Verhalten von mehrfachen Thetareihen bei Modulsubstitutionen*, Math. Ann. **116** (1939), 511–523.

39. *Séminaire Henri Cartan, 10e année* (1957/58). *Fonctions automorphes.* vols. 1, 2, Secrétariat Mathématique, Paris, 1958.

40. J.-P. Serre, *A course in arithmetic*, Springer, New York, 1973.

41. G. Shimura, *On modular correspondences for $Sp(n, \mathbb{Z})$ and their congruence relations*, Proc. Nat. Acad. Sci. U.S.A. **49** (1963), 824–828.

42. _____, *Introduction to the arithmetic theory of automorphic functions*, Princeton Univ. Press, Princeton, NJ, 1971.

43. _____, *On modular forms of half integral weight*, Ann. of Math. (2) **97** (1973), 440–481.

44. _____, *On the holomorphy of certain Dirichlet series*, Proc. London Math. Soc. (3) **31** (1975), 79–98.

45. M. Sugawara, *On the transformation theory of Siegel's modular group of the n-th degree*, Proc. Imp. Acad. Japan **13** (1937), 335–336.

46. _____, *An invariant property of Siegel's modular functions,*, Proc. Imp. Acad. Japan **14** (1938), 1–3.

47. T. Tamagawa, *On the ζ-functions of a division algebra*, Ann. of Math. (2) **77** (1963), 387–405.

48. E. Witt, *Eine Identität zwischen Modulformen zweiten Grades*, Abh. Math. Sem. Hansischen Univ. **14** (1941), 323–337.

49. K. Wohlfahrt, *Über Operatoren Heckescher Art bei Modulformen reeller Dimension*, Math. Nachr. **16** (1957), 233–256.

50. N. A. Zharkovskaya, *The Siegel operator and Hecke operators*, Funktsional. Anal. i Prilozhen. **8** (1974), no. 2, 30–38; English transl. in Functional Anal. Appl. **18** (1974).

51. _____, *The connection between the eigenvalues of Hecke operators and the Fourier coefficients of eigenfunctions for Siegel modular forms of genus n*, Mat. Sb. **96** (1975), no. 4, 584–593; English transl. in Math. USSR-Sb. **25** (1975).

52. V. G. Zhuravlev, *Hecke rings for a covering of the symplectic group*, Mat. Sb. **121** (1983), no. 3, 381–402; English transl. in Math. USSR-Sb. **49** (1984).

53. _____, *Euler expansions of theta-transformations of Siegel modular forms of half-integer weight and their analytic properties*, Mat. Sb. **123** (1984), no. 2, 174–194; English transl. in Math. USSR-Sb. **51** (1985).

List of Notation

$\mathbf{N}$ is the set of natural numbers

$\mathbf{N}_{(q)}$ is the set of natural numbers prime to q

$\mathbf{P}$ is the set of prime numbers in $\mathbf{N}$

$\mathbf{P}_{(q)} = \mathbf{P} \cap \mathbf{N}_{(q)}$

$\mathbf{Z}$ is the ring of rational integers

$\mathbf{Z}_q$ is the set of q-integral rational numbers

$\mathbf{Q}, \mathbf{R}, \mathbf{C}$ are the fields of rational, real, and complex numbers, respectively

$\mathbf{C}_1 = \{z \in \mathbf{C}; |z| = 1\}$

$\mathbf{F}_p = \mathbf{Z}/p\mathbf{Z}$ is the residue field modulo the prime p

K is a commutative ring with unit

K^* is the multiplicative group of invertible elements of K

$M_{m,n}(K)$ is the set of $m \times n$-matrices with entries in K

$M_n(K) = M_{n,n}(K)$

$S_n(K)$ is the set of symmetric matrices in $M_n(K)$

$M_{m,n} = M_{m,n}(\mathbf{Z}), M_n = M_n(\mathbf{Z}), S_n = S_n(\mathbf{Z})$

$\mathbf{E}_n$ is the set of matrices in S_n having even diagonal

$A > 0$ $(A \geqslant 0)$ means that A is a positive definite (semidefinite) matrix in $S_n(\mathbf{R})$

$\mathbf{A}_n = \{A \in \mathbf{E}_n; A \geqslant 0\}, \mathbf{A}_n^+ = \{A \in \mathbf{E}_n; A > 0\}$

$GL_n(K) = \{M \in M_n(K); \det M \in K^*\}$

$\Lambda = \Lambda^n = GL_n(\mathbf{Z})$

$SL_n(K) = \{M \in M(K); \det M = 1\}$

$E_n = (e_{\alpha\beta})$ is the $n \times n$ identity matrix

$J_n = \begin{pmatrix} 0 & E_n \\ -E_n & 0 \end{pmatrix}$

^tM is the transpose of the matrix M

$M^* = {}^tM^{-1}$ for a nonsingular square matrix M

$Q[M] = {}^tMQM$

$S_K^n = GSp_n^+(K) = \{M \in M_{2n}(K); J_n[M] = r(M)J_n, r(M) > 0\}$

$Sp_n(K) = \{M = GSp_n^+(K); r(M) = 1\}$

$\Gamma^n = Sp_n(\mathbf{Z})$

$\mathfrak{G}$ is the covering of the symplectic group $GSp_n^+(\mathbf{R})$

$P: \mathfrak{G} \to GSp_n^+(\mathbf{R})$ is the canonical epimorphism

$\mathbf{H}_n$ is the Siegel upper half-plane of degree n

$\mathbf{H}_n(\varepsilon) = \{Z = X + iY \in \mathbf{H}_n; Y \geqslant \varepsilon E_n\}$

$M\langle Z \rangle = (AZ + B)(CZ + D)^{-1} \quad \left(M = \begin{pmatrix} A & B \\ C & D \end{pmatrix} \right)$

$a|b$ $(a \nmid b)$ means that a divides (does not divide) b

$a|q^\infty$ means that every prime factor of a divides q

(a, b) is the greatest common divisor of a and b

$\left(\frac{a}{p}\right)$ is the Legendre symbol

$\langle n \rangle = n(n+1)/2$

$\sigma(A)$ is the trace of a square matrix A

$e\{A\} = \exp(\pi i \sigma(A))$

$r_p(A)$ is the rank of the matrix A over the field $\mathbf{F}_p$

$T(S) = \begin{pmatrix} E_n & S \\ 0 & E_n \end{pmatrix}$

$U(r, V) = \begin{pmatrix} rV^* & 0 \\ 0 & V \end{pmatrix}$, $U(V) = U(1, V)$

(Γg) is a left coset modulo the group Γ

$(g)_\Gamma = \displaystyle\sum_{g_i \in \Gamma \backslash \Gamma g \Gamma} (\Gamma g_i)$

$\Gamma_{(g)} = \Gamma \cap g^{-1} \Gamma g$

$D_K(\Gamma, S)$ is the Hecke ring of the pair (Γ, S) over the ring K

$D(\Gamma, S) = D_\mathbf{Z}(\Gamma, S)$

$H^n = D_\mathbf{Q}(\Lambda^n, G^n) = D_\mathbf{Q}(GL_n(\mathbf{Z}), GL_n(\mathbf{Q}))$

$L^n(q) = D_\mathbf{Q}(\Gamma_0^n(q), S_0^n(q))$, $L^n = L^n(1)$

$L_0^n = D_\mathbf{Q}(\Gamma_0^n, S_0^n)$ is the triangular Hecke ring

L_p, $\underline{L}$, E are the local, integral, and even subrings of the Hecke ring L

$\overline{L} = L \otimes_\mathbf{Q} \mathbf{C}$ is the complexification of the Hecke ring L

$\widehat{L}$ is the Hecke ring obtained by lifting the ring L

$\mathbf{L}$ is the image of the Hecke ring L in the ring $\mathbf{L}_0^n$

$\mathbf{T}$ is the image of an element $T \in L$ in the ring $\mathbf{L}_0^n$

Other Titles in This Series

(*Continued from the front of this publication*)

(See the AMS catalog for earlier titles)